215
Topics in Current Chemistry

Springer

Berlin
Heidelberg
New York
Barcelona
Hong Kong
London
Milan
Paris
Singapore
Tokyo

Glycoscience

Epimerisation, Isomerisation and Rearrangement Reactions of Carbohydrates

Volume Editor: Arnold E. Stütz

With contributions by
M. Albert, S. J. Angyal, K. Dax, B. Eder, M. H. Fechter, R. J. Ferrier, H. Häusler, Z. Hricovíniová, I. Lundt, R. Madsen, S. Osanai, L. Petruš, M. Petrušová, A. de Raadt, A. E. Stütz, J. Thiem, B. Werschkun, T. M. Wrodnigg

Springer

The series *Topics in Current Chemistry* presents critical reviews of the present and future trends in modern chemical research. The scope of coverage includes all areas of chemical science including the interfaces with related disciplines such as biology, medicine and materials science. The goal of each thematic volume is to give the non-specialist reader, whether at the university or in industry, a comprehensive overview of an area where new insights are emerging that are of interest to a larger scientific audience.

As a rule, contributions are specially commissioned. The editors and publishers will, however, always be pleased to receive suggestions and supplementary information. Papers are accepted for *Topics in Current Chemistry* in English.

In references *Topics in Current Chemistry* is abbreviated Top. Curr. Chem. and is cited as a journal.

Springer WWW home page: http://www.springer.de
Visit the TCC home page at http:/link.springer.de/series/tcc/
or http://link.springer-ny.com/series/tcc/

ISSN 0340-1022
ISBN 3-540-41383-9
Springer-Verlag Berlin Heidelberg New York

Library of Congress Catalog Card Number 74-644622

Springer-Verlag Berlin Heidelberg New York
a member of BertelsmannSpringer Science+Business Media GmbH

Cover design: Friedhelm Steinen-Broo, Barcelona; MEDIO, Berlin
Typesetting: Fotosatz-Service Köhler GmbH, 97084 Würzburg

SPIN: 10753427   02/3020 ra – 5 4 3 2 1 0 – Printed on acid-free paper

Volume Editor

Prof. Dr. Arnold E. Stütz
Institut für Organische Chemie
Technische Universität Graz
Stremayrgasse 16
8010 Graz, Austria
E-mail: stuetz@orgc.tu-graz.ac.at

Editorial Board

Topics in Current Chemistry
Now Also Available Electronically

For all customers with a standing order for Topics in Current Chemistry we offer the electronic form via LINK free of charge. Please contact your librarian who can receive a password for free access to the full articles by registration at:

http://link.springer.de/orders/index.htm

If you do not have a standing order you can nevertheless browse through the table of contents of the volumes and the abstracts of each article at:

http://link.springer.de/series/tcc

There you will also find information about the

- Editorial Board
- Aims and Scope
- Instructions for Authors

Preface

Carbohydrate chemistry is a rapidly growing research and development area with great potential for future achievements due to the eminent importance of carbohydrates in biochemistry and biology and the strong impact of glycosciences in these fields. One of the most important methods of carbohydrate transformation in terms of commercial significance is the conversion of sugars into high-added-value epimers or isomers such as the enzymatic transformation of D-glucose into D-fructose or the base induced isomerisation of lactose into lactulose. In addition to these industrial processes, quite a range of methods has become available to transform simple sugars into valuable intermediates, building blocks and sophisticated products. Consequently, a compilation of important but quite diverse methods in one book appeared to be a worthwhile task. Quite a few of the authors have made significant discoveries in the area under consideration. Others have contributed important and innovative extensions taking one or the other method to even higher levels of sophistication.

The classical method of epimerisation, the Lobry de Bruyn – Alberda van Ekenstein rearrangement has been treated by S. Angyal, the expert in the field of free sugars and their interconversions. L. Petruš and co-authors have contributed a first-hand experience based chapter on the Bílik reaction, which was discovered in Bratislava and is an important approach to rare sugars on a commercial basis. Another, more recently developed method, the epimerisation of free sugars catalysed by nickel (II)/ethylene diamine complexes, has been reviewed by S. Osanai.

Robin Ferrier, the discoverer of the allylic rearrangement of glycals (generally known in organic chemistry as the "Ferrier I" reaction) as well as of the transformation of hex-5-enopyranosides to highly functionalised cyclohexanones (coined "Ferrier II" reaction) has provided accounts on these milestones of carbohydrate chemistry. A chapter on base-catalysed transformations of aldonolactones has been provided by I. Lundt, the leading expert in this area, in collaboration with R. Madsen and an account on Claisen rearrangement reactions in carbohydrate chemistry has been contributed by B. Werschkun and J. Thiem.

From the Graz groups, K. Dax and M. Albert have gone through the efforts to categorise rearrangements occurring during (attempted) nucleophilic displacement reactions at various ring positions in monosaccharides. T. Wrodnigg and B. Eder have reviewed the Amadori and Heyns rearrangements. In addition, a contribution on D-xylose isomerases and related, preparatively useful enzymes some of which have been used in this laboratory has also been included. A short section on reactions related to the general topic under consideration but not treated in the other chapters in this book has been compiled by A. de Raadt and M. H. Fechter.

We all hope that you, the readers, will find this collection interesting to read and a useful source of information for further studies and rewarding applications.

Graz, January 2001

Arnold E. Stütz

Contents

Contents of Volume 186

Glycoscience

Synthesis of Oligosaccharides and Glycoconjugates

Volume Editors: H. Driguez, J. Thiem

ISBN 3-540-62033-8

Contents of Volume 187

Glycoscience

Synthesis of Substrate Analogs and Mimetics

Volume Editors: H. Driguez, J. Thiem

ISBN 3-540-62032-X

The Lobry de Bruyn-Alberda van Ekenstein Transformation and Related Reactions

Stephen J. Angyal

School of Chemistry, University of New South Wales, Sydney 2052, Australia
E-mail: s.angyal@unsw.edu.au

Recent examples of the use of a very old synthetic method are discussed. Discovery of the cation-catalyzed epimerization explains some of the previously controversial results. The mechanism of the reaction is discussed. Suggestions are made for improvements in yield and in the method of work-up.

Keywords: Epimerization, Synthesis of ketoses, Enediol intermediates

1 Introduction

Until the middle of the 20th century, carbohydrate chemistry was regarded as a separate, minor branch of organic chemistry; despite the tremendous importance of carbohydrates in nutrition, the food industry, in biology and medicine and as organic raw materials, they were regarded by many organic chemists with some dubiety. The sugars are not soluble in organic solvents, only in water; they cannot be distilled and frequently will not crystallize; and – worst of all – when a reducing sugar is dissolved in water, it produces a solution containing at least six different compounds. Not like organic compounds at all.

This view has since changed. Eminent organic chemists have extended widely the range of reactions applied to carbohydrates and, in turn, carbohydrates have been used widely as starting materials for the synthesis of a great variety of organic compounds, especially chiral ones. Nevertheless, some reactions of carbohydrates, especially in aqueous solutions, remain untypical of organic reactions; in particular, epimerization is not often carried out during organic syntheses (and, if carried out, would not be called "epimerization"). Epimerization, i.e. change of the configuration of C-2 in aldoses (for formulae, see Osanai, this vol.), is one of the main topics of this volume. A short review on this subject has recently been published [1].

Many sugars occur in Nature in large quantities and are therefore readily available; others have never been detected in natural sources. In particular, L-sugars are rare, only L-sorbose, L-rhamnose and L-arabinose are readily available. The rare or unavailable sugars are usually prepared from the common ones (although there have been lately quite a few syntheses of sugars from simpler chemicals). Since all the sugars are stereoisomers of several other sugars, it is tempting to prepare the rare ones from common ones by rearrangements. Much of this volume deals with such rearrangements.

There is a problem, however. If molecules rearrange – without additional energy being supplied by adding groups and removing them – the reactions are reversible and the outcome is an equilibrium, the position of which is determined by the relative free energies of the starting sugars and the products. Nature prefers for its operations compounds with low free energy [2]: D-glucose – the most stable of all hexoses – is the most common monosaccharide in Nature. D-Galactose and D-fructose are also widespread. However, the less common sugars have higher free energies and hence are found as minor components in the rearrangement equilibria. For example, the very expensive D-talose (**1**) is obtained by epimerization of the very cheap D-galactose (**2**) but, naturally, the yield will be low because talose – with two *syn*-axial hydroxyl groups – has a much higher free energy. Methods in Carbohydrate Chemistry [3] describes four instances of sugar syntheses based on rearrangements and the yield in each case is less than 20%. At a time when organic chemists are aiming at yields close to 100% in most reactions, these do not appear satisfactory. Yet, often they are. The reasons for satisfaction are: plentiful cheap starting materials, reagents and solvents, only one reaction step, and simple work-up procedures. Only occasionally have there been attempts to shift the position of the equilibria.

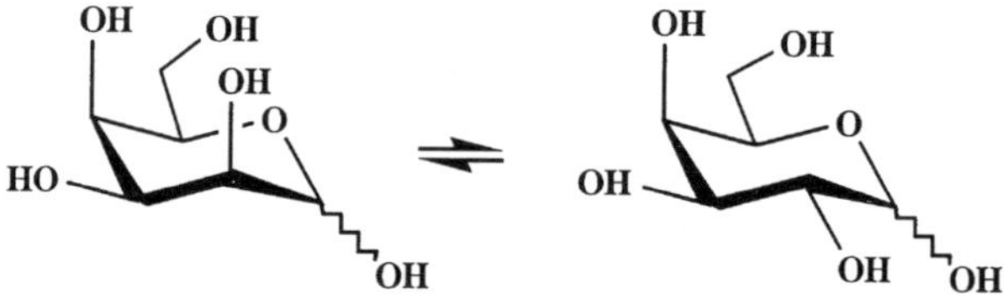

1 **2**

It has often been stated that most inventions in chemistry are due to accidents; they were not planned or sought for. This applies to most of the rearrangements discussed in this book. The first such rearrangement was discovered [4–6] by Cornelis Adriaan Lobry von Trostenburg de Bruyn, professor at the University of Amsterdam, and his collaborator Willem Alberda van Ekenstein, chemist at the Government Sugar Laboratory. The reaction was named after them; however, their names being so cumbersome – this is the longest name given to any chemical reaction [7] – it will be referred to here as the "LdB-AvE rearrangement". LdB and AvE were studying the degradation of D-glucose in alkaline solution in 1885 when they became aware, to their surprise, of the presence of other sugars in their reaction mixture. These were isolated and identified as D-mannose and D-fructose. The researchers also established that the same three sugars were present if mannose or fructose was used as the starting material. Thus the term "LdB-AvE rearrangement" includes two reactions: epimerization and aldose-ketose interconversion. In most cases, they occur simultaneously. During the 20th century, the LdB-AvE reaction has been used for the synthesis of many sugars. Its mechanism has been extensively studied but remains controversial. Speck [8] published an extensive review in 1958 on this reaction; the reader is referred to this for details that are not repeated here.

2
Experimental Procedures

The LdB-AvE reaction is usually conducted in basic solution; acids have a similar effect but that is of little practical importance. Sodium and calcium hydroxide in aqueous solution are the reagents commonly used. However, they cause a complex series of reactions: following epimerization and aldose-ketose interchange, further similar reactions occur along the chain, e.g. inversion on C-3 and migration of the keto group to C-3. Seven sugars (psicose, fructose, galactose, gulose, glucose, tagatose and sorbose) were found [9] in the mixture formed by the reaction of galactose or tagatose with potassium hydroxide. Some of the products were probably formed not through an enediol but by degradation to glyceraldehyde, followed by aldol condensation. After a prolonged period, all of the aldopentoses were found [9] amongst the reaction products formed from any of the aldopentoses. These reactions are followed by slower decomposition of all of the active species; more than 50 compounds (mostly acids) have been identified in the reaction mixture formed from D-glucose and aqueous sodium hydroxide [10]. To minimize such side reactions, the period of the reaction is reduced to a minimum, that is, equilibrium is seldom reached; and the concentration of the base is kept low. A marked improvement was the introduction of pyridine both as a solvent and a base [11] in 1927; pyridine has much less tendency to induce side reactions and makes it unnecessary to remove inorganic reagents. A good example is the preparation of D-*glycero*-tetrulose from D-erythrose in 39% yield [12]; another is the preparation of D-psicose from D-allose, which gives a 39% yield [13] and 50% after recycling the recovered allose. A more recent example is the preparation of 6-deoxy-L-fructose from L-rhamnose in 63% yield [14]. Ammonia and triethylamine have also been employed as bases in aqueous solution [15]. Anion-exchange resins in their hydroxide or aluminate form have been used as catalysts; the formation of fructose from glucose and from mannose has been studied by this method [16]. It is of industrial importance, as witnessed by numerous patents [17]. Borate was found, by chance, to promote the LdB-AvE reaction [18]: lactulose was formed from lactose during chromatography on a borate resin. A more efficient method is to use borate together with triethylamine [19].

When a new application of the LdB-AvE reaction is investigated, it is important to determine the best time to terminate the reaction. This is because the amount of the desired product will reach a maximum and then decline owing to the appearance of other isomers which are formed less rapidly, and to the general degradation of all the sugars. Mendicino [20] found that the presence of sodium borate slows down the degradations, hence a better yield of the product may be obtained. The proportion of fructose or psicose amongst the products increased in the presence of borate [21]. Benzeneboronate has a similar effect [22]. It appears that this observation has not been widely utilized.

Various sugars epimerize at different rates. For example, mannose is converted much slower than glucose or fructose [9] and it is also formed more slowly. Idose reacts very rapidly in the presence of bases, altrose rapidly but allose slowly [23]. The most investigated case is, of course, the glucose-mannose-fructose system. The products are formed in a ratio of approximately 4:1:4. More accurate figures have been quoted but the composition will depend on the nature of the base and of the starting material. True equilibrium, as already stated, is probably never obtained [9].

3
Epimerization

Of the two parallel reactions, epimerization is now of minor importance. The epimerized aldose is usually obtained in much smaller yield than the ketose, and has seldom been isolated. However, if C-2 does not carry a free hydroxyl group – and therefore a ketose cannot be formed – epimerization becomes a practical method of synthesis; examples are 2,4,6-tri-*O*-methyl-D-mannose from the corresponding D-glucose derivative [24] with barium hydroxide in 50% yield and *N*-acetyl-D-mannosamine from *N*-acetyl-D-glucosamine with calcium hydroxide in 20% yield [25]. A very useful synthetic application is the isomerization of 2,3-*O*-isopropylidene-aldoses in the open-chain form [26]. Curiously, if O-2 is replaced by sulfur, the LdB-AvE epimerization proceeds very easily: 2-*S*-ethyl-2-thio-D-glucose is obtained from the corresponding mannose derivative by the use of sodium bicarbonate [27]. Apparently, the presence of an electron-attracting group on C-2 facilitates the formation of the enolate anion. Another example is the epimerization of 2-thiosophorose (2-*S*-β-D-glucopyranosyl-2-thio-D-glucose) (**3**) to the *manno*-compound **4** by sodium bicarbonate [28]; this latter isomer, which is a syrup, had to be kept in a quartz tube because it slowly isomerized in a glass (presumably soda-glass) sample tube. Under the exceptionally facile conditions of these epimerizations, the other possible reactions do not seem to interfere.

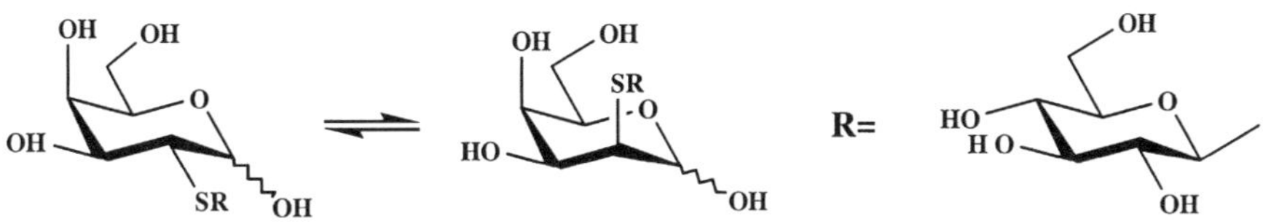

3 4

Closely related to this reaction is the base-catalyzed epimerization of aldonic acids and their lactones (see Lundt and Madsen, this vol.). This reaction is, of course, even older than the LdB-AvE process, and was first used by Emil Fischer [29]. Potassium hydroxide and tertiary amines (pyridine, quinoline) have been used as bases. The reaction is much slower than the epimerization of sugars and requires prolonged heating, but the aldonic acids are much more stable to the action of bases than the sugars, and no side reactions occur. The reaction proceeds even if the hydroxyl group at C-2 is methylated [30]. The kinetics and the mechanism of the interconversion of the aldopentonic acids in potassium hydroxyde solution have been studied; this reaction also occurs via the enediol and the removal of H-2 is the rate-determining step [31]. It is of industrial importance, as indicated by numerous patents [32].

When an alduronic acid is submitted to these conditions, reaction may occur at both ends of the molecule. Carlsson et al. [33] studied the reaction of D-glucuronic acid, as its potassium salt, at 100 °C for 3 h and then found 35% of D-*lyxo*-5-hexulosonic acid (of the configuration of fructose), 15% of D-*ribo*-hexulosonic acid (of the psicose configuration), 8% of D-mannuronic accid, 5% of D-altruronic acid and 1% of D-alluronic acid; the conversion of the aldehyde to the keto group is therefore predominant. However, in an earlier study, under practically the same conditions, epimerization of C-5 was found to be the main reaction [34]. In working out a successful synthesis of L-iduronic acid from D-glucuronic acid, it was found necessary to mask the reducing end of the latter by forming a 1,2-*O*-isopropylidene derivative [35].

4
Syntheses of Ketoses

With the development of newer methods of epimerization – the Bílik reaction (Petruš, Petrušová and Hricovíniová, this vol.) and the nickel-amine catalyzed rearrangement (see Osanai, this vol.) – the LdB-AvE epimerization of aldoses is no longer the best choice. On the other hand, the synthesis of ketoses from aldoses by the LdB-AvE reaction is of great interest. Already in their second paper, LdB and AvE described [5] the isolation and characterization of D-tagatose, a ketose previously not known, from D-galactose. The reaction is of industrial importance, as testified by numerous patents [17]. After equilibration, more ketose is usually found than aldose. For example, the equilibrium constant of D-fructose/D-mannose is 2.45 [36]. (The equilibrium constants of pentuloses/pentoses are all well below 1 but that is readily explained: the former cannot form pyranoses only furanoses and hence are less stable.) In the 1950s several enzymes became available [37] which were capable of catalyzing the aldose-ketose interconversion; such a specific enzyme is preferable to the chemical reaction since the epimerized aldose and other by-products are not formed (see Häusler and Stütz, this vol.). Speck stated [8]: "For some unknown reason, the best yields have consistently been obtained in the formation of D-*gluco*-heptulose from D-*glycero*-D-*gulo*-heptose". The reason is now quite obvious: the *gulo* configuration (**5**) of the aldose is of high energy, owing to its two axial hydroxyl groups, whereas the ketose formed has the most stable, all-equatorial configuration (**6**). Idose is of even higher free energy than gulose. Storing it in Jena glass tubes will convert it gradually into sorbose [38]. Idose is the only aldohexose which has not been found in any natural source, neither the D, nor the L form.

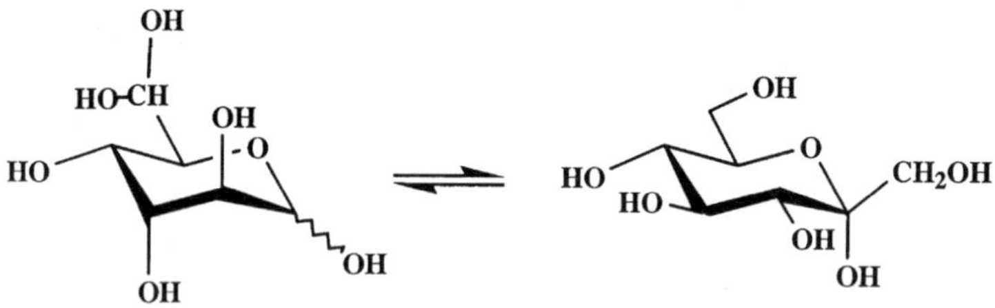

5 **6**

A ketose is rarely used as the starting material for the LdB-AvE transformation. The reaction in that case is even more complicated: besides formation of aldoses, epimerization of the ketose may occur and, if the keto group is at C-3, it may occur on both sides of the keto group. Thus, Schaffer [39] obtained D-*gluco*, D-*manno*- and D-*altro*-heptuloses from D-*manno*-3-heptulose by the action of calcium hydroxide. The involvement of 3,4-enediols has been proven, even when aldoses were used as starting materials [40]. One instance of the epimerization of a ketose is the preparation of D-psicose from D-fructose [15]; triethylamine was used as the base. In this case, the often cumbersome separation of starting material and products was achieved by the use of a cation-exchange column in its calcium form [41]. The size of this column could have been reduced to half by the use of lanthanum as the cation [42], or even further by the use of neodymium [43]. This technique should be used more often because frequently the desired sugar, with a higher number of axial hydroxyl groups, will complex more strongly with cations [44]. In fact, the first application of chromatography on a cation-exchange column was for the separation of an LdB-AvE mixture: galactose, tagatose and talose, eluted in that order [45]. A column with lead cations was used for the analysis of mixtures formed in the LdB-AvE reaction [9] but lead is not particularly suitable for preparative separations [44]. One might suggest that, in the case of D-psicose, addition of a salt of a complexing cation to the reaction mixture might have shifted the equilibrium in favor of the product which readily forms a complex. Addition of such a salt would not interfere with the separation on the column if the same cation were to be used for the reaction and for the column.

This preparation of D-psicose illustrates the possible economy of a reaction proceeding in poor yield. The yield is only 8% but the starting material and the reagents are so cheap that it is worth carrying out this one-step reaction on a large scale. With standard laboratory equipment, 10 g can be prepared in 2–3 days. Another preparation of D-psicose, from D-allose in boiling pyridine [13], would also benefit from the use of a cation-exchange column, particularly because psicose has never been obtained in the crystalline state. The column gives pure psicose without crystallization.

There appears to be no description of the synthesis of an aldose from a ketose by the LdB-AvE reaction on the preparative scale.

5
Catalysis by Cations?

In the long history of the LdB-AvE transformation [8], the suggestion has appeared several times that the reaction is catalyzed by calcium ions, but the proof has often been inconclusive. There are several points on which different authors disagree. Let us examine some of them. Kusin reported [46] that when D-glucose reacted in calcium hydroxide solution at 25 °C no fructose was formed whereas, in a solution of sodium hydroxide, fructose was detected under the same conditions. This is contrary to the observations of LdB and AvE [4], Sowden and Schaffer [47] and Topper and Stetten [48], who all isolated fructose from the reaction in calcium hydroxide solutions.

Sowden and Schaffer [49] also detected what appeared to be a catalytic effect of calcium ions. They studied the reaction of mannose and found that it disappeared much faster in a solution of calcium hydroxide than it did in barium hydroxide and, in turn, in the latter medium faster than in sodium hydroxide. They followed the reaction by polarimetry and the results proved rather puzzling: in sodium hydroxide solution the initial rotation was positive (as it should be for D-mannose) and declined over time; but in calcium hydroxide the rotation was negative and increased as the reaction proceeded. The behavior in barium hydroxide was intermediate. Clearly, something more than mere catalysis was involved here. Incidentally, in LdB-AvE's original work [4], the greatest amount of mannose was produced in a concentrated solution of calcium hydroxide.

Kusin's results were recently confirmed, by chance and unexpectedly. During studies on the nickel-amine catalyzed epimerization (see Osanai, this vol.), calcium hydroxide plus amine was also tried [50] and proved successful. However, calcium cations do not form complexes with amines; and when calcium hydroxide was tried without the amine, mannose was still formed from glucose but not fructose [51]. It was found that the reaction did not proceed through an enediol but by carbon-carbon bond migration within a calcium complex.

Hence it appears that calcium ions indeed exert a catalytic effect but on another reaction, not on the LdB-AvE transformation. The two reactions run simultaneously but the calcium-catalyzed one is faster – hence if interrupted at an early stage, the amount of ketose formed is minimal.

The controversial results can now be explained. Kusin [46] used one equivalent of calcium hydroxide, sufficient to convert most of the glucose into a complex; thereby the LdB-AvE reaction is impeded. However, the earlier workers used only sufficient calcium hydroxide to raise the pH to a value where the LdB-AvE reaction ran smoothly, in order to minimize the side effects. Only a small proportion of the sugar formed a complex and the calcium-catalyzed epimerization was negligible. It is also clear now that the negative rotation observed by Sowden and Schaffer [49] was due to a calcium-mannose complex, not to mannose alone. Barium forms a weaker complex and sodium none.

6
The Calcium-Catalyzed Epimerization

Yanagihara et al. [51] claimed that calcium was the only metal cation which catalyzed the epimerization; they tried many others but only strontium had a weak effect. A closer investigation [52] showed cations that complex well with polyols [44] all catalyze the epimerization; those metals tested by Yanagihara et al. were poor complex formers. Calcium hydroxide is best suited for this reaction because it has reasonable solubility in water and is readily available. The reaction can also be conducted in methanol in which both calcium hydroxide and the sugars have only low solubility; but when they are both present, the solvent dissolves many of them, owing to complex formation.

The calcium-catalyzed epimerization is clearly different from the LdB-AvE transformation. It does not proceed through the enediol: when it is conducted in deuterium oxide, there is no incorporation of deuterium; and, if C-1 is substituted by ^{13}C, the substituent shifts to become C-2. The mechanism is therefore the same as in the Bílik and in the nickel-amine catalyzed epimerization, a carbon-carbon migration. This mechanism is explained [52] by the formation of a complex between calcium cations and the anionic form of the sugar; this holds the sugar in a conformation suitable for the migration of the bond from C-2 to C-3, just as molybdic acid and nickel-amine do. To convert most of the sugar to this complex, an amount of calcium hydroxide equivalent to the sugar is required, or even more; this is where the reaction conditions differ from those of the LdB-AvE reaction.

Curiously, the calcium-catalyzed epimerization does not occur with galactose. It was found to require a *threo* configuration of O-3 and O-4. To explain this, it was postulated [52] that calcium complexes with four oxygen atoms, namely O-1, O-2, O-3 and O-4. Structure **7** was therefore suggested for the complex. If the configuration of C-4 were reversed, such a complex would have a very unfavorable conformation. Cations usually complex with only three hydroxyl groups on a carbon chain; however, an ionic form of the sugar would have a

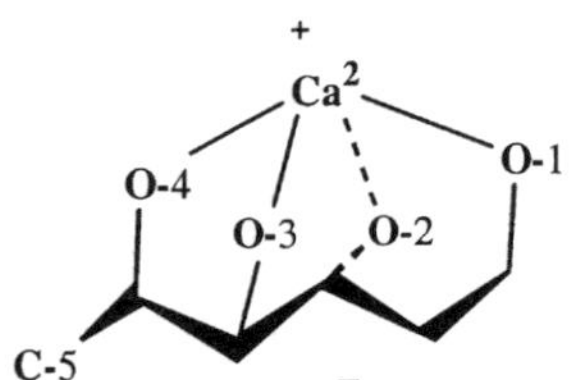

7

greater attraction to the cation and four of its oxygen atoms could replace the usual water molecules surrounding the cation. The distribution of the hydrogen atoms and the negative charge are not known and are therefore not shown in structure 7.

Owing to the large amount of calcium used, the reaction lends itself to the exploitation of complexation to increase the yield. Mannose (owing to its *cis* hydroxyl groups) forms a stronger complex than glucose does and, in the presence of a substantial amount of calcium ions, the equilibrium is shifted in favor of the former. Complex formation is stronger in methanol than in water [44] and in methanolic solution a yield of 55% of mannose can be obtained [51]. In water, in the absence of cations, the equilibrium mixture contains only 25% of mannose.

If this reaction is applied to a ketose it should result not in epimerization but in the formation of a branched-chain sugar, a 2-*C*-(hydroxymethyl)aldose, because the carbon atom which migrates carries a hydroxymethyl group. Preliminary results show that this is indeed the outcome of the reaction [53].

The interconversion of *N*-acetyl-D-glucosamine and *N*-acetyl-D-mannosamine in a saturated solution of calcium hydroxide [25], which leads to a 4:1 equilibrium, also proceeds by the calcium-catalyzed mechanism. It occurs very much faster in calcium hydroxide than in sodium hydroxide or barium hydroxide solution.

This reaction, being essentially restricted to the glucose/mannose and xylose/lyxose system and related equilibria, is not of wide applicability. It could be useful for the heptoses and for isotopically labeled or substituted sugars where the potentially higher yields make it desirable. However, the reaction is of interest because the carbon-carbon migration occurs with such ease (10 min at 65 °C is sufficient to almost reach equilibrium; the Bílik reaction requires 4 h at 90 °C), and a detailed study of its mechanism would be of interest.

7
The Mechanism of the LdB-AvE Transformation

It is generally agreed that the LdB-AvE reaction proceeds by an intermediate enediol structure; that is, the bond between the hydrogen atom and C-2 is broken and is then reinstated in another configuration or to another carbon atom. Wohl and Neuberg [54] were the first to suggest the enediol intermediate. It was confirmed by the fact that when the reaction was conducted in deuterium oxide, glucose, mannose and fructose were all found to be deuterium substituted; that is, deuterium from the solvent entered the molecules [55]. In subsequent experiments, sugars substituted with deuterium [47] or labeled with tritium [56, 57] were employed. Some of the label was lost during the LdB-AvE transformation. In illustrating the reaction, different authors have interpolated various supposed intermediates (or transition states) but essentially the reaction path is as shown in Scheme 1; full details are discussed at the end of this section.

However, during the last 60 years, there has been much dispute about this mechanism: numerous papers have appeared and various modifications of the simple enediol mechanism have been proposed. Many of these are reported in Speck's review [8]. There is poor agreement between the results of various au-

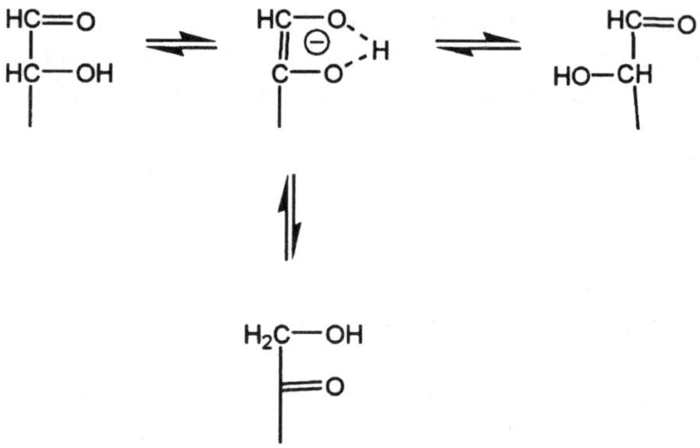

Scheme 1

thors and in some cases they are contradictory. It has sometimes been suggested that another concurrent reaction path exists; this has now been confirmed by the recognition of the cation-catalyzed epimerization. Most of the early investigations were carried out with D-glucose in solutions of calcium hydroxide. The extent to which the cation-catalyzed reaction would have contributed to the outcome of the reaction would therefore have been dependent on the amount of the calcium hydroxide used. This is particularly evident in the results of Fredenhagen and Bonhoeffer [55], where the outcome of the reaction clearly depended on the concentration of the calcium hydroxide. It is not possible, however, to correlate the results of the different investigators because they used different cations, concentrations, temperatures and reaction times. In particular, the authors refer to solutions saturated with calcium hydroxide but it is not stated whether saturation was carried out before or after the addition of sugar; this would have made a big difference because the solubility of calcium hydroxide is greatly increased by the presence of polyols. It is clear, for example, that with a large amount of calcium hydroxide present, the calcium-catalyzed reaction would have prevailed and there would have not been any incorporation of deuterium. However, equilibrium is rapidly established and, if the reaction is not then terminated, the slower LdB-AvE reaction will proceed and the ultimate result will be the formation of fructose and deuterium substitution.

Several modifications of the mechanism have been proposed in order to explain some of the incongruous results. It was suggested that mannose is not directly formed from glucose but from fructose. This is clearly inconsistent with the fact that sugars that cannot form a ketose (e.g., 2-*O*-methyl derivatives) are also converted into their epimers. Sowden and Schaffer [47] proved that fructose is not a necessary intermediate in the epimerization. They also suggested that there were two forms of the enediol, *cis* and *trans*, each giving rise to only one aldose. This seems very unlikely: the two forms would interchange very rapidly, and the *cis* form would be much more stable, being a resonance hybrid stabilized by an intramolecular hydrogen bond [58]. Only this form is shown in Scheme 1. Berl and Feazel had already suggested the resonance structure [59] and Bamford and Collins [60] proved that only one di-enol anion is involved in the reaction.

To explain some of the anomalous results in the distribution of tritium in D-arabinose formed from (2-^{3}H)ribose, Gleason and Barker [56] suggested that some of the reaction occurred by transfer of the C-2 hydrogen atom with its electrons to C-1, rather than by enolization. This appears unlikely to be energetically favored. Isbell [61] found that hydride transfer was insignificant in the conversion of D-glucose to D-fructose. These results can now be explained. The "tritium migration" could have occurred by the cation-catalyzed epimerization path; actually, C-2 would have migrated, carrying the tritium atom with it. Hence migration was found during the epimerization but not during the aldose-ketose interconversion (which is not catalyzed by cations).

The facile epimerization of aldoses that carry a sulfur atom on C-2 does not involve any other mechanism except enediol formation [28]; there is complete exchange of H-2 for deuterium when the reaction is carried out in heavy water. The epimerization of aldonic acids also proceeds via the enediol; here also complete exchange of H-2 occurs in heavy water [31]. Similarly, the enzyme-catalyzed epimerization proceeds via the enediol but, recently, a case has been described for which a path involving carbon-carbon migration was postulated [62]. de Wit et al. [58] carried out a more detailed study of the enolization process by using ultraviolet (UV) spectroscopy; reaction rate constants were determined. These data suggest that the neutral open-chain aldehyde form is not involved in the process. The base removes the proton from O-1 of the cyclic form, this being the most acidic proton in the molecule; then the ring of the sugar anion opens, the charge moving to O-5. It is the anionic O-5 that extracts the proton from O-2, thereby producing the enediol anion. This is the rate-determining step of the enolization. The essential steps are shown in Scheme 2 (without the steric details which were carefully worked out by the authors). The UV spectra show that the concentration of the enediol anion is up to 0.6% of the starting sugar. Calcium ions increase the rate of formation of the enediol anion [63].

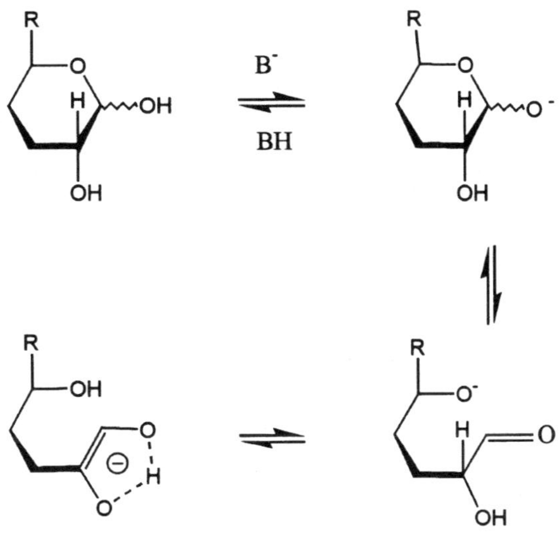

Scheme 2

Isbell and his group have carried out extensive investigations on the LdB-AvE reaction [9, 12, 14, 23, 57, 61] dealing mainly with the enolization step. However, the last of their papers [23] presents some puzzling results. The rate of disappearance of all of the hexoses was studied in aqueous potassium hydroxide solution at pH 11.5. Owing to the formation of acidic by-products, the pH dropped and had to be occasionally readjusted to 11.5 by the addition of potassium hydroxide. The results are useful in showing that the rate of the reaction varies considerably with the configuration of the sugar. Thus, idose reacts very rapidly, the reaction of altrose and talose is fast, that of glucose slower and that of allose and mannose very slow. Ketoses react faster, particularly tagatose.

The proportion of glucose, mannose and fructose ultimately approaches that at equilibrium, no matter which is the starting material. However, the disappearance of allose comes to a stop at about 75%, much higher than the equilibrium value; from altrose only 5% of allose is produced. The final proportion in some other cases is also far from that at equilibrium. The proposed explanation for these results, formation of hydrogen-bonded, unreactive ions or molecular complexes, is clearly unsatisfactory. Maybe the formation of acidic by-products from the less stable sugars creates conditions unfavorable for the LdB-AvE reaction even if the pH is kept constant. There are, indeed, many by-products: after 5 days at 55 °C in a solution of potassium hydroxide at pH 11.5, the total monosaccharides remaining in a solution of D-mannose accounted for only 16% of the starting sugar.

Glyceraldehyde and dihydroxyacetone interconvert at higher temperatures in aqueous solution even in the absence of a base. The reaction has recently been studied by infrared spectroscopy [64] and, here also, the enediol proved to be the intermediate but in its undissociated form. It has been suggested that it also has a hydrogen bond but it connects OH-3 to OH-1, the enediol being *trans*.

To summarize: the main reaction path of the LdB-AvE transformation is undoubtedly the one involving the anion of the enediol. A small proportion of the products, however, may be formed by the cation-catalyzed mechanism, by degradation and recombination, and possibly by other ways. There is no evidence of the occurrence of hydride migration, except when the reaction is catalyzed by acids [65]. The reaction path has always been illustrated as starting with the aldehydo form of the sugars and leading to the aldehydo or keto form of another sugar but it is by no means certain that these open-chain forms are intermediates in the LdB-AvE reaction; the enediol anion may be formed directly from a cyclic form and be converted into a cyclic form of another sugar.

8
Concluding Remarks

After more than a century, the LdB-AvE transformation is unlikely to find new uses or be developed further; in many cases it has been displaced by newer reactions. However, if the old procedures were to be applied, they could probably be improved in two ways. In some instances the equilibria could be shifted in favor of the desired product by the addition of complexing agents: either boric acid or boronates, or an excess of cations, like Ca^{2+}, thereby improving the yield. In other cases, the work-up could be simplified by the use of chromatography on cation-exchange resins. This would result in a better separation and improved yield of the product.

9
References

1. De Raadt A, Ekhart CW, Stütz AE (1995) Advances in detailed reaction mechanisms 4:175
2. Eliel EL, Allinger NL, Angyal SJ, Morrison GA (1966) Conformational analysis. Wiley, New York
3. Whistler RL, Wolfrom ML (1962/63) Methods in carbohydrate chemistry, vols 1–2. Academic Press, New York
4. Lobry de Bruyn CA, Alberda van Ekenstein W (1895) Recl Trav Chim Pays-Bas 14:203
5. Lobry de Bruyn CA, Alberda van Ekenstein W (1896) Recl Trav Chim Pays-Bas 15:92
6. Lobry de Bruyn CA, Alberda van Ekenstein W (1895) Recl Trav Chim Pays-Bas 14:156; ibid (1897) 16:262, 274, 282; (1899); ibid 18:147; ibid (1900) 19:1
7. Quadbeck-Seeger H-J, Faust R, Knauss G, Siemeling U (1999) World records in chemistry. Wiley-VCH, p 278
8. Speck JC Jr (1958) Adv Carbohydr Chem 13:63
9. El Khadem HS, Ennifar S, Isbell HS (1987) Carbohydr Res 169:13
10. Young BY, Montgomery R (1996) Carbohydr Res 280:27
11. Fischer HOL, Taube G, Baer E (1927) Ber 60:479
12. Linek K, Fedoronko M, Isbell HS (1972) Carbohydr Res 2:326
13. Steiger M, Reichstein T (1936) Helv Chim Acta 19:184
14. Ennifar S, El Khadem HS (1989) Carbohydr Res 193:303
15. Beveridge RJ, Davis M, Morris JL, Hoogenraad NJ (1982) Carbohydr Res 101:348
16. Rendleman JA, Hodge JE (1979) Carbohydr Res 75:83
17. Langlois DP, Larson RF (1956) US Patent 2,746,889 [CA 50:15110b]; Carubelli R (1970) US Patent 3,505,309 [CA 73:151169t]; Haack E, Braun F, Kohlor K (1966) US Patent 3,256,270 [CA 60:14598A]; Parrish FW (1969) US Patent 3,431,253 [CA 71:30672d]; Guth JH (1970) US Patent 3,546,206 [CA 74:100810k]; Tumerman L, Guth JH (1974) US Patent 3,822,249 [CA 81:63927a]
18. Carubelli R (1966) Carbohydr Res 2:480
19. Hicks KB, Parrish FW (1980) Carbohydr Res 82:393
20. Mendicino JF (1960) J Am Chem Soc 82:4975
21. de Bruijn JM, Kieboom HPG, van Bekkum H (1986) Rec Trav Chim Pays-Bas 105:176
22. Barker SA, Sommers PJ, Hatt BW (1973) Carbohydr Res 26:41; Ger Offen 2229064 [CA 80:83524z]
23. El Khadem HS, Ennifer J, Isbell HS (1989) Carbohydr Res 185:51
24. Prentice N, Cuendet LS, Smith F (1956) J Am Chem Soc 78:4439
25. Sugai T, Kuboki A, Hiramatsu S, Okazaki H, Ohta H (1995) Bull Chem Soc Jpn 68:3581
26. Lee AVM, Martin VS, Masamune S, Sharpless KB, Walker FJ (1982) J Am Chem Soc 104:3515
27. Horton D, Berrang B, Norris P (1996) Carbohydr Res 283:53
28. Petrušová M, Lattová E, Matulová M, Petruš L, BeMiller JN (1996) Carbohydr Res 283:73
29. Fischer E (1890) Ber 23:799; ibid (1891) 24:2136
30. Haworth WN, Long CW (1929) J Chem Soc 345
31. Kalman M, Sokolowski J, Szafranek J, Lönnberg H (1987) J Carbohydr Chem 6:587
32. Sternbach LH (1952) US Patent 2,438,881 [CA 42:5048e]; Schmidt W, Paust J (1980) Ger Offen 2,852,720 [CA 93:186747x]; 2,852,721 [CA 93:186746w]
33. Carlsson B, Samuelson O, Popoff T, Theander O (1969) Acta Chem Scand 23:261
34. Fischer FG, Schmidt H (1959) Chem Ber 92:2184
35. Baggett N, Smithson A (1982) Carbohydr Res 108:59
36. Palleroni MJ, Doudoroff M (1956) J Biol Chem 218:535

37. Cohen SS (1953) J Biol Chem 201:71
38. Vargha L (1954) Ber 87:1351
39. Schaffer R (1964) J Org Chem 29:1473
40. King-Morris M, Serianni AS (1986) Carbohydr Res 154:29
41. Angyal SJ, Bethell GS, Beveridge RJ (1979) Carbohydr Res 73:9
42. Angyal SJ, Mills JA (1985) Aust J Chem 38:1279
43. Angyal SJ, Craig DC (1993) Carbohydr Res 241:1
44. Angyal SJ (1989) Adv Carbohydr Chem Biochem 47:1
45. Jones JKN, Wall RA (1960) Can J Chem 38:2290
46. Kusin A (1936) Ber 69:1041
47. Sowden JC, Schaffer R (1952) J Am Chem Soc 74:505
48. Topper YJ, Stetten D Jr (1951) J Biol Chem 189:191
49. Sowden JC, Schaffer R (1952) J Am Chem Soc 74:499
50. Yamauchi TG, Fukushima K, Yamagihara R, Osanai S, Yoshikawa S (1990) Carbohydr Res 204:233
51. Yanagihara R, Saeda K, Shiina S, Osanai S, Yoshikawa S (1997) Bull Chem Soc Jpn 66:2268
52. Angyal SJ (1997) Carbohydr Res 300:279
53. Hricovíniová-Biliková Z, Hricovíni M, Petrušová M, Serianni AS, Petruš L (1999) Carbohydr Res 319:38
54. Wohl A, Neuberg C (1900) Ber 33:3095
55. Fredenhagen H, Bonhoeffer KF (1938) Z Physik Chem (Leipzig) A181:392
56. Gleason WB, Barker R (1971) Can J Chem 49:1433
57. Isbell HS, Frush HL, Linek K, Hepner KE Jr (1971) Carbohydr Res 9:319
58. de Wit G, Kieboom APG, van Bekkum H (1979) Carbohydr Res 74:157
59. Berl W, Feazel CE (1951) J Am Chem Soc 73:2054
60. Bamford CH, Collins JR (1955) Proc Royal Soc (London) A228:100
61. Isbell HS (1973) Adv in Chem Series 117:70
62. Arigoni D, Giner J-L, Sagner S, Wungsintaweekul J, Zenk MH, Kis K, Bacher A, Eisenreich W (1999) J Chem Soc Chem Commun 1127
63. de Bruijn JM, Kieboom APG, van Bekkum H (1987) Rec Trav Chim Pays-Bas 106:35
64. Yaylayan VA, Harty-Majors S, Ismail AA (1999) Carbohydr Res 318:20
65. Harris DW, Feather MS (1975) J Am Chem Soc 97:178

The Bílik Reaction

Ladislav Petruš[1], Mária Petrušová[2], Zuzana Hricovíniová[3]

Institute of Chemistry, Slovak Academy of Sciences, 84238 Bratislava, Slovakia
[1] *E-mail: chemlpet@savba.sk,*
[2] *E-mail: chemlaci@savba.sk,*
[3] *E-mail: chemhric@savba.sk*

The introduction of the Bílik reaction, the molybdic acid catalyzed interconversion of epimeric aldoses, is an important milestone in carbohydrate chemistry. The essentials of this unique, stereospecific carbon-skeleton rearrangement of epialdoses are presented. Emphasis is laid on the latest developments in the area, namely the mutual interconversion of 2-ketoses and 2-*C*-(hydroxymethyl)aldoses, a reaction that is exploited for the preparation of some important representatives of these reducing sugars. Mechanistic studies with isotopically substituted D-fructoses are also described.

Keywords: Bílik reaction, Molybdic acid catalysis, Epimerization, Carbon-skeleton rearrangement, Aldoses, 2-Ketoses, 2-*C*-(Hydroxymethyl)aldoses, Hamamelose, Sedoheptulose

1 Introduction

In 1972, a remarkable carbohydrate reaction, the molybdic acid catalyzed epimerization of aldoses, was reported [1], filling an existing gap in the synthesis of carbohydrates. While the facile preparation of ketoses from common aldoses by the Lobry de Bruyn-Alberda van Ekenstein isomerization reaction has been known since the end of the 19th century [2], several aldoses, formerly called rare aldoses, were only available rather infrequently, as their preparation procedures were very demanding. It is indubitable that the simple aldose interconversion, named the Bílik reaction after its discoverer, substantially helped to ignite the carbohydrate research boom that started at approximately the same time. Namely, an early commercialization of the Bílik reaction has made such aldoses as L-glucose, L-ribose, L-lyxose, D-lyxose, D-talose, or D-mannose easily available in one-step reactions starting from their more common respective epimers. Over the decades, different quantities of the pertinent sugars, ranging from grams up to hundreds of kilograms, have streamed to carbohydrate investigators in a broad array of research fields, through several chemical suppliers, chiefly from a Bratislava pilot plant facility.

In the meantime, other catalytic systems that enable similar interconversions of aldoses have been described. However, these have not been as widely exploited as the reaction catalyzed with molybdic acid. Despite the initial interest, approximately 15 years after the reaction had first been reported, investigations of the Bílik reaction and other related interconversions, in the main centered on their scope and mechanisms, had apparently exhausted all avenues of interest. In the mid-1990s, however, new, interesting applications of the reactions appeared which have subsequently facilitated easy obtainment of, e.g., some branched-chain aldoses, thus suggesting further possible developments in the area. Although the aim of this overview is to provide a survey of these latest developments, a brief description of the earlier stages of the research, as well as a necessary description of the mechanism of the Bílik reaction, are also included since there is still no available comprehensive review on the topic.

2 Essentials of the Bílik Reaction

2.1 Discovery

Similarly to many other great discoveries, the molybdic acid catalyzed interconversion of epimeric aldoses is the result of serendipity. Thus, when D-talose, obtained from another, lesser-known reaction also discovered by V. Bílik, namely the stereoselective, molybdate-catalyzed hydroxylation of glycals [3], was repeatedly recrystallized (in the presence of a catalyst that had not been completely removed), it was not possible to obtain it in a pure condition, as the aldose slowly epimerized to D-galactose [4]. A simple verification of the explanation of this originally undesirable interconversion, the treatment of D-galactose with molybdenum(VI) oxide in water at increased temperature, gave rise to D-talose. In spite of these primary observations of the molybdate-catalyzed epimerization of aldoses, the first two papers published describing the new interconversion of the epimeric aldoses were devoted to the glucose to mannose (and vice versa) mutual transformations in both the D and L series [1a,b], as no parallel reactions occurred during the interconversions performed under mild reaction conditions.

2.2 Substrate Scope and Limitations

As a result of extensive studies performed with as many epimeric pairs of aldoses as possible, the scope and limitations of the Bílik reaction (Scheme 1) can be summed up as follows. The treatment of an aldose, not shorter than a tetrose, in a mild acidic solution in the presence of molybdate ions (usually by heating an aqueous solution containing 10–20% of starting aldose and 0.1–0.2% molybdic acid for 2–6 h at 70–90 °C) gives rise to an equilibrium mixture of epimeric aldoses, without formation of the complementary 2-ketose, and independent of which aldose is used as starting material. In every case, a conformationally more stable aldose prevails in the thermodynamic mixture of the epimeric aldoses.

Since one aldose of any epimeric pair is usually more easily available than the other, either directly isolable from its natural source or in combination with a chain-length modification, the application of the Bílik reaction enables the other, less-available aldose to be obtained. Thus, the following equilibria can be obtained:

D-Threose/D-erythrose	4:3	[5]
5-Deoxy-L-arabinose/5-deoxy-L-ribose	3:1	[6]
D-Xylose/D-lyxose	7:5	[7]
L-Arabinose/L-ribose	2:1	[8]
D-Glucose/D-mannose	3:1	[1a]
D-Galactose/D-talose	4:1	[9]
D-Allose/D-altrose	3:2	[10]
L-Quinovose/L-rhamnose	3:2	[11]
L-Fucose/6-deoxy-L-talose	4:1	[6]
D-*glycero*-L-*gluco*-Heptose/D-*glycero*-L-*manno*-heptose	4:1	[12]
D-*glycero*-D-*gulo*-Heptose/D-*glycero*-D-*ido*-heptose	4:1	[13]
D-*glycero*-D-*galacto*-Heptose/D-*glycero*-D-*talo*-heptose	4:1	[13]
7-Deoxy-L-*glycero*-L-*galacto*-heptose/ 7-deoxy-L-*glycero*-L-*talo*-heptose	4:1	[6]

Similar thermodynamic equilibria were also obtained with a few other higher aldoses (D-*erythro*-L-*galacto*-octose, D-*erythro*-L-*talo*-octose, D-*threo*-L-*galacto*-octose, D-*threo*-L-*talo*-octose [14], D-*erythro*-L-*gluco*-octose and D-*erythro*-L-*manno*-octose [15]) and some 1–6 linked reducing disaccharides (melibiose (6-*O*-α-D-galactopyranosyl-D-glucose) and epimelibiose (6-*O*-α-D-galactopyrano-

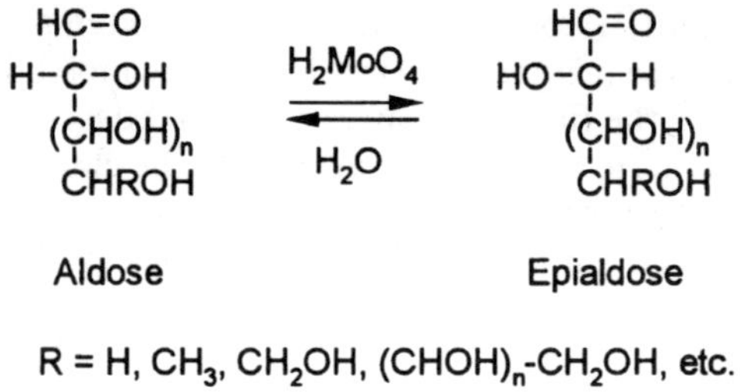

Scheme 1

syl-D-mannose) [16]) on treatment with molybdic acid. Again, the prevailing epimers were those with the *galacto* or *gluco* configurations at their C-2 to C-5 atoms. While common 1–4 linked reducing disaccharides such as lactose (4-*O*-β-D-galactopyranosyl-D-glucose) and maltose (4-*O*-α-D-glucopyranosyl-D-glucose), as well as their epimers (epilactose and epimaltose), do not undergo molybdic acid catalyzed epimerization [17], celtrobiose (4-*O*-β-D-glucopyranosyl-D-altrose) does to give epiceltrobiose (4-*O*-β-D-glucopyranosyl-D-allose) [18].

The importance of the presence of the OH group at C-4 for an unambiguous course of the molybdate-catalyzed epimerization was further shown in studies with trioses [19, 20], 4-deoxy-D-*lyxo*-hexose and several other 4-*O*-substituted derivatives of aldoses [20]. Much more important is the OH group at C-3 which is essential for the Bílik reaction. Its absence causes a quite different structural change in both 3-deoxy-D-*arabino*-hexose (**1**) and 3-deoxy-D-*ribo*-hexose (**2**) under otherwise the same reaction conditions; both sugars on treatment with molybdic acid are irreversibly transformed to 3-deoxy-D-*erythro*-hex-2-ulose (**3**, Scheme 2) [21].

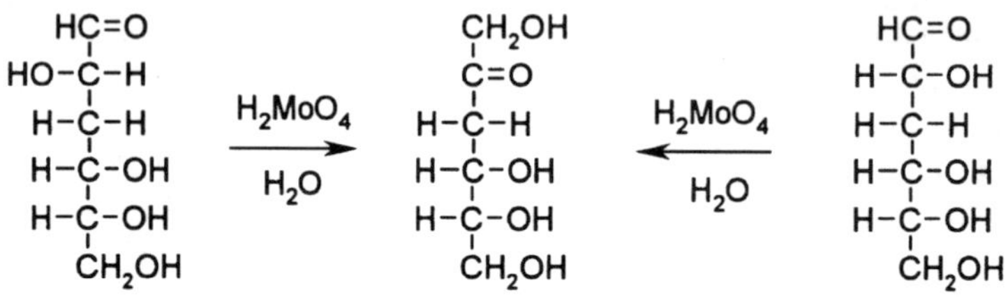

1 **3** **2**

Scheme 2

The essentiality of the OH group at C-2 for the epimerization of aldoses is obvious. The same is valid for the carbonyl group of aldoses since alditols do not undergo epimerization changes in molybdic acid [20, 21].

The most stable epimeric pair of aldoses, D-glucose and D-mannose, has been often used as the model system for optimization, kinetic and mechanistic studies of the reaction.

2.3 Catalysts

The rate of the molybdate-catalyzed epimerization reaction is highest in the pH range 1.5–3.5 and is up to 20 times higher than at pH 5.9, where the cyclic species of aldoses form the most stable molybdate complexes [21, 22]. At pH higher than 6.0 and below pH 0.1 the molybdate-catalyzed epimerization of aldoses does not occur [22, 23].

Practically any molybdenum(VI) compound can be used for the molybdate-catalyzed epimerization of aldoses, provided that the pH of the reaction mixture is adjusted with an acid to the aforementioned optimum value. There are a few restrictions: oxalic acid, which inhibits the Bílik reaction [22] apparently due to the formation of its bidentate molybdenum(VI)-oxalate complexes [24], and some hydroxycarboxylic acids which behave similarly cannot be considered for use. In practice, ammonium heptamolybdate (giving aqueous solutions with pH ~ 5.7) in combination with a mineral acid is widely used. This is, however, not the most convenient solution, as the removal of the catalyst by deanionization with an anion-exchange resin must be immediately followed by decationization because the basic medium can, in some cases, spoil the result of the molybdic acid catalyzed transformation. The same is valid for ammonium dimolybdate (giving aqueous solutions with pH ~ 4.5) which is, in some chemical catalogues, improperly named as molybdic acid. There are two forms of the real molybdic acid [25]; both of them possess a polymeric structure (the "white" molybdic acid, $MoO_3 \cdot H_2O$, is an infinite double chain while the yellow molybdic acid, $MoO_3 \cdot 2\,H_2O$, is an infinite double plain). The acids crystallize from fresh precipitates obtained on strong acidification of aqueous ammonium polymolybdate solutions. The fresh precipitates [26] are very soluble in water, give solutions of pH ~ 2.9, which is an optimum value for the Bílik reaction, and the catalyst is easily removed by a single deanionization with a strongly basic anion-exchange resin. The same effect can be obtained even with molybdenum(VI) oxide that forms a three-dimensional polymer network [27]. Although the oxide is poorly soluble in water, under the conditions of the Bílik reaction it quickly dissolves and functions properly [23].

Two more sophisticated applications of the molybdate catalyst are worthy of mention. An anion-exchange resin immobilized heptamolybdate has provided an effective catalysis while the same resin activated with mononuclear sodium molybdate was completely ineffective [23]. In contrast, a patent application claims a successful activation with the same sodium molybdate for an effective epimerization of D-glucose to D-mannose [28]. Therefore, attention has to be paid when performing such applications as the equilibria between the (poly)molybdate species in solution are very complex and extremely pH-sensitive [29].

Dioxobis(2,4-pentanedionato-*O,O′*)molybdenum(VI) has been reported as a catalyst for molybdate-catalyzed epimerizations in polar aprotic solvents such as dimethylformamide (DMF) [30]. In spite of more interesting reaction equilibria, the method is not as convenient or as economical as the classical aqueous one due to necessary additional purification procedures.

2.4
Mechanism

Since the first observation of the molybdic acid catalyzed interconversion of epimeric aldoses it has been clear that the mechanism of the reaction is different to the enediol pathway of the Lobry de Bruyn-Alberda van Ekenstein (LdB-AvE) isomerization reaction of aldoses that provides 2-ketoses as primary major products [31]. Moreover, the interconverting aldoses do not incorporate deuterium if the molybdic acid catalyzed reaction is performed in deuterium oxide [32]. Furthermore, using specifically the tritium-labeled aldoses that were available in the 1970s, it has been proved that during the interconversion D-[1-^{3}H]glucose transforms to D-[2-^{3}H]mannose and vice versa. The results suggested that the process is intramolecular and involves a simultaneous transfer of the hydrogen atoms linked to the carbon atoms C-1 and C-2.

All these unusual observations have suggested that the reaction must involve molybdate complexes of the interconverting aldoses. Molybdate complexes of saccharides [33] and other polyhydroxy compounds [27] are known and several types of structures have been proposed and proved in which the ligands can be of bidentate, tridentate or tetradentate nature. To explain the simultaneous hydrogen transfer suggested, the idea of bidentate aldopyranose ligands in linear oligomolybdate chains present in solutions at pH $\leq$ 5 was adopted [32]. Typical examples of such bidentate ligands in molybdate complexes are dimolybdates $(NH_4)\{[MoO_2(C_6H_4O_2)_2]O\} \cdot 2\ H_2O$ [34] and $K_2\{[MoO_2(C_2O_4)H_2O]_2O\}$ or polymolybdate $\{NH_4Na[MoO_3(C_2O_4)] \cdot 2\ H_2O\}_n$ [24]. In every case the bidentate ligands are planar. The aldopyranose bidentate ligand **4** suggested could adopt such a planar structure to be efficiently bound only after a deformation of the C-3-C-2-C-1-O-5 segment of its chair conformation with both original C-2-OH and C-1-OH hydroxyl groups in the equatorial positions. In such a half-chair ${}^{4}H_{5}$-conformation of the aldose moiety of the complex **5**, which must be in an equilibrium with the ${}_{4}H^{5}$-conformer **6**, the parallel, *trans*-oriented C-2-H and C-1-H bond orbitals can efficiently overlap to form two transitory tricentric bonds (Scheme 3). This transition state could mediate the intramolecular hydrogen transfer between the carbon atoms C-1 and C-2 causing simultaneous configurational changes there, i.e. epimerization at C-2 in the final aldopyranose **7** (the configurational change at C-1 is not observable due to a fast anomeric equilibration) [32].

In 1982, however, a paper from a laboratory involved in the synthesis of saccharides enriched with stable isotopes appeared [20] which fundamentally changed the interpretation of the former results and significantly contributed to the completion of the understanding of the mechanism of the Bílik reaction. Namely, when aldoses ^{13}C-isotopically substituted at C-1 or C-2 were subjected to molybdate-catalyzed epimerization in order to prepare the corresponding epimers, the C-1, C-2 carbon-skeleton rearranged products were isolated instead. Thus, D-(1-^{13}C)mannose, D-(1-^{13}C)talose, D-(1-^{13}C)galactose, D-(1-^{13}C) glucose, 6-deoxy-L-(1-^{13}C)talose, 6-deoxy-L-(2-^{13}C)talose, D-(1-^{13}C)xylose and D-(1-^{13}C)arabinose gave D-(2-^{13}C)glucose, D-(2-^{13}C)galactose, D-(2-^{13}C)talose, D-(2-^{13}C)mannose, L-(2-^{13}C)fucose, L-(1-^{13}C)fucose, D-(2-^{13}C)lyxose and D-(2-

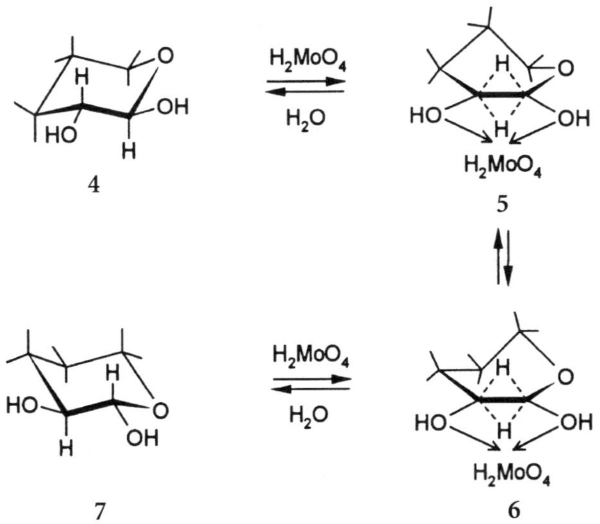

Scheme 3

^{13}C)ribose, respectively. Based on these observations, and the structures of the molybdate complexes of D-mannitol and D-lyxose determined by X-ray crystal structural analysis (which have appeared since) [35, 36], the acyclic, aldehyde forms of interconverting aldoses in the respective zigzag and sickle conformations, linked in the dimolybdate anionic complexes as tetradentate ligands via their carbonyl (O-1) and neighboring three hydroxylic (O-2, O-3 and O-4) oxygen atoms, were suggested as the active species mediating the stereospecific carbon-skeleton rearrangement, formally resulting in the C-2 epimerization of aldoses [20]. The interpretation scheme suggested by the authors is still accepted except for the participation of the aldehyde oxygen in the molybdate complexes of the interconverting aldoses. A close analysis of the structure of the molybdate complexes of aldoses [27, 35–37] reveals strict stereochemical rules, according to which all the participating oxygen atoms must come from the hydroxyl groups. Namely, two hydroxyl groups of the sugar, linked in a tetradentate fashion in the dimolybdate complex, are formally esterified (each of them linked to a different molybdenum atom) while the other two hydroxyls are coordinated to both molybdenum atoms and release their protons to give a formal (–II) charge to the complex [33c]. This is in accordance with the general equation of the formation of dimolybdate complexes of polyols, as shown in Eq. (1).

$$C_nH_m(OH)_k + H_2Mo_2O_7 \rightleftarrows \{[(MoO_2)O][C_nH_mO_4(OH)_{(k-4)}]\}^{2-} + 2\,H_3O^+ \tag{1}$$

The decrease in the acidity observed on mixing polyols with, for example, ammonium heptamolybdate is about $\Delta pH \approx 0.5$ [38].

Thus, from all the experimental data available, it follows that the aldoses must enter the catalytically active molybdate complexes in their aldehydrol forms, and this has subsequently been unambiguously proved by ^{1}H, ^{13}C and ^{95}Mo NMR spectroscopy [39]. The NMR studies also detected zigzag conformations for the dimolybdate complexes of the C-1 to C-4 moieties of the acyclic, hydrated forms of the C-2, C-3 *threo*-configured aldoses but sickle conformations for those of the C-2, C-3 *erythro*-oriented aldoses. Based on the scheme originally proposed by the authors [20] corrected by these NMR data and applying the strict stereochemical rules for the dimolybdate complexes (briefly, in every dimolybdate complex, two octahedron corners, opposite to both shortest, multiple Mo = O bonds, always being *cis* arranged, are occupied with ligands providing both longest coordination bonds while both single, usually the Mo-O bonds connect the remaining opposite corners of the MoO_6 octahedron [27]), the mechanism of the Bílik reaction generally accepted today is represented in Scheme 4.

The hydrated, acyclic form of an aldose complexed by its four adjacent hydroxyl groups, HO-1, HO-2, HO-3, and HO-4, in the tetradentate binuclear molybdate complex **8** undergoes delocalization to give the transient state **9**. At this point both the C-2–C-3 and the C-1–C-3 bond, the former one disrupting whilst the latter one is newly formed, are equivalent. Moreover, the process is

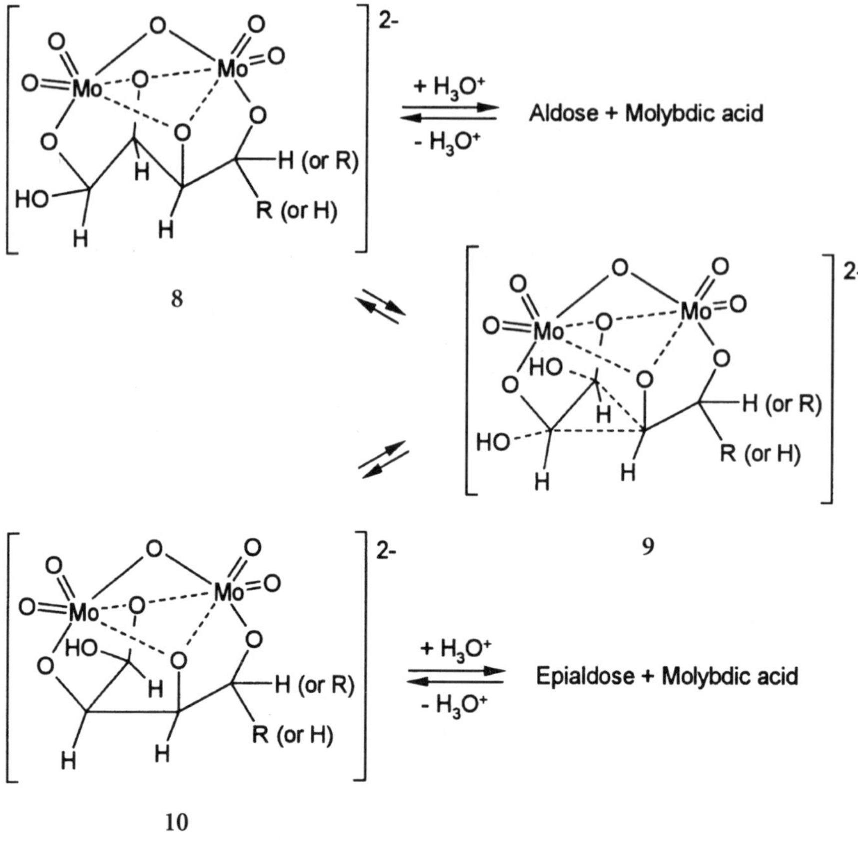

Scheme 4 R=H, CH_3, CH_2OH, $(CHOH)_n$–CH_2OH, etc.

accompanied by a simultaneously occurring dehydration and rehydration of the respective carbon(yl) atoms C-1 and C-2. Thus, the process results in the formation of the molybdate complex of the epimeric aldose **10**. The epialdose, however, in comparison with the starting aldose, has C-1 and C-2 mutually exchanged, together with their substituents (hydrogen atoms and hydroxyl groups). Therefore, the process, if strictly classified, should not be denoted as an epimerization, i.e., the process including the configuration change takes place only at one (C-2) carbon atom, but as a stereospecific carbon-skeleton rearrangement which, only in cases where all the aldose atoms are not isotopically distinguished, results in the epimerization of the unsubstituted aldoses. As the Bílik reaction utilizes only a catalytic amount of the catalyst (if expressed as dimolybdate, usually 1–2 mol % per aldose), it results in the formation of the thermodynamic equilibrium of the rearranged epimeric aldoses.

2.5 Secondary Process

A close analysis of the reaction mixtures obtained on treatment of any aldopentose or higher aldose with molybdic acid reveals the presence of the complementary, C-3 epimeric pair of aldoses. The amount of these secondary products is however different for each pair of epimers. Under gentle conditions of epimerization, practically no complementary aldoses are formed except in the glucose/mannose homomorphous series [1 a,b, 11, 12, 15] and the D-allose/D-altrose interconversions [10]. The lowest energy barrier of the formation of the C-3 epimers in the Bílik reaction has been observed for aldopentoses, where it is practically impossible to reach a thermodynamic equilibrium for any C-2 epimeric pair without contamination by the corresponding C-3 epimers [1a, 8]. Under more forcing reaction conditions involving higher amounts of the catalyst (over 10 mol %) and higher temperatures (over 95 °C), even the thermodynamic equilibria of all the C-2, C-3 diastereoisomers can be reached, as shown with aldopentoses [7] (Scheme 5) or the C-4, C-5 D-*erythro* aldohexoses [40] (Scheme 6).

The former example has been conveniently applied for the preparation of all D-[U-^{14}C]aldopentoses [7], while the latter enables D-allose and D-altrose to be obtained in interesting yields in one step from cheap D-glucose [40].

On treatment with molybdic acid, in addition to the primary C-2 epimeric products, the C-4, C-5 *threo* aldoses also similarly provide the complementary C-3 epimers. However, a proportion of these C-2, C-3 diastereoisomers, especially idoses and to some extent guloses and taloses, are transformed into more stable 2-ketoses, namely to sorboses and tagatoses in the case of hexoses [9] and their 6-deoxy derivatives [6] or to *gluco*-glyco-2-uloses and *manno*-glyco-2-uloses in the case of higher aldoses [6, 13]. The formation of these 2-ketoses from their parent aldoses is not associated with the molybdate catalysis and is obviously ascribed to the general acidity [41] of the reaction medium.

All the experimental data strongly suggest that the mechanism of the secondary process that operates in parallel under the conditions of the Bílik reaction may proceed via a transition state consisting of two tricentric C-3–H–C-2 bonds of an aldopyranose bidentately linked via C-3–OH and C-2–OH in a molybdate

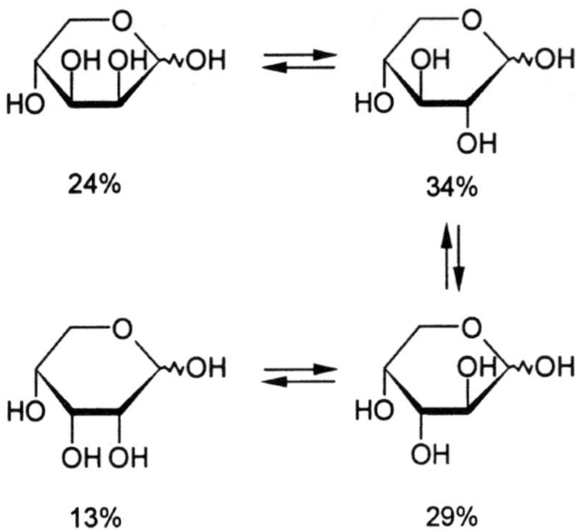

Scheme 5. Thermodynamic equilibrium of all D-aldopentoses obtained on treatment of D-[U-^{14}C]arabinose or D-[U-^{14}C]xylose (2 mg, 7.5 μmCi) with molybdic acid (2 mg) in water (2 mL) at 98–100 °C for 10 and 14 h, respectively [7]

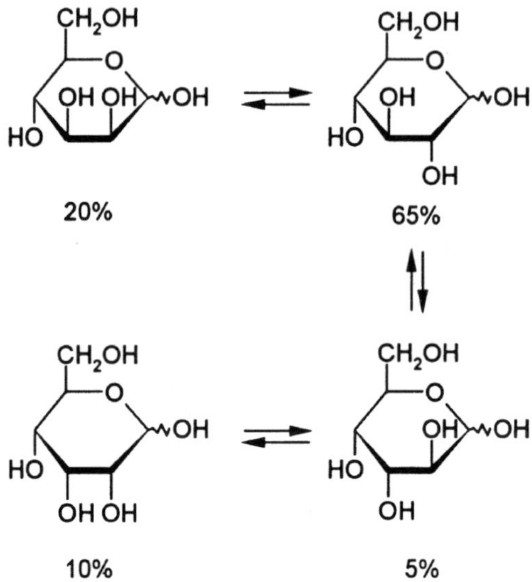

Scheme 6. Reaction equilibrium of the C-4, C-5 D-*erythro*-hexoses obtained on 2 h treatment of 2 M D-glucose in 0.004 M ammonium heptamolybdate in 0.1 M aqueous acetic acid at 120 °C [40]

complex, thus making possible a mutual reversal of configuration at the pertinent carbon atoms, i.e., in a similar manner to that originally suggested for a partially revealed pathway of the primary process [32] (cf. Scheme 3). In spite of this no direct evidence exists of the pathway proposed; observations such as, e.g., the formation of D-(1-^{13}C)gulose, D-(2-^{13}C)idose and D-(2-^{13}C)galactose from D-(1-^{13}C)talose [20], or similarly D-(2-^{13}C)arabinose, D-(2-^{13}C)xylose and D-(1-^{13}C)lyxose from D-(1-^{13}C)ribose [23] on the one hand, and the formation of D-(1-^{13}C, ^{2}H)lyxose essentially as the only product from D-(2-^{13}C, ^{2}H)xylose on the other [20], do admit the explanation. Analogously to the reactions that involve proton abstractions or hydride transfers, a significant isotope effect would be expected also for the mutual hydrogen exchange proposed. The substitution of the hydrogen atom with deuterium at C-2 should significantly diminish the reaction rate involving the transfer of the deuterium atom from C-2 to C-3 resulting in failure to form D-(2-^{13}C, 3-^{2}H)arabinose [and, subsequently, by the primary process, D-(1-^{13}C, 3-^{2}H)ribose] as expected.

Generally, the mechanism proposed to explain the secondary process includes the aldopyranose bidentate ligand **11** which, after linking in a bidentate molybdate complex via the oxygen atoms of its hydroxyl groups at C-2 and C-3 in the equatorial positions of the $^{4}C_{1}$ conformation of D-xylopyranose (**11a**) [alternatively of D-glucopyranose (**11b**) or D-galactopyranose (**11c**)], adopts a planar structure of its C-4–C-3–C-2–C-1 segment (Scheme 7). In such a half-chair conformation $^{0}H_{5}$ of the aldose moiety of complex **12**, which is in equilibrium with the conformer $_{0}H^{5}$ (**13**), the parallel, *trans*-oriented C-3–H and C-2–H bond orbitals efficiently overlap and form two tricentric bonds. In this transition state, intramolecular hydrogen transfer between C-2 and C-3 occurs caus-

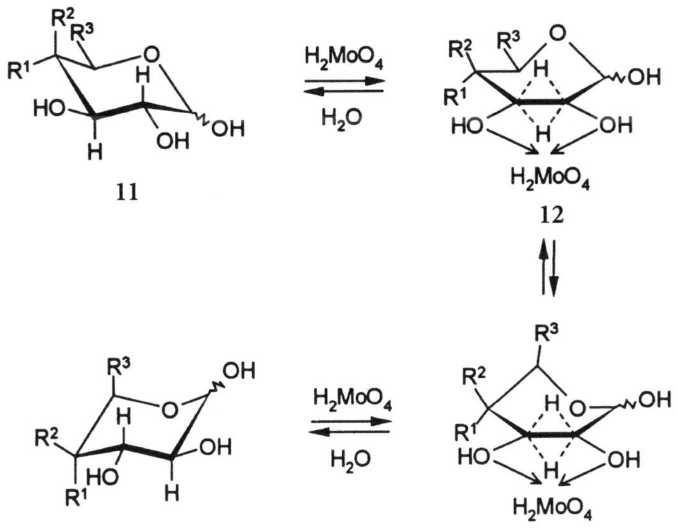

11 **12** **14** **13**

	R¹	R²	R³
a	OH	H	H
b	OH	H	CH_2OH
c	H	OH	CH_2OH

Scheme 7

ing simultaneous configurational changes on both the carbon atoms, resulting in the formation of the $^{1}C_4$ conformation of D-arabinopyranose (**14a**) [alternatively of D-altropyranose (**14b**) or D-idopyranose (**14c**)].

The high susceptibility of pentoses to the secondary process, with the proposed interface of the interconversion D-xylopyranose-$^{4}C_1$ $\Leftrightarrow$ D-arabinopyranose-$^{1}C_4$ between their two epimeric pairs, is probably connected with the conformational readiness of both aldoses for the reaction [42]. This susceptibility is less pronounced for higher aldoses because of a lower population of $^{1}C_4$ conformers of both D-altropyranose and D-idopyranose when compared with that of D-arabinopyranose.

Other indirect support for the pyranose half-chair structure of the transition state of the secondary process is provided by the behavior of alditols [21] and furanoid tetroses or 5-deoxypentoses [5, 6] under the conditions of the Bílik reaction. The former, which although able to form planar bidentate complexes are without the deformation necessary for the hydrogen exchange, do not undergo any structural change while the latter, sterically much more demanding furanoses (aparently unable to form bidentate molybdate complexes via their *trans*-oriented 2,3-hydroxyls), do not undergo racemization.

To a very limited extent, the vicinal hydrogen exchange mechanism might operate even within the primary process as originally suggested (Scheme 3). This is supported by the formation of ~0.2% of D-(1-^{13}C)glucose [besides the major product, D-(2-^{13}C)glucose] from D-(1-^{13}C)mannose, or even 1.2% of D-(1-^{13}C)arabinose [besides D-(2-^{13}C)arabinose] from D-(1-^{13}C)ribose [23]. Unfortunately, there are no data available for the formation of such isotopomers from the side of the C-2, C-3 *threo* aldopyranoses which are much more populated in the form of the C-1, C-2 diequatorial anomers, and thus should be much more susceptible to form such isotopomers by the hydrogen exchange mechanism than the aforementioned C-2, C-3 *erythro* aldopyranoses.

To suppress the secondary process that leads to undesired isotopomers, which is necessary when preparing regiospecifically isotope-substituted or isotope-labeled aldoses, a combined heptamolybdate-formate resin was developed [23]. Using the resin as a catalyst, the secondary process is practically negligible and can be reduced to < 0.1%.

2.6
Other Catalytic Systems Causing Skeletal Rearrangement of Epimeric Aldoses

In addition to molybdic acid, there have been reports of several other catalytic systems that cause mechanistically relative transformations of aldoses involving a 1,2-shift of their carbon skeleton. They include the nickel(II)-ethylenediamine [43] and cobalt(II)-ethylenediamine complexes [44] as well as the calcium(II), strontium(II), lanthanum(III) and neodymium(III) cations in both aqueous and alcoholic alkaline solutions [45, 46]. Unlike the Bílik reaction, all these transformations exploit rather equimolar amounts of the catalysts so that they result in thermodynamic equilibria of the catalytic complexes with the epimeric aldoses. Moreover, due to steric restrictions, the aforementioned alkaline earth and rare earth cations, which form mononuclear tetradentate complexes with aldoses, catalyze the carbon-skeleton rearrangement of aldoses with the C-3, C-4 *threo* configuration only. Although no secondary process (viz. a mutual hydrogen exchange between C-2 and C-3 leading to the formation of 3-epimers) analogous to that occurring in the Bílik reaction has been described for the other known catalytic systems, the necessary alkaline medium often brings about the formation of undesired 2-ketoses by the competitive LdB-AvE reaction. In other words, this is the main reason why these 1,2-carbon shift mediators cannot be exploited as real catalysts in the sense of the quantities applied. Due to the distinct restrictions and difficulties none of these alkaline complex catalyzed skeletal rearrangements has been commercialized and all of them are rather of theoretical interest.

3
Application of the Bílik Reaction for the Mutual Interconversion of 2-Ketoses and 2-*C*-(Hydroxymethyl)aldoses

Following the recognition of the unusual carbon-skeleton rearrangement that occurs in the Bílik reaction, new questions have been raised, in particular, whether the carbon atom 1,2-shift is able to transfer substituents other than the hydrogen atoms (i.e. protium, deuterium or tritium) (Scheme 8). This was shown by mechanistic studies with the hydrogen isotopomers of D-glucose (**15a**; X = ^{3}H, Z = H, *n* = 2) and D-mannose (**16a**; X = ^{3}H, Z = H, *n* = 2) [32], as well as of D-xylose (**15b**; X = H, Z = ^{2}H, *n* = 1) and D-lyxose (**16b**; X = H, Z = ^{2}H, *n* = 1) [20].

An immediately applicable example was the interconversion of 2-ketoses and 2-*C*-(hydroxymethyl)aldoses predictable from the analysis of the mechanism of the primary process of the Bílik reaction (Scheme 4). However, the primary studies performed with hex-2-uloses [47] and pent-2-uloses [48] did not provide results consistent with those expected according to the mechanism revealed later, as no 2-*C*-(hydroxymethyl)aldose was detected in the reaction mixtures. Based on the results of the primary studies, as well as on the assumption that the thermodynamic equilibrium of a pertinent 2-ketose and 2-*C*-(hydroxymethyl)aldose might be shifted totally in favor of the former, the investigation of the interconversion was approached from the side of the latter sugar. Some inspiration might be provided also by the analytical studies of the transformation of 2-ketoses to the corresponding 2-*C*-(hydroxymethyl)aldoses catalyzed by nickel(II)-ethylenediamine complexes [49] (see Osanai, this vol.).

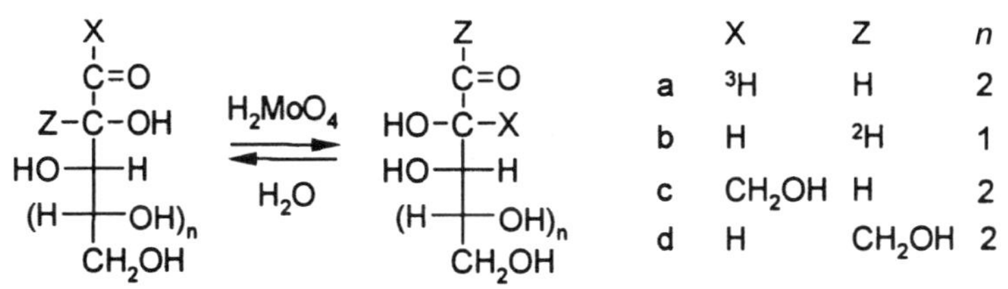

Scheme 8

3.1 Preliminary Studies

The D-glucose/D-mannose system, the most convenient for such model studies, was chosen again; however, with the necessary modifications. While 2-*C*-(hydroxymethyl)-D-mannose (**16c**; X = CH_2OH, Z = H, *n* = 2, Scheme 8) is conveniently available via aldolization of 2,3;5,6-di-*O*-isopropylidene-D-mannose with formaldehyde [50], this is not the case for 2-*C*-(hydroxymethyl)-D-glucose (**15d**; X = H, Z = CH_2OH, *n* = 2, Scheme 8). Therefore, both 2-*C*-(hydroxymethyl)aldoses were prepared by the addition of nitromethane to D-fructose (**17**) followed by transformation of the intermediates, epimeric 1-deoxy-2-*C*-(hydroxymethyl) alditol-1-nitronates **18** and **19** by the Nef reaction (Scheme 9) [51]. In spite of its apparent simplicity, it was the first synthesis of 2-*C*-(hydroxymethyl)aldoses using this sequence. Thus, a 1:1 mixture of the branched-chain aldoses **15d** and **16c** was obtained in 18% yield.

The individual 2-*C*-(hydroxymethyl)aldoses were obtained after a combined separation of the epimers via their phenylhydrazones followed by chemisorption chromatography on a cation-exchange resin in the Ba^{2+} form [50]. Then, on treatment with molybdic acid at increased temperature, 2-*C*-(hydroxymethyl)-D-mannose (**16c**) rearranged to D-*gluco*-hept-2-ulose (**15c**) as expected. Under similar conditions, 2-*C*-(hydroxymethyl)-D-glucose (**15d**) was transformed to D-*manno*-hept-2-ulose (**16d**). The analysis of the reaction equilibria by ^1H NMR spectroscopy revealed the presence of small quantities of the respective starting branched-chain aldoses **15d** (3.5%) and **16c** (8%). The same equilibria were also obtained from the side of the model ketoses **15c** and **16d**. Thus, for example, the equilibrium of the sugars **15c** and **16c** at their total 2% concentration in a 0.2%

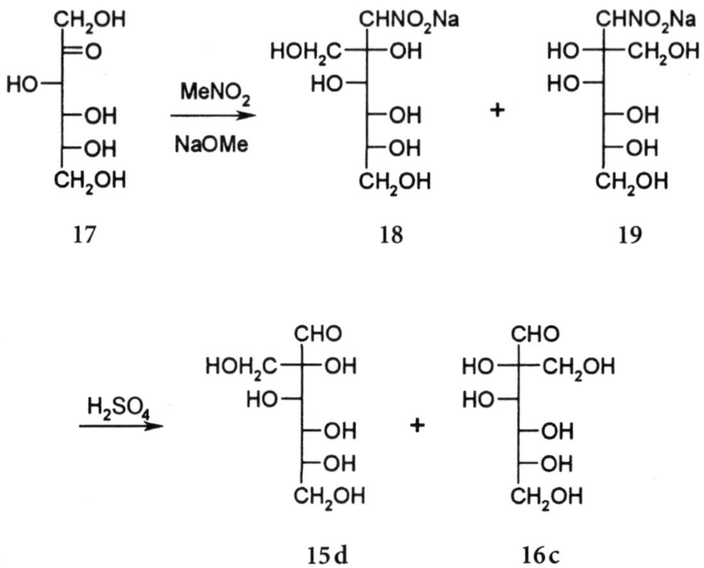

Scheme 9

aqueous solution of molybdic acid at 80 °C is obtainable in 1 h from the side of **16c** while in 20 min from the side of **15c**. This is in accordance with the different reaction rates observed also for the C-2 epimeric aldoses [20, 21], as the sugars with the *erythro* arrangement of their OH groups at two neighboring (α and β) carbon atoms to the carbonyl carbon atom formed, to a much higher extent, cyclic, catalytically inactive complexes than those with the *threo* arrangement at the pertinent moiety [39].

The data obtained suggest that conversions of the branched-chain aldoses **15d** and **16c** to the respective ketoses **15c** and **16d** are not irreversible but their reaction equilibria are strongly shifted to the side of the ketoses. Thus, between the interconverting pairs of sugars, **15c** ⇔ **16c** and **16d** ⇔ **15d**, respective thermodynamic equilibrium ratios of 23:2 and 55:2 were found. If the pairs of sugars **15c** and **16c** as well as **15d** and **16d** are considered to be substituted D-glucoses and D-mannoses (cf. Scheme 8), the mutual interconversion within the pairs can be regarded as further structural proof of the mechanism of the Bílik reaction.

3.2 Preparation of 2-*C*-(Hydroxymethyl)aldoses

While 2-*C*-(hydroxymethyl)aldoses of the C-2, C-3 *erythro* configuration are easily accessible via aldolization of their 2,3-*O*-alkylidene derivatives with formaldehyde, there is no simple and general method available for the synthesis of their C-2 epimers. The application of the Bílik reaction to 2-ketoses, however, has offered a much more universal method for their synthesis regardless of their stereochemistry.

3.2.1 *D-Hamamelose and Other 2-C-(Hydroxymethyl)pentoses*

D-Hamamelose [2-*C*-(hydroxymethyl)-D-ribose, **20**], a naturally occurring branched-chain monosaccharide with remarkable biological properties [52], has been synthesized by several, relatively complex procedures [53]. In spite of its poor yield (6.5%), the one-step synthesis of the sugar from D-fructose (**17**) by the Bílik reaction [54] (Scheme 10) makes the procedure preparatively interesting as the starting sugar can be easily removed by fermentation. The by-product D-sorbose (**21**), that apparently originates from the secondary process that

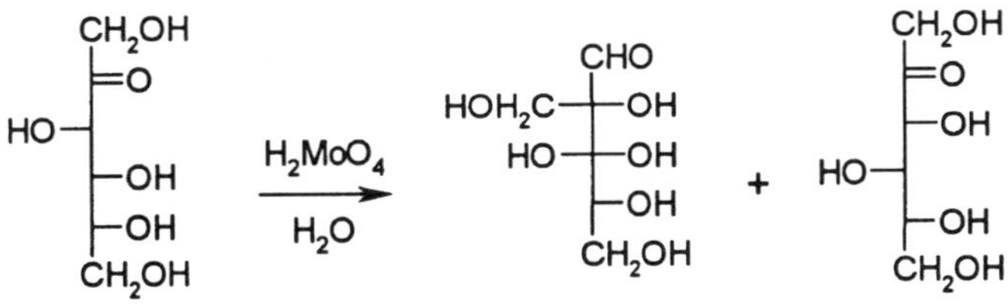

17 20 21

Scheme 10

operates in parallel, is separable by chemisorption chromatography on a cation-exchange resin in the Ba^{2+} form. The same thermodynamic equilibrium of D-fructose and D-hamamelose can also be obtained from the side of the branched-chain aldose; however, this reaction takes six times longer.

Addition of four equivalents of boric acid to the starting ketose improves the original 6.5% yield of D-hamamelose up to 20% [55]. Thus, the product of the molybdic acid catalyzed rearrangement, D-hamamelose, is apparently being removed from its thermodynamic equilibrium with D-fructose by a competitive complex formation with boric acid.

In a similar manner, L-sorbose, on treatment with a catalytic amount of molybdic acid, equilibrates with its rearranged isomer, 2-*C*-(hydroxymethyl)-L-lyxose, again accompanied by a side product of a secondary process, L-fructose [55]. Analogous co-application of boric acid shifts in this case the thermodynamic yield of 2-*C*-(hydroxymethyl)-L-lyxose from 3% to a preparatively more interesting 12%.

Molybdic acid catalyzed isomerization of D-tagatose (without boric acid) provided a 19:1 ratio (as estimated by ^{1}H NMR spectroscopy) of the ketose and 2-*C*-(hydroxymethyl)-D-xylose. Surprisingly, D-psicose on treatment with molybdic acid under otherwise identical reaction conditions, even after prolonged reaction times, did not lead to the formation of 2-*C*-(hydroxymethyl)-D-arabinose as expected. The co-application of boric acid, either in this case or in the case of D-tagatose, does not shift the equilibria towards the corresponding 2-*C*-hydroxymethyl)aldoses.

3.2.1.1
Mechanistic Studies

Regardless of the aforementioned structural-mechanistic considerations (cf. Scheme 8), classical mechanistic studies with two isotopically substituted D-fructoses, namely D-(2-^{13}C)fructose and D-(3-^{13}C)fructose, were also performed [54, 55] in order to unambiguously confirm the stereospecific carbon-skeleton rearrangement of 2-ketoses to 2-*C*-(hydroxymethyl)aldoses and vice versa. The study was also important in respect to gaining more detailed knowledge on the topology of the secondary process which occurs concomitantly with the prevailing primary process of the rearrangement. Thus, after treatment with 0.2% aqueous molybdic acid at 80 °C for 5 h, the reaction mixture in the first case also contained, in addition to the starting D-(2-^{13}C)fructose (89%), D-(2-^{13}C)hamamelose (6.5%) and D-(2-^{13}C)sorbose (2.5%), and, in the second, D-(1-^{13}C)hamamelose (7%) and D-(3-^{13}C)sorbose (3%) besides the starting D-(3-^{13}C)fructose (88%). The pertinent positions of the ^{13}C isotopic substitution in each isolated sugar were traced by ^{13}C NMR spectroscopy (Fig. 1) [54, 55].

Product analysis revealed that the branching C-2 atom of D-hamamelose, formed on treatment with molybdic acid, was the former C-2 atom of D-fructose, while the C-1 atom of D-hamamelose was originally the C-3 atom of the ketose. Isotope tracing tracked the origin of the only minor product, D-sorbose, to a different process, as both of its topologically most significant carbon atoms, namely C-2 and C-3, remain at the same positions as in the starting D-fructose.

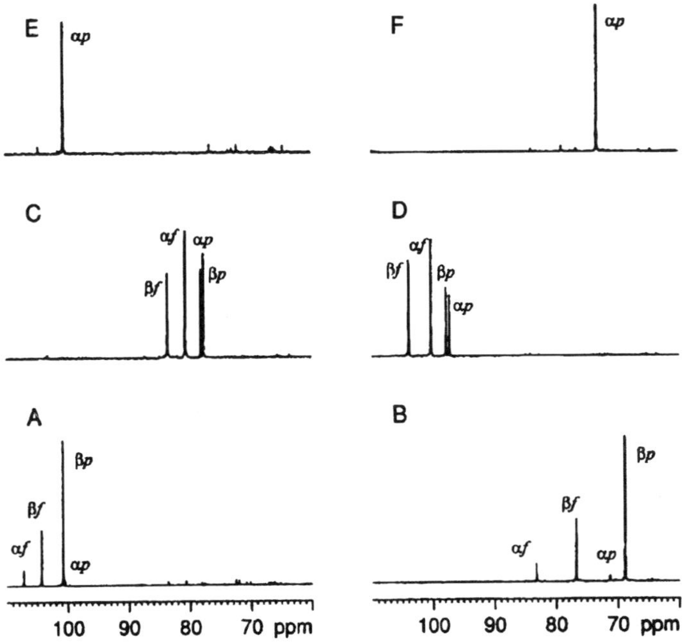

Fig. 1 A–F. The 75.5 MHz ^{1}H-decoupled ^{13}C NMR spectra in aqueous solution at 40 °C of the starting ^{13}C isotopically substituted compounds (**A, B**) and the synthetic products (**C-F**) obtained from isomerization reactions catalyzed with molybdic acid. D-(2-^{13}C)Hamamelose (**C**) and D-(2-^{13}C)sorbose (**E**) obtained from D-(2-^{13}C)fructose (**A**); D-(1-^{13}C)hamamelose (**D**) and D-(3-^{13}C)sorbose (**F**) obtained from D-(3-^{13}C)fructose (B) (Reprinted from Carbohydrate Research, Vol. 319, Zuzana Hricovíniová-Bíliková, Miloš Hricovíni, Mária Petrušová, Anthony S. Serianni, Ladislav Petruš, Stereospecific molybdic acid-catalyzed isomerization of 2-hexuloses to branched-chain aldoses, Pages 38–46, Copyright 1999, with permission from Elsevier Science.)

If this D-sorbose had come from a process similar to the LdB-AvE reaction, then such a final reaction mixture would also have to contain other 2-ketoses and 3-ketoses. The absence of these complementary ketoses suggests that D-sorbose is formed from D-fructose by a secondary process also catalyzed with molybdic acid, similarly to D-xylose from D-arabinose and vice versa. The other direction of the interconversion by the secondary process of both 2-ketoses was shown above by the isolation of L-fructose as the only minor product [besides the main product 2-*C*-(hydroxymethyl)-L-lyxose] from a reaction mixture obtained on treatment of L-sorbose with molybdic acid.

Based on the results of studies with isotopically substituted D-fructoses, and following the stereochemical rules of molybdate complexes, the mechanism of the molybdic acid catalyzed mutual interconversion of 2-ketoses and 2-*C*-(hydroxymethyl)aldoses (referred to as the primary process) [54, 55] is shown in Scheme 11.

The ketohydrol form of a 2-ketose (exemplified as D-fructose) enters the catalytically active tetradentate dimolybdate complex **22** which stereospecifically

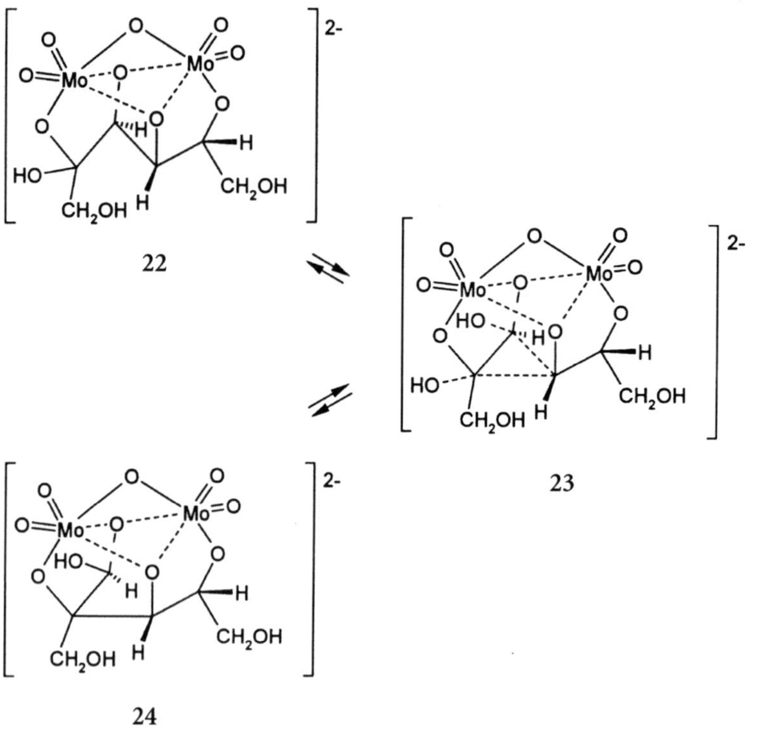

Scheme 11

rearranges via the transient state **23** into the corresponding complex **24** of the aldehydrol form of such a 2-*C*-(hydroxymethyl)aldose (exemplified as D-hamamelose) which has an opposite configuration at C-2 compared to that at C-3 of the starting 2-ketose. This is the result of a new C-2–C-4 bond formation with concomitant cleavage of the original C-3–C-4 bond (the numbering according to 22) while all the crucial atoms and their substituents of the four-carbon sugar moiety involved in the rearrangement remain fixed in space due to the chelation in the dimolybdate complex. The fixed topology of the atoms with their bond recombinations also results in differences in the conformational shapes of the interconverting species. Thus, while the four-carbon moiety of the starting D-fructose linked in the active complex **22** has a zigzag (or *anti*-periplanar) conformation, that of the product D-hamamelose in complex **24** is changed to a sickle (or *syn*-periplanar) one. In accordance with the stereochemistry of the molybdate complexes, both 2-ketoses and 2-*C*-(hydroxymethyl)aldoses should be complexed in the resulting complexes as the hydrates of their acyclic forms. In the molybdenum coordination octahedrons, two of the four hydroxyls, formally esterified, are placed opposite the bridging Mo-O-Mo oxygen atom, while the other two hydroxyl groups are coordinated to both molybdenum atoms and constitute, together with the bridging Mo-O-Mo oxygen atom, the joint triangular wall of the two molybdenum coordination octahedrons of the dimolybdate complex. The coordinated OH groups are located opposite to all the four multiple Mo=O bonds (actually of the multiciplity two and a half, resulting in a stable, 18-electron outer layer) in both, by the octahedrons fused at three corners and release their acidic protons to give a formal (–II) charge to the complex.

The mechanism of the secondary process (Scheme 12) that accompanies the molybdic acid catalyzed mutual interconversion of 2-ketoses and 2-*C*-(hydroxymethyl)aldoses is illustrated as the interconversion of D-fructopyranose (**25**) and D-sorbopyranose (**28**). The most probable structure of the active complex that catalyzes the secondary process is a dimolybdate of a joint-cornered dioctahedron structure such as, for example, $K_2\{[MoO_2(C_2O_4)H_2O]_2O\}$ [24]. It has been postulated [54] that, in such complexes, the 2-ketoses are involved as the bidentate hexopyran-2-ulose ligands exploiting their *trans*-arranged C-3–OH and C-4–OH hydroxyl groups (**25, 28**).

In accordance with the stereochemical rules, the hydroxyl group positioned opposite the bridging Mo-O-Mo oxygen atom (i.e., to the connecting corner of both octahedrons) is formally esterified, while the other hydroxyl group is coordinated and releases its acidic proton. Only the planar bidentate ligands that enable a maximum overlap with the *d* orbitals of the molybdenum(VI) atom are known to be components of stable complexes [27]. Apparently, the complex force field causes such a planar deformation of the C-2–C-3-C-4–C-5 segments of the complexed hexopyran-2-uloses **26** and **27** in their respective $^6H_0$ and $^0H_6$ conformations. Thus, the respective C-3–H and C-4–H bonds (numbering

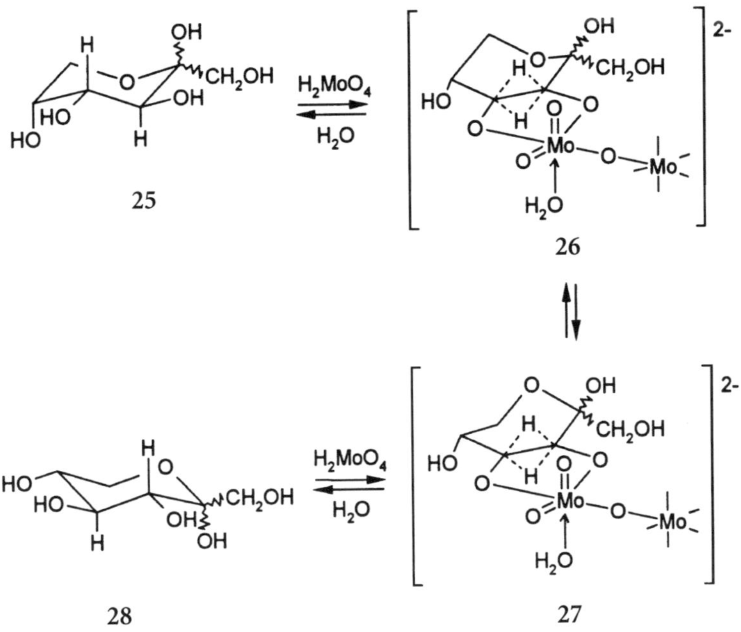

Scheme 12

according to **25**), originally *trans*-diaxially oriented, become strongly deformed, sticking below and above the plane of the sp^2-rehybridized carbon atoms C-3 and C-4 in an antiparallel manner, enabling them to efficiently overlap thus creating two tricentric bonds. The mediation of the mutual hydrogen atom exchange between C-3 and C-4, that occurs simultaneously below and above the plane, results in configurational changes at both carbon atoms, and, hence, in the formation of D-sorbopyranose (**28**) from D-fructopyranose (**25**) and vice versa (Scheme 12).

Although not yet proven, some important co-functioning during the secondary process must come also from the hemiacetal carbon atom, as no configurational changes in cyclitols such as, for example, *myo*-inositol, are observed under the conditions of the Bílik reaction [56].

To summarize the mechanistic and structural studies with hex-2-uloses and 2-*C*-(hydroxymethyl)pentoses performed under the conditions of the Bílik reaction, the following conclusion can be made. Molybdic acid catalyzes two types of interconversions between the sugars shown in Scheme 13. D-Fructose (**17**), D-sorbose (**21**) and D-tagatose (**29**) when treated with the catalyst are subjected to highly stereospecific carbon-skeleton rearrangements to produce thermodynamic equilibrium mixtures with the respective 2-*C*-(hydroxymethyl)-D-ribose (D-hamamelose, **20**), 2-*C*-(hydroxymethyl)-D-lyxose (**30**), and 2-*C*-(hydroxymethyl)-D-xylose (**31**). (For simplicity, all the sugars are represented in their acyclic non-hydrated forms in spite of which some of their interconversions proceed in the acyclic hydrated structures, while others proceed in cyclic hemiacetal ones. All of the interconversion relationships are schematically represented in the D series, despite the fact that some of them were experimentally performed with the L series.) Probably because of extensive formation of unproductive com-

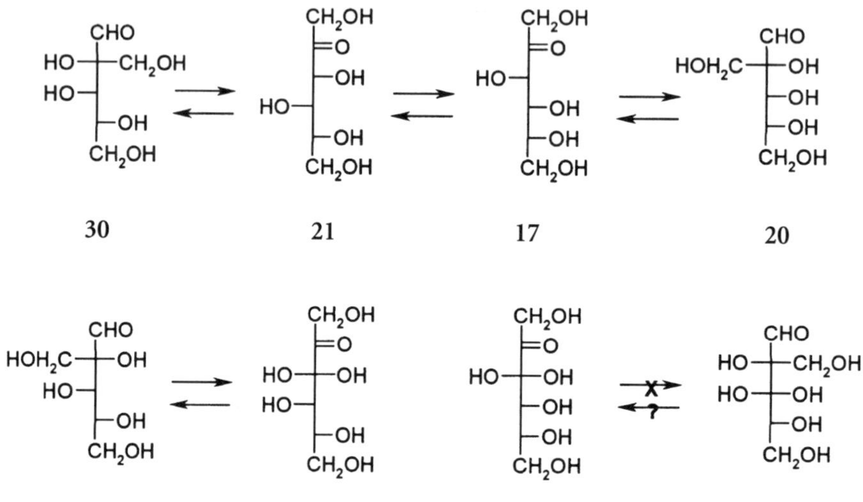

30 21 17 20

31 29 32 33

Scheme 13

plexes [57], D-psicose (**32**) is not transformed to the expected 2-*C*-(hydroxymethyl)-D-arabinose (**33**); the rearrangement, however, very probably proceeds from the side of **33** and remains to be studied. In addition to interconversions of the first type, referred to as the primary process, a competitive interconversion of a second type occurs, namely between ketoses **21** and **22**. This process is referred to as the secondary process and is accompanied *not* by a carbon-skeleton rearrangement but by an intramolecular hydrogen exchange.

3.2.2
2-C-(Hydroxymethyl)tetroses

The mutual interconversion between pent-2-uloses and 2-*C*-(hydroxymethyl)-aldoses has been studied from the side of the former sugars only. The process has been exploited for the preparation of 2-*C*-(hydroxymethyl)-D-erythrose (**35**) from D-xylulose (**34**) (Scheme 14) and 2-*C*-(hydroxymethyl)-D-threose (**37**) from D-ribulose (**36**) [58] (Scheme 15). The thermodynamic equilibria of the respective interconverting pairs of sugars, **34** ⇔ **35** and **36** ⇔ **37**, are 10:1 and 11:1, respectively. As both pairs can be efficiently separated on a column of cation-exchange resin (Ba^{2+} form), the branched-chain aldoses **35** and **37** can be obtained in 7 and 6% yield, respectively. The co-application of boric acid increases the yield of **35** up to 14%; however, the yield of **37** decreases to only 3% [58].

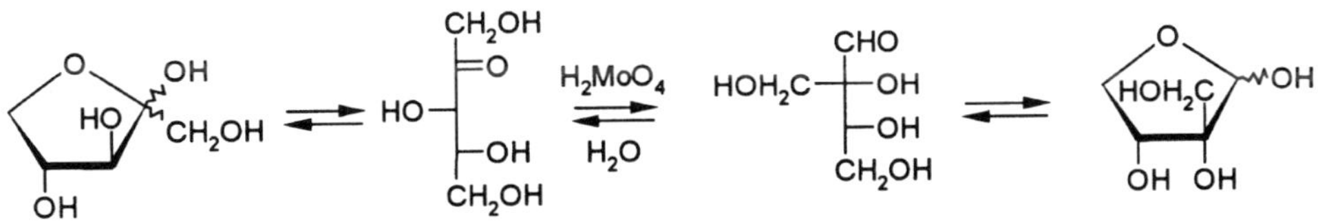

34 35

Scheme 14

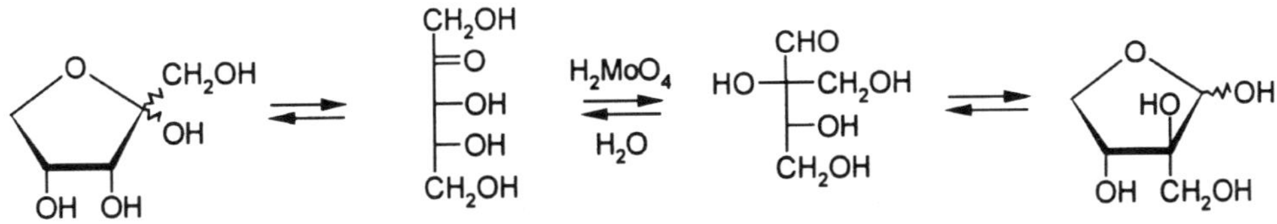

36 37

Scheme 15

3.3
Preparation of 2-Ketoses

In cases where a 2-*C*-(hydroxymethyl)aldose is easily available via base-catalyzed aldolization of a 2,3-*O*-alkylidene-aldofuranose with formaldehyde, the carbon-skeleton rearrangement operating in the Bílik reaction can also be conveniently exploited for preparation of 2-ketoses. The method is especially advantageous for synthesis of heptuloses and octuloses as (1) in special cases the 2-*C*-hydroxymethyl side chain construction is simpler than the classical aldose chain elongations, and (2) the equilibrium of a 2-*C*-(hydroxymethyl)aldose and its corresponding 2-ketose in the Bílik interconversion is much more favorably shifted to the side of the latter sugar (always >85%) than the LdB-AvE transformation of the pertinent unbranched aldose.

3.3.1
Sedoheptulose

Sedoheptulose (D-*altro*-heptulose) is a natural ketose that plays an important role in metabolism. Its content in natural sources, however, rarely exceeds 1% [59] so that its synthesis rather than its isolation is preferred. Until now, biochemical and biotechnological methods have been exploited to perform the synthesis [60].

D-Allose can be easily obtained in a one-step reaction with 10% yield from D-glucose via the secondary process of the Bílik reaction [40], and, hence, also 2-*C*-(hydroxymethyl)-D-allose (**40**) via base-catalyzed aldolization of 2,3:5,6-di-*O*-isopropylidene-D-allofuranose (**38**) with formaldehyde followed by deprotection of the aldol **39** [60, 61].

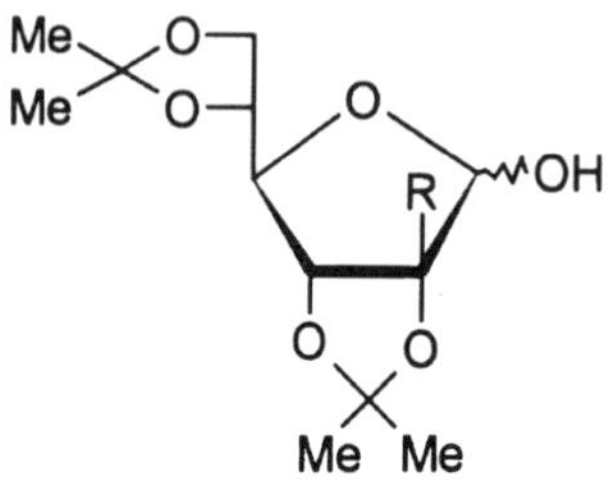

	R
38	H
39	CH_2OH

Molybdic acid treatment of the branched-chain aldose **40** results in a 1:12 equilibrium mixture with sedoheptulose (**41**), a proportion of which is present in the form of the 2,7-anhydro derivative **42**. For isolation purposes, **41** is completely converted in 0.5 M H_2SO_4 into the commercially available form of the sugar **42** (Scheme 16), which is thus obtained in 63% yield (in relation to the starting **40**).

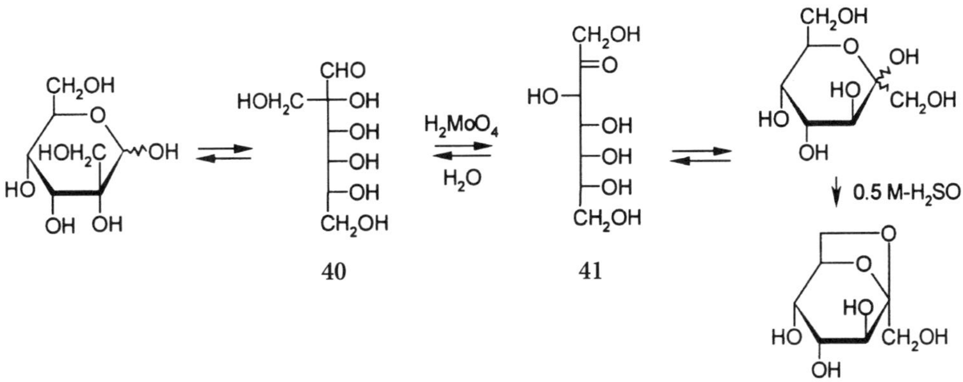

Scheme 16

3.3.2
D-*glycero*-D-*ido*-Octulose

This mysterious ketose occurs in *Craterostigma plantagineum*, a member of a small group of drough-tolerant higher plants. The free sugar represents almost 90% of the carbohydrates present in the fully hydrated leaves of the plant. A peculiarity of D-*glycero*-D-*ido*-octulose is its metabolic conversion into sucrose when dehydration occurs. The process is reversible as, on rehydration, sucrose is converted back to the octulose. This exceptional drought adaptation has attracted attention, and pertinent structural and mechanistic studies have been undertaken on the ketose isolated from its natural source [62].

The synthetic design based on the molybdic acid catalyzed carbon-skeleton rearrangement of the branched chain aldose as applied above to prepare sedoheptulose has also been exploited for the synthesis of D-*glycero*-D-*ido*-octulose [63]. The necessary branched-chain aldose **45**, 2-*C*-(hydroxymethyl)-D-*glycero*-D-*gulo*-heptose, was obtained by the potassium carbonate catalyzed aldolization of the D-*glycero*-D-*gulo*-heptose derivative **43** [64], which has an anchored configuration at C-2, by 2,3-*O*-isopropylidenation with formaldehyde and by subsequent deprotection of the aldolization product **44**.

Molybdic acid treatment of **45** resulted in a 1:7 equilibrium mixture of the starting sugar and D-*glycero*-D-*ido*-octulose (**46**) (Scheme 17). Under the condi-

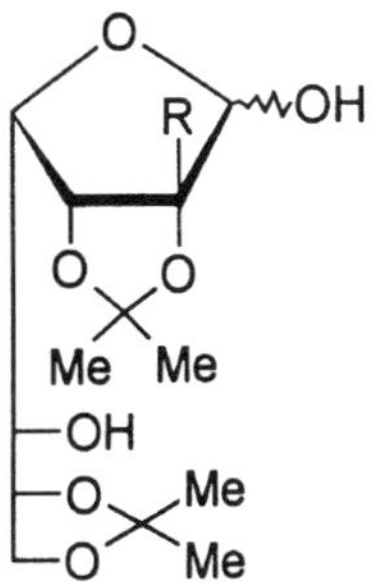

	R
43	H
44	CH_2OH

tions of the interconversion, ketose **46** is partly transformed and in 0.5 M H_2SO_4 completely transformed into its 2,7-anhydrofuranose and 2,7-anhydropyranose forms **47** and **48** in a ratio of 2:1.

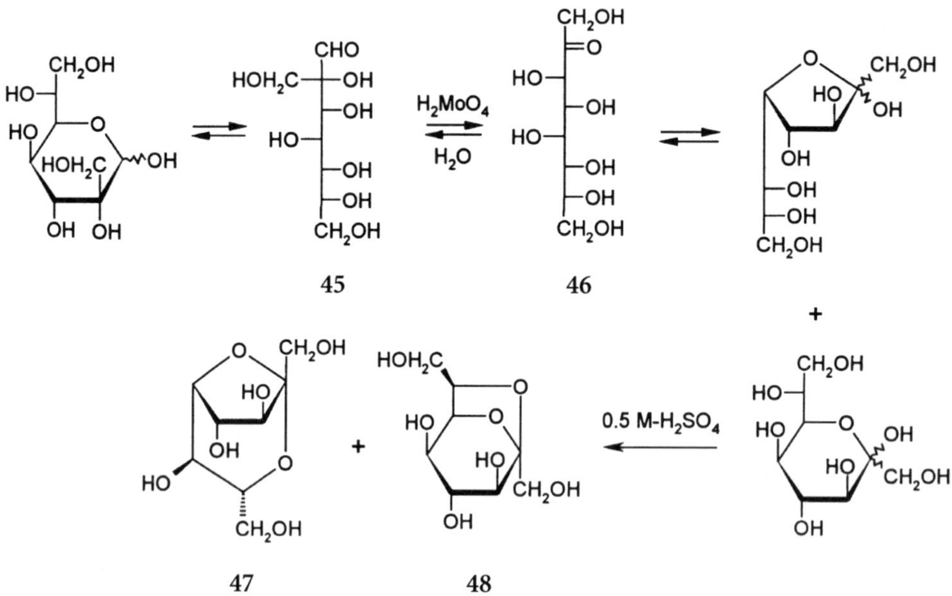

Scheme 17

3.4 Conclusions

The extension of the unique, stereospecific carbon-skeleton rearrangement of epimeric aldoses that occurs during the Bílik reaction to produce 2-ketoses and 2-*C*-(hydroxymethyl)aldoses has provided new, one-step preparative procedures of representatives of both these groups of reducing sugars by simple mutual interconversions. In accordance with the epimeric aldoses ratio representation of the Bílik reaction equilibria, the pertinent ratios of the interconverting sugars at their thermodynamic equilibria are as follows:

D-Xylulose/2-*C*-(hydroxymethyl)-D-erythrose	10:1	[58]
D-Ribulose/2-*C*-(hydroxymethyl)-D-threose	11:1	[58]
D-Fructose/2-*C*-(hydroxymethyl)-D-ribose (D-hamamelose)	14:1	[54]
L-Sorbose/2-*C*-(hydroxymethyl)-L-lyxose	32:1	[55]
D-Tagatose/2-*C*-(hydroxymethyl)-D-xylose	19:1	[55]
D-*gluco*-Hept-2-ulose/2-*C*-(hydroxymethyl)-D-mannose	23:2	[51b]
D-*manno*-Hept-2-ulose/2-*C*-(hydroxymethyl)-D-glucose	55:2	[51b]
D-*altro*-Hept-2-ulose/2-*C*-(hydroxymethyl)-D-allose	12:1	[61]
D-*glycero*-D-*ido*-Oct-2-ulose/ 2-*C*-(hydroxymethyl)-D-*glycero*-D-*gulo*-heptose	7:1	[63]

The mechanistic studies with isotopically substituted D-fructoses have unambiguously proved the topology of the carbon-skeleton rearrangement that occur during the interconversions, namely that they proceed in such a manner that the carbonyl, i.e. C-1 of the product branched-chain aldose, is the former C-3 carbon atom of the starting 2-ketose, while C-2 of the product sugar is the C-2 carbon atom of the starting ketose. In addition to this unambiguous regioselectivity, the process is also strictly stereospecific, as the configurations at the neighboring carbonyl carbon atoms of the interconverting sugars are opposite.

A competitive interconversion of the diastereoisomeric C-3, C-4 *threo* 2-ketoses proceeds via a different mechanism which is very probably a mutual hydrogen exchange between the C-3 and C-4 carbon atoms.

4 Perspectives

The development of knowledge concerning the mechanism of the Bílik reaction of epimeric aldoses and its understanding has initiated the utilization of the reaction for the mutual interconversion of 2-ketoses and 2-*C*-(hydroxymethyl)-aldoses. We are convinced that exploitation of this unique carbohydrate reaction is still in its infancy and that further, significant advances can be expected in the field if the right parameters of the reaction are properly extrapolated. There is also justification for investigating whether the reaction or similar transformations take place in Nature [65], as it is most definitely apparent that this unique chemical-bond rearrangement, the result of an immense force field, does not only take place in molybdate complexes of reducing sugars. In any event, however, the results of such studies as, for example, the transformations of 1-deoxy-2-ketoses or 3-ketoses under the conditions of the Bílik reaction, can be expected in the very near future.

Acknowledgements. Preparation of this chapter was supported by grant No. 2/7144 of the Slovak Academy of Sciences and by the sponsorship of CMS Chemicals.

5 References

1. (a) Bílik V (1972) Chem Zvesti 26:183; (b) Bílik V (1972) ibid 26:187; (c) Bílik V (1972) ibid 26:372
2. (a) Lobry de Bruyn CA, Alberda van Ekenstein W (1895) Recl Trav Chim Pay-Bas 14:156; (b) Lobry de Bruyn CA, Alberda van Ekenstein W (1895) ibid 14:203
3. (a) Bílik V, Kučár Š (1970) Carbohydr Res 13:311; (b) Bílik V (1972) Chem Zvesti 26:76
4. Bílik V (personal communication)
5. Bílik V, Stankovič L' (1973) Chem Zvesti 27:544
6. Bílik V, Petruš L, Stankovič L', Linek K (1978) Chem Zvesti 32:372
7. Bílik V, Petruš L, Farkaš V (1978) Coll Czech Chem Commun 43:1163
8. Bílik V, Čaplovič J (1973) Chem Zvesti 27:547
9. Bílik V, Voelter W, Bayer E (1974) Liebigs Ann Chem 1162
10. Bílik V (1973) Chem Zvesti 29:114
11. Bílik V, Voelter W, Bayer E (1971) Angew Chem 83:967
12. Bílik V, Petruš L (1976) Chem Zvesti 30

13. Petruš L, Bílik V, Anderle D, Janeček F (1979) Chem Zvesti 33:636
14. Bílik V, Petruš L, Alföldi J (1976) Chem Zvesti 30:698
15. Bílik V, Alföldi J, Matulová M (1986) Chem Zvesti 40:763
16. Bílik V, Jurčová E, Sutoris V (1978) Listy Cukrov 94:88 [CA 89:7948m]
17. Bílik V, Jurčová E, Sutoris V (1978) Chem Zvesti 32:252
18. Angyal SJ (personal communication)
19. (a) Königstein J (1978) Coll Czech Chem Commun 43:1152; (b) Königstein J, Bílik V (1992) Chem Pap 46:257
20. Hayes ML, Pennings NJ, Serianni AS, Barker R (1982) J Am Chem Soc 104:6764
21. Bílik V, Petruš L, Zemek J (1978) Chem Zvesti 32:242
22. Bílik V, Kněžek I (1990) Chem Pap 44:89
23. Clark EL, Hayes ML, Barker R (1986) Carbohydr Res 153:263
24. Cotton FA, Morehouse SM, Wood JS (1964) Inorg Chem 3:1603
25. (a) Krebs B (1972) Acta Cryst B28:2222; (b) Böeschen I, Krebs B (1974) ibid B30: 1795
26. Bauer G (1962) Handbuch der präparativen anorganischen Chemie, vol 2. Enke, Stuttgart, p 1233
27. Porai-Koshits MA, Atovmyan LO (1974) Kristallokhimiya i stereokhimiya koordinatzionnykh soedinenii molibdena. Nauka, Moscow
28. Swedo RJ, Arena BJ, Firth BE (1989) US Patent 4,815,445, [CA 111:59887h]
29. Jahr KF, Fuchs J (1966) Angew Chem 78:725
30. Abe Y, Takizawa T, Kunieda T (1980) Chem Pharm Bull 28:1324
31. (a) Speck JC Jr (1958) Adv Carbohydr Chem 13:63; (b) Isbell HS (1973) Adv Chem Ser 117:70
32. Bílik V, Petruš L, Farkaš V (1975) Chem Zvesti 29:690
33. (a) Bourne EJ, Hutson DH, Weigel H (1960) J Chem Soc 4252; (b) Weigel H (1963) Adv Carbohydr Chem 18:61; (c) Angus HJF, Bourne EJ, Weigel H (1965) J Chem Soc 21
34. Atovmyan LO, Tkachev VV, Shishova TG (1972) Dokl AN SSSR 205:609
35. Godfrey JE, Waters JM (1975) Cryst Struct Commun 4:5
36. Taylor GE, Waters JM (1981) Tetrahedron Lett 22:1277
37. Ma L, Liu S, Zubieta J (1988) Polyhedron 8:157
38. Bílik V, Matulová M, Kněžek I (1990) Chem Pap 45:829
39. (a) Matulová M, Bílik V (1990) Chem Pap 44:77; (b) Bílik V, Matulová M (1990) ibid 44:257; (c) Matulová M, Bílik V (1992) ibid 46:253; (d) Matulová M, Bílik V (1993) Carbohydr Res 250:203
40. Bílik V, Kněžek I, Bíliková K (1988) Chem Pap 42:401
41. Wolfrom ML, Hanessian S (1962) J Org Chem 27:1962
42. (a) Durette PL, Horton D (1971) Adv Carbohydr Chem Biochem 26:49; (b) Angyal SJ (1969) Angew Chem Int Ed Engl 8:157
43. (a) Tanase T, Shimizu F, Yano S, Yoshikawa S (1986) J Chem Soc Chem Commun 1001; (b) London RE (1987) ibid 661; (c) Tanase T, Shimizu F, Kuse M, Yano S, Hidai M, Yoshikawa S (1988) Inorg Chem 27:4085
44. (a) Tanase T, Shimizu F, Kuse M, Yano S, Yoshikawa S, Hidai M (1987) J Chem Soc Chem Commun 659; (b) Tanase T, Takei T, Hidai M, Yano S (1992) J Chem Res (S) 252
45. (a) Kusin A (1936) Ber 69:1041; (b) Yanagihara R, Osanai S, Yoshikawa S (1990) Chem Lett 2273; (c) Yanagihara R, Soeda K, Shiina S, Osanai S, Yoshikawa S (1993) Bull Chem Soc Jpn 66:2268
46. Angyal SJ (1997) Carbohydr Res 300:279
47. Bílik V, Tihlárik K (1974) Chem Zvesti 28:106
48. Stankovič Ľ, Bílik V, Fedoroňko M, Königstein J (1975) Chem Zvesti 29:252
49. (a) Yanagihara R, Osanai S, Yoshikawa S (1992) Chem Lett 89; (b) Yanagihara R, Egashira J, Yoshikawa S, Osanai S (1995) Bull Chem Soc Jpn 68:237
50. Ho PT (1979) Can J Chem 57:381
51. (a) Petruš L, Hricovíniová Z, Petrušová M, Matulová M (1996) Chem Pap 50:373; (b) Hricovíniová Z, Hricovíni M, Petrušová M, Matulová M, Petruš L (1998) ibid 52:238

52. (a) Shafizadeh F (1957) Adv Carbohydr Chem 11:267; (b) Masaki H, Atsumi T, Sakurai H (1995) J Dermatol Sci 10:25
53. (a) Ho PT (1979) Tetrahedron Lett 19:1623; (b) Overend WG, Williams NR (1965) J Chem Soc 3446; (c) Burton JS, Overend WG, Williams NR (1965) ibid 3433
54. Hricovíniová Z, Hricovíni M, Petruš L (1998) Chem Pap 52:692
55. Hricovíniová Z, Hricovíni M, Petrušová M, Serianni AS, Petruš L (1999) Carbohydr Res 319:38
56. Petruš L unpublished results
57. Matulová M, Bílik V (1990) Chem Pap 44:97
58. Hricovíniová Z, Hricovíni M, Petruš L (2000) J Carbohydr Chem 19
59. Richtmyer NK (1962) Methods Carbohydr Chem 49:167
60. Kohn BD, Kohn P, Dubin A (1971) Carbohydr Res 18:349
61. Hricovíniová-Bíliková Z, Petruš L (1999) Carbohydr Res 320:31
62. (a) Bianchi G, Gamba A, Murelli C, Salamini F, Bartels D (1991) Plant J 355; (b) Howarth OW, Pozzi N, Vlahov G, Bartels D (1996) Carbohydr Res 289:137
63. Hricovíniová Z, Hricovíni M, Petruš L unpublished results
64. Brimacombe JS, Tucker LCN (1968) J Chem Soc (C) 562
65. Petruš L (1999) Enantiomer 4:369

Nickel(II)-Catalyzed Rearrangements of Free Sugars

S. Osanai

Department of Applied Chemistry, Faculty of Science and Technology,
Keio University, Kohhoku-ku, Yokohama 223-8522, Japan
E-mail: osanai@applc.keio.ac.jp

To the memory of Sadao Yoshikawa, Ph.D. (1925-1998)

C-2 Epimerization of aldoses and ketoses catalyzed by the Ni(II) complex is described. Both epimerization and the preparation of side-chained sugars proceed in the ternary nickel complex through stereospecific rearrangement of the carbon skeleton, that is, from (1-^{13}C)-D-glucose into (2-^{13}C)-D-mannose and from (2-^{13}C)-D-fructose into (2-^{13}C)- D-hamamelose, and vice versa. The structure of the ligand in the complex plays an important role during the epimerization. The epimerization proceeds more smoothly in both methanolic and aqueous solutions with a more hydrophobic ligand such as *N,N'*-higher alkylated ethylenediamine. Due to their ability to self-aggregate in water, the organized metallomicelles can provide an appropriate reaction system for the epimerization of aldoses. The chirality of the ligand contributes to the shift of the equilibrium between the two C-2 epimers.

Keywords: C-2 Epimerization, Hydroxymethylated sugar, Stereospecific rearrangement, Nickel/diamine complex, Metallomicelle, Chirality of ligand

1
Introduction

If sugars are regarded as polyhydroxycarbonyl compounds, a new aspect is developed by observing how hydroxyl groups are positioned in these molecules. Although the individual hydroxyl groups are simple ones, their diversity in steric relationships seems to govern the complexity of the sugar chemistry.

It is well known that sugars can be complexed with metals in a wide variety of ways according to the difference of metals and structures of the ligands, such as aldoses and ketoses or mono- and polysaccharides. These subjects have been previously reviewed [1, 2].

The multiformity due to the slight differences in the stereochemical and electronic adaptabilities provides various subjects with regard to the coordination mode with metals. It is a possible that significant and effective action is shown for the recognition of the sugars. For these reasons, the interaction of carbohydrates with metals has been an attractive subject, and, therefore, it has been extensively studied in connection with various fields of chemistry for a long time.

In particular, the interaction between transition metals and sugars provides many kinds of applications. They are observed not only in reactions such as catalytic chemistry [3–9], coordination and chelation chemistry [10–15], but also in analytical chemistry, such as during electrophoresis [16–18], NMR studies [19–22], and in many other fields [23–29].

For example, aldoses are converted to their sugar acids with high selectivity in the presence of Pt or Rh during the dehydrogenation reaction [3, 4]. In an alkaline solution, the reducing sugars undergo the Cannizzaro reaction with assistance from Ni or Pt [5–7]. In every case, it is recognized that coordinational interaction between the metal and the sugar plays an important role during the reaction.

As mentioned above, the coordinating behavior of nickel(II) with carbohydrates has been studied in detail [30–38]. Yoshikawa and Yano clarified that the nickel complexes that coordinate ethylenediamine have the ability to isomerize aldoses into their corresponding epimers at C-2 along with stereospecific rearrangement [39, 40, 43]. They applied the coordinating abilities of nickel, diamine and sugar in a sophisticated reaction, and established a distinct reaction system that is different from the traditional "enediol" epimerization of aldoses by alkaline solution.

Based on these findings, we were encouraged to develop further the potential of this reaction system. In this chapter, a survey of the conditions of epimerization by nickel complexes is provided. It should be mentioned that the structures of the ethylenediamines are not the only important factors in their complexation during a epimerization with stereospecific rearrangement; the importance of hydrophobicity of the nickel complex must also be stressed when the reactions are carried out in an aqueous solution and not in methanol. This suggests an important new concept concerning the effects of aggregation of metal complexes in addition to the influence of the chirality of the ligand diamine.

2
Principal Aspects of the C-2 Epimerization of Aldoses

2.1
Epimerization by Reversible Enolization

It is well known that aldoses are susceptible to epimerization along with isomerization under alkaline conditions. The equilibrium between glucose, mannose and fructose in an alkaline media is the best-known example of such a case. This results in the interconversion of two aldoses and the corresponding constitutionally isomeric ketose is an equilibrium mixture of the three [41]. An aldose is

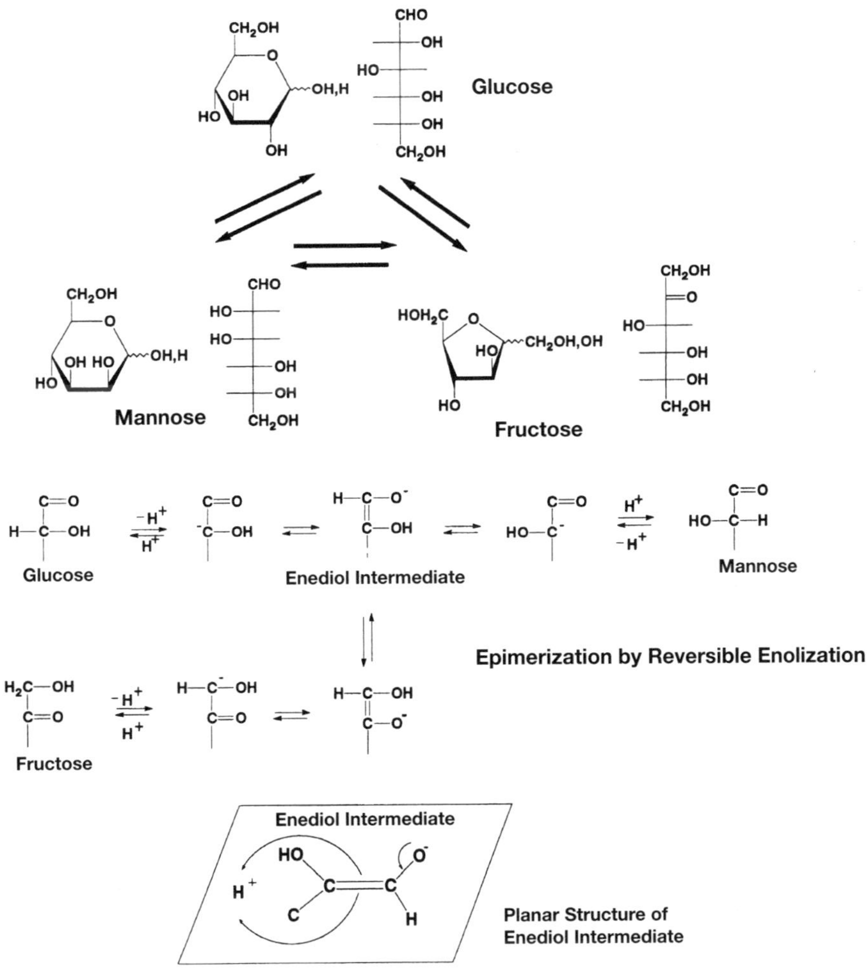

Fig. 1. Interconversion of glucose, mannose and fructose; epimerization by reversible enolization (LdB-AvE rearrangement)

partially converted via a 1,2-enediol intermediate. The mechanism involves abstraction of a proton α to the carbonyl group. Subsequent migration of electrons leads to an ionized enediol, as shown in Fig. 1. The ratios of C-2 epimers in the equilibrium are shown and they are estimated from the thermodynamic equilibrium [2]. The results are due to the reversible enolation that is well known as the "Lobry de Bruyn–Alberda van Ekenstein (LdB-AvE) rearrangement" (see Angyal, this vol.).

According to our work [42], when glucose and mannose are treated with 66.7 mM of aqueous NaOH solution at 65 °C for a period as short as even 5 min, the composition of the products Glc/Man/Fru are 74:3:23 and 4:87:9, respectively. The corresponding ketose, i.e. fructose, was preferentially formed rather than the corresponding epimers, mannose and glucose.

In these isomerization and epimerization reactions, as the mechanism shows, the carbon sequence of the sugar is held unchanged, that is, C-1 of glucose is C-1 of the epimerized mannose and fructose, respectively. As shown in Fig. 2 [42], this is proved by the ^{13}C NMR spectrum. In this study, glucose and mannose whose C-1 carbons had been substituted with ^{13}C were used as the starting substrates. They are denoted as 1*-Glc and 1*-Man, respectively.

2.2 Epimerization by Nickel(II)/Diamine

In an extensive series of studies, Yoshikawa, Yano and co-workers have confirmed that various kinds of aldose can be complexed with nickel and amino compounds such as diamines and amino acid in a methanolic solution [43]. They studied the nickel(II) complexes coordinating *N*-glycosides derived from diamine and monosaccharide. The structure of the complex was determined from three-dimensional X-ray data [43d]. Their further studies elucidated that aldoses can be epimerized by reaction systems containing a nickel/diamine complex. This novel epimerization proceeds through an alternative mechanism that is definitely distinct from the epimerization mechanism by reversible enolization that was introduced in Sect. 2.1. It results in the formation of an equilibrium mixture of the two epimers. According to their study [39], three significant peculiarities of this epimerization are:

1. The reaction involves an exchange of C-1 and C-2 atoms by inversion of the C-1–C-2 fragment in the carbon skeleton.
2. As an active catalyst, a nickel complex composed of a balanced coexistence of nickel(II) and diamine is indispensable for this epimerization.
3. The free C-1 carbonyl group and the C-4 hydroxyl group are essential for the epimerization. This is ascertained by the fact that methyl glycosides (methyl α-D-glucopyranoside and methyl α-D-mannopyranoside) and 4-*O*-substituted disaccharides (D-lactose and D-cellobiose) were not epimerized under conditions where simple pyranoses and furanoses are able to undergo epimerization.

A ^{13}C NMR study using enriched D-glucose and EXAFS (extended X-ray absorption fine structure) analysis revealed that the reaction proceeds via an interme-

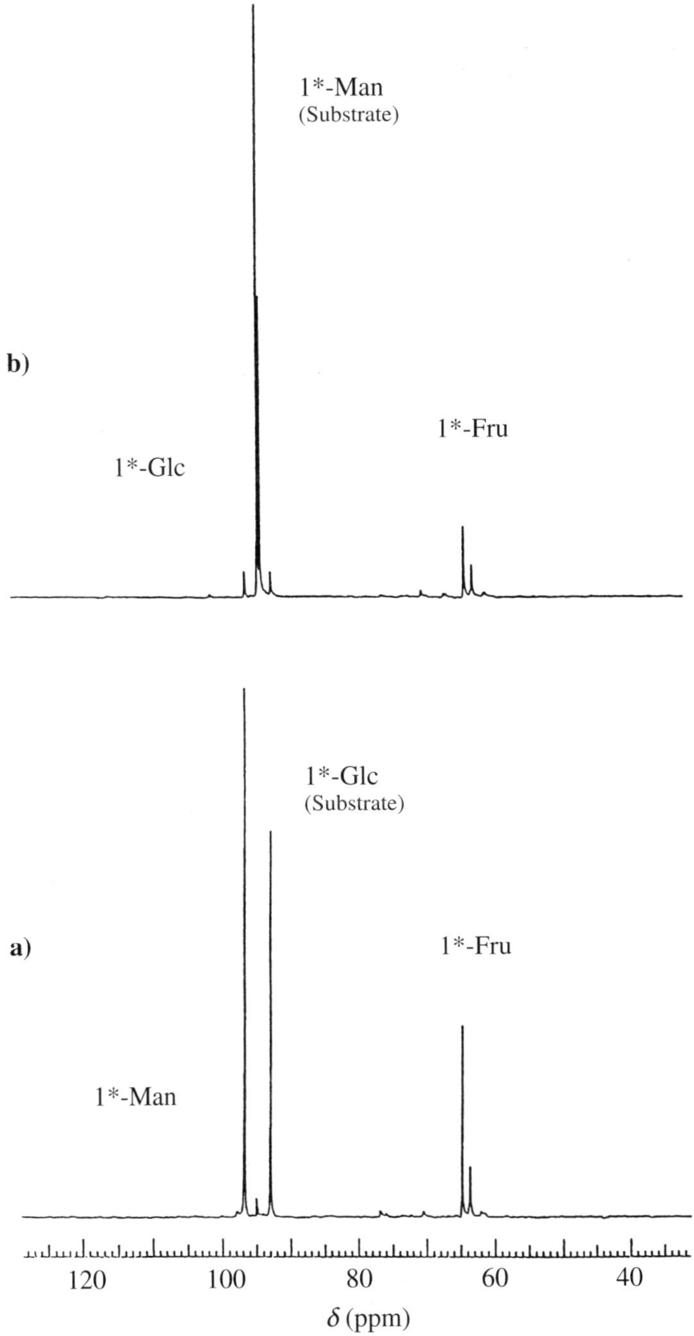

Fig. 2a, b. ^{13}C NMR spectrum (22.5 MHz) of the product mixture derived from 1*-glucose [(1-^{13}C)Glucose] (**a**) and 1*-mannose (**b**) with NaOH (66.7 mM). The ^{13}C NMR data are shown in Table 1

diate mononuclear nickel(II) complex. Figure 3 shows the outline of this reaction. C-1 and C-2 are exchanged on going from the substrate to the product. The reaction proceeds as shown in Fig. 4. The rearrangement takes place via a Ni(II)-involved five-membered chelate intermediate, the postulated glycoside complex, in which the leaving OH group and the migrating C-2–C-3 bond are arranged in an antiperiplaner relationship to create an *S*-configuration at C-2 by inversion. In other words, the sequence of carbons in the aldose changes from C-1–C-2–C-3 to C-2–C-1–C 3. ^{13}C NMR spectroscopy is an eloquent witness to this mechanism, as shown in Fig. 5 [40]. The ^{13}C NMR data of glucose, mannose and fructose are summarized in Table 1.

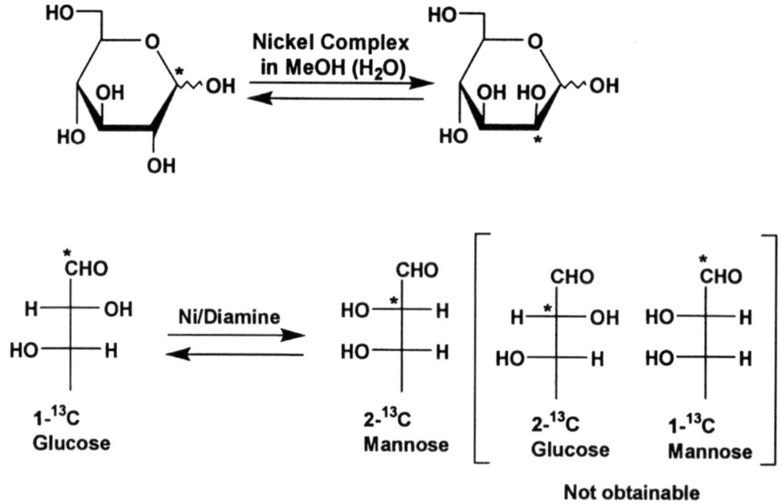

Fig. 3. C-2 epimerization of aldoses by a nickel complex

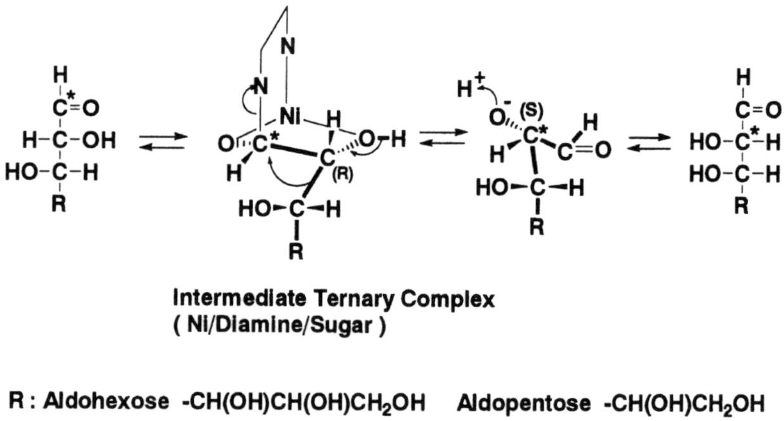

Fig. 4. Mechanism of the epimerization by a nickel complex

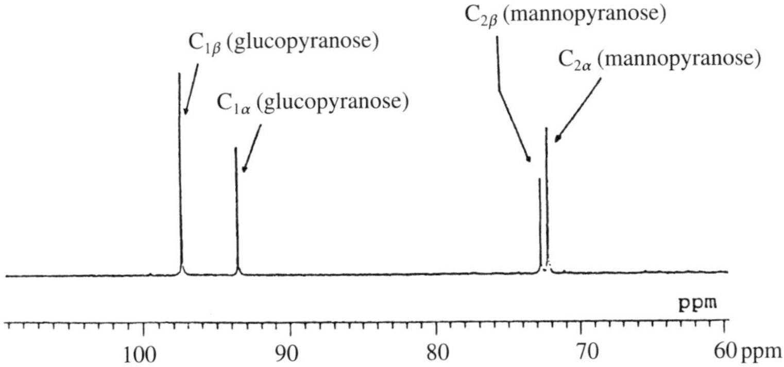

Fig. 5. ^{13}C NMR spectrum of the product mixture obtained from 1*-glucose with Ni^{2+} and 1,1,1′,1′-en

Table 1. ^{13}C NMR data of glucose, mannose and fructose

	C-1	C-2	C-3	C-4	C-5	C-6	Ref.
α-D-Glucopyranose	93.1	73.8	72.4	70.6	72.5	61.6	[66]
β-D-Glucopyranose	96.7	75.1	76.7	70.6	76.8	61.7	[67]
α-D-Mannopyranose	95.1	71.6	71.1	67.7	72.8	61.9	[68]
β-D-Mannopyranose	94.4	71.8	73.9	67.4	76.7	61.9	[68]
α-D-Fructopyranose	65.9	99.1	70.9	71.3	62.1	61.9	[69]
β-D-Fructopyranose	64.7	99.1	68.4	70.5	70.1	64.1	[69]
α-D-Fructofuranose	63.7	105.5	82.9	77.0	82.2	61.9	[69]
β-D-Fructofuranose	63.6	102.6	76.4	75.4	81.6	63.2	[69]

The epimerization and the resulting product analysis were carried out in the following ways [44]. Scheme 1 summarizes the actual processes of the epimerization. Aldose (180 mg, 1.0 mmol, 1 equiv for glucose) was added to a methanolic solution (15 mL) of $NiCl_2 \cdot 6H_2O$ (1 mmol) and ethylenediamine (2 mmol). This solution was incubated for 15 min at 65 °C with stirring. The mixture was then kept at pH 6.5 with 1 M HCl for 0.5 h at room temperature. Samples of the mixture were adjusted to pH 7 and de-ionized by consecutive use of a cation-exchange resin (Dowex 5W × 8, H^+ form) and an anion-exchange resin (Dowex 2 × 8, HCO_3^- form). After eluting 500 mL of water through the resin columns, evaporation of the eluent left a mixture of unreacted and epimerized sugars. Samples for GLC analysis were prepared by trimethylsilylation of the free sugars and injected using a Hamilton Model 7001. A GL Science Model GC-380 instrument obtained GLC analysis data with a FID (flame-ionization detector). A capillary column (0.25 mm ID × 25 m, OV-1 bonded) was set at 190 °C isothermally with the injector temperature higher by 50 °C than the column temperature. Under a given set of operating conditions, the retention time served as a method for qualitative identification.

Fortunately, the epimerization by the Ni(II) complex proceeds intelligibly and quantitatively. Reaction mixtures contain only two epimers including unreacted

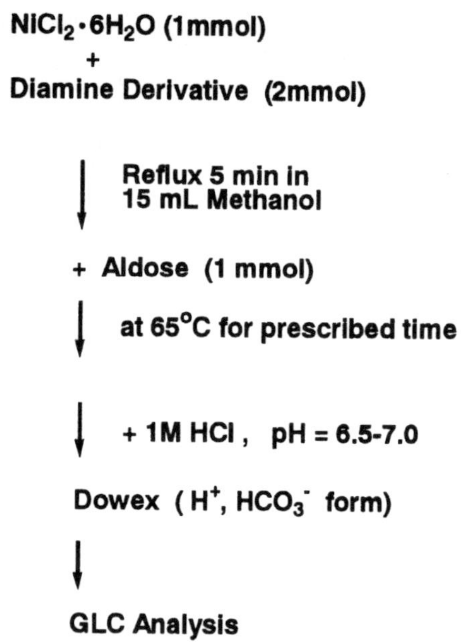

Scheme 1. Epimerization Procedure

substrate aldose. An ambiguous sugar mixture or any amines derived from the catalyst are not detected during the reaction. Consequently, complete interpretation has been formulated without difficulty based on an assignment of NMR spectra and estimation of GLC charts of the products. A representative GLC chart is shown in Fig. 6. Samples are a mixture of α- and β-anomers. Retention times of the product sugars were in fair agreement with those of the authentic glucose and mannose.

Bilik, Petruš and co-workers [45], as well as Heyes, Serianni and colleagues [46], have thoroughly investigated closely related reactions. Epimerization and isomerization of carbohydrates were studied using molybdate ions in a mild acidic medium (see Petruš, Petrušová and Hricovíniová, this vol.).

Yano and Yoshikawa examined the C-2 epimerization of carbohydrates using various metal ions and *N*-substituted diamines, in order to elucidate the role of the metal center and the diamine component [40]. Among the metal ions studied, only Ni^{2+}, Co^{2+}, Ca^{2+} and Sr^{2+} promoted C-2 epimerization of aldoses. The reaction was carried out in a manner similar to the procedure shown in Scheme 1, except for the type of metal ions. When H_2MoO_4 (1 equiv) was used as the metal center, C-2 epimerization did not occur in this short reaction time. The reaction using Ca^{2+} proceeds slowly compared with that using Ni^{2+}, but this has advantages for practical use. Yanagihara confirmed that Ca^{2+} possesses a similar ability to that of Ni^{2+} [42, 47]. The reaction did not proceed via an enediol but by carbon–carbon bond migration within the Ca^{2+} complex.

The metal ions Ni^{2+}, Co^{2+}, Ca^{2+} and Sr^{2+} are known to have a strong affinity for carbohydrates, so it is assumed that a certain interaction or complexa-

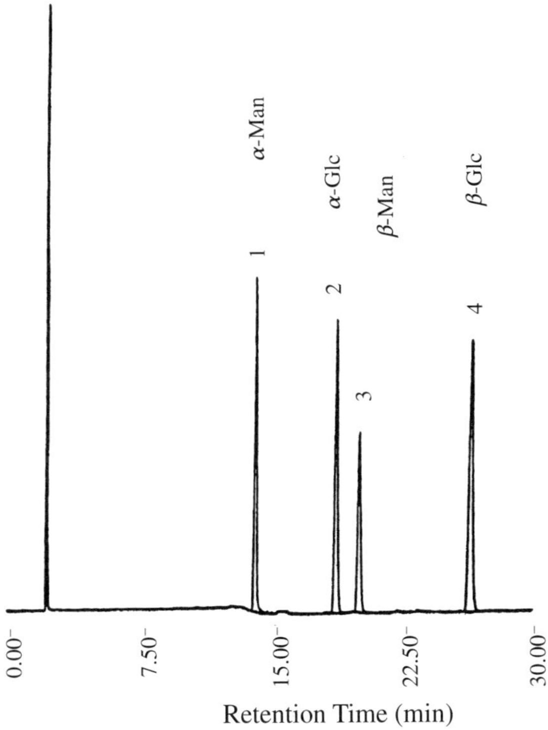

Peak	Rt (min)	Aldoses (D-sugars)	Conc (%)
1	13.85	α-Mannose	27.2
2	18.46	α-Glucose	29.7
3	19.96	β-Mannose	16.5
4	26.30	β-Glucose	26.5

Fig. 6. GLC chart of the reaction product from D-glucose (Table 2, run 8)

tion between metal ions and aldoses is indispensable to the present epimerization.

3 Influence of the Structure of the Ternary Nickel Complex

We know that functional groups are commonly the major factors controlling the reaction system. The reaction mechanism shown in Fig. 4 suggests that the intermediate ternary complex, nickel/diamine/sugar, governs the epimerization. It indicates that aldoses cannot be epimerized by a nickel ion or ethylenediamine alone. Formation of a ternary complex comprising of metal, sugar, and a diamine derivative is indispensable and a minimum requirement for the epimerization. We have carried out experiments to clarify whether an amine-type base possesses the ability to epimerize aldoses. Amine is one of the representative organic bases. If it works as a base like sodium hydroxide, epimerization of aldoses will take place by the LdB-AvE rearrangement through an ionized enediol. When glucose and mannose are treated with amine only, each aldose remains perfectly unchanged. Similarly, aldoses cannot be epimerized by an aqueous solution of $NiCl_2$ alone. This shows how important the simultaneous presence of nickel and diamine is. Experimental results describe that the epimerization systems rely heavily on mutual interactions between the three components in terms of the stability of the ternary complex. A very important observation of its chemistry is that the steric hindrance is strongly dependent on the nature and arrangement of atoms attached to the amino group. The coordination would normally proceed in the direction of the most stable form, and, consequently, the coordinating property determines the products as well as the equilibrium between the two epimers.

Structural effects of the complex on reactivity are divided into two categories: those of the diamines and those of the aldoses. Here, they are considered in terms of their epimerizing activity.

3.1 Influence of the Diamine Ligand

Firstly, the influence of the amine structure should be considered. The structure of the diamine ligand is as significant as that of the sugar as it coordinates to nickel. We place emphasis on developing new ligands that fit formation of the ternary complexes. How much of the epimerized aldose will be obtained from a starting sugar will depend on the stability of the intermediate ternary complex.

In order to develop a suitable catalytic system for nickel/diamine complex assisted epimerization of aldoses, various kinds of diamines were prepared and their influences on epimerization studied [44a, c]. One of the most effective chemical alterations is *N*-alkylation of ethylenediamine. Due to the variety of alkyl chain lengths and degree of alkylation around nitrogen atoms, the stability of the complexes should be profoundly changed according to the steric interactions between the sugar and the ligand in the complex.

The effects of ligand structure, such as *N*-alkylated chain length, degree of *N*-substitution, and the distance between the two amino groups, that is, methylenediamine (mn), ethylenediamine (en), trimethylenediamine (tn) and some kinds of polymethylenediamine (bn, hn), were extensively studied. In addition, ethylenediamine derivatives containing ether linkages and polyamines such as triamine and tetramine are also mentioned. Scheme 2 depicts the relationships between the abbreviations and the structure of representative ligands studied in this work.

Firstly, mn, en, tn, bn and hn denote methylenediamine, ethylenediamine, tri-, tetra- and hexamethylenediamine, respectively. The numbers before en and tn refer to the number of carbons in the *N*- or *N′*-alkylated groups. For example, 4-en and 1,1,1′,10′-en signify *N*-butylethylenediamine and *N,N,N′*-trimethyl-*N′*-decylethylenediamine, respectively. Ph denotes a phenyl group. Similarly, *N,N′*-

n=1 mn; n=2 en; n=3 tn; n=4 bn; n=6 hn	:	$H_2N\text{-}(CH_2)_n\text{-}NH_2$	[Diamine]
4-en	:	$C_4H_9NHCH_2CH_2NH_2$	[Diamine]
8,8'-en	:	$(C_8H_{17})NHCH_2CH_2NH(C_8H_{17})$	[Diamine]
1,1,1',10'-en	:	$(CH_3)_2NCH_2CH_2N(CH_3)(C_{10}H_{21})$	[Diamine]
(0-2)(0-2)'-en	:	$HOC_2H_4NHCH_2CH_2NHC_2H_4OH$	[Hydroxyl Diamine]
(2-3)(2-3)'-en	:	$C_2H_5OC_3H_6NHCH_2CH_2NHC_3H_6OC_2H_5$	[Ether Diamine]
8,8'-tn	:	$(C_8H_{17})NHCH_2CH_2CH_2NH(C_8H_{17})$	[Diamine]
12,12'-[2,2]	:	$(C_{12}H_{25})NHCH_2CH_2NHCH_2CH_2NH(C_{12}H_{25})$	[Triamine]
12,12'-[2,4,2]	:	$[(C_{12}H_{25})NHCH_2CH_2NHCH_2CH_2\text{-}]_2$	[Tetramine]
Cyclene	:	1,4,7,10-Tetraazacyclododecane	[Tetramine]

Scheme 2. Abbreviation of polyamine ligands

dioctyltrimethylenediamine is abbreviated as 8,8′-tn. Ethylenediamine derivatives indicated in parentheses possess ether linkages in their alkyl groups (Table 3). The symbol (2–2–2–3) means that there are three ether linkages, which are expressed with the aid of hyphens. They connect ethyl (2), ethylene (2), ethylene (2), and trimethylene (3) groups through the ether oxygen atoms [$C_2OC_2OC_2OC_3$-]. Amines with [ ] denote polyamines containing more than three amino groups. Numbers in [ ] show the carbon number between two amino groups, that is, [2, 3, 2] means $NH_2(CH_2)_2NH(CH_2)_3NH(CH_2)_2NH_2$. Numbers in front of [ ] denote the carbon number of the alkyl group bonded to each of two terminal primary amino groups.

In general, recoveries of the total yields of epimerized and unepimerized sugars are quantitative. They vary between 90 and 100% in every experiment carried out under the conditions mentioned in Scheme 1. Tables 2 and 3 summarize the epimerization carried out by nickel complexes coordinating the various kinds of diamines mentioned above. Run 1 eloquently states that the existence of nickel along with ethylenediamine is indispensable for the epimerization using this reaction system. Further, ethylenediamine must be alkylated with an appropriate chain length. The catalytic activities of the complex coordinating non-substituted ethylenediamine are negligibly small, as shown in run 2.

As runs 8–13 in Table 2 show, some diamine ligands fit the purpose of the epimerization under consideration here very well. The complex composed of nickel and *N,N′*-dialkylethylenediamine is expected to catalyze the conversion of aldohexoses into their corresponding C-2-epimers with a high degree of specificity. On the other hand, the catalytic activity of *N,N*-dialkylethylenediamine is extremely low. For some *N,N,N′*-trialkyl-substituted ethylenediamines, every li-

Table 2. Epimerization of glucose and mannose by nickel complexes of ethylenediamine derivatives

Run	Ligand (cf. Scheme 2)	Conversion (%) from glucose		Conversion (%) from mannose	
		Glc	Man	Glc	Man
1	None	100	0	0	100
2	en	100	0	0	100
3	1-en	95	5	2	98
4	4-en	90	10	1	99
5	1,1-en	86	14	1	99
6	1,4-en	84	16	9	91
7	1,1′-en	66	34	28	72
8	2,2′-en	56	44	55	45
9	1,4′-en	60	40	58	42
10	4,4′-en	59	41	59	41
11	8,8′-en	63	37	61	39
12	12,12′-en	67	33	66	34
13	14,14′-en	64	36	64	36
14	18,18′-en	86	14	3	97
15	Ph,Ph′-en	100	0	0	100
16	1,1,1′-en	53	47	53	47
17	1,1,4′-en	52	48	54	46
18	1,4,1′-en	59	41	53	47
19	1,4,4′-en	65	35	59	41
20	4,4,1′-en	92	8	6	94
21	4,4,4′-en	96	4	1	99
22	1,1,1′,1′-en	50	50	48	52
23	1,1,1′,4′-en	87	13	9	91
24	1,1,4′,4′-en	99	1	0	100
25	1,4,1′,4′-en	100	0	0	100
26	1,4,1′,4′-en	100	0	0	100
27	4,4,4′,4′-en	100	0	0	100
28	tn	92	8	1	99
29	4-tn	74	26	25	75
30	8-tn	74	26	24	76
31	12-tn	85	15	32	68
32	4,4-tn	100	0	2	98
33	8,8-tn	99	1	0	100
34	12,12-tn	98	2	0	100
35	4,4′-tn	93	7	5	95
36	8,8′-tn	89	11	11	89
37	12,12′-tn	80	20	13	87
38	1,1,1′,1′-tn	97	3	0	100
39	1,1,1′,1′-mn	100	0	34	66
40	1,1,1′,1′-en	50	50	48	52
41	1,1,1′,1′-tn	97	3	0	100
42	1,1,1′,1′-bn	100	0	0	100
43	1,1,1′,1′-hn	97	3	0	100

Table 3. Epimerization of glucose and mannose by nickel complexes of polyamine derivatives containing ether linkages

Run	Ligand (cf. Scheme 2)	Conversion (%) from glucose				Conversion (%) from mannose			
		Glc	Man	Fru	Others	Glc	Man	Fru	Others
44	4-Amine (*n*-butylamine)	100	0	0	0	0	100	0	0
45	4,4-Amine (di-*n*-butylamine)	100	0	0	0	0	100	0	0
46	8-Amine	100	0	0	0	0	100	0	0
47	8,8-Amine	100	0	0	0	0	100	0	0
48	12-Amine	100	0	0	0	0	100	0	0
49	12,12-Amine	100	0	0	0	0	100	0	0
50	Ethanolamine	100	0	0	0	0	100	0	0
51	(12–3)(12–3)′-en	62	36	2	0	53	38	1	8
52	(8–3)(8–3)′-en	58	36	4	2	54	37	5	4
53	(4–3)(4–3)′-en	59	35	5	1	53	37	6	4
54	(2–3)(2–3)′-en	56	34	7	3	53	38	7	2
55	(2–2-3)(2–2-3)′-en	55	37	8	0	52	36	9	3
56	(6–2-3)(6–2-3)′-en	53	34	3	10	53	38	4	5
57	(2–2-2–3)(2–2-2–3)′-en	53	37	8	2	51	37	9	3
58	(2–2)(2–2)′-en	63	27	10	0	58	29	12	1
59	(2–2)(2–2)(2–2)′(2–2)′-en	94	0	5	1	0	100	0	0
60	(0–2)(0–2)′-en	98	0	1	1	0	100	0	0
61	12,12′-[2,2]	100	0	0	0	0	100	0	0
62	12,12′-[2,2,2]	100	0	0	0	0	100	0	0
63	12,12′-[2,3,2]	100	0	0	0	0	100	0	0
64	12,12′-[2,4,2]	78	21	1	0	20	80	0	0
65	12,12′-[2,6,2]	67	32	1	0	65	33	2	0
66	12,12′-[3,2,3]	100	0	0	0	0	100	0	0
67	12,12′-[2,2,2,2]	100	0	0	0	0	100	0	0
68	12,1,1,12-[2,2,2]	100	0	0	0	0	100	0	0
69	Cyclene	100	0	0	0	0	100	0	0

gand tested provides an excellent reaction system except for the ligands possessing two butyl groups at one amino group. Among the tetraalkylated diamines, tetramethylated ethylenediamine (run 22) is the only ligand that shows effective and excellent epimerizing ability. When these effective ligands are used, the final product compositions are almost identical when either D-glucose or D-mannose is used as the starting sugar. This means that the nickel(II) complex coordinating an adequately alkylated diamine promotes epimerization at C-2 of aldoses to provide a near-equilibrium mixture of 2-epimeric aldoses with considerable speed under mild conditions. When the epimerization proceeded smoothly and reached equilibrium, the ratio of the two epimers, Glc/Man, was approx. 60 ± 5:40 ± 5. This equilibrium value lies somewhat on the side of mannose compared with that calculated from the thermodynamic equilibrium state, i.e. 70:30 [48, 49]. Other varieties of epimeric pairs are discussed in more detail in Sect. 3.2.

Table 3 summarizes the epimerization carried out by nickel complexes coordinating monoamines (runs 44–50), ethylenediamines containing ether linkages (runs 51–60), and polyamines such as triamine, tetramine, pentamine and cyclic tetramine (runs 61–69).

The introduction of ether oxygen atoms enhances the solubility of the *N*-alkylated ethylenediamine. Moreover, the ether linkage exhibits a strong interaction with hydroxyl groups of aldoses. This could have an influence on the preparation of the ternary complex including a steric arrangement of the aldoses. Runs 51–60 give us some information to strengthen this argument [44c]. Ligands used in runs 51 and 52 (Table 3) can be regarded as homologs of *N,N′*-didodecylethylenediamine, which is introduced in run 12 in Table 2. Ligand 51 possesses the trimethyleneoxy group inserted between the dodecyl group and the nitrogen atom of ethylenediamine. In ligand 52, an ether oxygen atom is substituted for C-4 of the dodecyl group, dividing the dodecyl group into C_8 and C_3 alkoxyl groups. Little epimerization of glucose/mannose resulted when *N,N′*-dioctadecylethylenediamine was used as the ligand (run 14). On the other hand, the epimerization proceeded readily when ligand 51, 52 or another ether ethylenediamine was used. Both the increases in epimerization and in the solubility of the complex appear to be due to introduction of the ether linkage in the long hydrophobic alkyl chain.

There is little difference between the ligands 52, 55, and 57 with regard to the epimerization. These three diamines can be considered as polyoxygenated analogs of *N,N′*-didodecylethylenediamine. This result indicates that, once the ligand imparts a degree of solubility, the influence of the number of ether linkages of the ligand on epimerization is small, as long as the diamine is an ethylenediamine derivative.

When polyamines which possess more than two amino groups are used as ligands (runs 61–69), the epimerization activity of the corresponding nickel complex is negligible. Ligand 66 (12,12′-[3, 2, 3]) is similar to ligand 51 but the oxygen atom of the ether group is replaced by nitrogen. Epimerization occurs through incorporation of the aldose into the Ni/ethylenediamine complex, that is, the ternary complex formation is prerequisite for epimerization. Ligand 12 (12′-[3, 2, 3]) is known to complex strongly with Ni(II), coordinating its four amino groups. This complex is so stable that it will not accept further incorporation of the substrate aldose. A similar state would occur for the ligand (0–2)(0–2)′-en (run 60), as stated in Fig. 7.

This epimerization system has the further advantage of permitting simple washing to regenerate the ion-exchange resin used for the de-ionization. When ligands containing long alkyl chains are used, the resin may be poisoned because

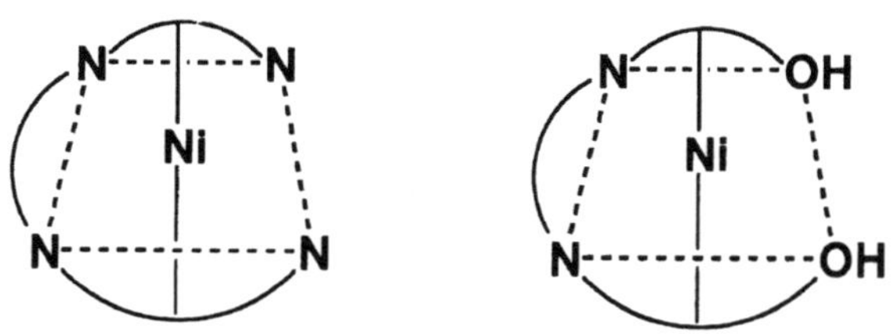

Fig. 7. Postulated structures of the tetradendate ligand complex

it is not possible to remove the adhesive and viscous alkylethylenediamine by simple washing. In contrast, the ether diamines are readily removed from the used resin by elution with methanol. Equally good results for the epimerization are achieved by using the cleaned, regenerated resin.

These suitable *N*-alkylated diamine ligands show high levels of epimerization. On the other hand, if an ethylenediamine or its derivative containing one or two free primary amino groups is used as the ligand, the epimerization capacity of the system vanishes.

Where does this difference originate from? Previously, Yano and Yoshikawa et al. prepared and isolated the ternary nickel complex coordinating α-D-glucosamine and ethylenediamine. Its structure was established by X-ray crystallography [43d]. This experimental fact is of considerable significance when considering epimerization by Ni complexes. It is convenient to imagine that the complex comprising free ethylenediamine and sugar is so stable that it could be isolated. It means that this ternary complex is too stable to accept further bonds for interconversion of the corresponding two epimers.

Looking at it from another angle, it would be feasible to state that there are two prerequisites to accomplish the epimerization of aldoses by a nickel complex. As the first requirement, the Ni(II)/diamine complex must have the ability to incorporate sugar and lead to the ternary complex. In this context, simple diamines such as free ethylenediamine and methylated ones have the potential to produce the ternary complex. As the second prerequisite, the resulting ternary complex must be appropriately unstable to undergo a mutation and a transition in the complex.

No matter how smoothly the ternary complex is produced, when the resulting complex fails to undergo further stereoselective skeleton interconversion of the carbon of C-1 and C-2 in the aldose, the epimerization cannot be realized, as demonstrated in Fig. 4.

Judging from this point of view, some diamine derivatives are confirmed to be fit for the epimerization. Among tetramethylated polymethylenediamines, only the ethylenediamine derivative works as a functional ligand. It is suited for the formation of a cyclic intermediate, through the bond of Ni-*N*-$(C)_n$-*N*-C-*O*-Ni, as shown in Fig. 4. If the two amino groups are apart by more than three methylene groups, the diamine cannot function as a diamine. It acts as a monoamine and is not able to produce the intermediate needed for a key cyclic complex.

3.2 Influence of the Configuration of the Sugar

In Sects. 2.2 and 3.1, the mechanism of this novel, regiospecific rearrangement of epimers such as glucose and mannose, and the influence of the diamine ligands, were clarified. In this section, the influence of the configuration of the sugar on the epimerization, that is, the influence of the position of the hydroxyl groups in aldoses, will be examined. For convenience and consistency, the Ni(II)/1,1,1′,1′-en complex was employed as the catalyst for all the experiments listed in Table 4.

Table 4. Epimerization of various kinds of aldoses by nickel complexes of 1,1,1′,1′-en

Run	Substrates	Composition of products (%)		
		Glc	Man	Fru
1	D-Glucose [0.02]	54 (45)	45 (55)	1
2	D-Mannose [0.045]	52 (34)	47 (66)	1
		Gal	Tal	Tag
3	D-Galactose [0.04]	29	51	20
4	D-Talose [0.27]	27	55	18
		Qui	Rha	Ketose
5	D-Quinobose	66 (55)	33 (45)	1
6	L-Rhamnose	65 (55)	34 (45)	1
		Ara	Rib	Ribu
7	D-Arabinose [0.06]	38	61	1
8	D-Ribose [0.25]	39	60	1
9	L-Arabinose	63	36	1
		Xyl	Lyx	Xylu
10	D-Xylose [0.03]	41 (49)	58 (51)	1
11	D-Lyxose [0.08]	43 (50)	56 (50)	1
12	L-Xylose	43	56	1
13	L-Lyxose	43	56	1

[ ]: electrophoretic mobility of sugars [16]; ( ): data from [39].

We carried out an extensive survey of the C-2 epimerization of various aldoses and ketoses [44b]. Because of the variety of the possible aldose configurations, the stability of the complexes should vary accordingly. In order to understand the effects of configuration on the epimerization of aldoses, a variety of aldohexoses, aldopentoses, and 6-deoxyaldohexose were studied along with four ketoses. The stereospecific and regiospecific rearrangement of these ketoses will be studied in detail in Sect. 4.

We determined the epimerization efficiency of the systems by measuring the amount of epimers as a percentage of the total amount of aldose. Table 4 summarizes the results of the aldose epimerization conducted in the nickel complex system. Losses of the original sugar through the reaction process were small and more than an 85% recovery was readily available in every instance. Table 4 shows that the resultant ratio of a pair of epimers is almost equal, irrespective of which sugar is epimerized. This tendency was observed with all the sugars studied that included aldohexose pairs such as glucose/mannose, galactose/talose and with 6-deoxyaldohexoses (D-quinobose, L-rhamnose), as well as aldopentoses (ribose/arabinose, xylose/lyxose). These results indicate that the nickel complex promotes the C-2 epimerization of aldoses to rapidly provide a near-equilibrium mixture of 2-epimeric aldoses under mild conditions.

These results are discussed in terms of the mutual interaction between nickel and the sugars. Angyal et al. clarified the correlation of the configuration and conformation of sugars with the electrophoretic mobilities in solutions of calcium ions [16]. They concluded from their studies that sugars containing an *ax-eq-ax* sequence of three hydroxyl groups in a pyranose-ring, or a *cis-cis* sequence in a five-membered ring, readily form complexes with alkaline earth metal ions [2c]. The complex formation is directly proportional to the electrophoretic mobility.

Some significant similarities are recognized with respect to the coordination of sugars to nickel and mobility, as shown in Table 4. For example, the equilibrium position of D-galactose/D-talose and D-arabinose/D-ribose lies almost in favor of D-talose and D-ribose, respectively, which possess the *ax-eq-ax* sequence of three hydroxyl groups at C-1, C-2 and C-3. Therefore, they exhibit a high electrophoretic mobility among the pair of two epimers. This suggests that the more readily formed complex is preferentially present in the reaction system containing the nickel(II) ion (Table 5).

3.3 Influence of the Chirality of the Diamine Ligand

As discussed in previous sections, during our studies of the epimerization of free aldoses and ketoses using the Ni(II)/diamine system, it was elucidated that the coordinating state of the complex consisting of the three units, nickel, diamine and sugar, played an important role in this reaction. When the epimerization by the nickel complex was carried out at low temperature, the equilibrium state of the epimerization was successfully controlled using an optically active diamine as the ligand [50]. The objective of this section is to clarify the influence of the chirality of the ligand in the nickel complex during the epimerization and how this difference in optical isomerism affects the equilibrium between the two epimers, glucose and mannose. In addition, the influences of ligand structure, reaction temperature and reaction time on the kinetics and thermodynamics of the reaction are examined.

Table 5. Complexation of sugars with a nickel ion

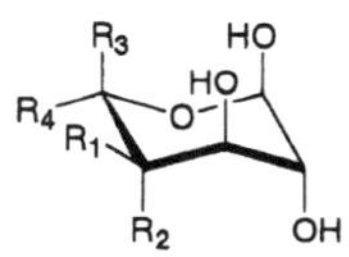

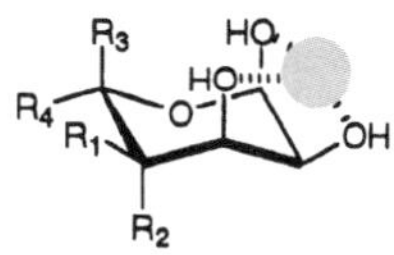

	R_1	R_2	R_3	R_4		R_1	R_2	R_3	R_4
Glucose	H	OH	CH_2OH	H	Mannose	H	OH	CH_2OH	H
Galactose	OH	H	CH_2OH	H	Talose	OH	H	CH_2OH	H
Quinobose	H	OH	CH_3	H	Rhamnose	H	OH	CH_3	H
Xylose	H	OH	H	H	Lyxose	H	OH	H	H
Arabinose	OH	H	H	H	Ribose	OH	H	H	H

The epimerization and analysis of the reaction mixture were conducted in a manner similar to that described in a previous section, except for the reaction temperature and time. In this study, different kinds of optically active chiral diamines were used as ligands. Diamines considered as ethylenediamine derivatives were chosen as ligands because of their structural similarity to ethylenediamine (en) that had shown the highest epimerizing ability in previous studies. The optically active cyclohexanediamine was obtained by resolution of the diastereomers obtained by the reaction of L-tartaric acid and the primary cyclohexanediamine (chxn) [51]. 2,2′-Chxn was prepared by the *N*-acetylation of the optically resolved chxn, followed by reduction of the acetyl group. Three other types of ethylenediamine derivatives, 2,2′-Me-en, 2,2′-Bn-en and 2,2′-Ph-en, were synthesized from the amino acids, alanine, phenylalanine and phenylglycine, respectively. The optical purity of the chxn was confirmed by comparison of specific rotations with published values [52]. The results of the epimerization are summarized in Table 6 and Figs. 8 and 9.

As shown in Table 6, in reference to the influence of 2,2′-chxn, it was confirmed that at 65 °C the configuration of the employed ligand diamine had no significant effect on the epimerization. At 30 °C, there were small differences,

Table 6. Epimerization of glucose and mannose at various temperatures using racemic and optically active diamine ligands

Run	Ligand	Chirality of ligands	Solvent, Temp. (°C)	Time (min)	Conv. (%) from glucose		Conv. (%) from mannose	
					Glc	Man	Glc	Man
1	2,2′-chxn	*R,R*	MeOH 65	5	55.3	44.7	52.5	47.5
2	2,2′-chxn	Racemate	MeOH, 65	5	54.2	45.8	57.5	42.5
3	2,2′-chxn	*S,S*	MeOH, 65	5	53.3	46.7	51.7	48.3
4	2,2′-chxn	*R,R*	MeOH, 30	5	2.6	47.4	46.8	53.2
5	2,2′-chxn	Racemate	MeOH, 30	5	54.3	45.7	49.1	50.9
6	2,2′-chxn	*S,S*	MeOH, 30	5	58.4	41.6	53.3	46.7
7	2,2′-chxn	*R,R*	MeOH, 0	300	37.5	62.5	35.9	64.1
8	2,2′-chxn	Racemate	MeOH, 0	90	45.4	54.6	44.7	55.3
9	2,2′-chxn	*S,S*	MeOH, 0	60	60.4	39.6	59.9	40.1
10	2,2′-Me-en	*R*	EtOH, 20	300	47.2	52.8	46.8	53.2
11	2,2′-Me-en	Racemate	EtOH, 20	300	49.7	50.3	49.0	51.0
12	2,2′-Me-en	*S*	EtOH, 20	300	53.9	46.1	53.4	46.6
13	2,2′-Ph-en	*R*	EtOH, 35	120	34.3	65.7	34.0	66.0
14	2,2′-Ph-en	Racemate	EtOH, 35	120	45.0	55.0	45.0	55.0
15	2,2′-Ph-en	*S*	EtOH, 35	120	55.0	45.0	55.0	45.0
16	2,2′-Bn-en	*R*	EtOH, 25	240	40.0	60.0	39.2	60.8
17	2,2′-Bn-en	Racemate	EtOH, 25	240	47.0	53.0	47.0	53.0
18	2,2′-Bn-en	*S*	EtOH, 25	240	55.0	45.0	54.5	45.5

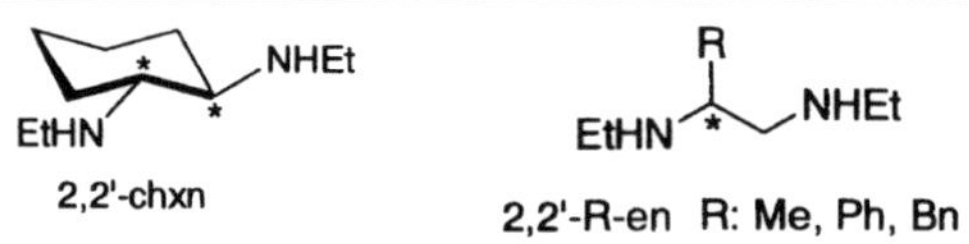

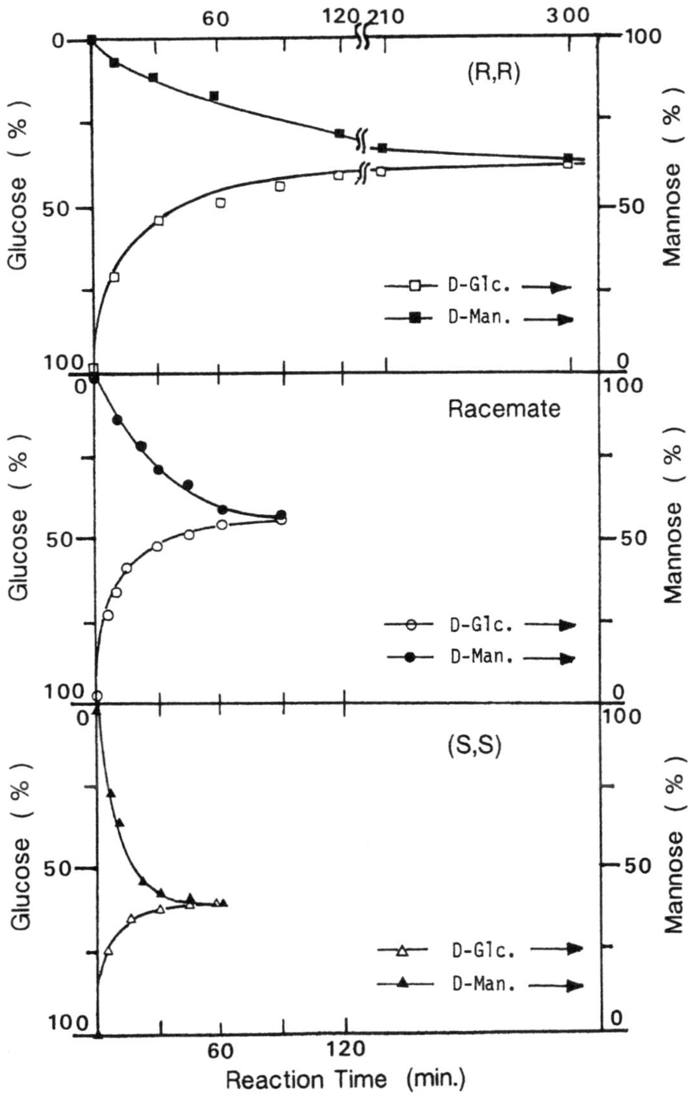

Fig. 8. Time course of the epimerization at 0 °C for Glc and Man with chxn ligands possessing (*R,R*), (*R,S* and *S,R*) as well as (*S,S*) absolute configurations (Table 6, runs 7–9)

but the influence of the configuration was not readily apparent. When the reaction was carried out at 0 °C, the differences in the rate of epimerization and equilibrium became more distinct.

Figure 8 shows that the epimerization of glucose at 0 °C employing chxn with (*R,R*) configuration was found to give a better yield of the mannose epimer, although a longer time was required to reach equilibrium. When the epimerization reaction was performed with the (*S,S*)-isomer, a smaller proportion of man-

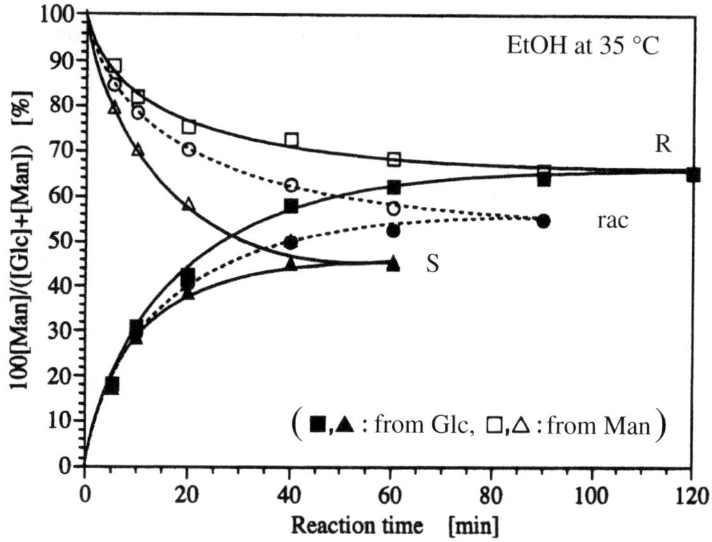

Fig. 9. Time course of the epimerization at 35 °C for Glc and Man with optically active ethylenediamine derivatives (Table 6, runs 13–15)

nose was formed but the reaction required a shorter time to reach equilibrium. Conversely, when mannose was used as the starting compound, the epimerizing activity of the reaction system was specifically stimulated by the (*S*,*S*)-isomer. In this particular case, the largest amount of epimer was obtained in the shortest time. Not surprisingly, the racemic ligand exhibited intermediate properties.

As for the mode of the epimerization by the optically active ethylenediamine derivatives whose chiral properties arise from a side chain between the two amino groups, similar results were observed. The higher yield of the mannose-type epimer was obtained with the (*R*)-isomer. The epimerization with (*R*)-2,2′-Ph-en (run 13) gave the corresponding epimer as the major product. With respect to the influence of the side-chain group structure, the phenyl group was more significant than the benzyl group. It is easily stipulated that it is associated with the proximity of the large aromatic ring to the ethylenediamine part.

The most significant result of this work is that the reaction system showed a molecular recognition ability, that is, the composition of the mixture of the two epimers could be regulated by nickel complex systems having the appropriate ligand. The equilibrium of the reaction system shifted in accordance with the configuration of the diamine ligand employed. This is the first observed example of molecular recognition of such steric effects during the coordination of these complexes that could be obtained in a relatively straightforward manner due to the unique geometry of the six-membered ring. The existence of two axial or equatorial positions, which are fixed by the ring structure, is a prime requisite for molecular recognition.

4 Preparation of 2-Hydroxymethylaldoses from Ketoses

As stated in Sect. 3, the nickel complex that coordinates the *N*- and *N′*-alkylated ethylenediamine ligand shows an excellent epimerizing ability in methanol with high selectivity. Based on this information, it is expected that 2-*C*-hydroxymethylaldopentose could be obtained in a manner similar to that of the formation of the epimerized aldose. If a 2-ketohexose could be treated as an aldopentose, which possesses a hydroxymethyl group at the carbonyl carbon, as illustrated in Fig. 10, a 2-hydroxylmethylaldopentose would be obtained. In agreement with this expectation, the ternary nickel complex readily proceeds with the rearrangement, and four different ketoses yield the corresponding 2-hydroxymethylated aldoses, as shown in Fig. 11 [53, 54].

In this section, a highly efficient synthetic system for the preparation of 2-*C*-hydroxymethylaldopentoses is discussed. The reactions of four different ketohexoses (D-psicose, D-fructose, L-sorbose and D-tagatose) were studied under various conditions in order to elucidate the relationship between the two key processes, the coordination of the sugar to the nickel complex, and the rearrangement of the carbon skeleton of the substrate ketose. The system was then compared with that of the epimerization of the aldose/nickel(II)/ethylenediamine complex.

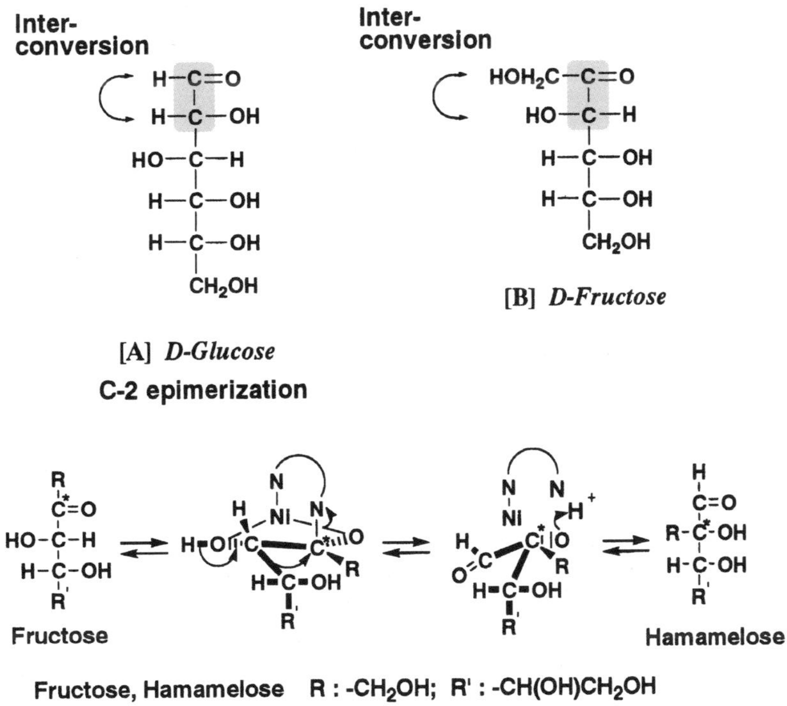

Fig. 10. Reaction mechanism of the formation of 2-*C*-hydroxymethylaldose (cf. Fig. 4)

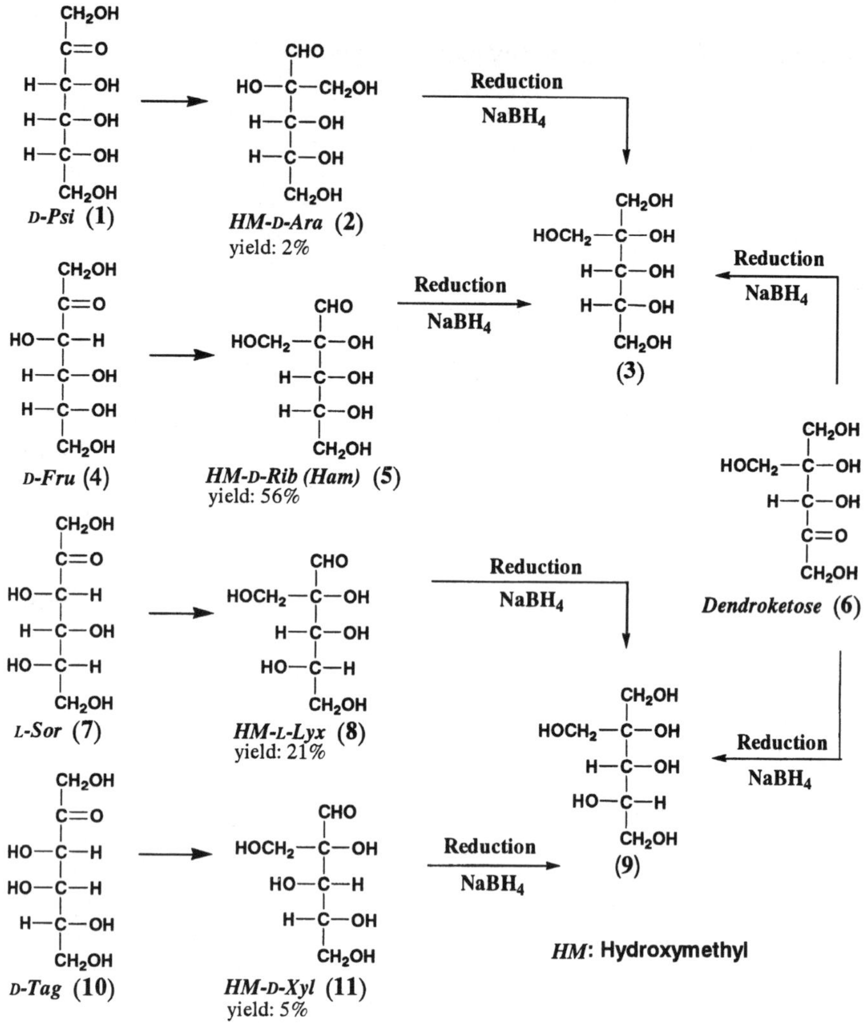

Fig. 11. Preparation of 2-*C*-hydroxymethylaldoses from the corresponding ketoses

The results for the rearrangement of D-fructose (4) into D-hamamelose (5) and L-sorbose (7) into 2-*C*-hydroxymethyl-L-lyxose (8) are summarized in Table 7. Each reaction was conducted in a manner similar to the epimerization of aldoses as shown in Scheme 1, except for the reaction temperature (30 °C) and the reaction time (40 min), employing methanol as the solvent.

Figure 12 demonstrates that the isomerization of the fructose to the corresponding branched 2-hydroxymethylated sugar proceeds through a sequence of stereospecific rearrangements. In this study, the C-2 carbon of fructose was substituted with a ^{13}C isotope. The carbon atom of the carbonyl group in fructose is converted into the C-2 carbon of the product, hamamelose. The δ values in the ^{13}C NMR spectrum agreed very closely with those of the authentic data shown

Table 7. Preparation of 2-*C*-hydroxymethyl-D-ribose (hamamelose) (%) from D-fructose and 2-hydroxylmethyl-L-lyxose (%) from L-sorbose

Ligand	Fru	Ham	Glc	Man	Others
en	97	1	0	0	2
2,2′-en	86	13	1	0	0
1,1,2′-en	68	19	2	2	9
1,1,1′,1′-en	88	12	0	0	0
2,2′-pn	69	27	0	0	4
chxn	93	2	0	0	5
1,1′-chxn	90	4	0	0	5
2,2'-chxn	59	41	0	0	0
3,3′-chxn	41	56	0	0	3
i4,i4′-chxn	79	21	0	0	0
8,8'-chxn	54	46	0	0	0
NaOH	42	0	39	9	10
Ligand	**Sor**	**2-HM-Lyx**			**Others**
2,2′-en	77	19			4
1,1,1′-en	72	10			18
3,3′-chxn	62	21			17
8,8′-chxn	67	20			13

Structure of *N*,*N*′-dipropyl-cyclohexanediamine (3,3′-chxn)

in Table 8. The hamamelose obtained consists of four constitutional isomers, α- and β-hamamelopyranoses and α- and β-hamamelofuranoses.

Taking into account the influence of the ligand diamine structures, a series of *N*,*N*′-dialkylated cyclohexanediamines (*R*,*R*′-chxn) were the most suitable ligands for the preparation of the 2-*C*-hydroxymethylated sugars. Each of the cyclohexanediamine derivatives is a racemic mixture of stereoisomers. Hamamelose (**5**) was obtained with high selectivity in a one-pot reaction in 56% yield employing the Ni(II)/3,3′-chxn complex. Very few by-products, such as glucose and mannose, were formed under the standard conditions for the reaction. As it is well known that monosaccharides are susceptible to alkali, and the solution containing the nickel/diamine complex is weakly basic, these minute amounts of aldoses could have been formed via an enediol intermediate in this basic environment.

Conversely, when the reaction mixture contained sodium hydroxide, fructose gave glucose and mannose in addition to a mixture of unidentified carbohydrates. The aldoses must have been formed by the LdB-AvE rearrangement of fructose via an enediol intermediate. In this case, the equilibrium between the two epimers favored glucose [41b, 48].

In a discussion on the influence of the diamine structure on the rearrangement of ketoses, information obtained from the epimerization of aldoses could be of interest. As pointed out in the context of the reaction mechanism, when a

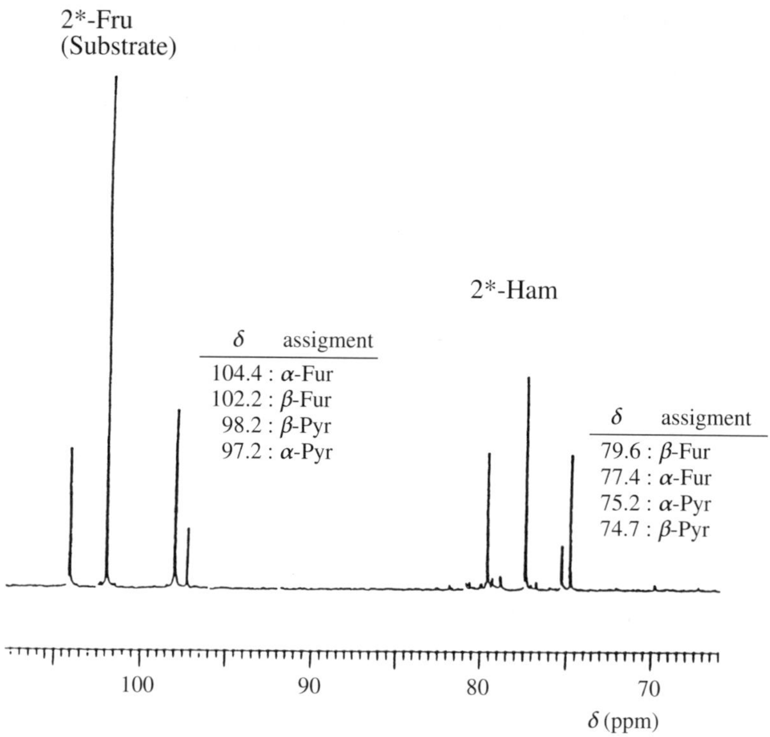

Fig. 12. ^{13}C NMR spectrum (22.5 MHz) of the product mixture derived from (2-^{13}C)fructose [2*-Fru]

Table 8. ^{13}C NMR data of fructose and hamamelose

	C-1	C-2	C-3	C-4	C-5	C-6	Ref.
α-D-Fructopyranose	65.9	99.1	70.9	71.3	62.1	61.9	[69]
β-D-Fructopyranose	64.7	99.1	68.4	70.5	70.1	64.1	[69, 70]
α-D-Fructofuranose	63.8	105.5	82.9	77.0	82.2	61.9	[69, 70]
β-D-Fructofuranose	63.6	102.6	76.4	75.4	81.6	63.2	[69, 70]
α-D-Hamamelopyranose	94.5	75.2	67.3	68.2	64.8	60.5	[71]
β-D-Hamamelopyranose	95.0	74.7	66.0	68.9	63.2	62.9	[71]
α-D-Hamamelofuranose	96.6	77.4	69.6	81.7	62.2	62.4	[71]
β-D-Hamamelofuranose	101.5	79.6	71.3	82.6	62.7	62.7	[71]

harmonic motion between the coordination of the sugar and release of the ligand into the reaction system occurs, the reaction proceeds smoothly. Whereas a primary amino group is unsuitable, a secondary or tertiary one substituted with a suitable alkyl group provides for the rearrangement of the ketose.

One can readily understand that the configuration and the conformation of a particular sugar has a significant effect on the formation of the ternary glycoside complex. For a series of ketohexoses, the contrasting yields can be explained in terms of the relative configuration of the two hydroxyl groups of the ketose substrate involved in the process. This suggests that a sugar/Ni complex interaction contributes in some way to the coordination during the reaction. Observing the influence of the substrate sugar's configuration, the following tendency was recognized. Among the yields of the four types of 2-*C*-hydroxymethylated aldopentoses including hamamelose, the yields of 2-hydroxymethyl-D-ribose (**5**) and 2-hydroxymethyl-L-lyxose (**8**) were 56 and 21%, respectively. Conversely, the yields of 2-hydroxymethyl-D-arabinose (**2**) and 2-hydroxymethyl-D-xylose (**11**) were extremely low (2 and 5%, respectively). The coordination of the sugar to a nickel complex is a very important process for preparing a branched-chain sugar in high yield from a ketose. The 2-*C*-branched-chain sugars were produced in relatively high yield from two ketoses, D-fructose (**4**) and L-sorbose (**7**), in which the two hydroxyl groups at C-3 and C-4 are aligned in a *threo* relationship. Two other ketoses, **1** and **10**, in which in the two hydroxyl groups at C-3 and C-4 are *erythro* positioned, gave the corresponding isomerized sugars in very low yields.

This indicates that the availability of the hydroxyl groups for coordination with the nickel/diamine complex is the principal factor which governs the extent of the ketose rearrangement. These results suggest that the coordination of **1** and **10** to the nickel/diamine complex is more difficult to achieve as compared with the interaction of **4** and **7**. This is also in keeping with the fact that the configuration of the hydroxyl groups in the sugar is a controlling factor of the aldose epimerization [44b].

In this context, it appears to be of relevance to briefly mention some comparable reaction systems for ketose epimerization. Petruš and his group, as well as Petruš, Serianni and co-workers, recently reported the stereospecific molybdic acid catalyzed isomerization of 2-hexuloses to branched-chain aldoses [55–57]. Upon treatment with a catalytic amount of molybdic acid in aqueous solution, the 2-ketohexoses, D-fructose, L-sorbose and D-tagatose, underwent a stereospecific intramolecular rearrangement to give the corresponding 2-*C*-hydroxymethylaldoses, 2-*C*-hydroxymethyl-D-ribose (D-hamamelose), 2-*C*-hydroxymethyl-L-lyxose and 2-*C*-hydroxymethyl-D-xylose, respectively (see Petruš, Petrušová and Hricovíniová, this vol.).

Hadwiger and Stütz reported the nickel(II)-catalyzed isomerization reactions of D-fructose derivatives modified at positions 5 and/or 6 [58, 59]. As already mentioned, this reaction system has attracted significant interest because of its possible extension to a wide range of carbohydrates.

5
Epimerization by a Hydrophobic Nickel Complex in Aqueous Solution

In recent years, reactions carried out in organized molecular assemblies such as micellar solutions have attracted much interest in many fields. This is because organized molecular aggregates can be applied as mimics of biological systems, that is, many reactions proceed smoothly as in living systems. Such reactions are observed, for example, as chemiluminescence [60], hydroxylation of benzene [61], debromination of bromocarboxylic acid [62], phosphate ester hydrolysis [63] and others, carried out in the aggregate.

In this section, the effect of aggregates such as metallomicelles comprised of nickel and hydrophobic diamines on the epimerization of aldoses is outlined. A homologous series of nickel/ethylenediamine complexes of various *N,N*-dimethyl-*N′*-alkylethylenediamines (1,1,n′-en) was prepared and examined [64]. The influence of the hydrophobicity of the ligand on the epimerization in aqueous media was assessed. The objective of this work was to clarify why differences in hydrophobicity affect the outcome of the epimerization and to characterize the nature of the aggregative metallomicelle.

5.1 Progress of the Epimerization

The epimerization reaction was carried out in a manner similar to the procedure shown in Scheme 1 (Sect. 2.2) except for the reaction medium. Water instead of methanol was used in this investigation. Analytical results and yields of the epimers are given in Table 9, and illustrated in Fig. 13. Data show the degree of epimerization at 80 °C after 20 min in an aqueous solution as a function of alkyl chain length of *N,N*-dimethyl-*N′*-alkylated ethylenediamine at a nickel complex concentration of 67 mM.

For the epimerization of aldohexoses (glucose and mannose), minute amounts of fructose are produced as the sole by-product. In the case of the aldo-

Table 9. Epimerization products from glucose, mannose, xylose and lyxose in an aqueous solution

Ligand	Products (%)				Products (%)				
	Substrate	Glu	Man	Fru	Substrate	Xyl	Lyx	Xylu	Others
1,1,1′-en	Glucose	95	2	3	Xylose				
	Mannose	3	90	7	Lyxose				
1,1,2′-en	Glucose	97	2	1	Xylose	84	7	5	4
	Mannose	2	95	3	Lyxose	5	75	10	10
1,1,6′-en	Glucose	97	2	1	Xylose	82	9	6	3
	Mannose	2	95	3	Lyxose	7	76	12	5
1,1,8′-en	Glucose	97	2	1	Xylose	75	11	6	8
	Mannose	3	95	2	Lyxose	10	75	9	6
1,1,9′-en	Glucose	87	11	2	Xylose	58	29	6	7
	Mannose	27	69	4	Lyxose	42	45	8	5
1,1,10′-en	Glucose	87	13	0	Xylose	55	33	5	7
	Mannose	34	63	3	Lyxose	53	37	3	7
1,1,12′-en	Glucose	80	20	0	Xylose	56	33	5	6
	Mannose	45	55	0	Lyxose	54	33	7	6
1,1,14′-en	Glucose	75	25	0	Xylose				
	Mannose	61	37	2	Lyxose				
1,1,16′-en	Glucose	69	29	2	Xylose	56	33	5	6
	Mannose	67	32	1	Lyxose	57	35	4	4
1,1,18′-en	Glucose	71	29	0	Xylose				
	Mannose	69	29	2	Lyxose				

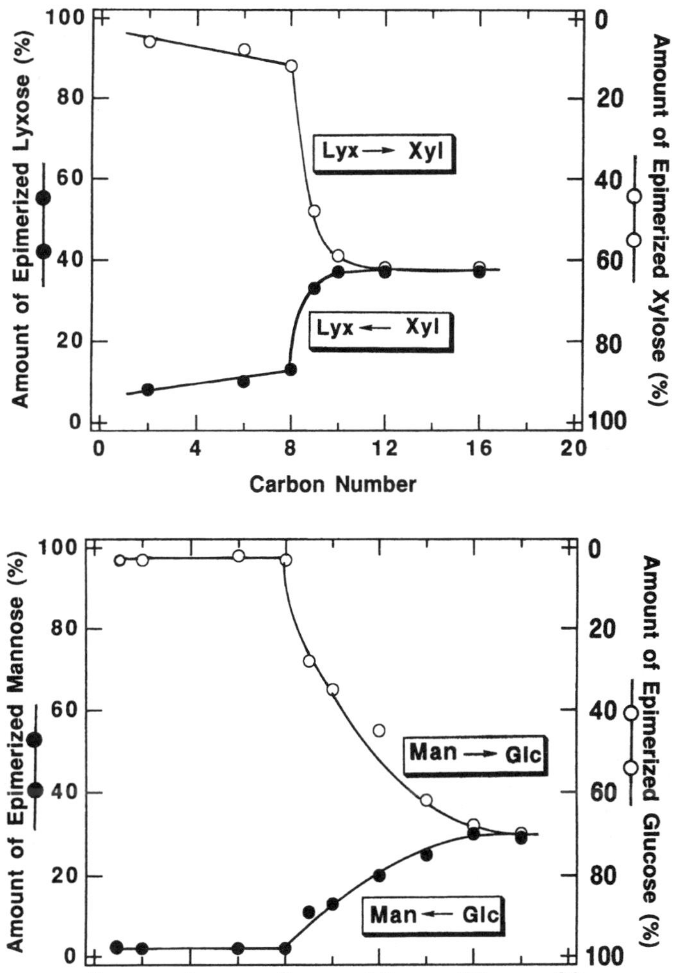

Fig. 13. Influences of the carbon number in a Ni(II)/1,1,n′-en complex on the epimerization between glucose and mannose (*bottom*) and between xylose and lyxose (*top*)

pentoses (xylose and lyxose), the product contains corresponding ketose (xylulose), other complex sugars, as well as unidentified decomposition products, in addition to starting material and the corresponding C-2 epimerized aldoses.

The nickel complex coordinating a less alkylated diamine lost much of its ability to epimerize although it showed a high epimerizing activity to attain the equilibrium in methanol, as discussed in Sect. 3.1. Although 1,1,1′-en shows an excellent epimerizing capacity in methanolic solution, its ability vanishes in water.

Conversely, an equilibrium between the C-2 epimers, with a composition almost identical for D-glucose and D-mannose as starting sugars, was smoothly attained when *N*-alkylated ethylenediamines with longer chains, such as 1,1,16′-en or 1,1,18′-en, were used as ligands. As Fig. 13 shows, the yields of the C-2 epimers increase with the carbon chain length of the diamine, irrespective of which aldose is subjected to the reaction conditions. The epimerization increases noticeably when the *N*′-alkyl group becomes longer than octyl (1,1,8′-en) and the epimerization curves shift to higher values to attain equilibrium with increasing alkyl chain length. A dramatic difference is observed between the yields of epimers, depending on the hydrophobicity of the diamine complex employed. Epimerization was observed in the hydrophobic complex system (ligands 1,1,9′-en and 1,1,18′-en) furnishing yields of 30–70%, which are approximately 15–35-fold better than observed in hydrophilic system (ligands 1,1,1′-en and 1,1,2′-en), both in the aldohexose and the aldopentose series.

The influence of the pH of the medium on the extent of epimerization was investigated. The pH values of the aqueous solutions containing nickel chloride and the ethylenediamine derivative were measured at 25 °C, and the results were as follows: 1,1,2′-en, pH 9.1; 1,1,8′-en, 8.6; 1,1,9′-en, 8.0; 1,1,12′-en, 7.4. The pH of the solution decreased as the alkyl chain length increased and a clear relationship between the rate of the epimerization and pH of the system could not be recognized.

According to the stereospecific epimerization mechanism (see Sect. 2.2), the nature of the coordination of the aldose to a Ni^{2+}/diamine complex is crucial to this reaction. A stereospecific rearrangement occurred in the ternary complex system composed of Ni^{2+}, diamine and sugar. It was demonstrated that the epimerization in aqueous media proceeds via a similar mechanism to that which occurs in methanolic solution. The rearrangement may take place via a Ni^{2+}-chelated five-membered intermediate, where the migration occurs with an antiperiplaner alignment to the leaving diamine ligand leading to inversion of configuration at C-2. Figure 14 shows the ^{13}C NMR spectrum of the product mixture derived from (1-^{13}C)glucose (top) and from (1-^{13}C)mannose (bottom). The signals at 92.8 and 96.9 ppm were assigned to C-1 of α- and β-glucose, and those at 72.2 and 72.8 ppm to C-2 of α- and β-mannose, respectively (Table 1). Epimerized mannose possesses a carbon substitution at C-2, showing that even in aqueous solution the carbon sequence of the aldose was rearranged.

The introduction of hydrophobicity into the aqueous reaction system seems to be important to attain effective epimerization. Aggregation of complex and hydrophobic interaction due to amphiphilic properties plays a key role in the coordination of aldose. Possible reasons for the large epimerizing ability of amphiphilic nickel complexes include enhanced concentration of nickel ion and ethylenediamine in a metal complex core structure on the micelle surface, which enhances the coordination of the aldose to the nickel complex. The integrated effect of an active site is responsible for the pronounced degree of epimerization.

In contrast, the nickel complexes that coordinate shorter-chain alkylated diamines show lesser degrees of epimerization, although they exhibit a high epimerizing activity to attain the equilibrium in methanol, as demonstrated in runs 16 and 17 in Table 2. They would be homogeneously dissolved in the matrix and would be strongly hydrated. The coordination of substrate sugar to the complex, which is a key process in the reaction, was inhibited by hydration.

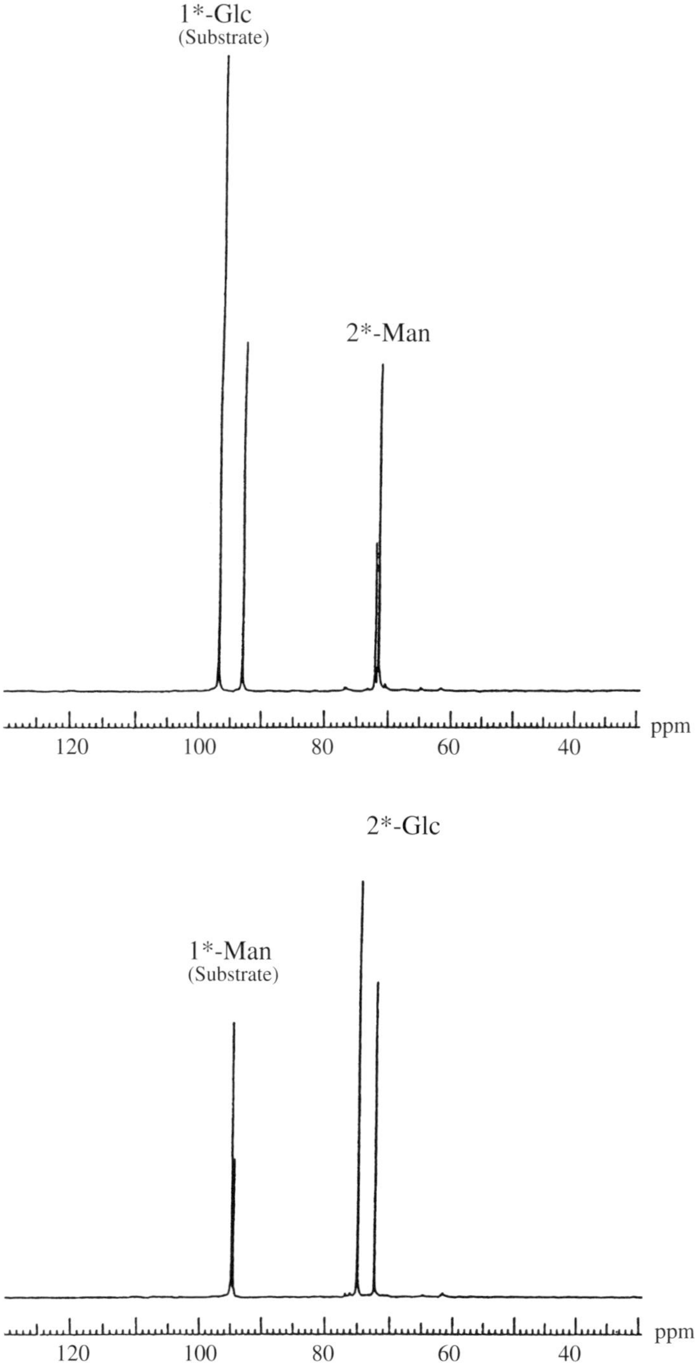

Fig. 14. ^{13}C NMR spectrum (22.5 MHz) of the product mixture derived from (1-^{13}C)glucose [1*-Glc] and from (1-^{13}C)mannose in aqueous solution

5.2 Formation of Metallomicelles

It has been suggested that significant enhancement of the epimerization was due to the formation of metallomicelles in the reaction system. To confirm the existence of the micelles, the surface tension of the reaction system was measured. It was confirmed that hydrophobic Ni/diamine complexes containing alkyl chains longer than decyl form micelles, whereas Ni/diamine complexes containing alkyl chains shorter than nonyl do not. It should be emphasized that such formation of micelles was in accordance with the occurrence of epimerization. The fact that a nickel complex that can aggregate to form metallomicelles shows excellent epimerizing ability is the most noteworthy point.

The dependence of the critical micelle concentration (cmc) on the chain length was also investigated. The results are shown in Fig. 15. In general, the cmc of amphiphiles in aqueous media decreases as the hydrophobic character of the amphiphile increases, and the relationships between the cmc and the number of carbon atoms (N) in the hydrophobic group can be expressed by the following equation:

$$\log C_{cmc} = A - BN$$

where A and B are constants reflecting the free energy changes involved in transferring the hydrophilic group and a methylene unit of the hydrophobic group,

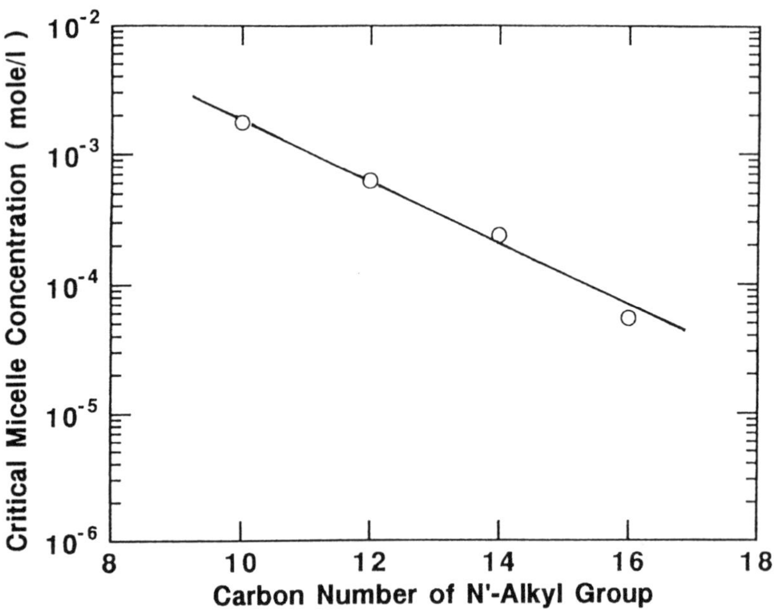

Fig. 15. Effect of length of the hydrophobic *N′*-alkyl group on the critical micelle concentration on the carbon number in ligands 1,1,n′-en

respectively, from an aqueous environment to the micelle. The values of A and B for the micelle of this metallic complex were −0.43 and 0.24, respectively. They are compatible with those for known commercial anionic and cationic surfactants [65].

5.3 Catalytic Activity of the Amphiphilic Complex System

The variation of the epimerization in aqueous media was also studied. The catalytic activity was presented by means of relative activity with changing molar ratio between glucose and nickel complex. The epimerization of glucose under the normal conditions was employed as a standard. The results are summarized in Fig. 16.

When a hydrophobic ligand was employed, the epimerizing ability of the nickel complex was not significantly reduced, even when a relatively large amount of aldose was applied. The Ni^{2+}/1,1,12′en complex, in particular, showed excellent epimerization properties even when the molar ratio of complex to aldose was increased by a factor of about 30. This would suggest interconversion between free aldose in a bulk aqueous phase and aldose coordinated to the nickel complex, and that the interconversion is accompanied during the epimerization by the skeletal rearrangement of the two epimers. In contrast, the hydrophilic nickel complex shows no catalytic epimerizing ability under standard conditions.

Furthermore, the relationship between concentration and catalytic activities was considered. The hydrophobic nickel complexes facilitate the epimerization

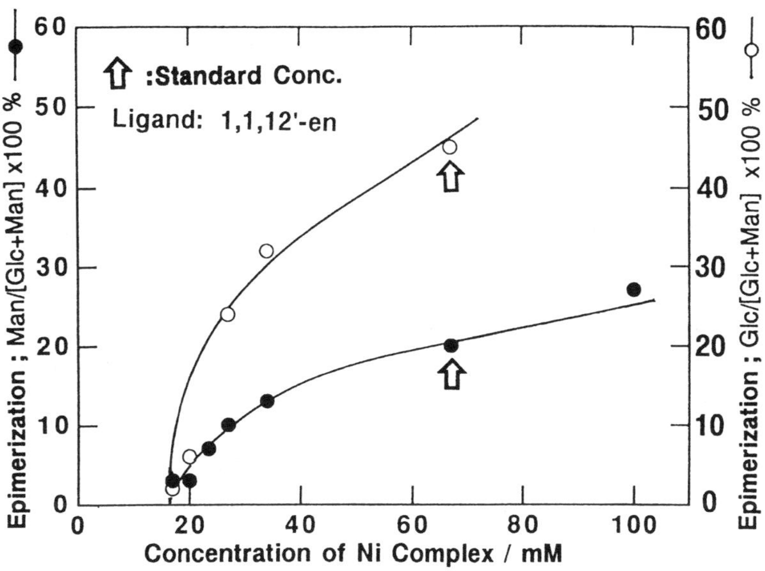

Fig. 16. Influences of the concentration of the Ni complex on the epimerization

by the assembly of the Ni^{2+} ion and diamine on the micelle. The effect of the concentration of the Ni/diamine complex on the epimerization of aldoses was discussed. The results are shown for $Ni_2Cl_2 \cdot 6H_2O$ and *N,N*-dimethyl-*N′*-dodecylethylenediamine, giving the concentration of nickel complex as indicated in Fig. 16. Under standard conditions, with the concentration of the nickel complex at 67 mM, glucose and mannose were smoothly converted into the corresponding epimers, whereas no epimerization took place at concentrations below 20 mM.

Comparison of these results may not be straightforward, since the surface tension and epimerization were measured at different temperatures (the former at 25 °C and the latter at 80 °C). However, it is interesting that the findings are consistent with the interpretation that aggregates were formed, giving additional evidence for metallomicelles.

6 References

1. (a) Bartok M (1985) Stereochemistry of heterogeneous metal catalysis. John Wiley and Sons, New York, chap 6; (b) Stoddart JF (1971) Stereochemistry of carbohydrates. Wiley-International, London, chap 5
2. (a) Angyal SJ (1989) Chem Soc Rev 9:415; (b) Angyal SJ (1972) In: Carbohydrates, vol IA. p 195; (c) Angyal SJ, Davies KP (1971) J Chem Soc Chem Commun 500
3. de Wit G, de Vlieger JJ, Kock-van Dalen AC, Heus R, Laroy R, van Hengstum AJ, Kieboom APG, van Bekkum H (1981) Carbohydr Res 91:125
4. de Wit G, de Vlieger JJ, Kock-van Dalen AC, Kieboom APG, van Bekkum H (1978) Tetrahedron Lett 1327
5. Delephine M, Horeau A (1937) Bull Soc Chim Fr 1524
6. Levene PA, Christman CC (1937) J Biol Chem 120:575
7. Bach AN, Alexejewa JP, Drewing WP (1938) Chem Zentralblatt I 626
8. Kaiwar SP, Raghavan MSS, Rao CP (1995) J Chem Soc Dalton 10:1569
9. Costamagna J, Barroso NP, Matsuhiro B (1998) Inorg Chim Acta 273:191
10. Kozlowski H (1990) Carbohydr Res 197:109
11. Escandar GM, Sala SF, Sierra MG (1994) Polyhedron 13:143
12. Virtanen POI, Oikarinen E, Nevala H (1989) J Carbohydr Chem 8:13
13. Loeb KE, Zaleski JM, Hess CD, Hecht SM, Solomon EI (1998) J Am Chem Soc 120:1249
14. Klufers P, Schuhmacher J (1994) Angew Chem Int Ed Engl 33:1742
15. Buzas N, Gajda T, Nagy L, Kuzmann E, Vertes A, Burger K (1998) Inorg Chim Acta 274:167
16. Angyal SJ (1980) Chem Soc Rev 9:415
17. Oefner PJ, Chiesa C (1994) Glycobiology 4:397
18. Bazzanella A, Bachmann K (1998) J Chromatogr A 799:283
19. de Wit G, Kieboom APG, van Bekkum H (1979) Recl Trav Chim Pays-Bas 98:355
20. Horitsu K, Gorin PA (1977) Agric Biol Chem 41:1459
21. Venema FR, Peters JA, van Bekkum H (1993) Recl Trav Chim Pays-Bas 112:445
22. Lehmann TE, Ming LJ, Rosen ME (1997) Biochemistry 36:2807
23. Matulová M, Bílik V (1993) Carbohydr Res 250:203
24. Sauvage JP, Verchere JF (1992) Carbohydr Res 237:23
25. Rodio RT, Pereira EM, Tavares MFM (1999) Carbohydr Res 315:319
26. Diez M, Barros FJG, Calahorro CV (1995) J Inorg Biochem 60:199
27. Ferrier F, Avezou A, Terzian G, Benlian D (1998) J Mol Struct 442:281
28. Frutos AA, Signorella S, Sala LF, Escandar GM, Salas Peregrin JM, Moreno V (1998) Polyhedron 17:3371
29. Amanokura N, Kanekiyo Y, Shinkai S (1999) J Chem Soc Perkin Trans 2 10:1995

30. Templeton DM, Sarcar D (1985) Biochem J 230:35
31. London RE (1987) J Chem Soc Chem Commun 661
32. Veres S (1987) Magy Kem Foly 93:199; CA 107:249257
33. Vijayalakshmi MA, Cadelis F, Tse SBB (1989) UCLA Symp Mol Cell Biol New Ser 80:315
34. Khandelwal S, Flora SJS, Tandon SK (1990) Chem Biol Interact 75:341
35. Deiana S, Gessa C, Piu P, Seeber R (1991) J Chem Soc Dalton Trans 5:1237
36. Gajda T, Nagy L, Rozlosnik N, Korecz L, Burger K (1992) J Chem Soc Dalton Trans 3:475
37. Jankiewicz B, Soloniwicz R (1990) Acta Pol Pharm 47:39; CA 117:239618
38. Kim KM, Song SC, Lee SB (1998) Inorg Chem 37:5764
39. Tanase T, Shimizu F, Yano S, Yoshikawa S (1986) J Chem Soc Chem Commun 1001
40. Tanase T, Shimizu F, Kuse M, Yano S, Hidai M, Yoshikawa S (1988) Inorg Chem 27:4085
41. (a) El Khadem HS (1988) In: Carbohydrate chemistry, monosaccharides and their oligomers. Academic Press, New York, p 102; (b) El Khadem HS, Ennifar S, Isbell HS (1989) Carbohydr Res 185:51
42. Yanagihara R, Osanai S, Yoshikawa S (1990) Chem Lett 2273
43. (a) Takizawa S, Sugita H, Yano S, Yoshikawa S (1980) J Am Chem Soc 102:7969; (b) Yano S, Takizawa S, Sugita H, Takahashi T, Tsubomura T, Shioi H, Yoshikawa S (1985) Carbohydr Res 142:179; (c) Shioi H, Yano S, Toriumi K, Ito T, Yoshikawa S (1983) J Chem Soc Chem Commun 201; (d) Yano S, Sakai Y, Toriumi K, Ito T, Yoshikawa S (1985) Inorg Chem 24:498; (e) Tsubomura T, Yano S, Toriumi K, Ito T, Yoshikawa S (1983) Polyhedron 2:123; (f) Tsubomura T, Yano S, Toriumi K, Ito T, Yoshikawa S (1984) Bull Chem Soc Jpn 57:1833; (g) Tsubomura T, Yano S, Toriumi K, Ito T, Yoshikawa S (1985) Inorg Chem 24:3218; (h) Yano S (1988) Coord Chem Rev 92:113; (i) Yano S, Takahashi T, Sato Y, Ishida K, Tanase T, Hidai M, Kobayashi K, Sakurai T (1987) Chem Lett 2153
44. (a) Fukushima K, Takahashi M, Nagano H, Osanai S, Yoshikawa S (1988) Nippon Kagaku Kaishi 585 (main text is written in Japanese; tables and figures in English); (b) Yamauchi T, Fukushima K, Yanagihara R, Osanai S, Yoshikawa S (1990) Carbohydr Res 204:233; (c) Osanai S, Inaba K, Yoshikawa S (1991) Carbohydr Res 209:289
45. (a) Bílik V, Stankovic L (1973) Chem Zvesti 27:544; (b) Bílik V, Caplovic J (1973) Chem Zvesti 27:547; (c) Bílik V, Petruš L, Farkaš V (1978) Collect Czech Chem Commun 43:1163; (d) Bílik V, Petruš L, Farkaš V (1975) Chem Zvesti 29:690; (e) Bílik V, Petruš L, Zemek J (1978) Chem Zvesti 32:242
46. Hayes ML, Pennings NJ, Serianni SA, Barker R (1982) J Am Chem Soc 104:6764
47. Yanagihara R, Soeda K, Shiina S, Osanai S, Yoshikawa S (1993) Bull Chem Soc Jpn 66:2268
48. Angyal SJ (1969) Angew Chem Int Ed Engl 8:157
49. Tewari YB, Goldberg RN (1986) Biophys Chem 24:291
50. Hataya K, Yanagihara R, Osanai S, Yoshikawa S (1991) J Chem Soc Chem Commun 1246
51. Asperger RG, Liu CF (1965) Inorg Chem 4:1492
52. Buckingham J (1982) In: Dictionary of organic compounds. Chapman and Hall, New York, c-03330
53. Yanagihara R, Osanai S, Yoshikawa S (1992) Chem Lett 89
54. Yanagihara R, Egashira J, Yoshikawa S, Osanai S (1995) Bull Chem Soc Jpn 68:237
55. Hricovíniová Z, Hricovíni M, Matulová M, Petruš L (1998) Chem Zvesti 52:238
56. Hricovíniová Z, Hricovíni M, Petruš L (1998) Chem Zvesti 52:692
57. Hricovíniová Z, Hricovíni M, Petrušová R, Serianni AS, Petruš L (1999) Carbohydr Res 319:38
58. Hadwiger P, Lechner A, Stütz AE (1998) J Carbohydr Chem 17:241
59. Hadwiger P, Stütz AE (1998) J Carbohydr Chem 17:1259
60. (a) Warner M (ed) (1991) Advances in multidimensional luminescence, vol 1. JAI Press, p 149; (b) Papadopoulos K, Chantoron A, Nikokavouras J, Hrbac J, Lasovsky J (1998) J Photochem Photobiol A Chem 116:153
61. Tamagaki S, Fukuda K, Tagaki W (1996) Chem Express 7:868
62. Takagi K, Miyake N, Nakamura E, Usami H, Sawaki Y (1988) J Chem Soc Faraday Trans I 84:3475
63. Menger FM, Gan LH, Johnson E, Durst DH (1987) J Am Chem Soc 109:2800

64. Osanai S, Yanagihara R, Uematsu K, Okumura A, Yoshikawa A (1993) J Chem Soc Perkin Trans 2:1937
65. Rosen MJ (1978) In: Surfactants and interfacial phenomena. John Wiley and Sons, New York, p 103
66. Johnson LF, Jankowski WC (1973) JEOL Carbon-13 spectra, vol 1, #197
67. Pfeffer PE, Valentine KM, Parrish FW (1979) J Am Chem Soc 101:1265
68. Perlin AS, Casu B, Koch J (1970) Can J Chem 48:2596
69. Wolff GJ, Breitmaier E (1979) Chem Ztg 103:232
70. Angyal SJ, Bethell S (1976) Aust J Chem 29:1249
71. Schilling G, Keller A (1977) Liebigs Ann Chem 1475

D-Xylose (D-Glucose) Isomerase and Related Enzymes in Carbohydrate Synthesis

Herwig Häusler, Arnold E. Stütz

Glycogroup, Institut für Organische Chemie der Technischen Universität Graz, Stremayrgasse 16, 8010 Graz, Austria
E-mail: stuetz@orgc.tu-graz.ac.at

Enzymatically catalysed isomerisation of free sugars is an important and economically viable alternative to chemical means for achieving such transformations. The conversion of D-glucose into D-fructose with D-xylose (D-glucose) isomerase (EC 5.3.1.5) has become a large-scale industrial application producing more than ten million tons of fructose syrup annually. Due to the remarkable tolerance of this and related enzymes, for example, L-fucose isomerase (EC 5.3.1.3), for epimers, as well as chemically modified substrate analogues bearing non-natural features, such as azido or fluoro substituents, a large variety of interesting and novel synthetic applications has become feasible for these remarkably useful and easy-to-handle biocatalysts.

Keywords: Aldose ketol isomerases, Enzymatic isomerisation, L-Fucose isomerase, D-Glucose (D-Xylose) isomerase, Non-natural substrates, L-Rhamnose isomerase

Topics in Current Chemistry, Vol. 215

1 Introduction

The tremendous importance of organometallic as well as bioorganic chemistry in contemporary organic synthesis and manufacturing is nicely mirrored in recent developments in carbohydrate chemistry, in particular when concerned with the preparation of commercial products ranging from commodities to sophisticated building blocks and high added value compounds. Long-known methods of conventional epimerisation and isomerisation of free sugars, such as the Lobry de Bruyn – Alberda van Ekenstein rearrangement (see Angyal, this vol.), as well as more recently discovered approaches via transition metal complex formation in protic solutions (see chapters by Petruš, Petrušová, and Hricovíniová and Osanai, this vol.), have been complemented by the discovery of a range of enzymes which interconvert aldoses into their epimers or into the corresponding ketoses with a high degree of selectivity. In some cases, very interesting and preparatively useful substrate tolerances with respect to a large variety of structural modifications of their natural substrates have been observed. These epimerases and isomerases belong to class 5 of EC numbers and feature a wide range of interesting properties.

1.1 Epimerases and Isomerases

Epimerases are a group of quite diverse enzymes of varying molecular masses whose catalytic mechanisms have not yet been fully investigated. They may utilise free sugars (Scheme 1) such as *N*-acetylglucosamine 2-epimerase (EC 5.1.3.8) [1], their corresponding terminal phosphates, for example, *N*-acetylglucosamine-6-phosphate 2-epimerase (EC 5.1.3.9, Scheme 1, compounds **1** and **2**) [2] and L-ribulose-5-phosphate 4-epimerase (EC 5.1.3.4, compounds **3** and **4**) [3], or 1-GDP or -UDP-"protected" sugars, for example, UDP-L-arabinose 4-epimerase (EC 5.1.3.5) [3] or GDP-D-mannose 3,5-epimerase (EC 5.1.3.18, compounds **5** and **6**) [3, 4]. In all cases of *O*-1 unprotected sugars, these give the corresponding epimers at C-2. No obvious trends are observed with open-chain ketose phosphates as well as UDP- or GDP-protected pyranoses.

Interesting preparative applications of epimerases could be envisaged, as exemplified by the preparation of D-sorbose (**8**) from D-tagatose (**7**) [5] or of D-

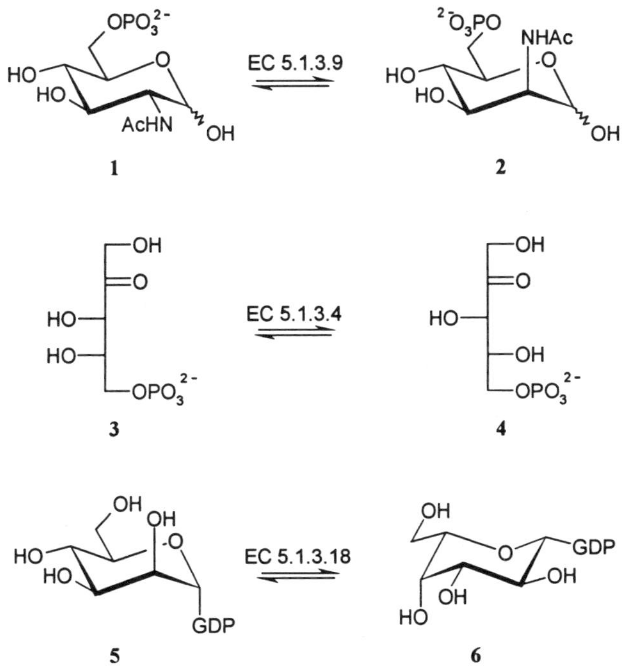

Scheme 1

psicose (**10**) from D-fructose (**9**) [6] with immobilised D-tagatose 3-epimerase (D-TE), just to mention a few recent examples (Scheme 2).

Isomerases interconvert free aldoses and the corresponding 2-ketoses. Other representatives, such as glucose 6-phosphate isomerase (EC 5.3.1.9), require phosphate groups at the terminal carbon of the respective sugar (Scheme 3, compounds **11** and **12**).

These enzymes, in particular D-xylose isomerase (D-glucose isomerase, EC 5.3.1.5) and L-fucose isomerase (EC 5.3.1.3), have attracted considerably more attention in terms of synthetic applicability, as compared with the epimerases,

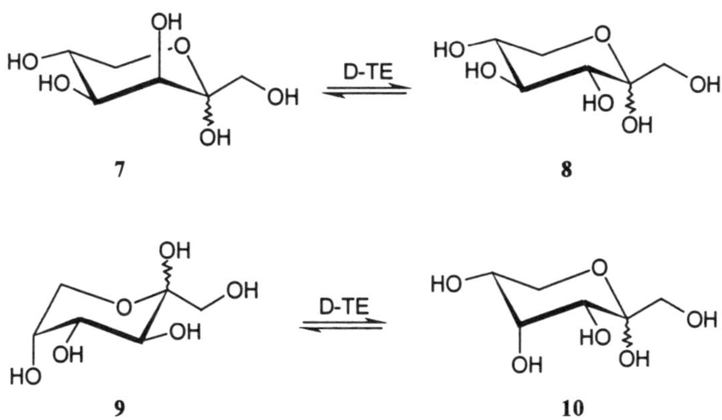

Scheme 2

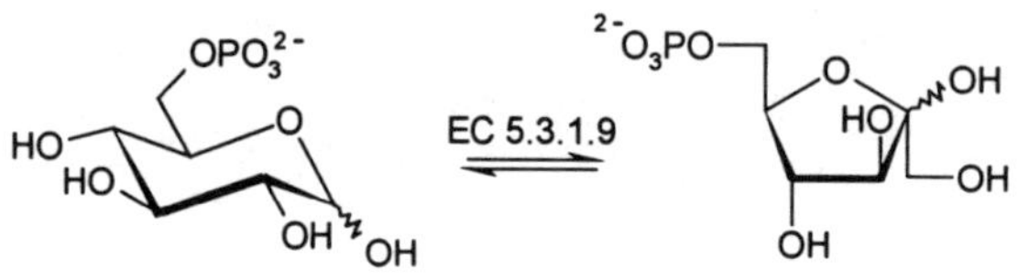

Scheme 3 **11** **12**

and several preparative applications, mostly in context with natural product synthesis, can be found in the literature.

2 D-Glucose (D-Xylose) Isomerase (EC 5.3.1.5)

2.1 History

In 1952, for the first time, an enzyme that had the ability to isomerise an unsubstituted free sugar, namely D-erythrose into D-*glycero*-tetrulose, was discovered [7]. Subsequently, it was found that *Escherichia coli* produces enzymatic activity that converts D-arabinose (**13**) into D-*erythro*-pentulose (**14**) and it was suggested that D-arabinose isomerase (EC 5.3.1.3) catalyses the isomerisation of L-fucose (**15**) into L-fuculose (6-deoxy-L-*lyxo*-hexulose, **16**) (Scheme 4) [8].

The conversion of D-glucose (**17**) into D-fructose (**9**) by a microbial enzyme (Scheme 5) was first reported in 1957 when Marshall and Kooi found glucose isomerase activity in cell-free extracts of *Pseudomonas hydrophila* [9]. This enzymatic activity was enhanced in the presence of arsenate. Soon thereafter, other arsenate-requiring enzymes were isolated from *Aerobacter* sp. as well as *Escherichia freundii* [10]. Enzymes required arsenate when D-glucose or D-fructose was the substrate but not when the corresponding 6-phosphates **11** and **12** were offered. Purification of the arsenate-dependent principle component from *Escherichia intermedia* allowed the conclusion that the enzyme was a glucose 6-phosphate isomerase (EC 5.3.1.9) that was able to isomerise free D-glucose when it was complexed with arsenate [11].

A true glucose isomerase that did not require arsenate for its activity was discovered in *Lactobacillus brevis* [12]. Such enzymes were later also found in the genus *Streptomyces* and this source was first reported in 1965 [13]. Subsequent studies demonstrated that the *Streptomyces* enzyme responsible for this conversion is inducible and is rather a D-xylose isomerase, as the value of K_M

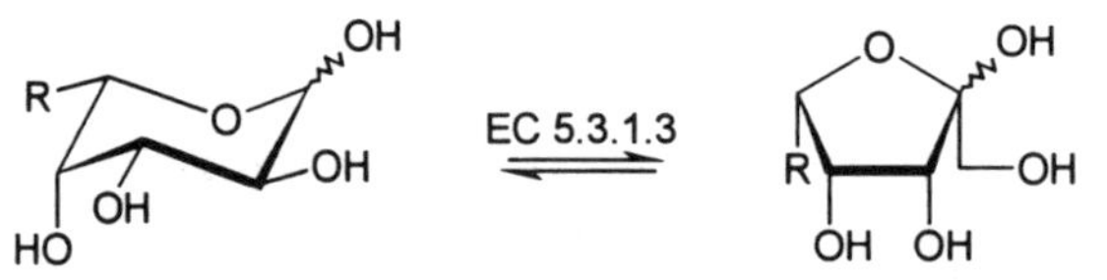

Scheme 4 R = H **13** R = Me **15** **14** **16**

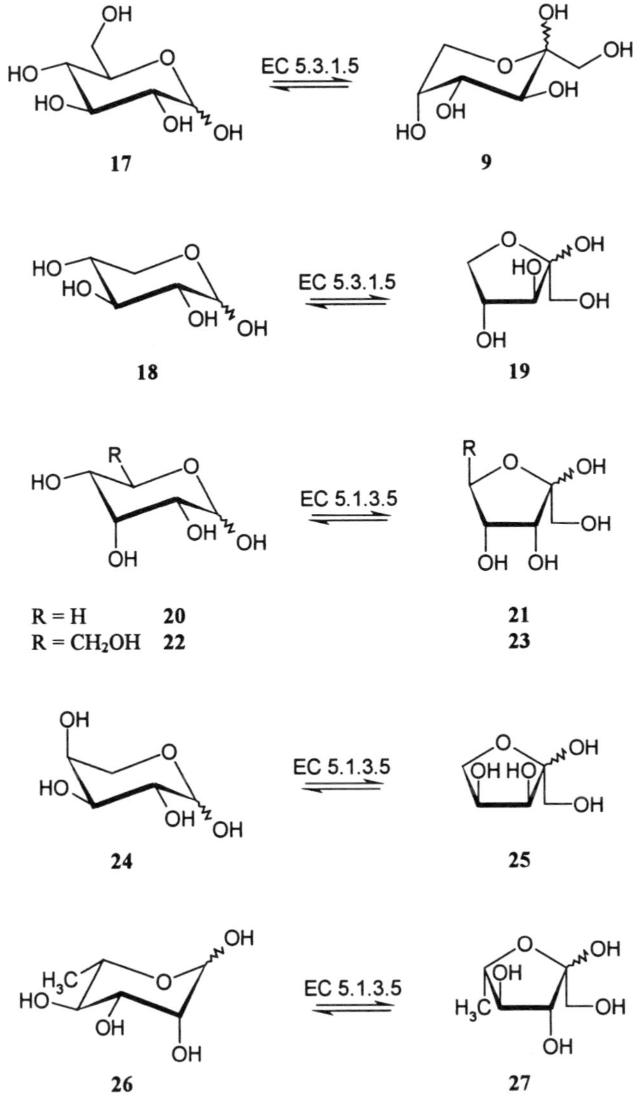

Scheme 5

is one to two orders of magnitude smaller with D-xylose (**18**)/D-*threo*-pentulose (D-xylulose, **19**) as the substrates [14]. D-Xylose isomerases found in other microorganisms, such as *Lactobacillus brevis* [15] and *Bacillus coagulans* [16], were also able to convert D-ribose (**20**) into D-ribulose (**21**). A further extension to the range of substrates was reported for D-xylose isomerase from *Streptomyces albus*, which was also able to isomerise D-allose (**22**), L-arabinose (**24**) and L-rhamnose (**26**) into the corresponding ketoses **23**, **25** and **27** (Scheme 5) [17].

It was discovered that several other strains of *Streptomyces*, *Bacillus* and *Lactobacillus* were able to produce these enzymes [18]. Their pH-optimum was determined around pH 8, but the range of activity reaches from pH 5 to pH 12. Enzymes from a variety of sources were found to be dependent on magnesium, manganese, or cobalt ions, while calcium and some divalent transition metal ions, such as copper, iron and nickel, were found to inhibit their activities. Optimum temperatures for the conversion range between 70 and 85 °C [3] as yields of D-fructose increase with temperature, but enzyme stability decreases dramatically in this temperature range. Consequently, industrial processes are conducted around 60 °C. Recently, enzymes that allow higher reaction temperatures of up to 95 °C have been reported [19].

Purification of enzymes from several sources allowed the determination of their molecular masses of around 120,000 [20] to around 200,000 [21] for the respective complete tetrameric structures. Enzymes from *Streptomyces olivochromogenes* and from *Bacillus stearothermophilus* were compared in respect to their physicochemical and enzymatic properties [20].

The lack of a supply of sucrose after the Cuban revolution in 1958 triggered in-depth research into glucose (xylose) isomerases and initiated their commercial importance in the USA. As early as 1967, immobilised D-xylose isomerase found its first commercial application in the United States based on the production of "high-fructose corn syrup" (HFCS) from corn-starch based on a Japanese patent [22] and, in 1984, Coca-Cola and Pepsi Cola approved quantitative substitution of sucrose by EFCS (enriched HFCS) in the USA. The annual world consumption of HFS is estimated at more than 10 million tons dry weight, making D-xylose isomerase the largest scale intracellular immobilised enzyme used in industry [23, 24]. The annual growth rate was recently estimated at 3–4% on a global basis [24]. Several methods for immobilisation have been made available and were recently reviewed [24]. Comprehensive surveys on various aspects including industrial scope and significance of xylose isomerases are available [23–25].

2.2 Structural and Mechanistic Studies

Many of the mechanistic aspects of glucose isomerase catalysed aldose-ketose interconversion have been under discussion for some time and are still not fully understood. By comparison with triose phosphate isomerase (TIM, EC 5.3.1.1) and glucose 6-phosphate isomerase (EC 5.3.1.9), the base-catalysed formation of an 1,2-enediol was invoked as the key step of the epimerisation based on the work of Rose and co-workers with tritium-labelled substrates [26]. An unexplained feature of the epimerisation process was that in contrast to isomerisations with triose phosphate isomerase no proton exchange with the medium could be observed with D-xylose isomerase, a fact that was attributed to the phosphate group of the former as a mediator for the exchange process [26]. Subsequently, additional important differences between triose phosphate isomerase and xylose isomerase were recognised. For example, D-xylose isomerase is apparently a very slow enzyme catalysing about five molecules per second per active site with an absolute requirement for divalent cations, while TIM does not need co-factors and operates at nearly 1000-fold the speed of D-xylose isomerase at the diffusion-controlled limit. For these reasons it can be called an ideal enzyme [27]. A detailed discussion of the reaction mechanism of xylose isomerase based on the enediol pathway was provided by Dyson and Noltmann [28] as well as Alworth [29].

Investigations into the anomeric specificity of the enzyme suggested that the α-xylopyranose was utilised exclusively, a fact demonstrated by comparison of initial reaction rates with pure and mixed anomers assuming that the rate of spontaneous mutarotation of each sugar is slow [30].

In a seminal contribution, Bock and co-workers conducted mechanistic studies on soluble as well as immobilised xylose isomerase from *Streptomyces murinus* sp. (Sweetzyme Q from NOVO) employing NMR-spectrometric methods. These studies confirmed the lack of deuterium exchange with solvent molecules of D-glucose and D-fructose regiospecifically *C*-deuterated at C-1 and C-2 [31]. It is noteworthy that they found the (1*S*)-diastereomer of D-(1-^{2}H)fructose (**29**) was formed with perfect selectivity by isomerisation of D-(1-^{2}H)glucose (**28**) and so was the corresponding (1*R*)-epimer (**31**) when D-(2-^{2}H)glucose (**30**) was employed as the substrate (Scheme 6).

No other products were detected in the reaction mixture and no isotopic scrambling occurred in the course of the reaction. These observations remained uncommented on but, in the light of later findings, do already suggest a highly ordered mechanism involving a hydride (deuteride) shift as opposed to the proton shift based enediol pathway involving solvent exchange as seen with triose phosphate isomerase. Applying ^{13}C-NMR spectrometry these workers also confirmed earlier findings [29] that the α-anomer of D-glucopyranose is the substrate and β-D-fructofuranose is released from the enzymatic process (Scheme 7).

Further NMR-spectroscopic investigations into the mechanism and reactive species of the D-glucose/D-fructose interconversion by Optisweet P (Miles Kali-Chemie) employing D-(1-^{13}C)glucose confirmed that the α-anomer of D-glucose Is the aldose substrate for the enzyme. However, in contrast to prior observations [21], α-D-fructofuranose (not β-D-glucofuranose) appeared to be the ketose substrate [32]. Very recently, these interpretations have been supported by Col-

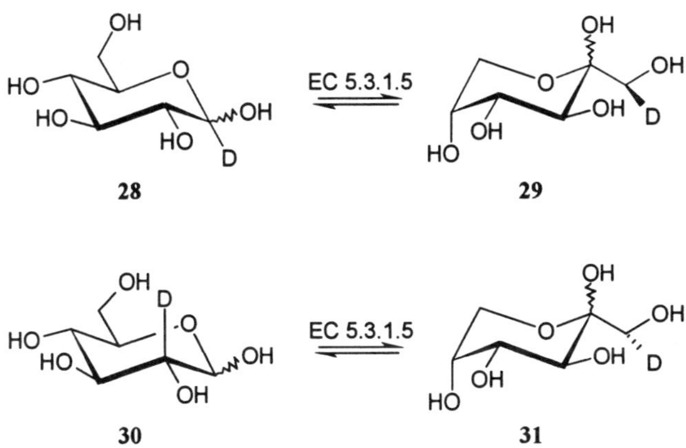

Scheme 6

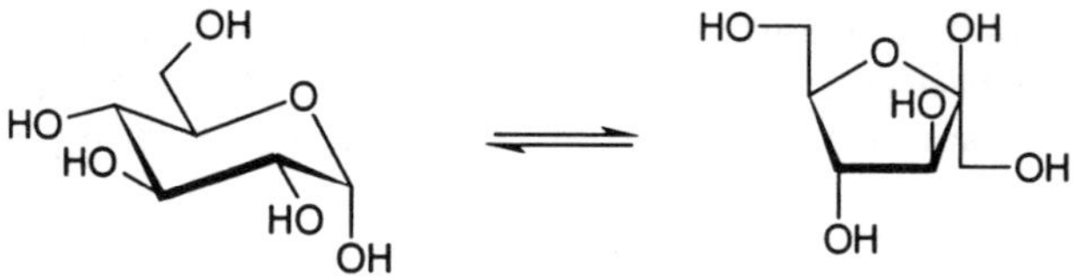

Scheme 7

lyer and co-workers from their interpretations of crystal structures of D-xylose isomerase with stable structural analogues of the putative reactive species [33].

Applying the "principle of least molecular motion" a reaction pathway (Scheme 8) based on the enediol mechanism was proposed [32] that also explained the stereospecific formation of (1*S*)- (**29**) and (1*R*)-D-(1-^{2}H)fructose (**31**) from D-(1-^{2}H)- (**28**) and D-(2-^{2}H)glucose (**30**), respectively [31].

In 1984 Carrell and co-workers published the first X-ray crystal structure (Scheme 9) of a D-xylose isomerase at 4 Å resolution [34]. This cobalt-dependent enzyme from *Streptomyces rubiginosus* had been shown to have an approximate molecular mass of 165.000 and to consist of four structurally identical peptide chains each containing around 370 amino acid residues [35]. It was established that the main section consisted of eight β-strand α-helix ($\beta\alpha$) units arranged in a configuration similar to the ones found in triose phosphate isomerase [36], pyruvate kinase [37], 2-keto-3-deoxy-6-phosphogluconate aldolase [38], as well as ten, possibly eleven [39] other enzymes [40], forming a characteristic closed α/β-barrel, while a smaller loop-shaped domain overlaps the larger domain of another subunit, so furnishing a tightly bound "embracing dimer". The whole, tetrameric structure consists of two such dimers. Two years later, a magnesium-dependent D-xylose isomerase from *Arthrobacter* sp. was crystallised [41]. The sequences and gene coding for D-xylose isomerases from various microorganisms, including *Bacillus subtilis* [42], *E. coli* [43], *Ampulariella* sp. [44] and *Streptomyces violaceoniger* [45], have been made available and aligned. It was shown [46] that a number of residues are conserved in all sequences implying significant roles in the structures or functions of the enzyme. It was recognised that two histidine residues, in particular, were conserved in the sequences of all species under consideration and their function was probed by site-specific substitution [47].

By this time, X-ray structures had been reported for several xylose isomerases. The previously reported [34] structure of the *Streptomyces rubiginosus* enzyme had been refined to a 3 Å resolution [48]. A structure of the same reso-

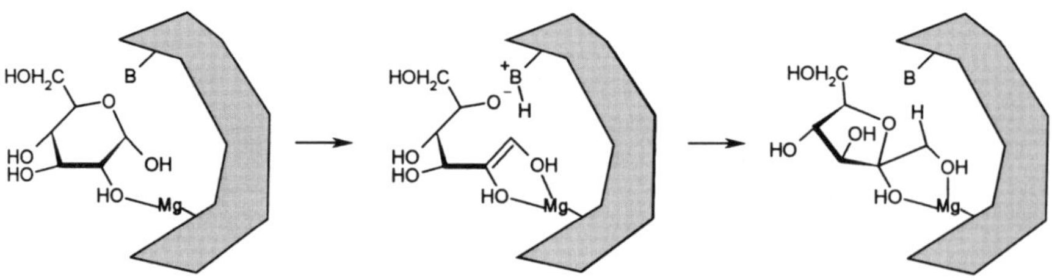

Scheme 8

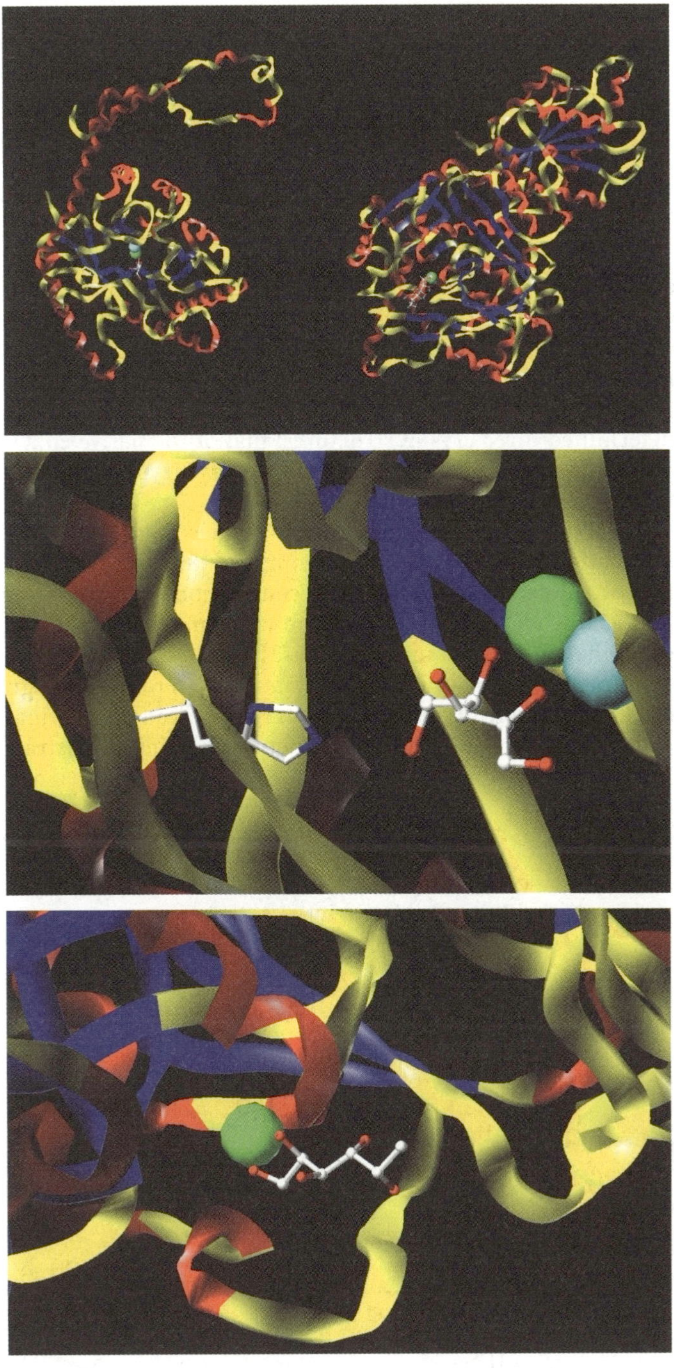

a b c

Scheme 9. **a** Shaded ribbon presentation of D-xylose (D-glucose) isomerase (EC 5.3.1.5) [61 a] with open-chain D-xylose/xylulose (capped sticks) coordinated to two metal ions in the active site (left) and L-fucose isomerase (EC 5.3.1.3) [94] with L-fucose and one metal ion (right). **b** Detail of the active site of D-xylose isomerase with histidine coordinating O-5 of D-xylose and the two metal ions interacting with O-1, O-2 and O-4 of the substrate. **c** Detail of the active site of D-fucose isomerase with the metal ion coordinating O-1 and O-2 of the substrate

lution had been determined [49] for xylose isomerase from *Streptomyces olivochromogenes*, the structure of the *Arthrobacter* B3278 isomerase was available at 2.5 Å resolution [41] and the structure of the enzyme from *Actinoplanes missouriensis* [50] as well as the one from *Streptomyces albus* [46, 51] had been discussed. They all exist as tetramers with 222 symmetry; the respective monomers in each case exhibiting the characteristic eight-stranded α/β-barrel structure, thus confirming the close relationship of these enzymes.

Initiated by a paper from Carrell and co-workers [52] reporting the structure of D-xylose isomerase from *Streptomyces rubiginosus* refined to 1.9 Å resolution, in which the authors claimed experimental support for the enediol mechanism, controversy on the true mechanism of the enzymatic catalysis arose. This group had designed the aldose analogue **32**, a suicide inhibitor of the isomerase, and had determined three crystal structures: that of the native enzyme, of the enzyme with bound substrate/product, and the enzyme after reaction with the mechanism-based inhibitor (Scheme 10).

The two metal cations were located, one of which exhibited nearly ideal octahedral coordination with oxygen atoms from Glu-181, Glu-217, Asp-245, Asp-287 as well as two molecules of water. The second cation was found to coordinate Glu-217, Asp-255, Asp-257, His-220 and one water molecule. The metals were 4.9 Å apart in the native structure and Glu-217 was established to coordinate both with different oxygens of its carboxylate group. From the magnitudes of temperature factors and from peak heights in the electron density maps the cations were inferred to be Mn^{2+}. From the structure containing substrate/product

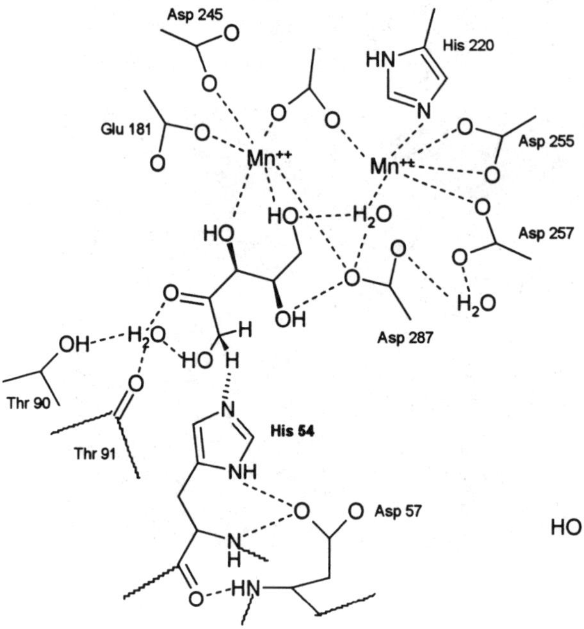

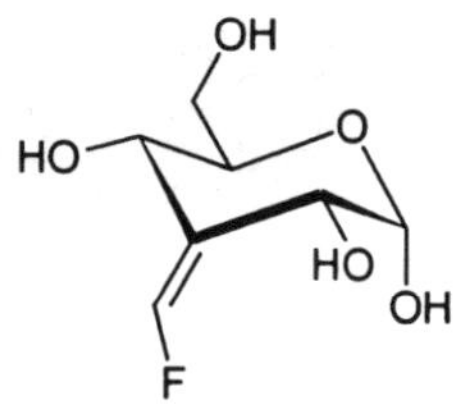

Scheme 10

32

it was apparent that the substrate molecule was bound in the open-chain form and coordinated to metal one by O-3 and O-5 while O-1 and O-2 were found to be coordinated by a water molecule in the proximity of Thr-90 and Thr-91. The mechanism-based inhibitor was found to alkylate His-54.

Because of the close interaction of C-1 of the substrate with His-54, indicating an incipient hydrogen abstraction through the *Re*-face of C-1, it was suggested that His-54 was the base catalyst facilitating the formation of the *cis*-configured enediol intermediate. The metal ions were assumed to be important for the structural integrity of the active site and for the coordination of hydroxyl functions of the substrate remote from the site of the isomerisation reaction. Alkylation of His-54 by the suicide inhibitor reinforced the interpretation of the role of this amino acid residue as the base catalyst.

In contrast, in their publication on crystallographic studies of xylose isomerase from *Streptomyces olivochromogenes*, Petsko and co-workers [53] came to the conclusion that a hydride shift mechanism (Scheme 11) agreed better with their findings from experiments also conducted under steady-state conditions by use of a flow cell, than the classical enediol pathway.

This technique is based on the constant flow of substrate over the crystal providing products being washed away. It was assumed that the time-averaged structure determined in such an experiment was dominated by the species that precedes the transition state of highest free energy (which as the authors clearly say does not need to be the same as that observed in an equilibrium experiment by soaking the crystals in the substrate).

The only base in the active site possibly working as the catalyst is a histidine, His-219. Yet, this residue is positioned to act as the ring-opening base. Other possible candidates for the role as ring-opening base could also be excluded, a fact that again pointed to the histidine as the base that catalyses the conversion of the cyclic into the open-chain form of the sugar. As the absence of label exchange with the medium is also consistent with a hydride shift, the divalent metal cation must serve as a Lewis acid stabilising the formation of a carbocation at the carbonyl carbon prior to the concerted transfer of a hydride. By coordination with O-1 and O-2, the cation could also stabilise a conformation favouring the hydride shift (Scheme 11). From chemical reactions involving a hydride shift, such as the Meerwein-Ponndorf-Verley-Oppenauer [54] and the Cannizzaro [55] reactions, the role of metal ions acting in this way was known.

This view on the reaction mechanism was strongly supported by Collyer and Blow [56] who compared their findings with the enzyme from *Arthrobacter* B3278 with Carrell's results [52]. In their view, the "mechanism-based inactivator", which is supposed to become "armed" upon isomerisation of the aldose to the α,β-unsaturated ketose, then alkylate His-53 (which corresponds to His-54 of the *Streptomyces rubiginosus* enzyme), does not bind in a "productive" conformation which would involve interaction of O-3 with one of the cations in the active site. This is not possible if this compound does not bear a hydroxy function in this position. They reasoned that the alkylation step could occur independently of the isomerisation as, in their opinion, His-53 is not involved in the latter. Furthermore, the authors disagreed on the orientation of the substrate D-

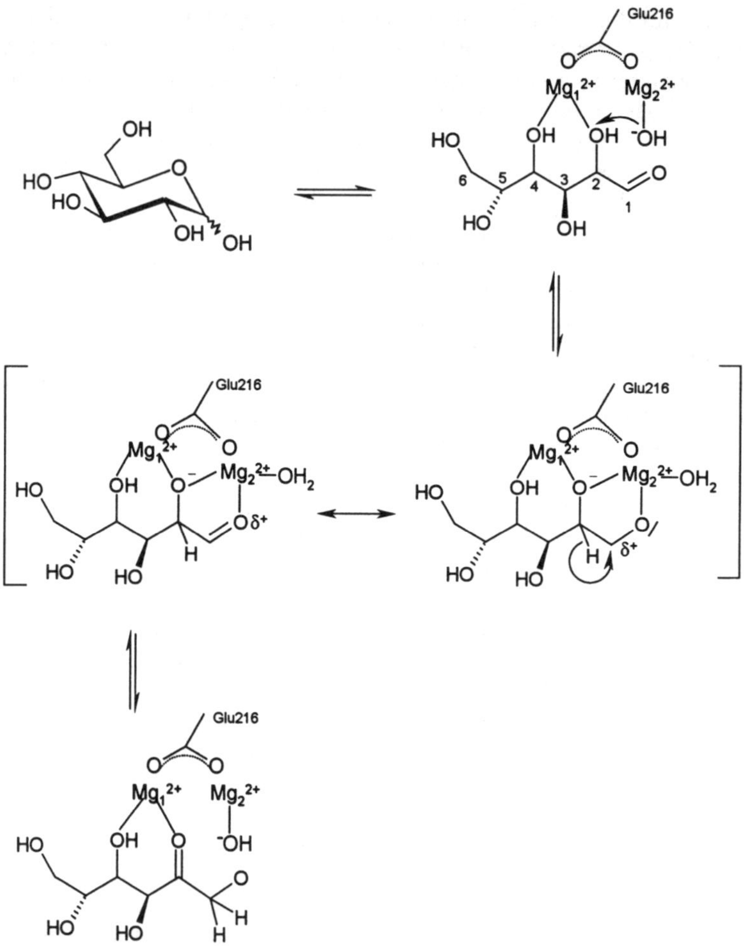

Scheme 11

xylose/D-xylulose bound to the active site "allowing opposite conclusions on the isomerisation mechanism". Following their arguments, His-53 serves to ring open the substrate. In this open-chain conformation C-1 and C-2 are close to the cation binding sites but there is no base in the right position to catalyse the formation of the enediol intermediate. The cation in site two moves slightly to coordinate O-1 and O-2 and, aided by cation 1 and a lysine moiety, Lys-182, becomes the electrophile that expels the proton from O-2 during the hydride ion shift. The substrate side where the hydride shift occurs is positioned in a strongly hydrophobic environment preventing the exchange of hydride with solvent confirming previous experimental observations [26, 33].

By site-directed mutagenesis of His-101 in xylose isomerase from *Clostridium thermosulfurogenes*, and identification of the rate-limiting step during catalysis by monitoring deuterium isotope effects on the V_{max} values of both wild-type and Gln-101 mutant enzymes, the hydride shift mechanism was further supported by Zeikus and co-workers [57]. It was found that D-(2-^{2}H)glucose (**30**) slowed the reaction rate of the wild-type as well as the mutant enzyme by a factor of four when compared with the reaction rate of D-glucose under the same conditions. These results suggested that the rate-determining step for both the wild-type enzyme and the mutant enzyme, that lack a base catalyst at position 101, is the hydrogen transfer and His-101 does not act as a simple base, since, in this case, inactive enzymes would have to be expected as the result of site-directed substitutions.

As it had been found that the enzyme specifically reacts with the α-anomer, but that the substrate detected in crystals of enzyme substrate complex was the open-chain form, the authors suggested, similar to Collyer and Blow, that the mechanism involves three major steps in the catalytic isomerisation of D-glucose: ring opening of the substrate α-D-xylopyranose aided by interaction of His-101 with O-5 of the aldose, isomerisation by a metal ion catalysed hydride shift, and ring closure of the product.

Detailed kinetic analysis of the catalytic mechanism of the *Arthrobacter* enzyme by Rangarajan and Hartley [58] also supported the hydride shift mechanism. However, these workers imply different reaction pathways for the Mg^{2+} and the Co^{2+} enzymes. The rate-limiting step was inferred to be the isomerisation and not the ring opening, in contrast to the results of Zeikus and co-workers. Reports from Withlow and co-workers [59] as well as Collyer and his group [60] provided further support for the hydride shift mechanism.

In a comprehensive investigation, Lambeir and co-workers added significant information to the knowledge of the catalytic mechanism of xylose isomerase [61]. In the first of a series of three papers, this group established the role of the two metal ions in the course of the reaction with the enzyme from *Actinoplanes missouriensis*. Metal site 1 is tetracoordinated and tetrahedral in the absence and becomes hexacoordinated and octahedral in the presence of substrate. During the interaction with metal ion 1, O-2 and O-4 of the sugar are coordinated. Metal ion 2 is always octahedral and changes its position by 0.7 Å when binding to O-1 and more than 1 Å when coordinating O-2 of the substrate. These bonds replace bonds to carboxylate ligands from the protein backbone. In the presence of open-chain ligands, metal site 1 was found to be virtually identical in *A. missouriensis* and *S. rubiginosus*. The two metal ions play an essential part in binding the substrate, stabilising the open-chain form and catalysing the hydride shift. The second paper identified two histidine residues (His-220 and His-54) as important for catalysis. His-220, however, is not a catalytic base but plays a role in the coordination of metal ion 2. The role of His-54, previously believed to be the base catalyst facilitating the ring opening of the bound substrate, despite its interaction with OH-1 of the cyclic and OH-5 of the acyclic substrate, respectively, was questioned but the authors came to the conclusion that this residue governs the anomeric specificity of the enzyme for steric reasons.

In the last account of the series the previously observed conformational flexibility [59, 62] of Asp-255 was suggested to be an integral part of the catalytic process. The residue was believed to support the movement of metal ion 2 during binding to the substrate and subsequent hydride shift as well as the shuttling of a proton between O-1 and O-2 with the aid of a water molecule in the proximity of both, metal 2 and Asp-255.

Subsequent investigations into the catalytic action of the isomerase from *Streptomyces olivochromogenes* further substantiated the putative hydride shift pathway and gave strong evidence against a proton transfer via an enediol intermediate as part of the reaction (Scheme 12). A reinvestigation of the original solvent isotope incorporation experiments of Rose and co-workers at temperatures of 15, 25 as well as 55 °C [63] showed that in D-fructose resulting from the isomerisation reaction, less than 0.6% of hydrogen was substituted by tritium convincingly confirming Rose's original report.

Neither 3-deoxy-3-fluoro-D-glucose (**33**) [63, 64] nor the corresponding D-*allo*-configured epimer **35** were observed to release fluoride by β-elimination when exposed to the enzyme which would have to be expected upon proton abstraction at C-2 (Scheme 13). Furthermore, both deoxyfluoro sugars failed to act as substrates of the enzyme; this also held true for 3-deoxy-3-fluoro-D-fructose (**34**) [64]. It was concluded that His-53 (corresponding to His-54 in the *Streptomyces rubiginosus* enzyme) was not the ring-opening base as had been hypothesised previously but, in concert with the Mg^{2+} ions, acted as stabiliser of the open-chain tautomer by interaction with O-5 of the substrate prior to 1,2-hydride transfer [63].

Replacement of the tetracoordinated Mg^{2+} ion at binding site 1 by an amino acid employing site-directed mutagenesis furnished a catalytically incompetent mutant underscoring the crucial role of this catalytic entity [65]. Additional support for the hydride shift on the open-chain tautomer of the respective substrate was collected by employing 3-*O*-methyl-D-glucose (**36**) [66].

This sugar was observed to coordinate to metal site 1 of the Mg-dependent enzyme through O-2 and O-4 as well as to Mg-2 through O-1 and O-2 of the open-chain tautomer. His-53 was again found to stabilise the complex by interaction with O-5. This interaction, however, is not essential for catalysis as was demonstrated by other workers [69]. The catalytic process was again accompanied by a movement of Mg-2 towards Mg-1. In the reverse reaction, substrate activation of the ketose was accomplished by complexing of both metals with the carbonyl oxygen at C-2. These findings found additional support from X-ray crystallography of an enzyme-inhibitor complex of xylose isomerase with D-threonohydroxamic acid (**37**, Scheme 14) at 1.6 Å resolution [67].

To re-examine a proposal suggesting a hydride shift on a cyclic tautomer of the substrates [68], the influence of active site residues in the *Streptomyces rubiginosus* enzyme was probed by site-directed mutagenesis [69]. This particular reaction pathway would include binding of the α-anomer of D-xylopyranose, with O-1 coordinated to His-54 and O-3 as well as O-4 interacting with metal 1. Isomerisation would then occur via proton abstraction of O-2 by Asp-287 and concomitant shift of hydride from C-2 to C-1. Subsequently, negatively charged O-5 completes the reaction by ring closure to C-2 of the ketose.

Based on a wide range of crystallographic data of the mutant enzymes with D-xylose (**18**), 5-thio-D-glucopyranose (**38**) or 5-deoxy-D-xylulose (**39**) (Scheme 15) in their active sites, this view was not supported by the investigators. An in-

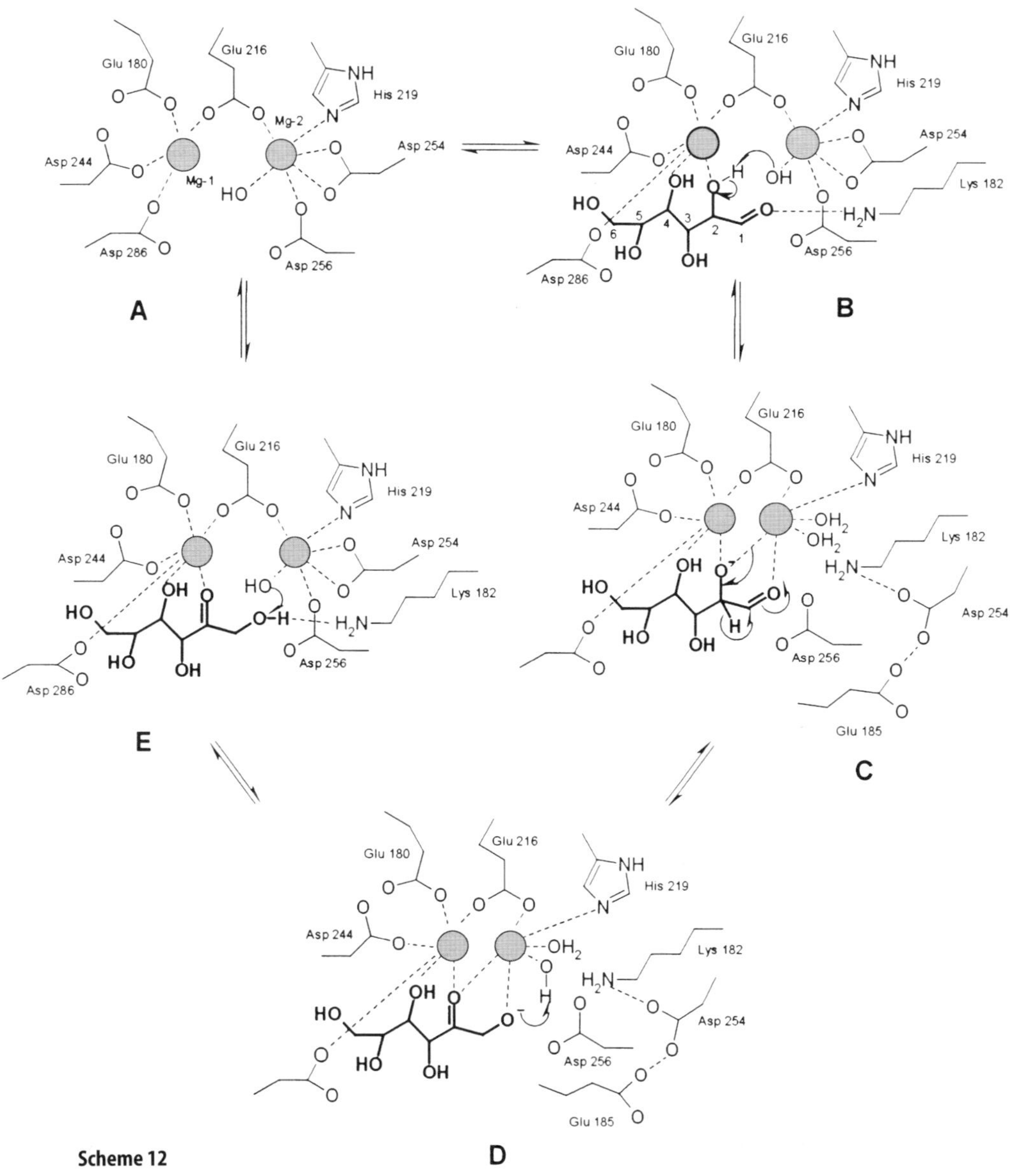

Scheme 12

vestigation into the isomerisation behaviour of glyceraldehyde in weakly alkaline aqueous solution revealed that under the conditions employed in this study (0.01 M KOD in D_2O at 25 °C) proton and hydride transfer occurred at similar rates [70]. Harris and Feather [71] observed, in acidic solutions, the isomerisation of D-glucose into D-fructose accompanied by a hydride shift. This result underscores the overwhelming evidence amassed for the hydride shift mechanism in D-xylose isomerases. Recent theoretical studies of the enzymatic reac-

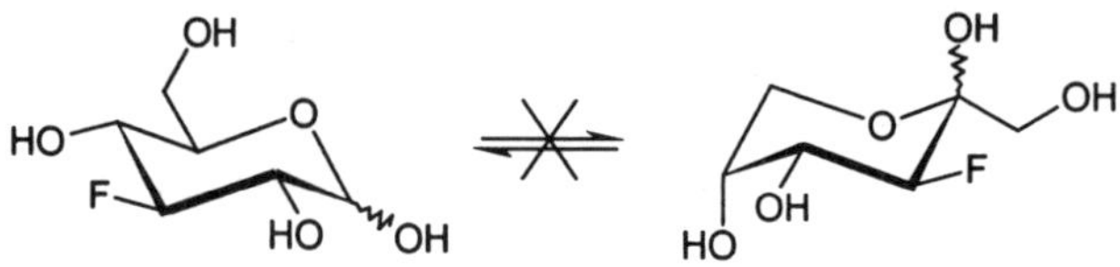

33 34

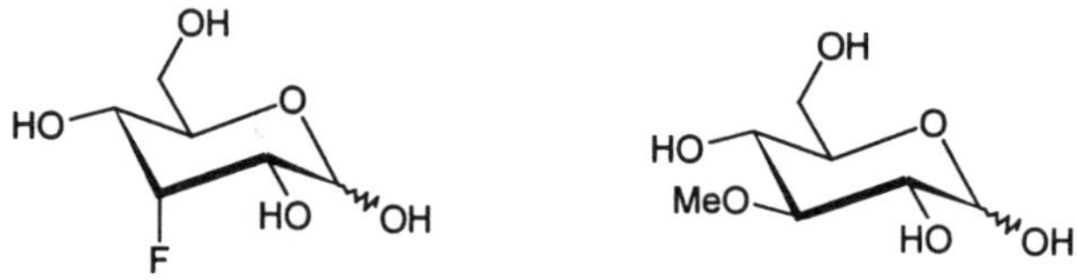

35 36

Scheme 13

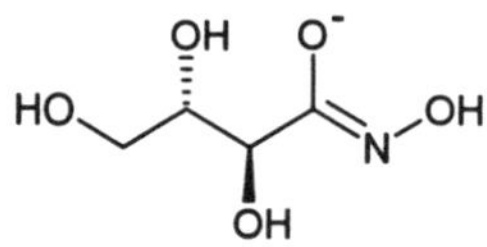

37

Scheme 14

38 39

Scheme 15

tion pathway employing a semiempirical molecular orbital method (PM3) also supported this reaction pathway [72] and excluded the enediol intermediate as well as a charge-relay mechanism during ring opening [73].

An interesting observation in the light of accumulated data supporting the hydride shift mechanism was made during the recently reported successful isomerisation of D-ribose (**20**), L-ribose (**40**), as well as D-lyxose (**41**), into the corresponding ketopentoses (**21**, **25**, and **19**, respectively, Scheme 16) [74]. It was found that the C-2 epimer of the respective aldopentose was also formed in the presence of high enzyme concentrations and upon extended reaction times of 12 h. For example, the respective enantiomer of arabinose was identified as the third component in the equilibrium of D- or L-ribose with the corresponding ketopentose. In the case of D-ribose/D-ribulose (**20/21**) isomerisation, the reaction mixture contained 40% of D-ribose and 30% each of D-ribulose and D-arabinose (**13**) after 12 h. With the L-sugars, the proportion of the "wrong" aldose epimer, L-arabinose, amounted to 2% after the same reaction period. Based on these results, the authors invoked a reaction pathway via the putative enediol

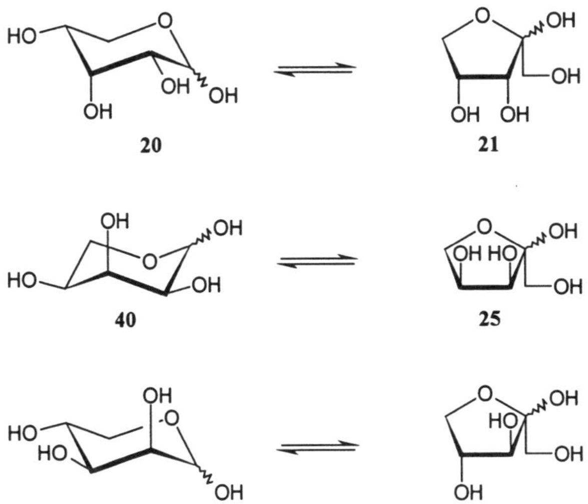

Scheme 16 — 20, 21, 40, 25, 41, 19

intermediate as initially proposed by Rose and co-workers [26]. Procedures were reported and the conditions employed by these workers practically rule out a non-enzymatic epimerisation, as reactions were run at pH 6.8 to avoid acid- or base-catalysed Lobry de Bruyn – Alberda van Ekenstein rearrangement reactions; blanks without enzyme, but containing Mg^{2+} or Co^{2+} ions as well as with metal-depleted enzyme, did not furnish isomerised products. Nevertheless, their conclusion is in marked contrast to the currently widely accepted hydride shift based mechanism for this type of enzyme. In keeping with the previously reported observations mentioned above that the corresponding enzymes from *Streptomyces olivochromogenes* and *Bacillus stearothermophilus* as well as *Bacillus* No. KX-6 accept and isomerise not only D-xylose (**18**) and D-glucose (**17**), but also D-ribose (**20**) as well as D-arabinose (**13**), the results neither contradict or support either of the two reaction pathways. L-Arabinose (**24**) has been reported to be a substrate for the enzyme from *Streptomyces albus* [17], which would explain this, previously unreported, very minor side activity in *Streptomyces rubiginosus.*

The formation of D-lyxose (**41**) from D-xylulose (**19**) could be rationalised on the basis of the previously observed quantitative conversion of L-erythrose (**42**) into the corresponding tetrulose (**43**) (Scheme 17) [87]. Clearly, the observations deserve further attention in order to establish whether the conversions of aldopentoses into their C-2 epimers result from interactions at the active site or with the basic enzyme surface at random. Either might occur due to the sterically less demanding, hence less tightly fitting, aldopentose structures under consideration (as opposed to D-glucose which only yielded D-fructose (**9**) – but no D-mannose (**44**) – under the same conditions).

Obviously, despite many efforts, the mechanism of D-xylose isomerase catalysed isomerisations of suitable pentoses and hexoses is far from rigorously established and need not be the same for sugars of different chain lengths and

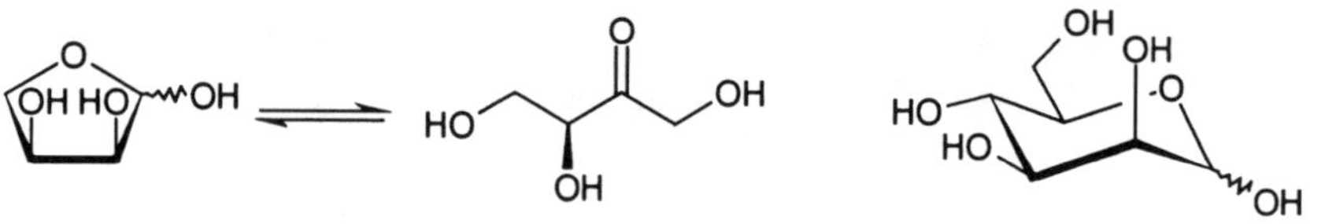

42 43 44

Scheme 17

configurations. As with other isomerases which only bear one metal ion per active monomer, and likely catalyse by enediol formation (for example, L-fucose isomerase, see Sect. 3.1), suitable pentoses might fit into the active site interacting with only one of the two catalytic metal sites, thus bypassing the hydride shift with the equally likely [70] enolisation pathway.

2.3 Non-Natural Substrates and Synthetic Applications

D-Xylose isomerases have been found in a wide range of microorganisms. As mentioned in Sect. 2.1, their substrate specificities are not confined to D-xylose (**18**) and D-glucose (**17**) but can include D- and L-arabinose (**13** and **24**) [17, 74], D-ribose (**20**) [15, 20, 75], and L-rhamnose (6-deoxy-L-mannose, **26**) [17], as well as D-allose (**22**) [17]. Very recent investigations already mentioned [74] have shown that L-ribose (**40**) as well as D-lyxose (**41**) are also accepted as substrates by some enzymes of the type under consideration. This would mean that of all eight aldopentoses, only L-lyxose and L-xylose have not yet been identified as substrates of D-xylose isomerases, demonstrating an intrinsic tolerance of these enzymes towards "unusual" substrates.

The first non-natural sugar derivative reported to be a substrate was 6-thio-D-glucose (**45**) which was found to be quantitatively converted into the corresponding D-fructopyranose **46** with sulfur in the ring (Scheme 18) [76]. In-depth investigations into the substrate tolerance of a D-xylose isomerase (Sweetzyme Q from Novo A/S, the enzyme from *Streptomyces murinus* sp.) were conducted by Bock and co-workers in 1983.

Results of this work showed that D-glucose derivatives modified at C-6, such as 6-deoxy-D-glucose (**47**) and 6-*O*-methyl-D-glucose (**49**), or C-3, for example, 3-*O*-methyl-D-glucose (**36**) and the corresponding 3-deoxy derivative **52**, are substrates (Scheme 19) while modifications at C-4, for example, 4-deoxy- (**54**), 4-*O*-methyl-D-glucose (**55**) as well as 4,6-di-*O*-methyl-D-glucose (**56**), epimers at C-2 (D-mannose, **44**), C-3 (D-allose, **22**) and C-4 (D-galactose, **57**) as well as L-

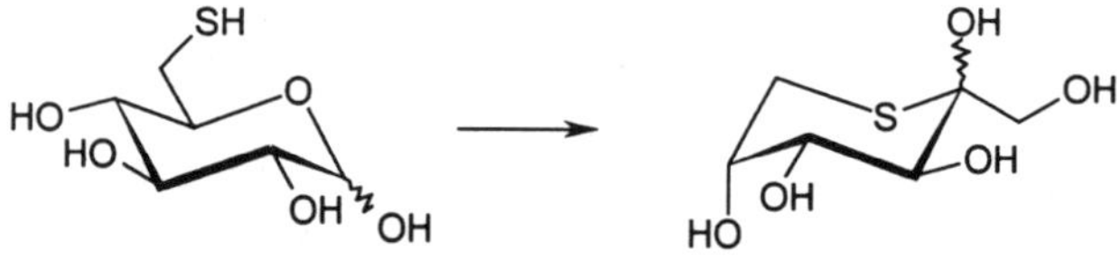

Scheme 18 45 46

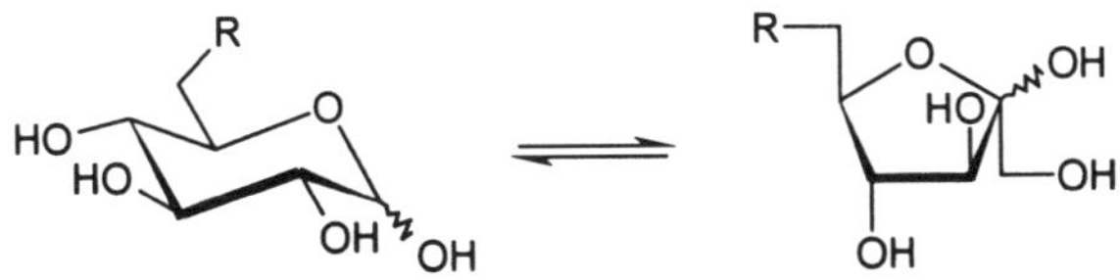

R = H 47 48
R = OMe 49 50

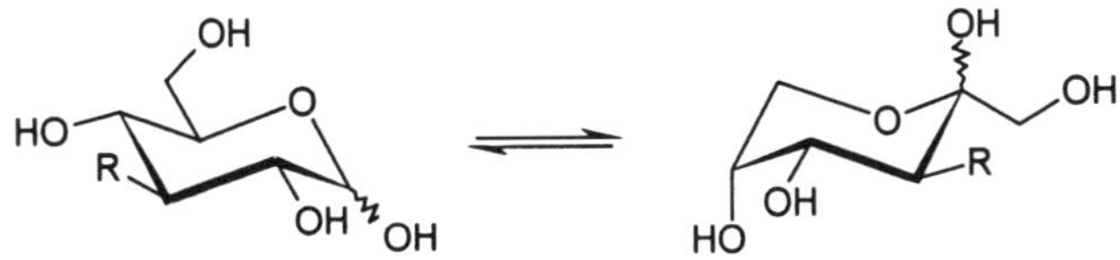

R = OMe 36 51
R = H 52 53

Scheme 19

glucose (**58**) were not accepted by the xylose isomerase employed in this study (Scheme 20).

5-Thio-D-glucose (**38**) did not serve as a substrate either. On the other hand, these workers discovered that 5-deoxy-D-xylohexofuranose ("5-deoxy-D-glucose", **59**) was quantitatively converted at pH 7.2 by the enzyme into the corresponding ketose (Scheme 21). They noted that incubation of the 5-deoxysugar at pH 8.5 in the absence of enzyme also gave 100% conversion but, under these conditions, unknown decomposition products were formed.

Subsequently, Card and co-workers found a conversion rate of 10–15% for the isomerisation of 6-deoxyfluoro-D-glucose (**61**) into the corresponding D-fructose derivative **62** (Scheme 22) [77].

These results were essentially confirmed by Wong and co-workers for the 6-deoxy (**48**), 6-deoxyfluoro (**62**) and 6-OMe (**50**) as well as for the 3-deoxy derivatives (**53**) of D-fructose employing immobilised xylose isomerase from *Flavobacterium aborescens* (Takasweet from MILES) [78] extending Bock's list of sub-

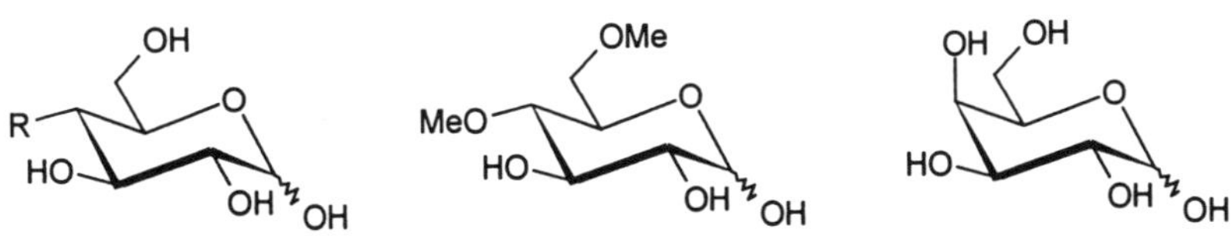

R = H 54 56 57
R = OMe 55

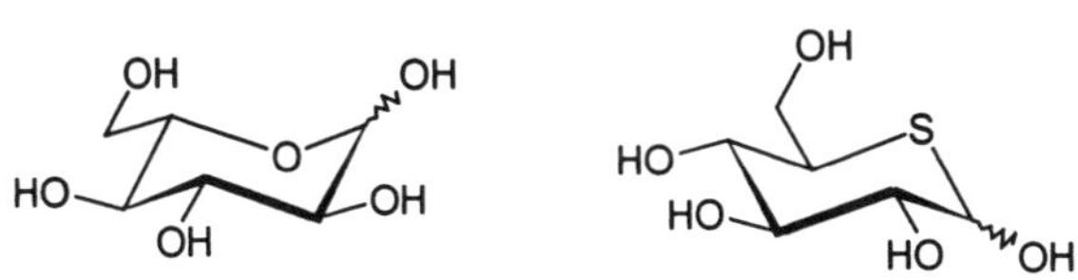

Scheme 20 58 38

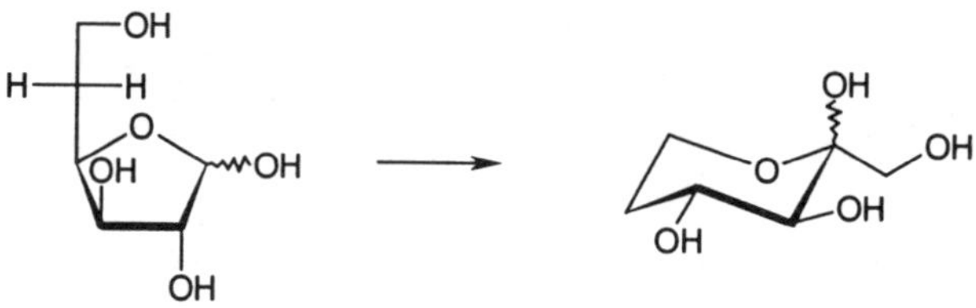

Scheme 21 59 60

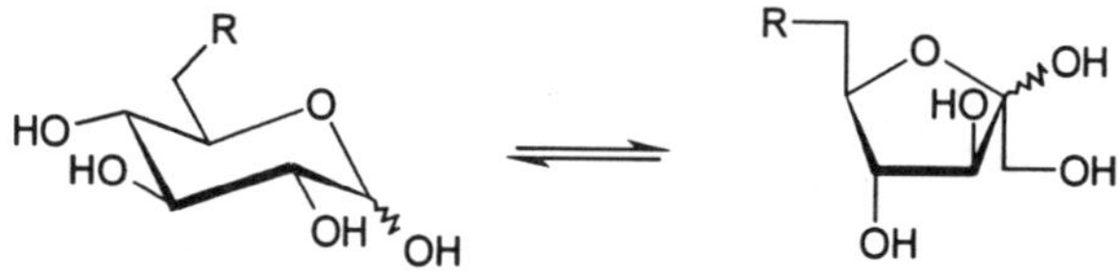

Scheme 22 R = F 61 / $R = N_3$ 63 — 62 / 64

strates by the 6-azidodeoxy derivatives of D-glucose (**63**) and D-fructose (**64**), respectively [79]. According to these authors, L-sorbose derivatives are not substrates for the enzyme [78b]. They also confirmed the previously reported quantitative conversion of 5-deoxy-D-xylohexofuranose into the corresponding fructopyranose for the *Flavobacterium* enzyme [78, 79b]. The availability of 6-azido-6-deoxy-D-glucose (**63**) from the corresponding fructose analogue **64**, which had been prepared by aldolase-catalysed addition of racemic 3-azido-2-hydroxypropanal to dihydroxyacetone phosphate, was exploited by the Wong group during an approach to 1,6-dideoxy-1,6-imino-D-glucitol (**65**, Scheme 23), a seven-membered iminoalditol exhibiting glycosidase inhibitory properties [79c].

The enzymatic conversion of 6-azido-6-deoxy-D-glucose into its D-fructose isomer was utilised as a key step in a five-step synthesis from sucrose [80] of the natural product and powerful D-mannosidase inhibitor 1-deoxymannojirimycin (1,5-dideoxy-1,5-imino-D-mannitol, **66**) [81]. Subsequently, it was shown that the immobilised enzyme from *Streptomyces murinus* sp. (Sweetzyme T from Novo A/S) was able to isomerise D-glucose derivatives with modifications at C-3 and C-6 such as 6-azido-6-deoxy-3-*O*-methyl-D-glucose (**67**) as well as the corresponding 3-deoxy derivative **69** (Scheme 24) [64, 82].

Following Bock's observation with the only substrate that had been converted quantitatively, 5-deoxy-D-*xylo*-hexofuranose ("5-deoxy-D-glucose", **59**), it was discovered [83] that any derivative of D-glucose not bearing a free hydroxyl function at C-5, such as the 5-azidodeoxy (**71**), the 5-deoxyfluoro (**73**), and the 5-*O*-

Scheme 23 65 66

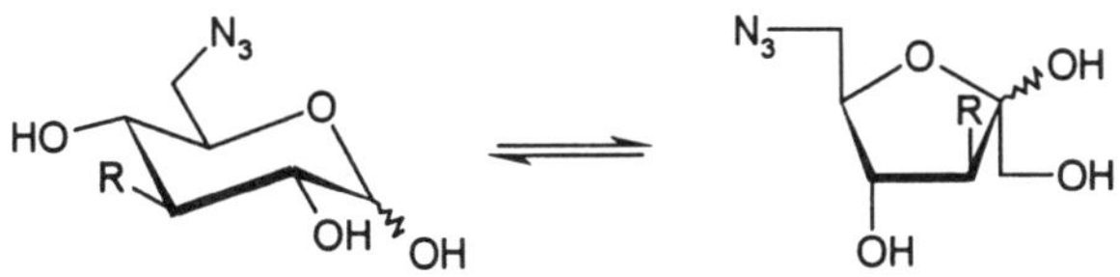

Scheme 24 R = OMe **67** **68**
R = H **69** **70**

benzyl (**75**) derivatives, was quantitatively converted to the corresponding D-fructopyranose (**72**, **74**, and **76**, respectively) by Sweetzyme T, a successor of the immobilised enzyme investigated by the Danish group (Scheme 25). Furthermore, it could be shown that the corresponding L-idofuranoses **77** and **79** were substrates as well, also being quantitatively converted into derivatives **78** and **80** of L-sorbose (Scheme 26).

Chain-extended compounds of both the D- (**81**) and the L-series (**83**, **85**) bearing an additional chiral centre were also isomerised (Scheme 27), albeit requiring longer reaction times, leading to interesting higher carbon ketoses (**82**, **84**, and **86**, respectively).

Attempts to isomerise C-5 modified analogues of D-xylose, for example, **87**, were successful leading to the corresponding open-chain D-xyluloses such as **88** as the predominant components in the equilibrium [64, 84]. When 5,6-dimodified derivatives of D-gluco- as well as L-idofuranose, such as **89** and **91**, were exposed to the enzyme, conversion to the corresponding open-chain D-fructose (**90**) and L-sorbose (**92**) analogues took place (Scheme 28) [64, 84]. At equilibrium, the reaction usually gave a 4:1 to 3:1 mixture in favour of the ketose.

This interesting finding was utilised in a synthetic approach to novel inhibitors of D-glucosidases [85]. A 6-*O*-(6-hydroxy)hexyl-substituted derivative (**93**)

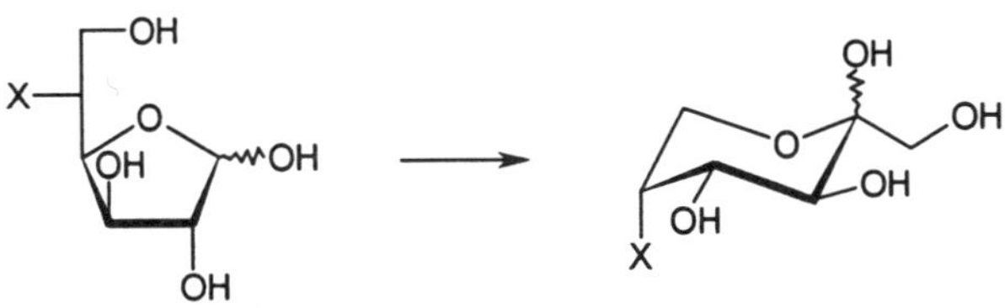

Scheme 25 X = N_3 **71** **72**
X = F **73** **74**
X = OBn **75** **76**

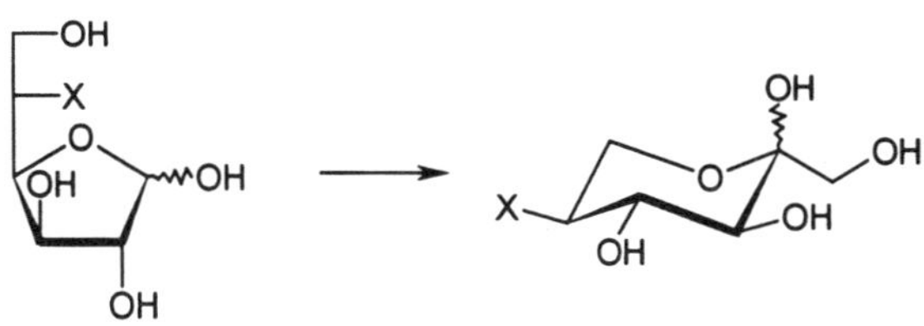

Scheme 26 X = N3 **77** **78**
X = F **79** **80**

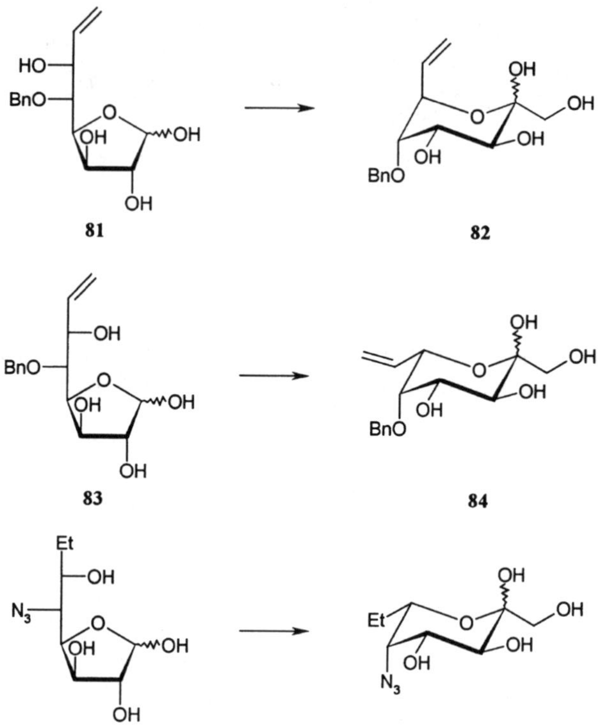

81 82

83 84

85 86

Scheme 27

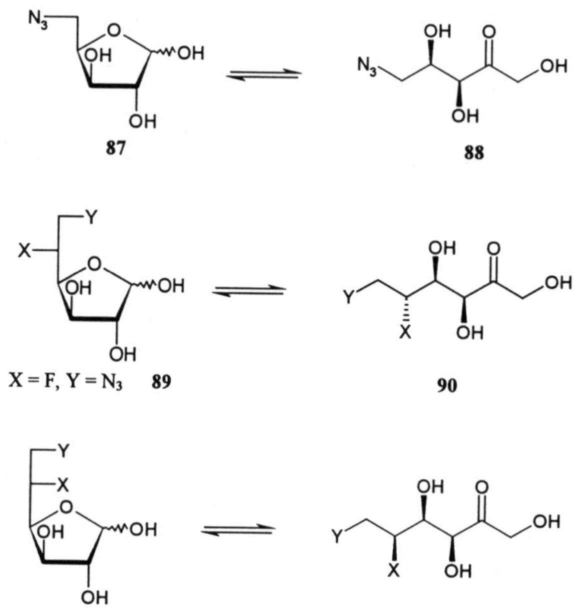

87 88

X = F, Y = N_3 89 90

X = N_3, Y = F 91 92

Scheme 28

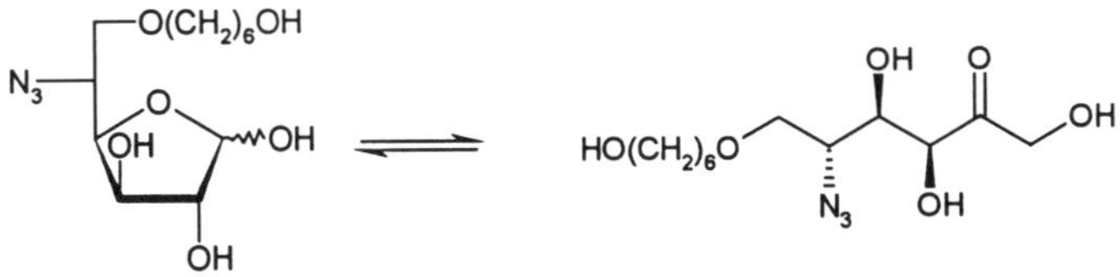

Scheme 29 **93** **94**

of 5-azidodeoxy-D-glucofuranose has recently been successfully employed as a substrate to give ketose **94** (Scheme 29) [86].

Based on the early observations that D-ribose and D-allose can act as substrates of xylose isomerases, the properties of (2*R*,3*R*)-configured sugars were also probed; it was found that D-erythrose (**95**) was quantitatively converted into the corresponding ketose **96** (Scheme 30).

D-Ribose derivatives such as 5-azidodeoxy- (**97**) and 5-deoxyfluoro-D-ribofuranose (**99**) were isomerised leading to 3:1 mixtures in favour of the open-chain ketoses **98** and **100** [87]. Interestingly, L-erythrose (**42**) was also accepted by Sweetzyme T and, akin to its enantiomer **95**, was quantitatively transformed into L-*glycero*-tetrulose ("L-erythrulose", **43**). On the other hand, no conversion was observed with 5,6-di-*O*-methyl-D-allofuranose (**101**, Scheme 31).

When the (2*R*,3*R*)-configured ketose was able to form a pyranose ring (Scheme 32), isomerisation was found to proceed, as observed with 5-deoxy-D-*ribo*-hexofuranose (**102**), 5-azido-5-deoxy-D-allofuranose (**104**), as well as the L-sugars 5-azido-5-deoxy- (**106**) and 5-deoxy-5-fluoro-L-talofuranose (**108**) [88].

The corresponding ketopyranoses (**103**, **105**, **107**, and **109**) as precursors to interesting glycosidase-inhibiting iminoalditols were formed in 3:2 ratios favouring the corresponding aldose with isolated yields ranging between 30–40%.

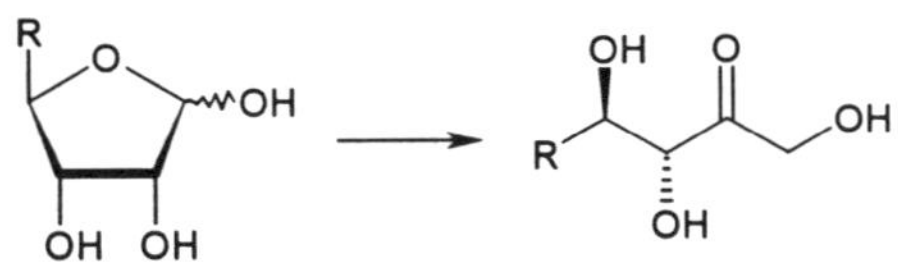

	R = H	**95**	**96**
	R = CH_2N_3	**97**	**98**
Scheme 30	R = CH_2F	**99**	**100**

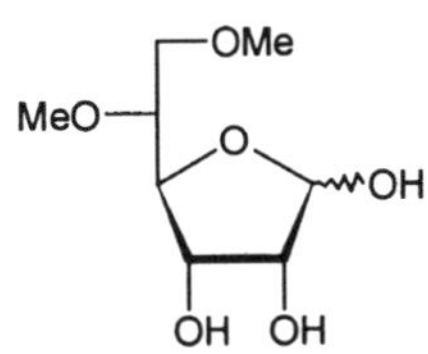

Scheme 31 **101**

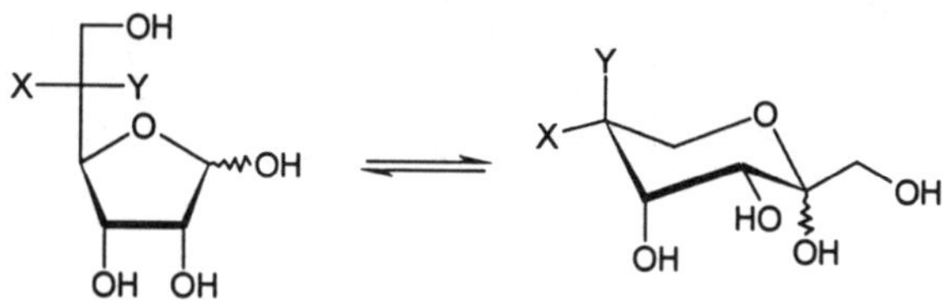

X = Y = H	**102**	**103**
X = N_3, Y = H	**104**	**105**
X = H, Y = N_3	**106**	**107**
X = H, Y = F	**108**	**109**

Scheme 32

3
L-Fucose (D-Arabinose) Isomerase (EC 5.3.1.3) and L-Rhamnose Isomerase (EC 5.3.1.14)

In 1953, in the cell-free extract of a D-arabinose-adapted strain of *Escherichia coli*, enzymatic activity was found which converted D-arabinose (**13**) into D-*erythro*-pentulose (D-ribulose, **14**) [8]. Moreover, this enzyme was able to convert L-fucose (6-deoxy-L-galactose, **15**) into the corresponding ketose **16** [89]. In-depth investigations by the same authors confirmed this finding and revealed that the equilibrium could be shifted from originally 11% of L-fuculose (**16**) to over 80% by in situ complexation of the ketose with excess borate [90].

Subsequently, isomerases from *Aerobacter aerogenes* strains which also converted L-xylose (**110**, Scheme 33) were purified [91, 92] and the inducible enzymes from two other *E. coli* strains were compared [93]. It was demonstrated that both preparations were tetramers with masses around 350 kDa and were active with L-fucose (**15**) and D-arabinose (**13**). However, this was not the case for the other aldopentoses and hexoses probed such as L-arabinose (**24**), L-rhamnose (**26**), D-xylose (**18**), D-ribose (**20**), D- (**17**) or L-glucose (**58**), D- (**44**) or L-mannose (**111**), D-galactose (**57**) or D-fucose (6-deoxy-D-galactose, **112**). Both enzymes were inhibited by alditols related to their natural substrates, exhibited their optimal catalytic activities in alkaline media, and were stimulated by the presence of Mn^{2+} as well as Co^{2+} but strongly inhibited by Cd^{2+}.

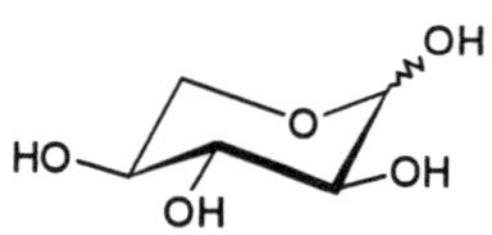

110

Scheme 33 **111** **112**

3.1 Structure and Mechanism

The structure of an L-fucose isomerase from *E. coli* was determined by X-ray crystallography at 2.5 Å resolution [94]. This manganese-dependent hexameric enzyme with subunits of 65 kDa is the largest aldose ketol isomerase of known structure. As with xylose isomerases, the open-chain tautomer of the substrate is processed and the enzyme was assumed to recognise the α-anomer prior to ring opening. The manganese ion was found to coordinate O-1 and O-2 of the substrate with two bases, Asp-361 and Glu-337, in close proximity to this part of the substrate structure. Based on the assumption that Glu-337 would be a superfluous residue in a hydride shift mediated isomerisation reaction, the enediol pathway was suggested for this particular isomerase.

3.2 Non-Natural Substrates and Synthetic Applications

Very interesting applications of the preparative potential of microbial ketol isomerases were reported by Fessner and co-workers, who gained access to preparative amounts of L-rhamnose isomerase (EC 5.3.1.14) as well as L-fucose isomerase (EC 5.3.1.3) from *E. coli* by cloning and over-expression [95]. In this way they were able to convert (Scheme 34) a variety of D- and L-pentoses and -hexoses into the corresponding 2-ketoses, such as L-mannose (**111**) (to L-fructose, **113**), D-gulose (**114**) (to D-sorbose, **8**) and D-allose (**22**) (to D-psicose, **10**), to mention just a few.

By combining the enzymatic isomerisation procedure with an L-rhamnulose kinase (EC 2.7.1.5) mediated phosphorylation step high yields ranging from 70 to 90% could be obtained [96]. Subsequently, Wong and co-workers [97] reported the cloning and over-expression of L-rhamnose isomerase and fucose isomerase. This group reported the conversion of L-fructose (**113**) that they had synthesised by aldolase-based methodology into L-glucose (**58**) in fair yield (Scheme 35) [98]. Isomerisation of L-fuculose (**16**) and C-6 modified analogues of L-tagatose such as compounds **115** and **117** led to the corresponding L-fucose

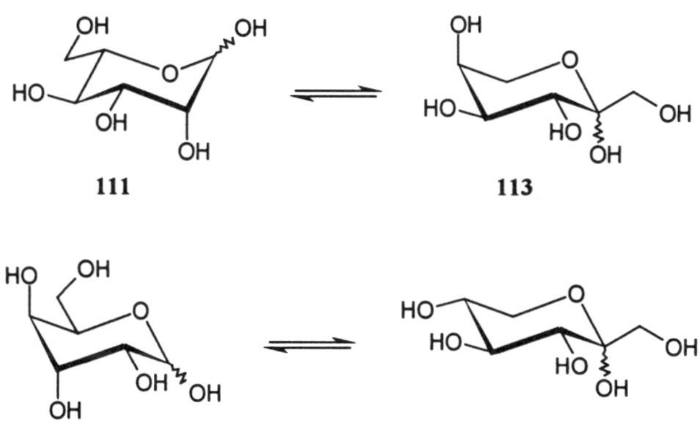

111 **113**

Scheme 34 **114** **8**

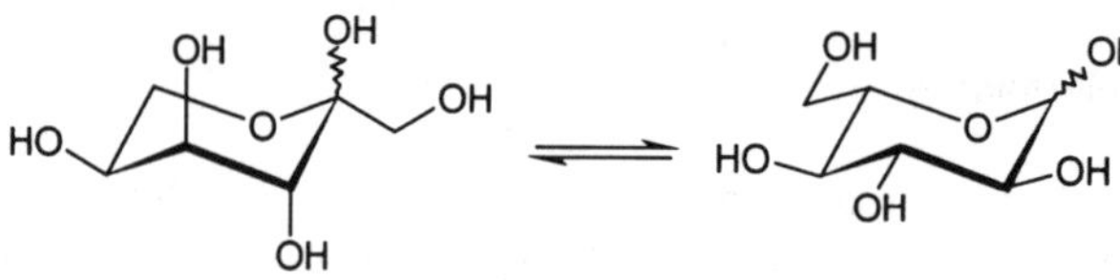

Scheme 35 **113** **58**

derivatives such as L-galactose, the corresponding 6-*O*-methyl derivative (**116**), and 6-azidodeoxy-L-galactose (**118**) (Scheme 36) [99].

Simple chain-extended analogues, for example, 7-deoxy-D-*glycero*-L-*galacto*-heptose (**120**) and the corresponding 6-deoxy compound (**122**), were also found to be available employing this methodology (Scheme 37) [99].

In context with a project aimed at structure-activity relationships of sialyl Lewis X epitope analogues, a range of new L-fucose derivatives with increased hydrophobicities of the C-5 substituents, such as compounds **124** and **126**, was recently synthesised [100] employing Fessner's proven L-fuculose 1-phosphate aldolase/L-fucose isomerase protocol (Scheme 38).

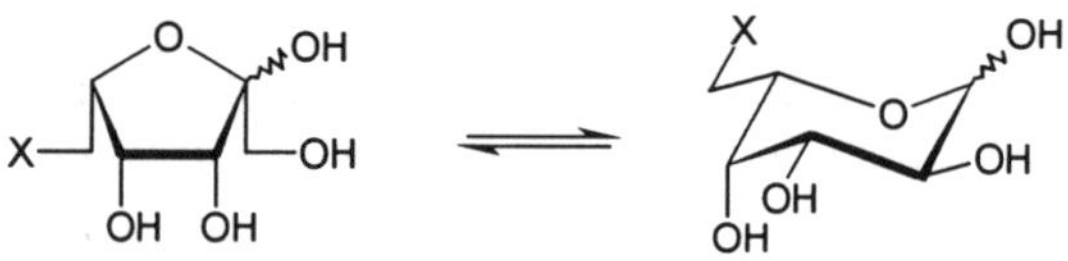

X = OMe **115** **116**
X = N_3 **117** **118**

Scheme 36

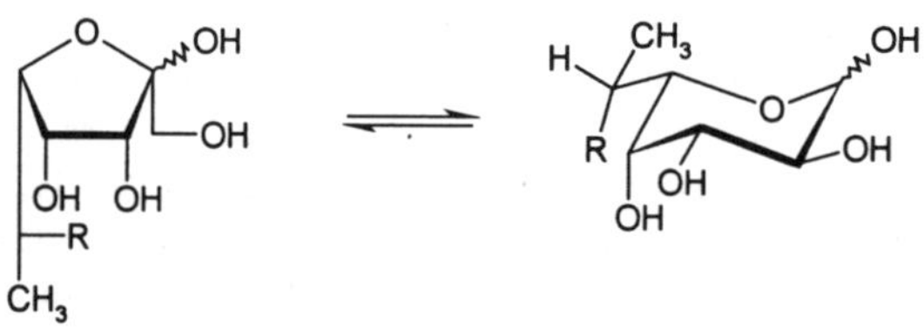

R = OH **119** **120**
R = H **121** **122**

Scheme 37

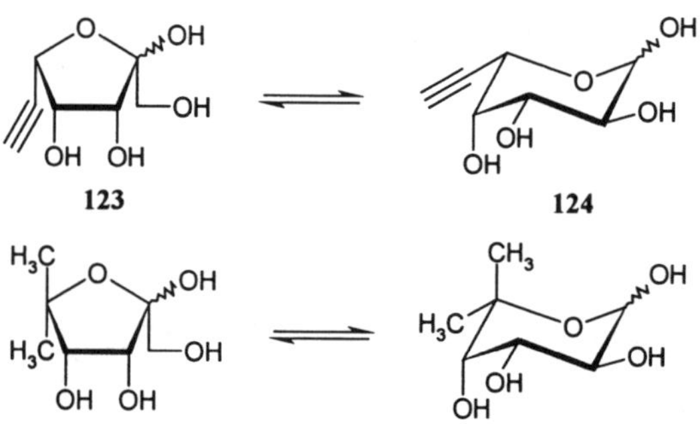

123 **124**

125 **126**

Scheme 38

4 Synthetic Perspectives

D-xylose (D-glucose) isomerase as well as L-fucose and L-rhamnose isomerases have found an increasing range of interesting synthetic applications in the recent past, in particular in the synthesis of iminoalditol-based glycosidase inhibitors as well as for the preparation of rare (highly substituted and/or open-chain) sugars. The application of glucose isomerase is based on the preparative availability of suitably modified aldoses which have been enzymatically transformed into the desired corresponding ketoses; these were either converted into iminoalditols or were found to be useful biochemical probes in their own right. Conversely, L-fucose and L-rhamnose isomerases have been employed to convert ketoses that had been synthesised with the aid of aldolase or transketolase methodology into the corresponding aldoses (Scheme 39).

Frequently, the glucose isomerase based approach to non-natural ketoses and the aldolase-aided synthesis of such ketoses have been found to be convergent and lead to the same product(s). Depending on the nature of the non-natural substituents, either of these two routes can be advantageous over the other. Whereas aldolase-catalysed ketose synthesis is superior in cases of simple modifications at C-5 and/or C-6 of the desired sugar, extensively derivatised analogues are clearly more conveniently accessible by chemical modification of the aldose and subsequent enzymatic isomerisation into the corresponding ketose. This conclusion is based on preparative restrictions such as the availability of the α,β-dimodified glyceraldehyde derivative to be coupled to dihydroxyacetone phosphate in the aldolase-catalysed carbon chain forming reaction. These are caused by the rapidly increasing overall number of synthetic steps necessary to access enantiomerically pure α,β-dimodified derivatives of glyceraldehyde bear-

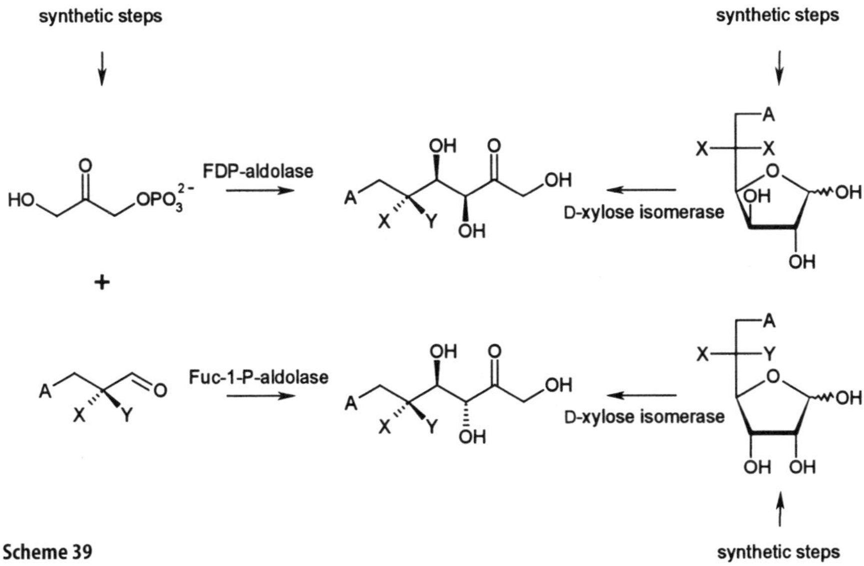

Scheme 39

ing different substituents, for example, a fluoro and an azido group at positions 2 and 3, or vice versa.

Interesting extensions of isomerase chemistry are "coupled" enzyme systems. These could be advantageous for the removal of the desired isomer from the sometimes unfavourable equilibrium to increase yields, as pointed out by Fessner and co-workers. Other applications could make rare sugars conveniently accessible from easily available starting materials, for example, in the preparation of D-psicose from D-fructose [6] by the combination of D-xylose isomerase with D-tagatose 3-epimerase. Clearly, the wide range of other available enzymes, for example, oxidoreductases, would allow for even more exciting opportunities to access unusual carbohydrates.

5 Conclusions

Enzymatic isomerisation of free sugars has been shown to be a useful and remarkably versatile entry to a wide range of non-natural and unusual aldoses and ketoses. These, in their own right, are interesting probes for sugar processing enzymes and, consequently, useful tools for glycobiology. Furthermore, they can be exploited as enantiomerically pure building blocks and precursors of modified natural products and sugar mimetics.

Ketose chemistry is, by far, not as advanced as the knowledge on aldose reactions and conversions and, consequently, the application of isomerases in this area will be an important contribution to open up this fascinating and underrated field of carbohydrate chemistry. Rare and potentially interesting aldoses, on the other hand, which have been resistant to conventional chemical means of access, can today be made available with a range of enzymatically catalysed reactions employing sugar modifying enzymes such as the ones under consideration in this account.

Currently, about ten sugar ketol isomerases are known [3], most of them produced from more than one known source. Similar to D-xylose isomerase and L-fucose isomerase, many of them are dependent on divalent cations, and some have not even been investigated in much detail at all. None has gained even a small fraction of the interest that has been paid to D-xylose isomerase due to its long known industrial importance. L-Fucose isomerase, yet another example, is one of a very few isomerases of current noteable synthetic importance.

In addition, a range of aldose phosphate isomerases, for example, D-mannose 6-phosphate isomerase, as well as a good selection of epimerases and oxidoreductases, are available.

It is certain that the future will open new avenues for the application of these enzymes and there are clearly many opportunities for chemists in investigating their properties and putting their highly appreciable preparative potential to use.

6 Tables of Substrates

6.1 Substrates of D-Xylose Isomerase

6.1.1 *Tetroses*

	Ratio	Time (h), Yield (%)	Ref.
D-erythro 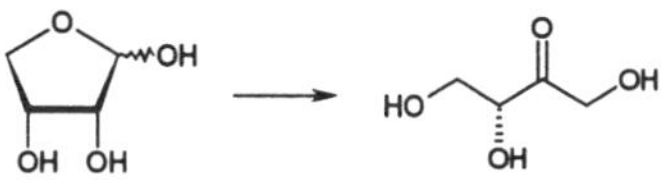	0:1	8, 66	87
L-erythro 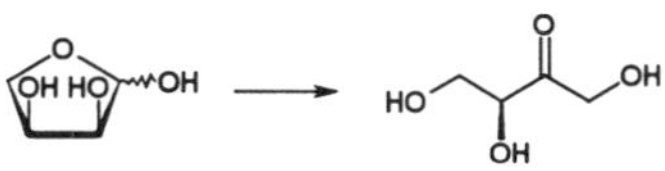	0:1	8, 39	87

6.1.2 *Pentoses*

	Ratio	Time (h), Yield (%)	Ref.
D-arabino 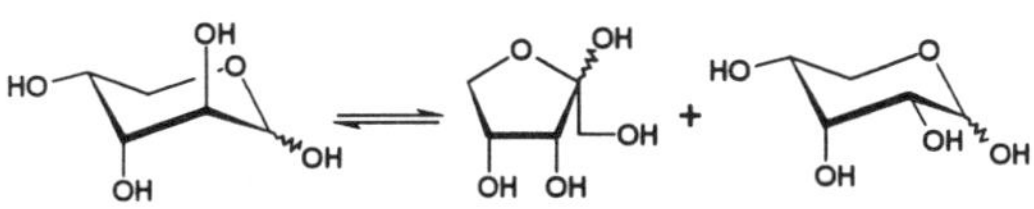	70:10:10	12	74
L-arabino 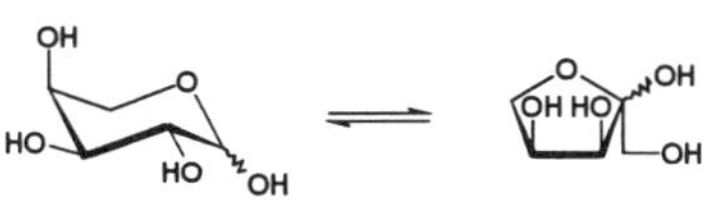	74:24	12	74
D-lyxo 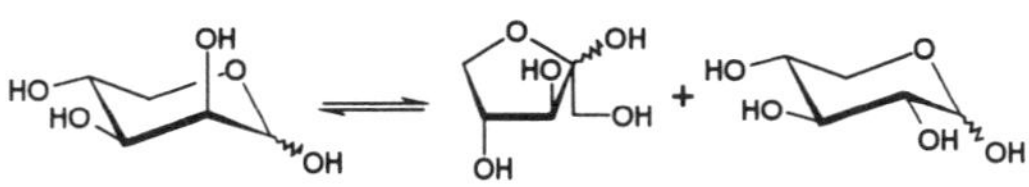	50:10:40	12	74

	Ratio	Time (h), Yield (%)	Ref.
D-ribo			
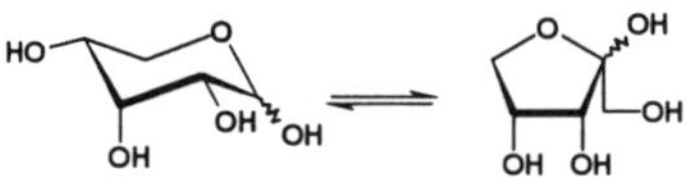	55:45	2	15, 16, 20, 74, 75
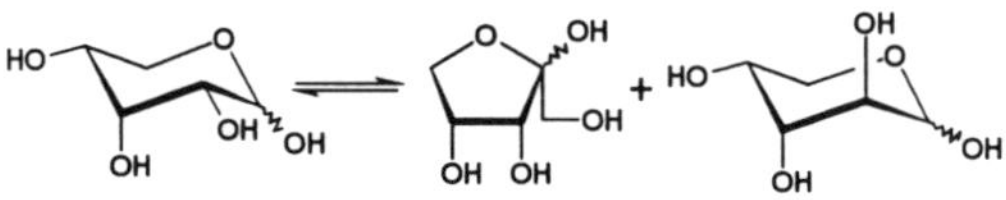	40:30:30	12	74
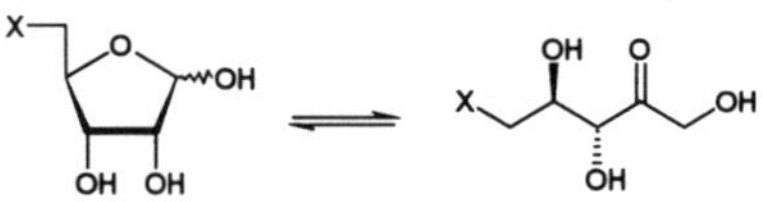 X = N_3, F	25:75	3–6, 65	87
D-xylo			
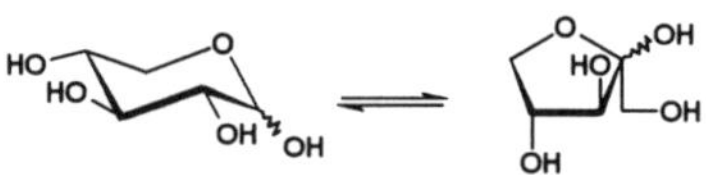	80:20	24	31
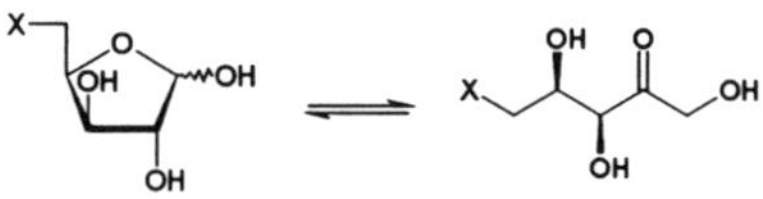 X = H, N_3, F	20:80	4–6, 50	64, 84

6.1.3
Hexoses

	Ratio	Time (h), Yield (%)	Ref.
D-allo/D-psico			
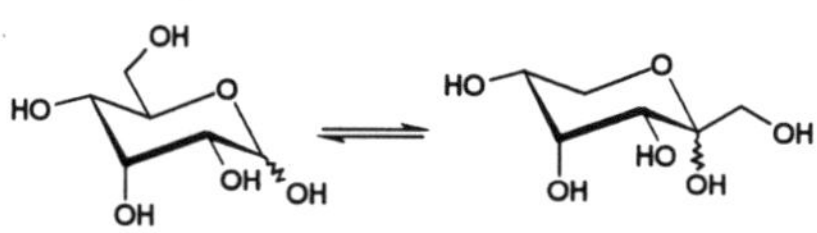			17
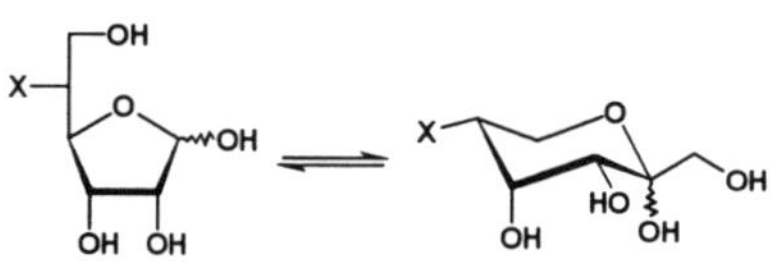 X = N_3, H	60:40	8, 33–37	88

	Ratio	Time (h), Yield (%)	Ref.
D-gluco/D-fructo			
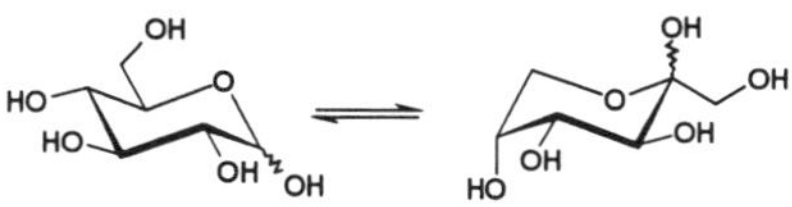	50:50 45:55	24	31 24
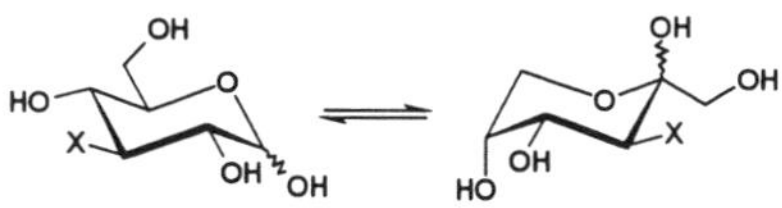 X = H, OMe	62:38 78:22	24	31 78
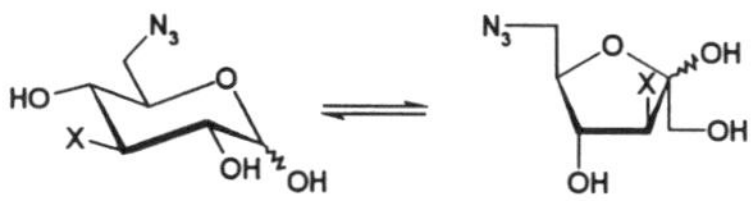 X = H, X = OMe	X = H 30:70 X = OMe 85:15	6	64, 82
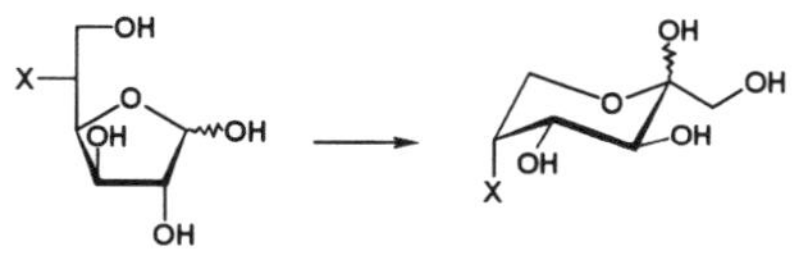 X = N_3, F, H, OBn	0:1, X = H X = N_3 X = F X = OBn	 24 4, 81 8, 78 4, 72	 31, 78 83
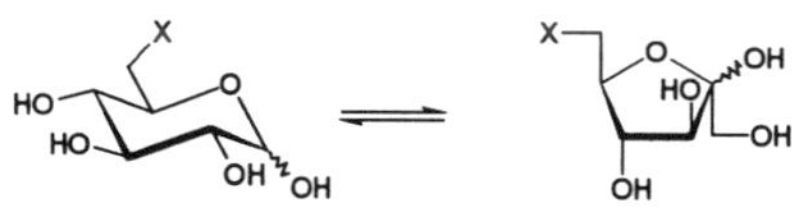 X = N_3, F, H, OMe, SH	X = N_3 H:85:15 X = F: 90:10 X = OMe: 79:21 X = SH: 0:1		31, 77 78–80 77, 78 31, 78 76
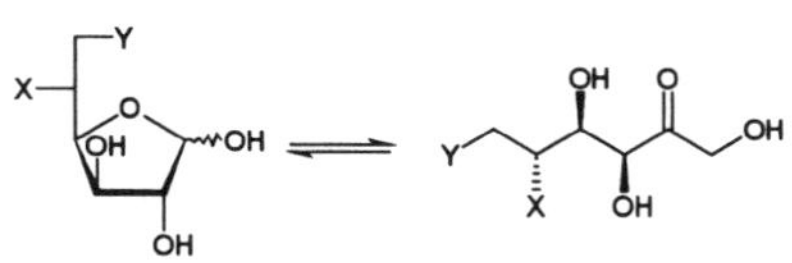 X = N_3, F, H; Y = N_3, F, H; X = N_3, Y = OMe; X = N_3, Y = $O(CH_2)_6OH$	25:75 to 20:80	6, 22–63	64, 84–86
L-ido			
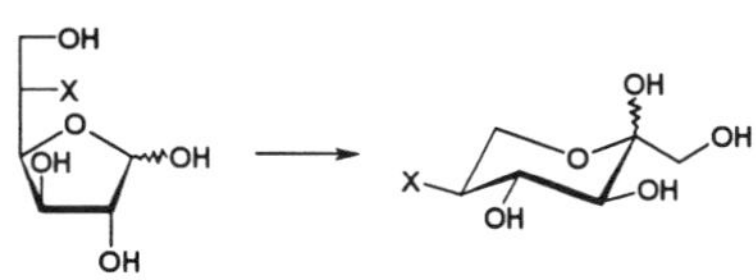 X = N_3, F	0:1 X = N_3 X = F	 4, 76 8, 73	83

	Ratio	Time (h), Yield (%)	Ref.
L-ido 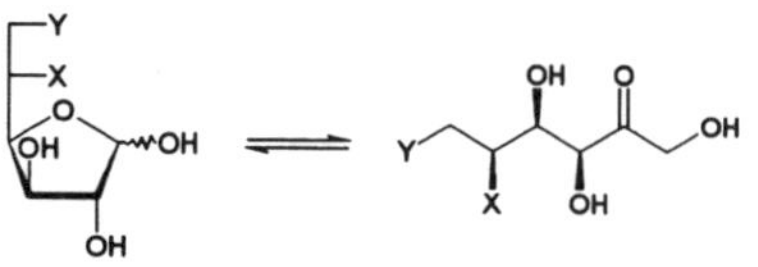 $X = F, Y = N_3$	20:80	8, 55	85
L-rhamno (6-deoxy-L-manno) 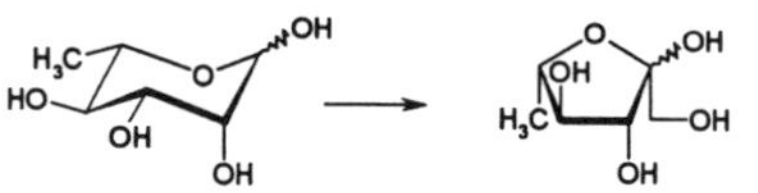			17
L-talo/L-tagato 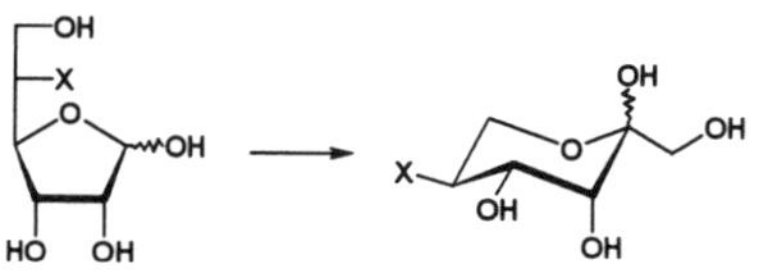	60:40 $X = N_3$ $X = F$	 8, 34 8, 35	 88

6.1.4
Higher Carbon Sugars

	Ratio	Time (h), Yield (%)	Ref.
D-glycero-D-gluco 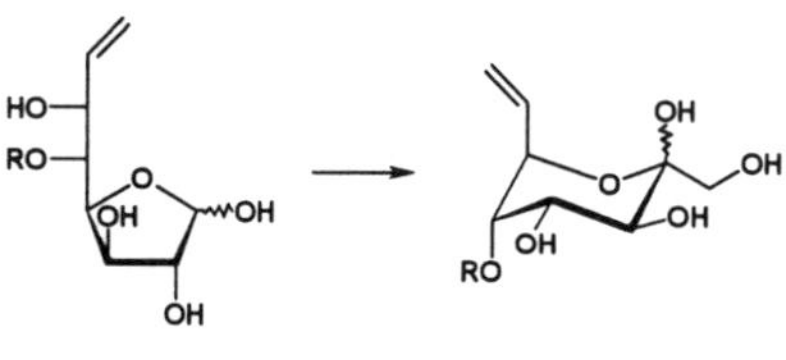 R = OBn	0:1	20, 55	83
L-glycero-D-gluco 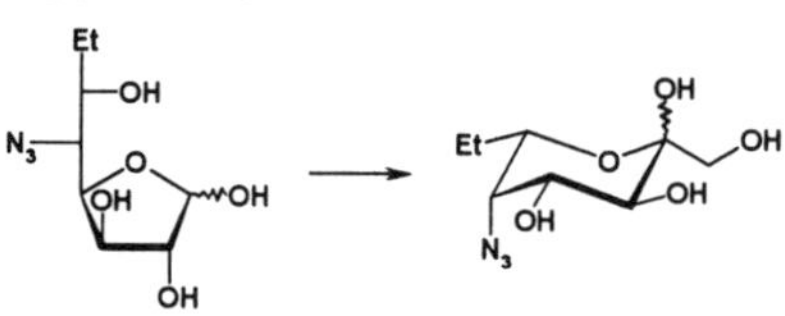	0:1	20, 61	83

	Ratio	Time (h), Yield (%)	Ref.
L-glycero-D-gluco 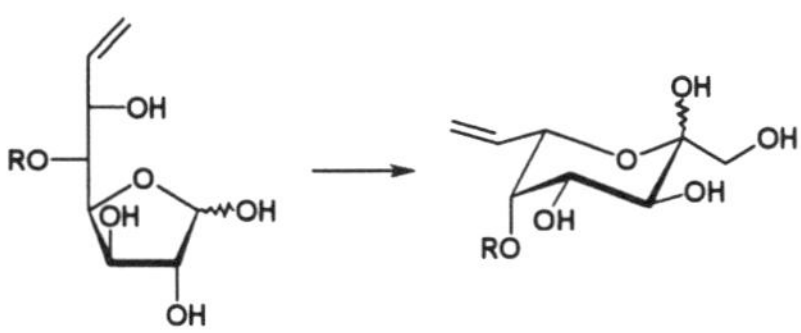 R = OBn	0:1	20, 48	83

6.2
Substrates of L-Fucose Isomerase

6.2.1
Pentoses

	Ratio	Time (h), Yield (%)	Ref.
D-arabino 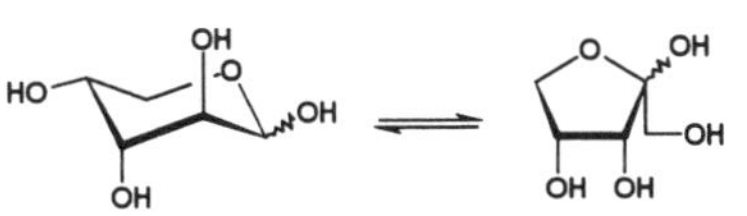	85:15	30	8, 93, 99
L-xylo 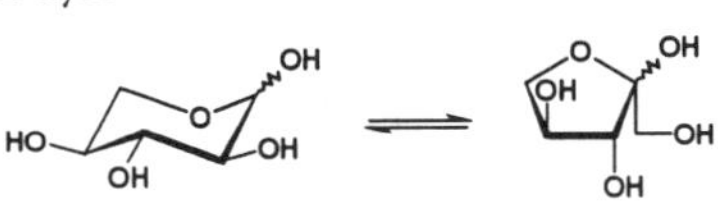			91, 92, 99

6.2.2
Hexoses and Higher Carbon Sugars

	Ratio	Time (h), Yield (%)	Ref.
L-fuco (6-deoxy-L-galacto) 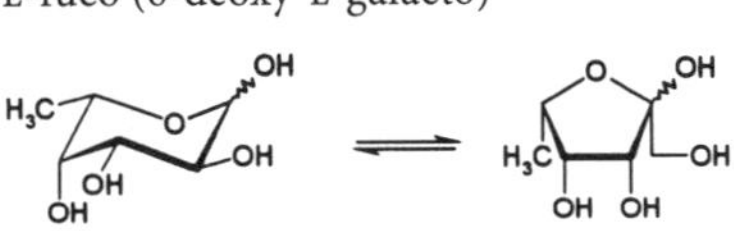	89:11	55% aldose, 78% ketose	89, 93, 99 90, 96

	Ratio	Time (h), Yield (%)	Ref.
L-galacto/L-tagato 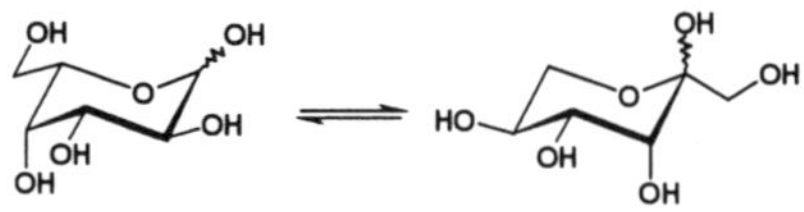		58%	99
L-gluco/L-fructo 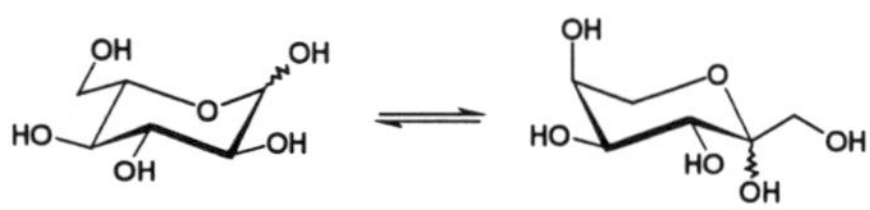		29% aldose	98
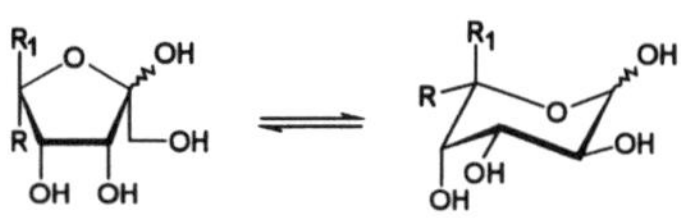 $R = CH_2N_3, CH_2OMe$; $R = C_2H_5, CH = CH_2, CCH; R_1 = H; R = R_1 = Me$	$R = CH_2N_3$ CH_2OMe R = Et 10:90 Others: 25:75	20% 25% 2–5 d, 83% 34–58%	 99 100

6.3 Substrates of L-Rhamnose Isomerase

6.3.1 *Pentoses*

	Ratio	Time (h), Yield (%)	Ref.
L-lyxo 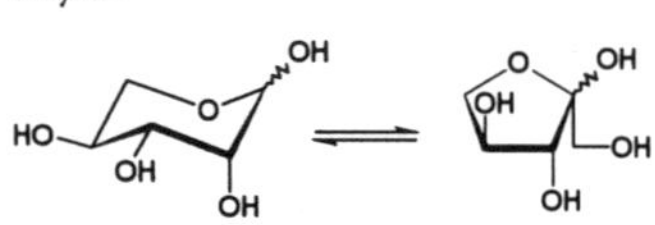		78% ketose	96
D-ribo 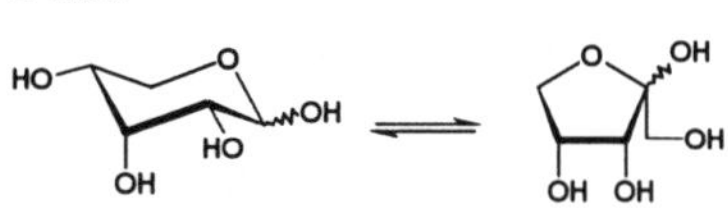		82% ketose	96

6.3.2
Hexoses

	Ratio	Time (h), Yield (%)	Ref.
D-allo/D-psico 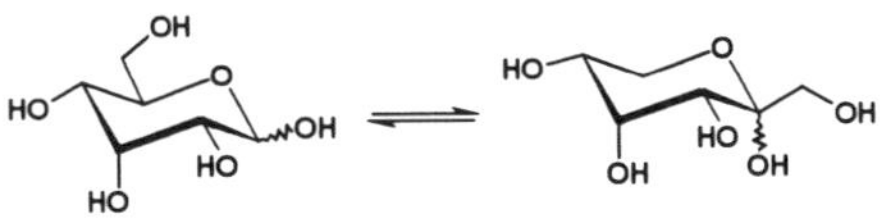		90% ketose	96
D-gulo/D-sorbo 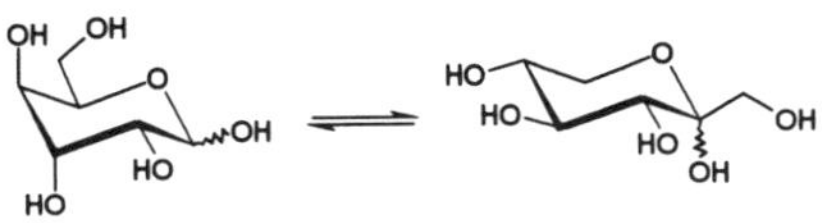		72% ketose	96
L-manno/L-fructo 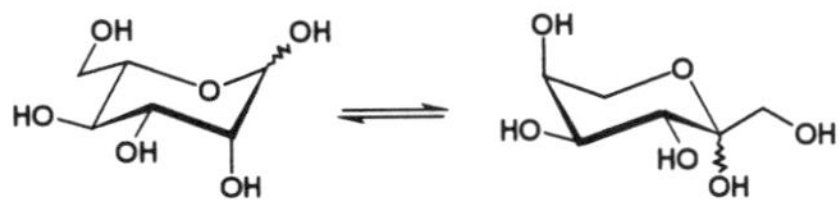		76% ketose	96
L-rhamno (6-deoxy-L-manno) 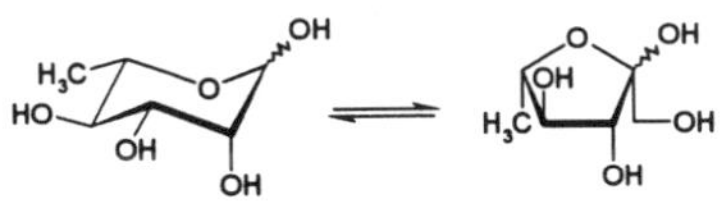		90% ketose	96
L-talo/L-tagato 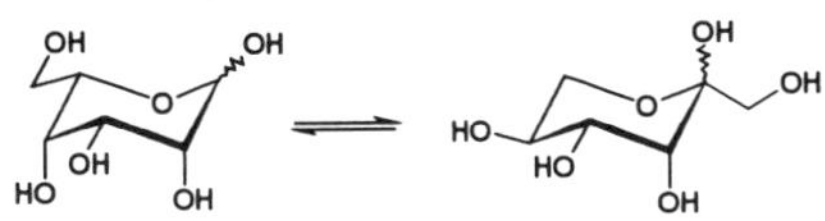		75% ketose	96

Acknowledgements. Glycogroup appreciates continuous support from the Austrian Fonds zur Förderung der wissenschaftlichen Forschung (FWF), Vienna (Projects 7335, 8415, 10067, 10805, and 13593 CHE). Novo Nordisk A/S is thanked for their gift of Sweetzyme T.

7
References

1. Ghosh S, Roseman S (1965) J Biol Chem 240:1531; Corfield AP, Rainey JB, Clamp JR, Wagner SA (1985) Biochim Biophys Acta 840:264 and references cited therein
2. Ghosh S, Roseman S (1966) Methods Enzymol 8:185
3. Schomburg D, Salzmann M (eds) (1990) Enzyme handbook, vol 2. Springer, Berlin Heidelberg New York
4. Goudsmit EM, Neufeld EF (1967) Biochem Biophys Res Commun 26:730; Barber GA, Hebda PA (1982) Methods Enzymol 83:522 and references cited therein

5. Itoh H, Sato T, Takeuchi T, Khan AR, Izumori K (1995) J Ferment Bioeng 79:184
6. Itoh H, Sato T, Izumori K (1995) J Ferment Bioeng 80:101
7. Akabori S, Uehara K, Muramatsu I (1952) Proc Jpn Acad 28:39
8. Cohen SS (1953) J Biol Chem 201:71
9. Marshall RO, Kooi ER (1957) Science 125:648
10. Tsumura N, Sato T (1960) Bull Agric Chem Soc Jpn 24:326
11. Natake M (1966) Agric Biol Chem 30:887; Natake M (1968) Agric Biol Chem 32:303
12. Yamanaka K (1963) Agric Biol Chem 27:265
13. Tsumura N, Sato T (1965) Agric Biol Chem 29:1129
14. Takasaki Y, Kosugi Y (1969) In: Perlman D (ed) Fermentation advances. Academic Press Inc, New York, p 561
15. Yamanaka K (1968) Biochem Biophys Acta 151:670
16. Danno G (1971) Agric Biol Chem 35:997
17. Sanchez S, Smiley KL (1975) Appl Microbiol 29:745
18. (a) Parker JH (1978) Appl Environ Microbiol 36:969; (b) Kwon HJ, Kitada M, Horikoshi K (1987) Agric Biol Chem 51:1983; (c) Yamanaka K, Takahara N (1977) Agric Biol Chem 41:1909
19. (a) Chang C, Park BC, Lee DS, Suh SW (1999) J Mol Biol 288:623; (b) Liu SY, Wiegel J, Gherardini FC (1996) J Bacteriol 178:5938; (c) Liao WX, Earnest L, Kok SL, Jeyaseelan K (1995) Biochem Mol Biol Int 36:401; (d) Vielle C, Hess JM, Kelly RM, Zeikus JG (1995) Appl Environ Microbiol 61:1867; (e) Quax WJ, Mrabet NT, Luiten RGM, Schuurhuizen PS, Lasters I (1991) Bio/Technol 9:738
20. Suekane M, Tamura M, Tomimura C (1978) Agric Biol Chem 42:909
21. Yamanaka K (1975) In: Woods W (ed) Methods in enzymology, vol XLI. Academic Press Inc, New York, p 466
22. Japanese Patent Appl (1965) 27,525; British Patent 1,103,394; (1967) CA 66:84776f
23. Hagen HH, Pedersen S (1990) In: Gerhartz W (ed) Enzymes in industry. VCH, New York, p 102
24. Pedersen S (1993) Bioprocess Technol 16:185
25. Bhosale SH, Rao MB, Desphande VV (1996) Microbiol Rev 60:280 and references cited therein
26. (a) Rose IA (1970) In: Boyer PD (ed) The enzymes, 3rd edn, vol II. Academic Press, New York, p 290 and references cited therein; (b) Hanson KR, Rose IA (1975) Acc Chem Res 8:1
27. Blacklow SC, Raines RT, Lim WA, Zamore PD, Knowles JR (1988) Biochemistry 27:1158
28. Dyson JED, Noltmann EA (1968) J Biol Chem 243:1401
29. Alworth WL (1975) Stereochemistry and its application in biochemistry. Wiley, New York, p 244
30. (a) Schray KJ, Rose IA (1971) Biochemistry 10:1058; (b) Feather MS, Desphande V, Lybyer MJ (1970) Biochim Biophys Res Commun 38:859
31. Bock K, Meldal M, Meyer B, Wiebe L (1983) Acta Chem Scand B 37:101
32. Makkee M, Kieboom APG, van Bekkum H (1984) Recl Trav Chim Pays-Bas 103:361
33. Collyer CA, Goldberg JD, Viehmann H, Blow DM, Ramsden NG, Fleet GWJ, Montgomery FJ, Grice P (1992) Biochemistry 31:12211
34. Carrell HL, Rubin BH, Hurley TJ, Glusker JP (1984) J Biol Chem 259:3230
35. Hogue-Angeletti RA (1973) J Biol Chem 250:7814
36. Banner DW, Bloomer AC, Petsko GA, Phillips DC, Pogson CI, Wilson IA, Corran PH, Furth AJ, Milman JD, Offord RE, Priddle JD, Waley SG (1975) Nature 255:609
37. Stuart DI, Levine M, Muirhead H, Stammers DK (1979) J Mol Biol 134:109
38. (a) Mavridis IM, Hatada MH, Tulinsky A, Lebioda L (1982) J Mol Biol 152:419; (b) Banner DW, Bloomer AC, Petsko GA, Phillips DC, Wilson IA (1976) Biochem Biophys Res Commun 72:146; (c) Phillips DC, Sternberg MJE, Thornton JM, Wilson IA (1978) J Mol Biol 119:329
39. Janecek S (1997) Int J Biol Macromol 21:277
40. Chothia C (1988) Nature 333:598 and references cited therein

41. Akins J, Brick P, Jones HB, Hirayuma N, Shaw P-C, Blow DM (1986) Biochim Biophys Acta 874:375
42. Wilhelm M, Hollenberg CP (1985) Nucleic Acids Res 13:5717
43. Lawlis VB, Dennis MS, Chen EY, Smith DH, Henner DJ (1984) Appl Environ Microbiol 47:15
44. Saari GC, Kumar AA, Kawasaki GH, Insley MY, O'Hara PJ (1987) J Bacteriol 169:612
45. Drocourt D, Bejar S, Calmels T, Reynes JP, Tiraby G (1988) Nucleic Acids Res 16:9337
46. Dauter Z, Dauter M, Hemker J, Witzel H, Wilson KS (1989) FEBS Lett 247:1
47. Batt CA, Jamieson AC, Vandeyar MA (1990) Proc Natl Acad Sci USA 87:618
48. Henrick K, Blow DM, Carrell HL, Glusker JP (1987) Protein Eng 1:467
49. Farber GK, Petsko GA, Ringe D (1987) Protein Eng 1:459
50. Rey F, Jenkins J, Janin J, Lasters I, Alard P, Claessens M, Matthyssens G, Wodak S (1988) Proteins 4:165
51. Dauter Z, Terry H, Witzel H, Wilson KS (1990) Acta Crystallogr B 46:833
52. Carrell HL, Glusker JP, Burger V, Manfre F, Tritsch D, Biellmann J-F (1989) Proc Natl Acad Sci USA 86:4440
53. Farber GK, Glasfeld A, Tiraby G, Ringe D, Petsko GA (1989) Biochemistry 89:7289
54. Shiner VJ, Whittaker D (1969) J Am Chem Soc 91:394; Warnhoff EW, Reynolds-Warnhoff P, Wong MYH (1980) J Am Chem Soc 102:5956
55. Okuyama T, Kimura K, Fueno T (1982) Bull Chem Soc Jpn 55:2285
56. Collyer CA, Blow DM (1990) Proc Natl Acad Sci USA 87:1362
57. Lee C, Bagdasarian M, Meng M, Zeikus JG (1990) J Biol Chem 265:19082
58. Rangarajan M, Hartley BS (1992) Biochem J 283:223
59. Whitlow M, Howard AJ, Finzel BC, Poulos TL, Winborne E, Gilliland GL (1991) Proteins 9:153
60. Blow DM, Collyer CA, Goldberg JD, Smart OS (1992) Faraday Discuss 93:67
61. (a) Jenkins J, Janin J, Rey F, Chiadmi M, van Tilbeurgh H, Lasters I, De Maeyer M, Van Belle D, Wodak SJ, Lauwereys M, Stanssens P, Mrabet NT, Snauwaert J, Matthyssens G, Lambeir A-M (1992) Biochemistry 31:5449; (b) Lambeir A-M, Lauwereys M, Stanssens P, Mrabet NT, Snauwaert J, van Tilbeurgh H, Matthyssens G, Lasters I, De Maeyer M, Wodak SJ, Jenkins J, Chiadmi M, Janin J (1992) ibid 31:5459: (c) van Tilbeurgh H, Jenkins J, Chiadmi M, Janin J, Wodak SJ, Mrabet NT, Lambeir A-M (1992) ibid 31:5467
62. Collyer CA, Henrick K, Blow DM (1990) J Mol Biol 212:211
63. Allen KN, Lavie A, Farber GK, Glasfeld A, Petsko GA, Ringe D (1994) Biochemistry 33:1481
64. de Raadt A, Ebner M, Ekhart CW, Fechter M, Lechner A, Strobl M, Stütz AE (1994) Catal Today 22:549
65. Allen NK, Lavie A, Glasfeld A, Tanada TN, Gerrity DP, Carlson SC, Farber GK, Petsko GA, Ringe D (1994) Biochemistry 33:1488
66. Lavie A, Allen KN, Petsko GA, Ringe D (1994) Biochemistry 33:5469
67. Allen KN, Lavie A, Petsko GA, Ringe D (1995) Biochemistry 34:3742
68. Meng M, Bagdasarian M, Zeikus JG (1993) Proc Natl Acad Sci USA 88:4015
69. Whitaker RD, Cho Y, Cha J, Carrell HL, Glusker JP, Karplus PA, Batt CA (1995) J Biol Chem 270:22895
70. Nagorski RW, Richard JP (1996) J Am Chem Soc 118:7432
71. Harris DW, Feather MS (1974) J Org Chem 39:724
72. Hu H, Liu H, Shi Y (1997) Proteins 27:545
73. Fuxreiter M, Farkas O, Naray-Szabo G (1995) Protein Eng 8:925
74. Pastinen O, Visuri K, Schoemaker H, Leisola M (1999) Enzyme Microb Techn 25:695
75. Lehmacher A, Bisswanger H (1990) Biol Chem Hoppe Seyler 371:527
76. Chmielewski M, Chen M-S, Whistler R (1976) Carbohydr Res 49:479
77. Card PJ, Hitz WD, Ripp KG (1986) J. Am Chem Soc 108:158
78. (a) Durrwachter JR, Sweers HM, Nozaki K, Wong C-H (1986) Tetrahedron Lett 27:1261; (b) Durrwachter JR, Drueckhammer DG, Nozaki K, Sweers HM, Wong C-H (1986) J Am Chem Soc 108:7812

79. (a) Durrwachter JR, Wong C-H (1988) J Org Chem 53:4175; (b) Wong C-H (1991) Microbial aldolases in carbohydrate synthesis. In: Bednarski MD, Simon ES (eds) Enzymes in carbohydrate synthesis. ACS Symposium Series 466, American Chemical Society, Washington, DC, p 23; (c) Morís-Varas F, Qian X-H, Wong C-H (1996) J Am Chem Soc 118:7647
80. de Raadt A, Stütz AE (1992) Tetrahedron Lett 33:189
81. Simmonds MSJ, Kite GC, Porter EA (1999) Taxonomic distribution of iminosugars and their biological activities. In: Stütz AE (ed) Iminosugars as glycosidase inhibitors. Wiley-VCH, Weinheim, p 8
82. Andersen SM, Ekhart CW, Lundt I, Stütz AE (2000) Carbohydr Res 326:22
83. Berger A, de Raadt A, Gradnig G, Grasser M, Löw H, Stütz AE (1992) Tetrahedron Lett 33:7125
84. Hadwiger P, Mayr P, Nidetzky B, Stütz AE, Tauss A (2000) Tetrahedron: Asymmetry 11:607
85. Andersen SM, Ebner M, Ekhart CW, Gradnig G, Legler G, Lundt I, Stütz AE, Withers SG, Wrodnigg T (1997) Carbohydr Res 301:155
86. Wrodnigg TM, Gaderbauer W, Greimel P, Häusler H, Sprenger FK, Stütz AE, Virgona C, Withers SG (2000) J Carbohydr Chem 19:975
87. Ebner M, Stütz AE (1998) Carbohydr Res 305:331
88. Fechter MH, Stütz AE (1999) Carbohydr Res 319:55
89. Green M, Cohen SS (1955) Bact Proc 127
90. Green M, Cohen SS (1956) J Biol Chem 219:557
91. Oliver EJ, Mortlock RP (1971) J Bacteriol 108:293
92. Izumori K, Yamanaka K (1974) Agr Biol Chem 38:267
93. Boulter JR, Gielow WO (1973) J Bacteriol 113:687
94. Seemann JE, Schulz GE (1997) J Mol Biol 273:256
95. Badía J, Gimenez R, Baldomá L, Barnes E, Fessner W-D, Aguilar J (1991) J Bacteriol 173:5144
96. (a) Fessner W-D, Badia J, Eyrisch O, Schneider A, Sinerius G (1992) Tetrahedron Lett 33:5231; (b) Fessner W-D, Schneider A, Eyrisch O, Sinerius G, Badía J (1993) Tetrahedron: Asymmetry 4:1183
97. García-Junceda E, Shen G-J, Alajarín R, Wong C-H (1995) Bioorg Med Chem 3:1349
98. Alajarín R, García-Junceda E, Wong C-H (1995) J Org Chem 60:4294
99. Wong C-H, Alajarín R, Morís-Varas F, Blanco O, García-Juneda E (1995) J Org Chem 60:7360
100. Fessner W-D, Gosse C, Jaeschke G, Eyrisch O (2000) Eur J Org Chem 125

The Amadori and Heyns Rearrangements: Landmarks in the History of Carbohydrate Chemistry or Unrecognized Synthetic Opportunities?

Tanja M. Wrodnigg, Brigitte Eder

Glycogroup, Institut für Organische Chemie der Technischen Universität Graz, Stremayrgasse 16, 8010 Graz, Austria
E-mail: wrodnigg@orgc.tu-graz.ac.at

The Amadori and Heyns rearrangements have been known to carbohydrate chemists for decades. Following an outline of their historical development, a survey of the biological aspects, applications and alternative approaches will be given. Due to their nature, both reactions suffer from a variety of preparative shortcomings; nonetheless, they are very useful bearing in mind that no protecting group manipulations are required. Naturally occurring rearrangement products can play very important biological roles. For example, in the Maillard reaction cascade, this type of rearrangement reactions appear to be involved in the pathological effects of diabetes, Alzheimer's disease and aging processes in general. Consequently, alternative means of synthetic access to the corresponding products have also been investigated. Both rearrangements appear to be highly underrated as useful methods for natural products synthesis.

Keywords: Amadori rearrangement, Heyns rearrangement, Amadori products, 2-Amino-2-deoxysugars, 1-Amino-1-deoxysugars, D-Glucosamine derivatives

1 Introduction

The Amadori [1] and Heyns [1b, 2] rearrangements are two well-known reactions in carbohydrate chemistry. Both were discovered several decades ago and are closely interrelated with respect to their mechanisms and their outcomes.

The Amadori rearrangement is a reaction between α-hydroxy aldehydes and suitable amines leading to α-amino ketones. When applied to aldoses, for example, D-glucose (**1**), it allows the introduction of an amino group at position C-1 with concomitant isomerization leading to 1-amino-1-deoxyketoses such as **2** (Scheme 1).

The Heyns rearrangement follows, in principle, the same pattern but employs α-hydroxy ketones as starting materials. Applied to free sugars, this reaction starts from a ketose such as D-fructose (**3**) and proceeds via a glycosylamine to the corresponding 2-amino-2-deoxyaldoses **4** and **5** (Scheme 2). Due to the nature of the intermediate, both epimers at position C-2 can be formed.

The two rearrangement reactions are known as the initial step in the Maillard reaction cascade [3], the non-enzymatic browning of food. This reaction sequence takes place during cooking, baking and preservation processes and is of great importance for taste, aroma, flavor and color of food [4]. The reacting components are reducing sugars and free amino groups of amino acids or proteins and the reaction occurs extensively during food preparation where it affects nutritional quality and may be accompanied by the formation of toxic substances [5]. Additional aspects of the Maillard reaction have arisen from the discovery that it also occurs in the human body (see Sect. 2.2).

A great variety of rearrangement products are accessible just by choosing the starting materials as they are as simple as free sugars and almost all sorts of amines, bearing in mind that no protecting group manipulations are required for this kind of reactions.

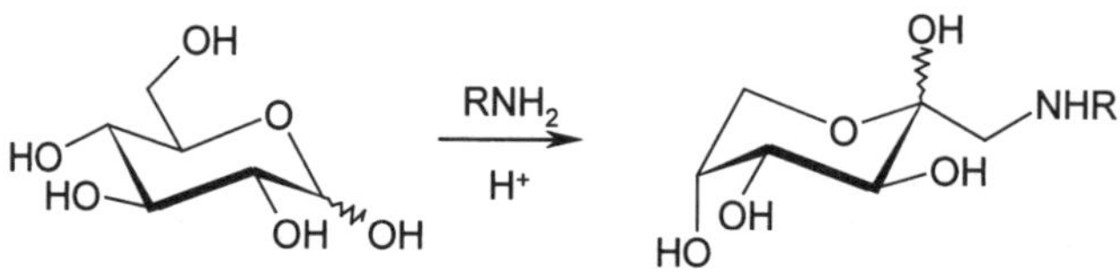

Scheme 1 1 2

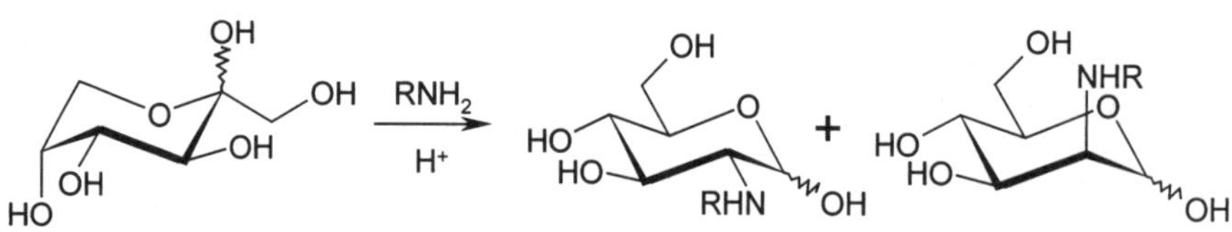

3 4 5

Scheme 2

2
Amadori Rearrangement

2.1
Historical Development

In the years 1925 to 1931, Amadori [6] claimed to have obtained two types of "composti", "A" (**6**, **7**) and "B" (**8**, **9**), from the reactions of D-glucose with *p*-toluidine, *p*-anisidine and *p*-phenetidine under different reaction conditions (Scheme 3). Referring to older literature, he suggested that his labile "composto A" was the same product as that previously isolated by Irvine and Gilmour [7] when they reacted D-glucose and *p*-toluidine in alcohol at the boiling point of the solvent. Amadori defined it as a mixture of α- and β-glucosyl amines **6** and **7**. Besides the expected *N*-phenyl-D-glucosylamines he observed isomeric products when he reacted D-glucose with the amines under consideration in substantia and at the time regarded the stable compounds obtained as related to Schiff bases **8** and **9**, having the same properties as the compounds described by Claus and Ree [8].

In the 1930s, Kuhn and Dansi reinvestigated the behavior of D-glucose with *p*-toluidine [9] in substantia as well as in alcoholic solution and postulated that the stable product obtained by Amadori was in fact the product of a novel rearrangement. Subsequently, Kuhn and Weygand established a mechanism (Scheme 4) [10], by which initially a reaction between the anomeric position of aldoses **1** with an amino group takes place leading to glycosylamines **10**. Ring opening occurs to form a Schiff base **12**, which is in equilibrium with its enol form **13** (enamine-aldimine tautomerism). This enol can be stabilized by formation of the 1-amino-1-deoxyketosugar **2**, which undergoes ring closure to the corresponding hemiacetal. This new reaction was coined the Amadori rearrangement by the authors.

A closer look into the history of the Amadori rearrangement reveals that the reaction between D-glucose and amines such as aniline had been explored by

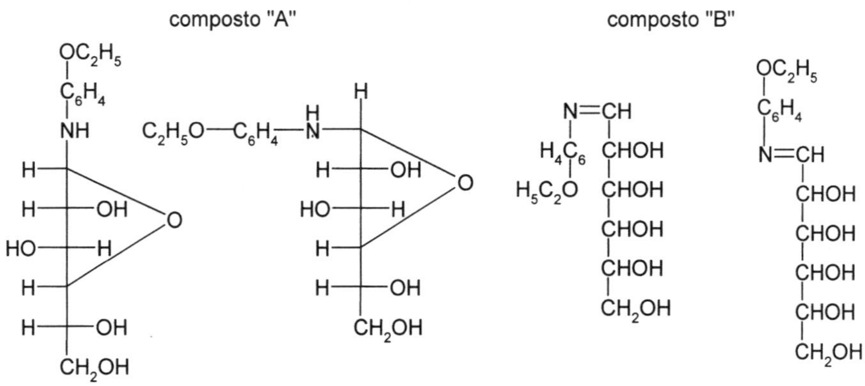

6 "α" **7** "β" **8** "sin" **9** "anti"

Scheme 3

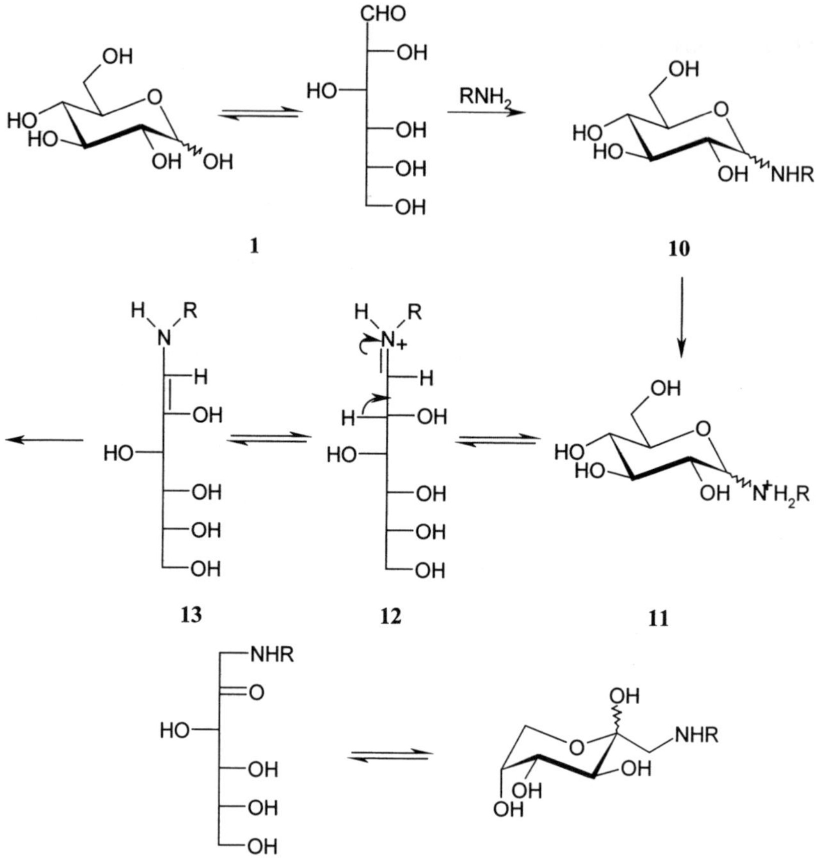

Scheme 4

several workers decades before Amadori's important contribution. This is not unexpected because the compounds employed were classical reagents and had been known and available for more than a century; eminent chemists such as Schiff, Fischer, and Sorokin, just to mention a few, had worked in this area of early carbohydrate chemistry.

Schiff, for example, had included carbohydrates in his studies on the formation of Schiff bases from aldehydes and amino compounds. He heated D-glucose with aniline or with *p*-toluidine in substantia [11], and described that these components, under loss of water, formed yellow glass-like condensation products. These supported his suggestions on the formation of Schiff bases from aldehydes and suitable aromatic amines [12].

Around the same time, Sachsse [13] investigated the reaction of aniline with lactose and reported that he obtained two different products or a mixture of both depending on the reaction conditions employed.

Subsequently, Sorokin reported [14] the isolation of products with different properties from those of Schiff, when he reacted D-glucose and aniline at lower temperatures in ethanol. Whereas Schiff had obtained amorphous materials by melting the components together, Sorokin isolated a crystalline compound [15], which he coined dextrose anilide. From the data supplied, which are typical for glycosylamines, it is safe to assume that he had isolated such a compound.

From these reports and by comparison with the conditions known to induce the Amadori rearrangement, it can be concluded that Schiff had already obtained the rearrangement products by carrying out the reaction without solvent and at elevated temperatures. Sorokin, on the other hand, preparing his products in an alcoholic solution and at lower temperatures, was able to isolate the comparably less stable glycosylamines. Sachsse would have found both, the unstable glycosylamine and the stable 1-aminodeoxyketose, depending on the reaction conditions employed.

In context with his interest in the formation of new products from sugars, Fischer focused his attention on phenylhydrazine as the basic component. He detected that the reactions of D-glucose on the one hand and D-fructose on the other with phenylhydrazine led to the same product [16]. When he reacted cane sugar with sulfuric acid followed by phenylhydrazine he initially isolated "phenylglucosazone" which he dissolved in ethanolic acetic acid to obtain the rearrangement product "isoglucosamine" (1-amino-1-deoxyfructose) [17]. The properties of this compound he found to be very closely related to "laevulose" (fructose) and he concluded that the relationship between "isoglucosamine" and fructose was the same as that between D-glucosamine and glucose. Consequently, the formation of this "isoglucosamine" had to take place involving a change from the dextrose to the laevulose series. Fischer also suggested that this was just one special case of a more general type of reaction.

Many years later, several groups conducted investigations into the mechanism of the osazone formation steps and concluded that an Amadori rearrangement is involved in this reaction cascade [18] (Scheme 5, **14** to **16**). Aldoses **14**

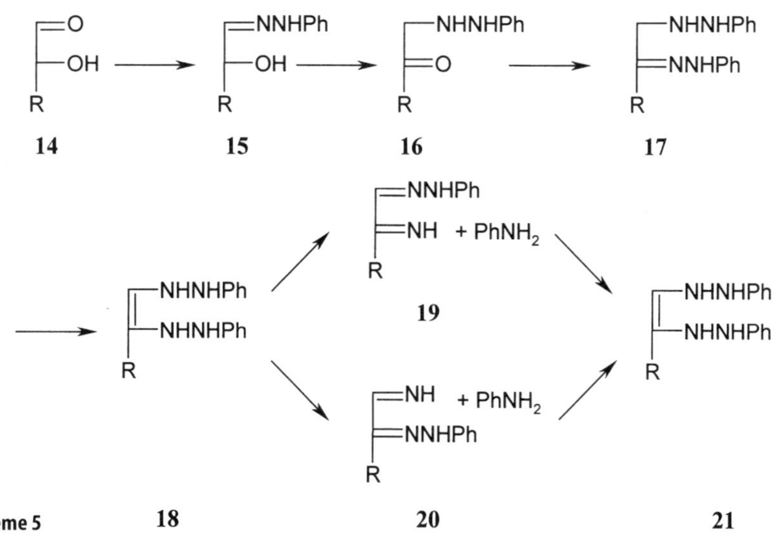

Scheme 5

react with phenylhydrazine to give phenylhydrazones **15**. These compounds can isomerize to Amadori rearrangement products **16**, whose ketone function may react with another molecule of phenylhydrazine to form **17**. The loss of aniline via intermediate **18** can take place either at position C-1 or C-2 to form both possible iminophenylhydrazones **19** and **20**. The last step of the sequence is the exchange of the imino moiety with a third molecule of phenylhydrazine to give osazone **21**. This view was strongly supported by Weygand's results with tritium-labeled components [19].

Remarkably, according to these pieces of information, Fischer had correctly interpreted his observations. An interesting fact in this context is that Amadori referred to the studies by Schiff, Sachsse, Irvine as well as those of Claus and Ree, but did not consider Fischer's important contribution. Amadori also had reported yields of around 70% of the *p*-phenetidino product but researchers following his work failed to reproduce these results. Improvements were made by Weygand [20], who was the first to recognize that both yields and product purity improved considerably when the reaction was performed under acidic conditions; a very important discovery in connection with this reaction.

In 1955, Hodge [1a] reviewed the available literature on the Amadori rearrangement, wherein he supplied experimental procedures and the physical properties of many Amadori rearrangement products, mainly from D-glucose and arylamines as well as some selected aliphatic amines. He also suggested that "this rearrangement may occasionally have gone unnoticed" in the older literature.

The influence of the basicity of the amine on the rearrangement was investigated by Rosen and co-workers [21]. They modified the conditions by employing aqueous pyridinium chloride as well as pyridinium acetate buffered glacial acetic acid as reaction media, and from these results concluded that the Amadori rearrangement had to be an example of general acid-base catalysis.

The influence of differently substituted arylamines on the outcome and the directing effects of *o*-, *p*- and *m*-substituents were reported by Micheel and Schleppinghoff [22].

At this point, except for a few special cases [1a], only rearrangement products from *N*-arylamines had been isolated. Mitts and Hixon [23] had failed in inducing an Amadori rearrangement using *N*-alkylamines under the reaction conditions known to be successful thus far. They could only isolate the glycosylamines. Subsequently, productive reaction conditions for this type of amines were found by Hodge and Rist [24], using a 1:1 mixture of ethanol/ethyl malonate as the solvent system. Micheel [25] employed dry oxalic acid as acidic component to obtain the corresponding ammonium oxalate salts. By this modification of the procedure, alkylamines were found to yield rearrangement products very easily.

In 1962, based on his experience with the reaction and results of other workers, Micheel [26] suggested a mechanism (Scheme 6) different from the one proposed by Kuhn and Weygand [10]. The initially formed glycosylamine **10** reacts with a second molecule of the amine to give a 1,1-bis-*N*-aminal **22**. In the following step, one of the nitrogens is protonated leading to intermediate **23**. Under loss of one molecule of the amine a carbonium ion **24** is generated, which, in turn, can undergo the rearrangement to the final product **2** via **25**.

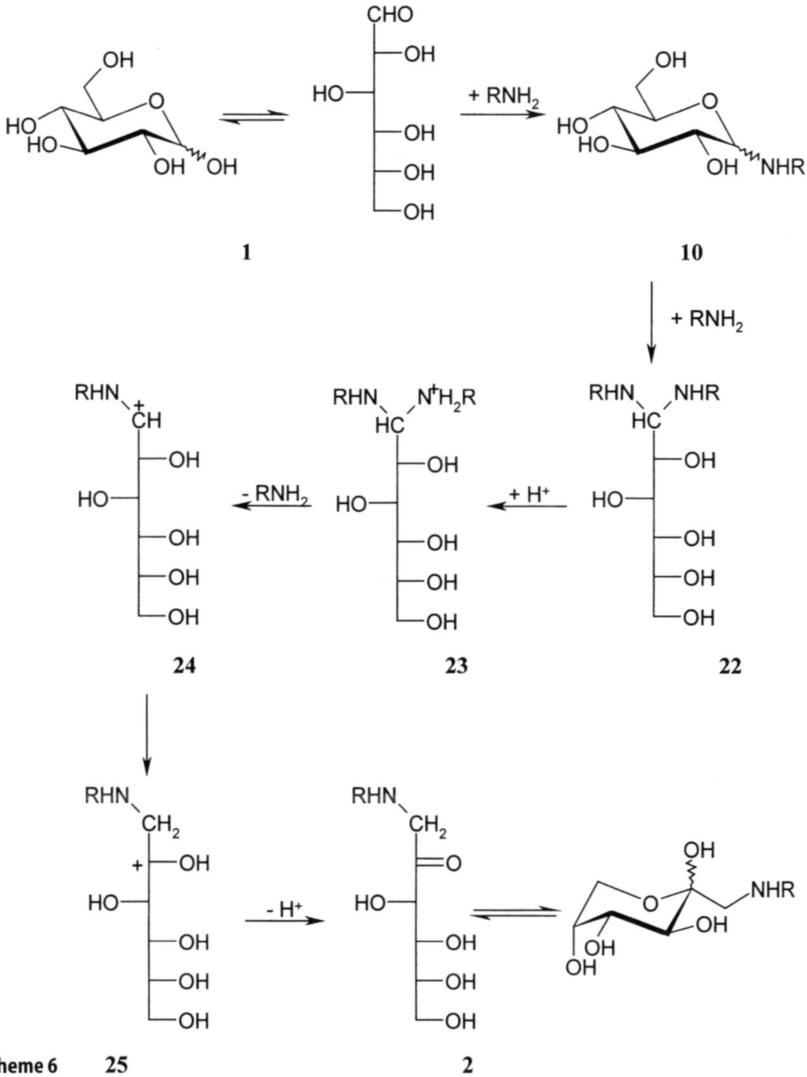

Scheme 6

This mechanism was challenged by Palm and Simon [18c, 27]. Employing tritium-labeled substrates these workers showed that the 1,1-bis-amino derivative **22** is not necessarily an intermediate of the rearrangement. The rate-determining step was found to be the abstraction of the proton at C-2 in the initially formed glycosylamine, supporting the observed general base and acid catalysis.

Following these observations, the influence of various borane-derived Lewis acids was probed by Yoshimura and co-workers [28]. Triethylborane exhibited no beneficial effect and difluoro(phenyl)borane was found to have a weaker catalytic activity than acetic acid. However, boron trifluoride and ethoxy difluoroborane showed stronger catalytic effects than acetic acid in the initial stage of the reaction, but colored side products were formed due to decomposition of the glycosylamine.

Nowadays, besides some special cases, typical reactions are performed in an alcoholic solution of the sugar in the presence of one equivalent of glacial acid as acidic catalyst.

2.2
Examples of Other Amines

2.2.1
Amino Acids

In addition to the investigations with aldoses and simple aryl- and alkylamines, the Amadori rearrangement was probed with a wide range of combinations of sugars with suitable amines. Gottschalk investigated the reaction conditions of the rearrangement between aldoses and amino acids and synthesized the first amino acid fructose adducts, by boiling the components in methanol, the carboxyl group of the amino acids providing the acidic catalyst for the Amadori rearrangement. This work opened a new and biologically important area [29].

Amino acids as amine components were also employed by Micheel and Frowein, by Heyns and co-workers, as well as by several other groups [30].

Furthermore, the Heyns group investigated the reactions of D-glucuronic acid with various amino acids and primary as well as secondary aliphatic amines [31]. They extended the range of amino components by employing taurine or aminomethansulfonic acids as basic components [32]. The reaction conditions were found to be different from the ones for the aminocarboxylic acids as, in the presence of oxalic acid as the acidic catalyst, only the sodium salts of the aminosulfonic acid could be successfully reacted to give the desired rearrangement products.

The synthesis and characterization of Amadori products with ω-amino acids was conducted by Feather [33] who employed them as model substances for biological studies.

The formation of amino acids as degradation products from the reaction of D-glucose and D-xylose with primary amines was investigated by Severin and co-workers [34]. They showed that the initially formed Amadori products either undergo oxidative degradation or give various amino acid compounds, such as *N*-propylalanine **26**, N^2-acetyl-N^6-(1-carboxyethyl)-L-lysine **27**, *N*-(carboxymethyl)- **28**, and *N*-(1-carboxyethyl)-alanine **29**, respectively, as well as *N*-substituted 4,5,6-trihydroxy-α-aminohexanoic acids **30** (Scheme 7).

Additionally, the amino groups of peptides [35] and, at a later stage, proteins [36] were found to react with D-glucose to give Amadori rearrangement products.

The Amadori rearrangement with amino acids as the basic component can be performed under self catalysis of the carboxylic acidic part of the molecule, but for better yields various additional acidic catalysts have also been used, depending on the starting materials.

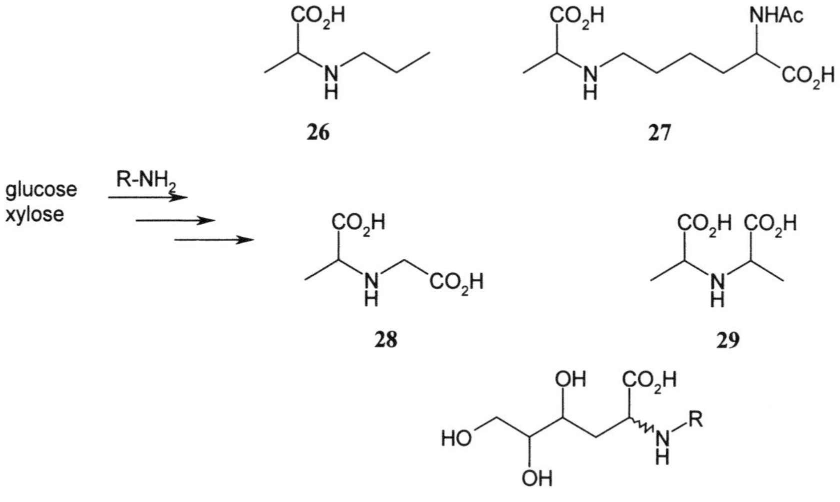

Scheme 7

2.2.2
Aminosugars

Micheel and co-workers employed D-glucosamine as the amine component (Scheme 8) [37]. 4,6-*O*-Benzylidene-D-glucose (**32**) was reacted with 1,3,4,6-tetra-*O*-acetyl-D-glucosamine (**31**) to give the expected glucosylamine **33**, which underwent rearrangement in ethanolic solution containing one equivalent of triethylamine and glacial acetic acid. The protecting groups were removed to obtain the free Amadori rearrangement product, an *N*-linked disaccharide **34**.

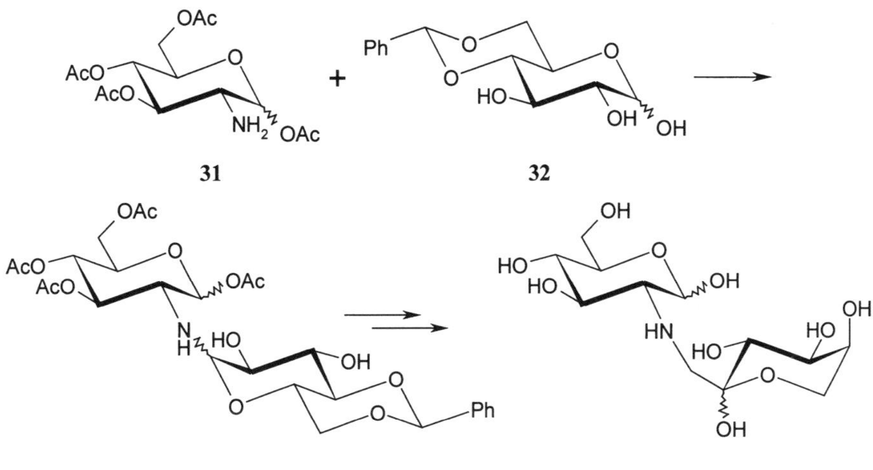

Scheme 8

Related disaccharides were synthesized by Klemer and Funcke in the reaction of D-glucosamine and D-galactosamine with D-*gluco*- and D-*galacto*-configured hexuronic acids. Insterestingly, good yields were obtained, although the rearrangement required ring contraction of the glycosyl amine leading to furanoid systems [38].

2.2.3
Polymeric Amines

Amadori products were also prepared from carbohydrates and polyvinylamine (**35**, Scheme 9) [39]. It was found that the initial products, such as **36**, undergo further reactions thus cross-linking the polymer strains intramolecularly or intermolecularly leading to structures like **37** and **38**, respectively.

2.3
Biological Aspects

Investigations into the formation and biological significance of 1-amino-1-deoxyketoses have become an interesting and important area, following the clarification of structure and behavior of the products. Gottschalk published

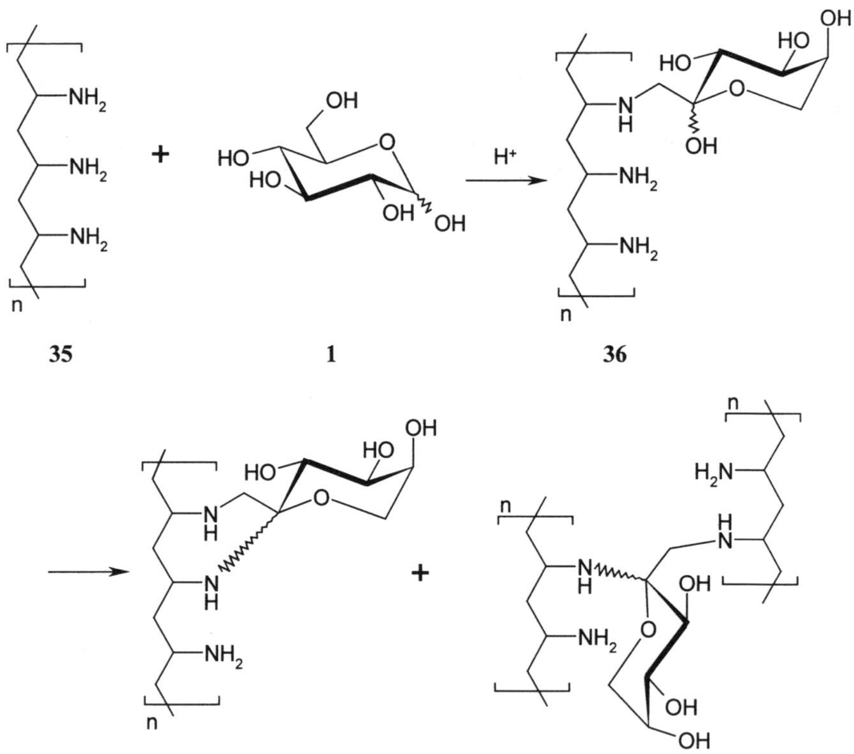

Scheme 9 **35** **1** **36** **37** **38**

evidence that the carbohydrate-amine linkage in the physiologically important mucoproteins is that of 1-amino-1-deoxy-2-ketoses [40].

Intermediates in the biosynthesis of tryptophans were also found to be Amadori rearrangement products of different aldoses with anthranilic acid such as **39**, derived from the corresponding phosphorylated forms **40** in cells (Scheme 10) [41]

An Amadori rearrangement was also found to take place as the first step in the reaction of D-glucose (**1**) with 2,4,5-triamino-6-hydroxypyrimidine (**41**) to give substituted pterine derivatives **42** (Scheme 11) [42]. Such compounds are important intermediates for the synthesis of folic acids and hydroxyalkylpterine and are of interest in context with the biogenesis of pterines **43** [43].

More recently, it has been shown that *N*-(1-deoxy-D-fructos-1-yl) derivatives of mammalian proteins are formed in vivo [44]. The process, which is called non-enzymatic glycation (formerly erroneously termed glycosylation), may be important in the pathology of diabetes [45] and of Alzheimer's disease [46], the formation of cataracts [47] as well as other aging processes [48]. The precise role of the Maillard reaction in the process of aging still remains a matter of specu-

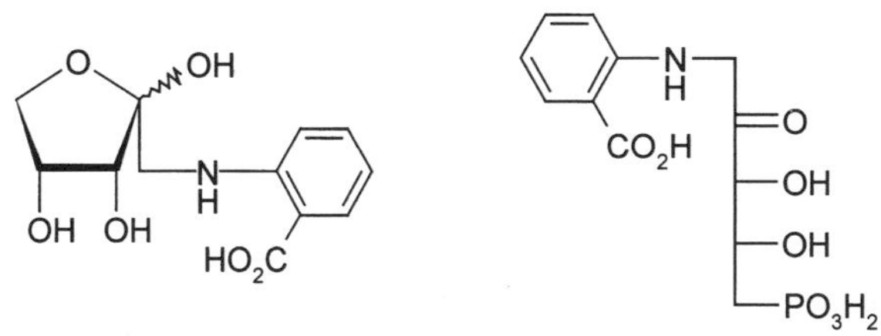

Scheme 10 **39** **40**

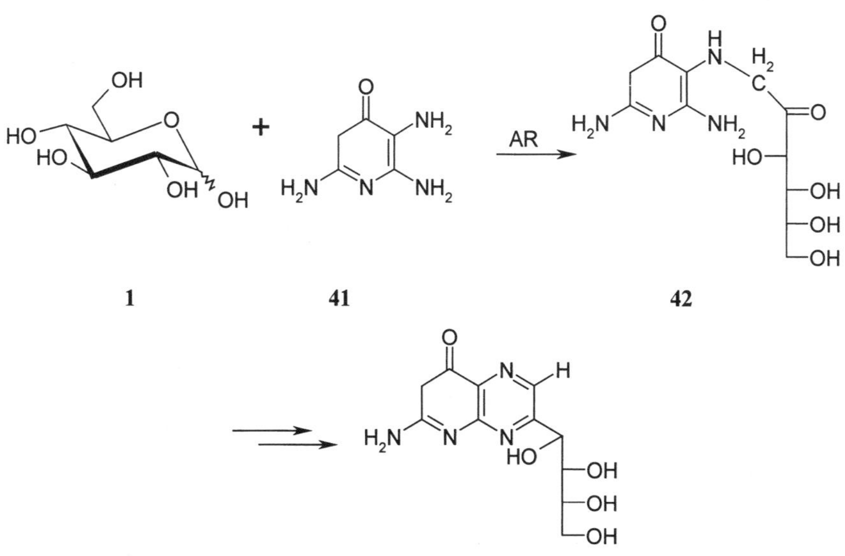

1 **41** **42** **43**

Scheme 11

lation, but the inter-relationship between the time-related appearance of Maillard reaction related cross-linking in proteins, cells and tissues and the overall aging process appears indisputable [49]. As not much is known about the chemical nature of the cross-linked units, several model reactions and different mechanisms for this process have been discussed [50]. For a better understanding of the impact which the Maillard reaction has, for example, on aging and diabetes, as well as for the development of effective therapeutic methods to prevent advanced glycation product accumulation in tissues, it is an absolute prerequisite to unambiguously establish the chemical nature of the major protein cross-links derived from this reaction [51].

Several Amadori products, notably those containing aromatic amino acid residues, have been found to behave as direct-acting mutagens in *Salmonella typhimurium* his^- strains [52] (i.e., in the Ames test), and some (e.g., nitrosated fructose-tryptophan and fructose-serotonin compounds) are known to induce DNA repair synthesis in cells of the human HeLa S3 cell line [53].

Furthermore, Amadori compounds and their degradation products have been reported to possess the ability to affect the adhesion and aggregation properties of cancer cells [54].

As a sequence of equilibrium reactions, the Amadori rearrangement is theoretically expected to be a reversible process [55], and this can have significant implications provided the factors promoting the reverse reaction can be identified and eventually controlled under physiological conditions. The tendency of the products to undergo a non-enzymatic reverse reaction into enolamines, and subsequently into free sugars and amino acids, has been demonstrated, when it was recently found from the data obtained on the kinetics of *N*-(5′-*O*-phosphono-β-D-ribosyl)anthranilate ketol isomerase (EC 5.3.1.24) that this enzyme involved in the biosynthesis of tryptophan catalyzes the reversible Amadori rearrangement reaction of anthranilic acid with ribose 5-phosphate [56]. This reverse reaction might limit protein glycation and prevent further tissue damage. The reversibility of the Amadori reaction was also supported by the identification of per-*O*-(trimethylsilyl) derivatives of aldoses generated from thermal decomposition of *N*-(1-deoxy-D-fructopyranos-1-yl)proline [57].

The non-enzymatic glycation of aminophospholipids supposedly plays an important role for lipid oxidation in vivo. Membranous functional lipids are vital for the maintenance of cellular integrity and survival and their non-enzymatic glycation could conceivably cause inactivation of receptors, cross-linking of aminophospholipids and proteins, membrane lipid peroxidation, and, consequently, cell death. Therefore, knowledge about both primary and secondary products from the reaction of D-glucose with aminophospholipids could provide a deeper insight into the mechanism of this advanced glycation end product initiated lipid oxidation. Related studies were conducted in an in vitro model system employing 1-deoxy-1-(2-hydroxyethylamino)-D-fructose. With the spectroscopic and chromatographic data obtained from this investigation it could be proven that Amadori products such as **44** (Scheme 12) are formed from D-glucose and phosphatidyl ethanolamine [58], which was also demonstrated by comparison with the outcome of the independent synthesis of aminophospholipid-linked Maillard products [59].

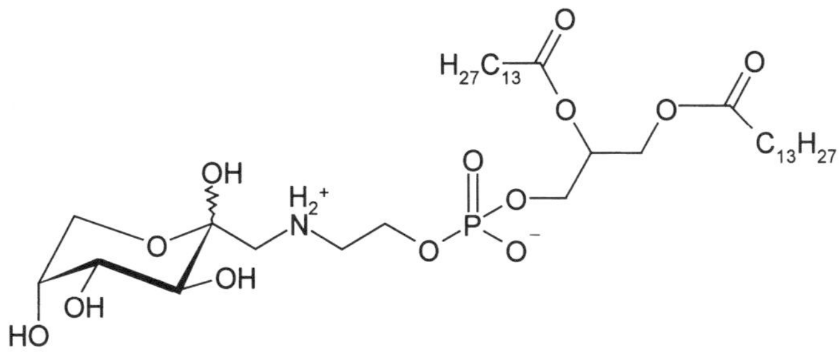

Scheme 12 **44**

Fructose analogs modified at positions C-1, C-2 or C-6 are of interest for studies of parasite glucose transporters, because the glycolytic metabolism of D-glucose is the unique source of energy for the parasite *Trypanosoma brucei*, the causative agent of African sleeping sickness. In contrast to the mammalian erythrocyte glucose transporter, which only recognizes D-glucose, the one of the parasite also accepts D-fructose as a furanose ring. C-1-Aminodeoxyfructose derivatives were synthesized via an Amadori rearrangement and subsequently marked with the fluorescent dansyl group and their affinities for the glucose transporter were examined [60].

The carcinogenic and mutagenic effects of heterocyclic compounds [61] such as **45** and **46**, which are formed during the Maillard reaction cascade, have been demonstrated (Scheme 13).

Other compounds can react with nitrite to nitroamine compounds, whereupon structures such as **47** are formed (Scheme 13), whose mutagenic and strongly carcinogenic effects are known [62]. This reaction may also occur in the stomach with nitrite from saliva.

2.4 Structural Investigations

The products resulting from Amadori rearrangement reactions, as free sugars, can exist in all of the typical sugar conformations; the acyclic form as well as the

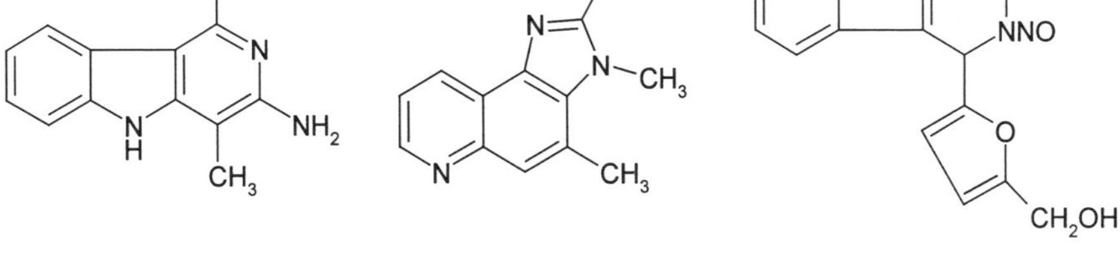

45 **46** **47**

Scheme 13

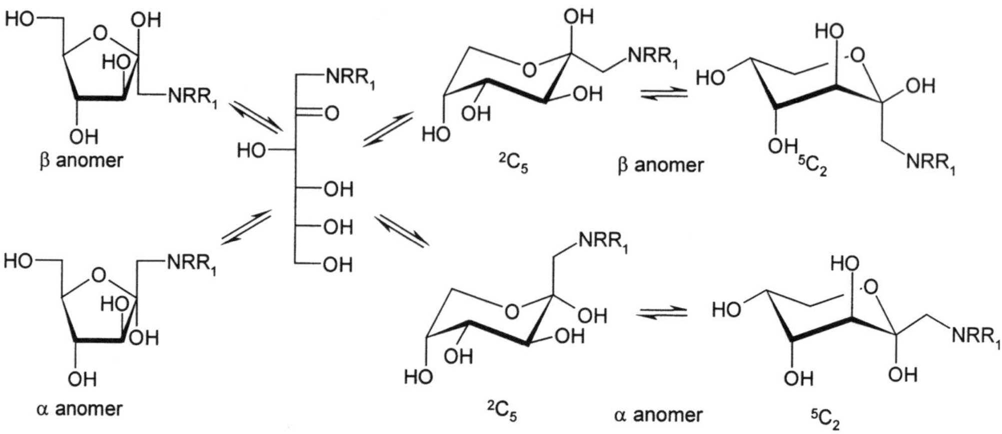

Scheme 14

furanoid and pyranoid $^5C_2$ and $^2C_5$ forms, and in addition, α and β configurations are possible (Scheme 14).

Early structural studies were conducted by Kuhn and co-workers who determined the ring size of 1-aminodeoxy-D-fructose derivatives by methylation and degradation experiments [63]. For the 1-(*p*-toluidinyl)fructose derivative they proved with this method the consequently most stable pyranoid conformation.

Qualitative and quantitative assessments of equilibrium mixtures of Amadori rearrangement products were performed employing ^{13}C-NMR spectrometry [64]. Substituents at C-6, C-4 and C-6 showed considerable influence on the equilibrium; the 4,6-*O*-benzylidene derivative, for example, was found to be in the open-chain form only, because ring closure of O-5 with the carbonyl function was energetically and sterically hindered. The conformation of compounds substituted at C-6 depended on the electron-withdrawing effect of the substituent. The lower the electron density at position C-5, the lower is the tendency of ring closure, so that the equilibrium is found on the side of the open-chain form.

^{1}H-NMR spectroscopy was subsequently investigated as a tool for the identification of Amadori products [65]. Isomeric mixtures of fructose derivatives of amino acid products in solution [66] as well as characterization of glycated proteins [67] were examined by exploiting this method.

2.5 Applications of the Amadori Rearrangement

An early application of the Amadori rearrangement reaction was the synthesis of lactulose **51** (Scheme 15) [68]. Reaction of lactose **48** with *p*-toluidine in pyridine/acetic acid furnished the corresponding rearrangement product **49**, which, after catalytic hydrogenolysis to 1-amino-1-deoxyketose **50** and subsequent deamination, gave ketose **51**. This was the first alternative approach to lactulose, which had been synthesized via a Lobry de Bruyn – Alberda van Ekenstein rearrangement [69].

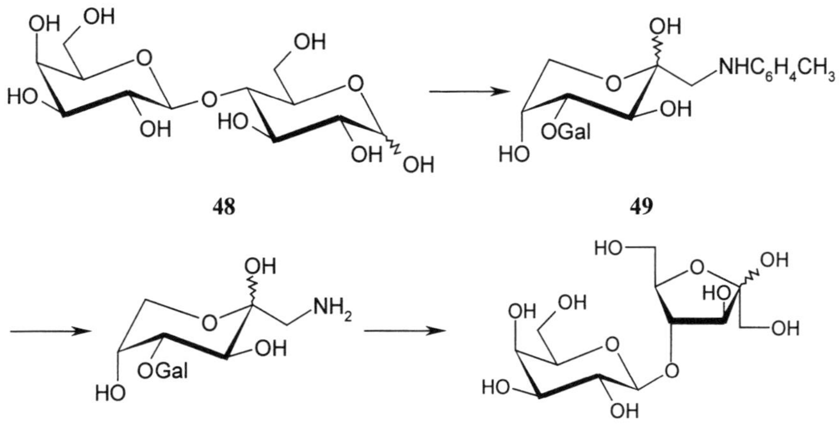

48 49

Scheme 15 50 51

A few years later Grünnagel and Haas used the same approach to synthesis D-tagatose from D-galactose via the Amadori rearrangement product 1-deoxy-1-dibenzylamino-D-tagatose [70].

Paulsen found that an intramolecular Amadori rearrangement of 5-amino-5-deoxy-D-xylose (**52**) under appropriate reaction conditions led to 1,5-dideoxy-1,5-imino-D-*threo*-pentulose (**55**), a direct precursor of 1,5-dideoxy-1,5-imino-xylitol (**56**) [71] (Scheme 16). When the temperature exceeded 0 °C or the acidic conditions became too strong, aromatization of the enamine-ol **54** took place, leading to 3-pyridinol. Release of strain by ring enlargement and the all-equatorial arrangement of substituents in the product after the rearrangement are strong driving forces of this intramolecular Amadori reaction. The sequence is a nice application of the rearrangement allowing access to a very important class of sugar analogs with basic nitrogen instead of the oxygen in the ring. Such compounds have been found to be powerful and highly useful glycosidase inhibitors [72].

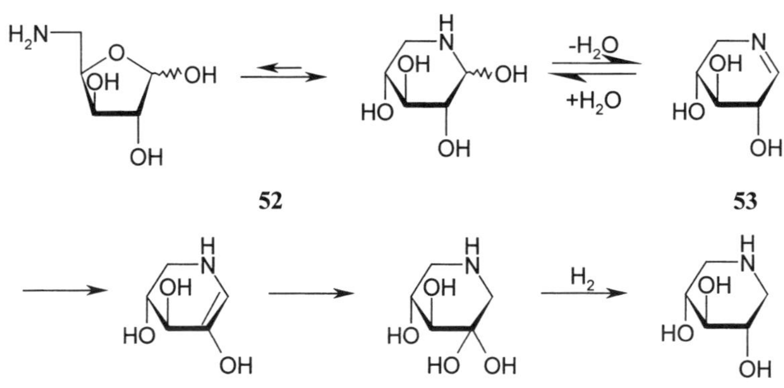

52 53

Scheme 16 54 55 56

Another example of an intramolecular Amadori rearrangement has been investigated with monosaccharide **57**, as well as of the corresponding D-*manno*- and D-*galacto*-configured sugars (Scheme 17). The aim was to determine the influence of the structure of the sugar component in the rearrangement products such as **58** on the overall course of these reactions and on the factors that contribute to the formation of the keto amine [73]. In particular, the knowledge on products from the reaction of, for example, D-galactose with endogenous peptides could provide a deeper insight into the mechanism of pathophysiology in hereditary galactosaemia; which was suggested to be related to the non-enzymatic galactose-dependent modification of proteins.

N-(1-Deoxyfructos-1-yl)dipeptides, such as **59** and **60** (Scheme 18), the rearrangement products of D-glucose and L-isoleucyl-L-aspartic acid and L-valoyl-L-aspartic acid, respectively, were found to be specific inhibitors of endopeptidase (EC 3.4.24.11) [74].

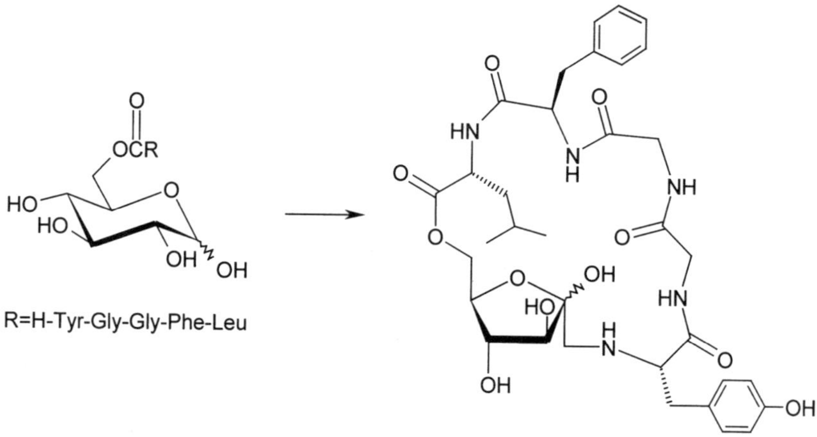

Scheme 17 **57** **58**

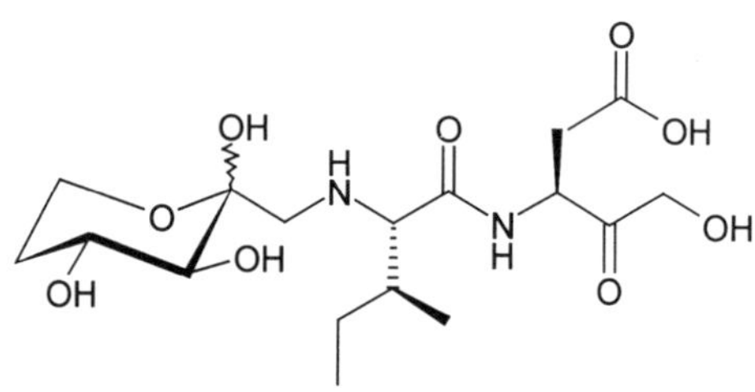

59

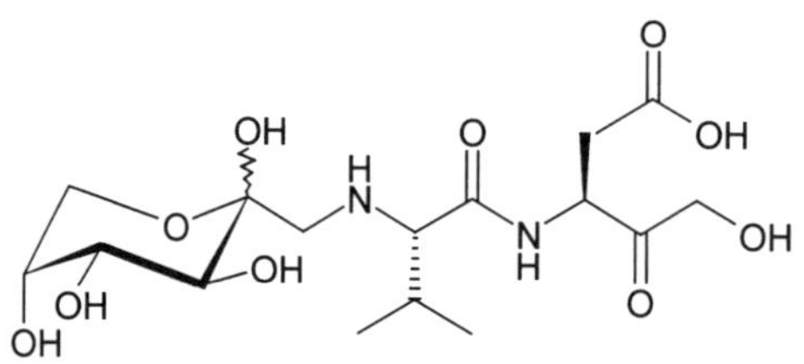

Scheme 18 **60**

This enzyme hydrolyses enkephalins, for example, tyr-gly-gly-phe-met or tyr-gly-gly-phe-leu, which are neuropeptides, and endogenous ligands of opiate receptors and are of importance for physiological supression of pain [75]. Compounds **59** and **60** were isolated after an Amadori rearrangement with the unprotected peptides in more than 70% yield and not only exhibited specific activity against metallopeptidases, but inhibited endopeptidase (EC 3.4.24.11) practically exclusively.

In the course of a project concerned with the synthesis of glycosidase inhibitors, interest arose in a simple, reasonably versatile, and efficient synthetic route to 1-aminodeoxy derivatives of the powerful D-glucosidase and invertase inhibitor 2,5-dideoxy-2,5-imino-D-mannitol [76] (Scheme 19). Bearing in mind that the release of ring strain in the five-membered ring **61** is a strong driving force for the quantitative isomerization [77] into the corresponding D-fructopyranose isomer, it was expected that an Amadori rearrangement reaction would introduce the desired amino group at C-1 with concomitant formation of the D-fructopyranose derivative. By reaction of glucofuranose **61** with a range of amines

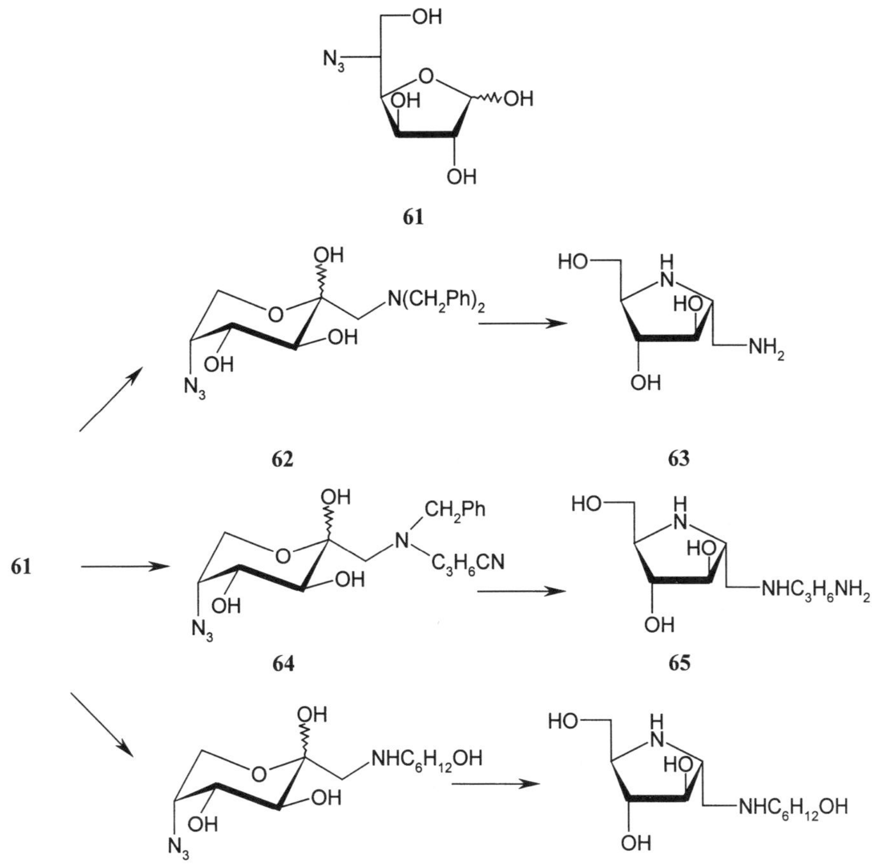

Scheme 19

(Scheme 19) the corresponding rearrangement products **62**, **64** and **66** were isolated in unusually high yields of more than 95%. These, in turn, were cyclized by intramolecular reductive amination to furnish inhibitors **63**, **65** and **67**, respectively, which exhibited very interesting biological activities [78].

Aminopolysaccharides have attracted considerable attention because of their structures. These are generally different from those of normal polysaccharides such as cellulose, chitin being one of the most prominent examples. The synthesis of such polymers is an important prerequisite to elucidating the biological functions of naturally occurring aminopolysaccharides. A thermal polymerization reaction (Scheme 20) by Amadori rearrangement of 6-amino-6-deoxy-D-glucose (**68**) was employed to yield the corresponding aminopolysaccharide **69** [79].

2.6
Preparative Limitations

The Amadori rearrangement reaction suffers from a few preparative drawbacks. In many cases, as well as the desired rearrangement product, mixtures with the isomeric glycosyl amine precursors are obtained and these are frequently difficult to separate. Furthermore, the reaction is very sensitive to hydrolytic conditions. Thus, the rearrangement of the glycosylamine and the hydrolysis of the latter back to starting material may compete depending on the acidic catalyst employed. In addition, products are not infinitely stable under the reaction conditions and decomposition reactions can take place. These very easily enter the Maillard cascade, leading to unwanted side products such as deoxyosones which can start a complex series of reactions, for example, dehydration, cyclization, oxidation, or fragmentation, resulting in an abundance of secondary products.

It is recognized that the reaction is dependent on four factors: i.e. the type of sugar and amine to be reacted, the reaction time, the temperature, as well as the catalyst employed.

Regarding the amine component, ammonia is a special case because it is a nucleophile and also a strong base. A survey of the reactions of free sugars with aqueous ammonia was provided by Kort in 1970 [80]. A large number of products can be found in such reaction mixtures, depending on the conditions employed. Short reaction times, low temperature, and the absence of catalyst arrest the reaction before heterocyclic compounds can be formed. The products obtained are the same as those formed by action of alkali on sugars, whereby a Lobry de Bruyn – Alberda van Ekenstein (LdB-AvE) rearrangement (see An-

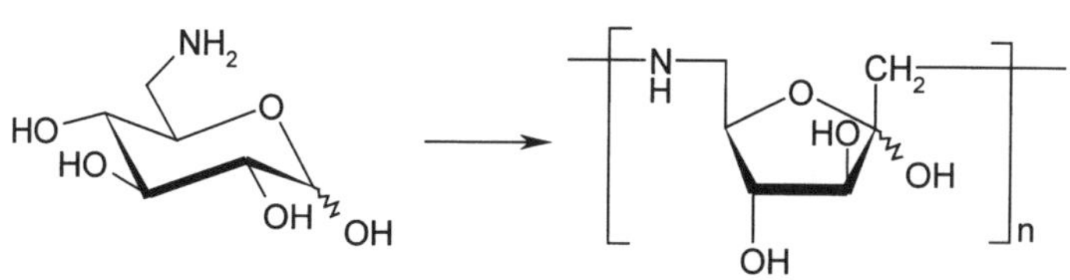

Scheme 20 **68** **69**

gyal, this vol.) takes place, leading to epimerization reactions. With prolonged reaction times, the reaction runs faster and becomes more complex and mainly degradation products are formed leading to substituted imidazoles and pyrazines. No Amadori rearrangement products can be detected with ammonia as basic component.

Similar problems can occur when the reaction is carried out with other strongly basic amines. The more basic the amino components are, the higher is the chance of epimerization at C-2. This leads to subsequent side reactions following the LdB-AvE isomerization resulting in unwanted degradation of the intermediates.

The longer the reaction times and the stronger the acid catalyst, the more Maillard decomposition reactions can take place. This was substantiated by the vacuum thermolysis of the Amadori product deriving from D-glucose and L-proline (1-deoxy-1-L-prolino-D-fructose), from which 22 different products were identified [81]. Amongst others, dihydrofurans, dihydropyrans, pyrons, a methylcyclopentenolone, pyrrolidine amides and formic, acetic, and propionic acid, substituted furfurylamines, a diazepine, proline, and pyrrolidines, maltol, pyrrolidine amides of formic, acetic and propionic acid, a γ-lactone, as well as 2-pyrrolino-substituted furans were found to have formed. Several bicyclic compounds have also been isolated from the reaction of D-glucose with methylamine [82] or glycine [83], including furans, pyrroles, pyridines, and a pyrogallol derivative. However, the thermal polymerization and degradation processes are not well elucidated due to the difficulty in obtaining reasonable quantities of the desired products via the Amadori rearrangement.

2.7
Alternative Synthetic Approaches to Amadori Rearrangement Products

Whereas in a few, unusual cases quantitative yields have been obtained from Amadori rearrangement reactions [78], they frequently range between 30 and 70%. Discouraged by the above-mentioned problems (Sect. 2.6), many workers interested in the products have chosen to synthesize such compounds by other, more reliable methods.

Xenakis and co-workers [84] converted 2,3:4,5-di-*O*-isopropylidene-β-D-fructopyranose (**70**) into triflate **71**. After displacement of the sulfonate with a nitrogen nucleophile, the corresponding protected Amadori compounds **72** were obtained in moderate to good yields (Scheme 21). The same approach was chosen by Lopez and Gruenwedel [85] when they needed larger quantities of pure products for biological and physiological studies designed to monitor the binding of aromatic Amadori compounds to DNA via intercalation. Winkel and co-workers [86] subsequently reported the deprotection sequence of the Amadori products thus obtained to form **73** .

Another access was exploited by Walton and McPherson (Scheme 22) [87]. Reductive amination of D-arabino-hexos-2-ulose (**74**) with various amines such as L-valine, L-leucine, L-methionine, L-phenylalanine or 6-aminohexanoic acid, in the presence of cyanoborohydride, resulted in the formation of the corresponding *N*-(1-deoxy-D-fructos-1-yl)-adducts **2**, but was accompanied by re-

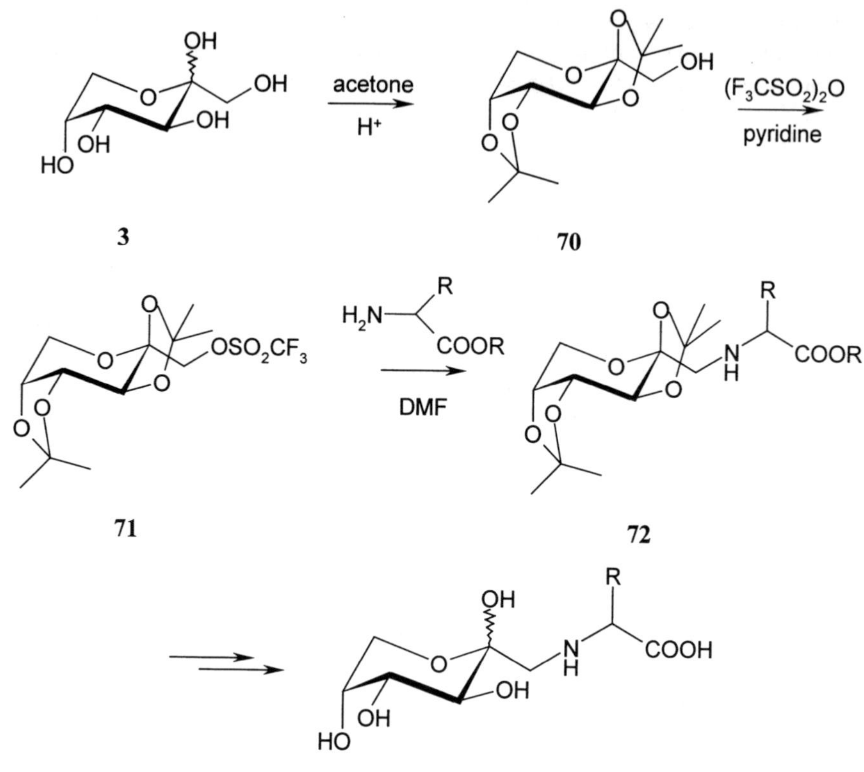

Scheme 21

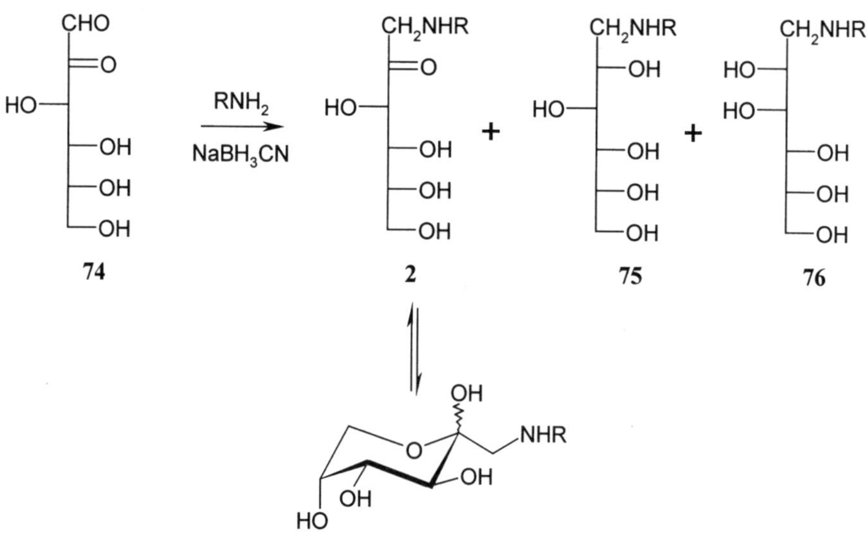

Scheme 22

duction of C-2 to give both epimers **75** and **76** as side products (Scheme 22). Separation was reported to be very difficult and could only be achieved by column chromatography on cellulose, providing the desired compounds in yields around 40%.

This problem was overcome by changing the starting material to the protected 2,3:4,5-di-*O*-isopropylidene-*β*-D-*arabino*-hexos-2-ulo-2,6-pyranose (**77**) (Scheme 23). Reductive amination employing the same amines circumvented the formation of side products improving the yield of the reduction step to around 75%. Removal of the protecting groups from **78** led to the desired products **2** in fair overall yields.

This method was also employed by Kojić-Prodić in the synthesis of the corresponding L-tyrosine derived Amadori product; the conformational properties were also examined [88].

Using a similar line of approach, Ohfune and his group produced aldehyde **77** from 2,3:4,5-di-*O*-isopropylidene-D-fructopyranose (**70**) by Swern oxidation (Scheme 24) [89]. Reductive coupling of this intermediate with unprotected or

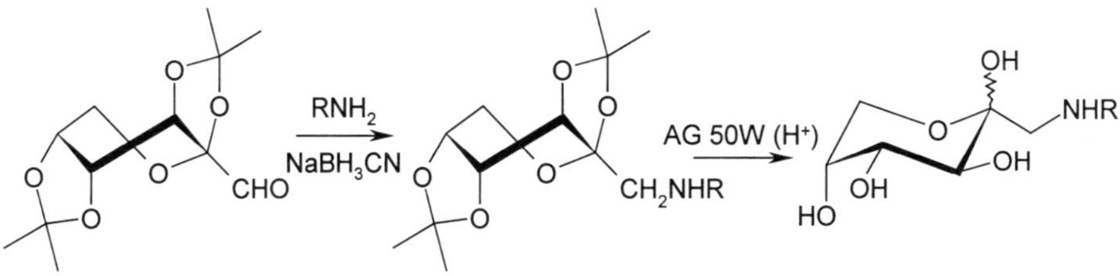

77 **78** **2**

Scheme 23

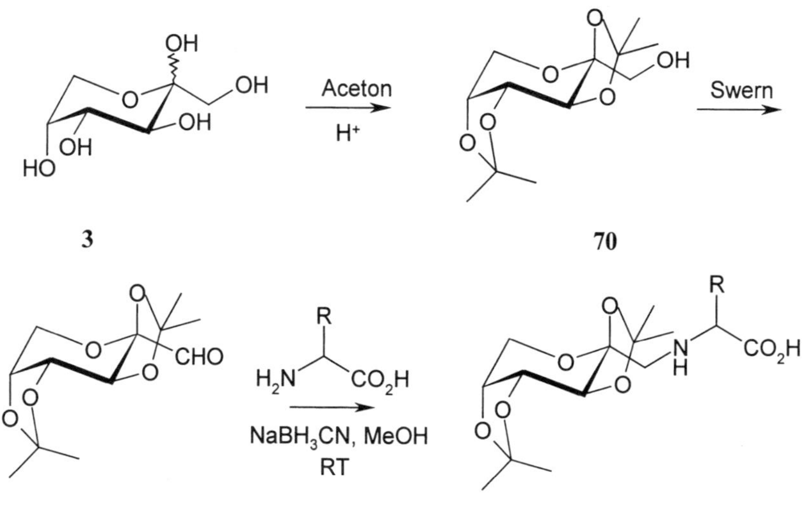

3 **70**

77 **79**

Scheme 24

partially protected amino acids was successfully achieved exploiting $NaBH_3CN$ as reducing agent to yield the desired Amadori products **79**.

A synthesis based on a novel Mitsunobu glycosylation procedure as the key step was recently reported by van Boom and co-workers [90]. They reacted the amino acid derived 2-nitrobenzenesulfonamides **81** with 2,3,4,6-tetra-*O*-acetyl-D-glucose (**80**) under Mitsunobu conditions (Scheme 25) to obtain fully protected glycosylamines **82**, which upon deprotection rearranged to the corresponding Amadori products **73** in overall yields of 70–80%.

1,2-Anhydro-3,4:5,6-di-*O*-isopropyliden-D-mannitol (**83**), which is available from D-mannitol in a few simple synthetic steps, can also be employed as starting material for the synthesis of Amadori products (Scheme 26) [91]. The epoxide was opened regioselectively at position C-1 to azide **84**, which was reduced to the amine **85**. Activated α-hydroxycarboxylic acids, such as L-lactic acid, can undergo nucleophilic attack in a S_N2 reaction leading to the substituted D-mannitol derivative **87**. Subsequent oxidation of the remaining free hydroxy group and deprotection furnished the Amadori compounds **73**.

The same epoxide **83** can be opened by the amino function of protected amino acids **88** with retention of configuration at this center leading to the Amadori product precursor **89**, which, in turn, after oxidation and deprotection, can be converted into the desired Amadori compounds, such as **73** (Scheme 27) [91].

3 Heyns Rearrangement

3.1 Historical Development

In 1898, Lobry de Bruyn mentioned that “chitosamine” [D-glucosamine (**4**)] was formed during the reaction of D-fructose **3** with ammonia but he was not able to isolate the syrupy product from the reaction mixture [92]. It was Breuer [93]

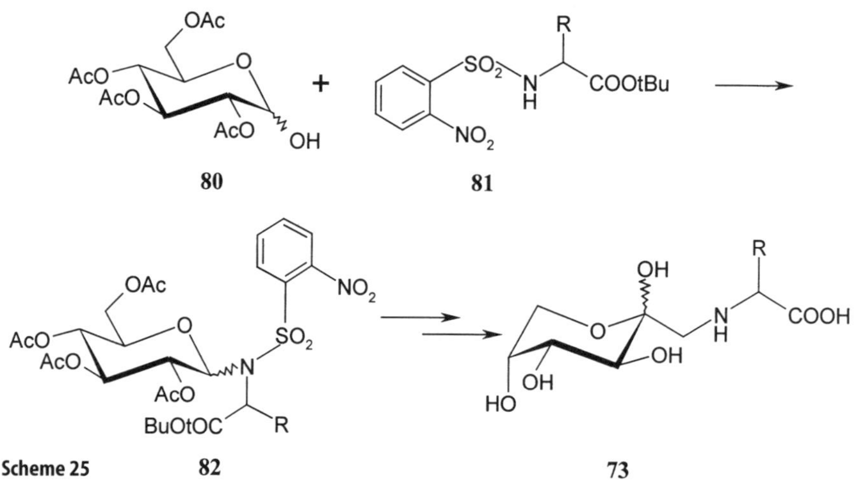

Scheme 25

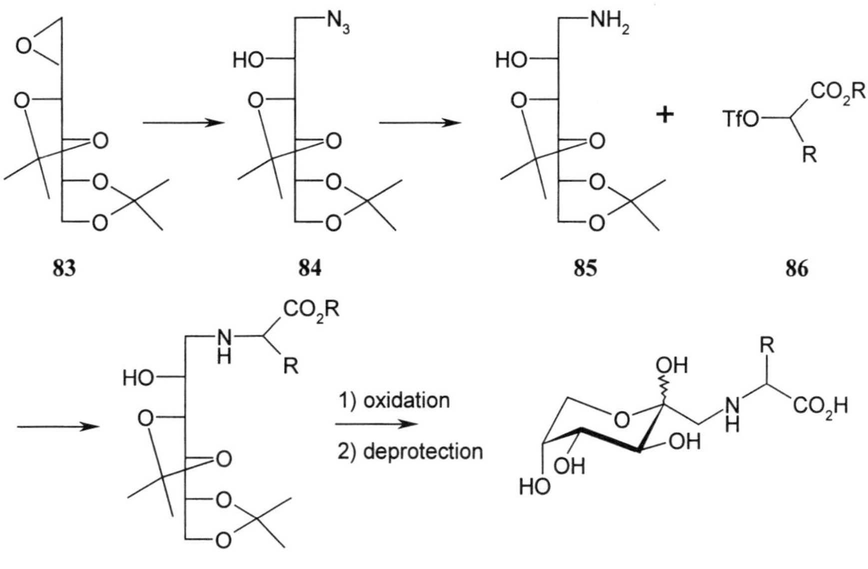

Scheme 26

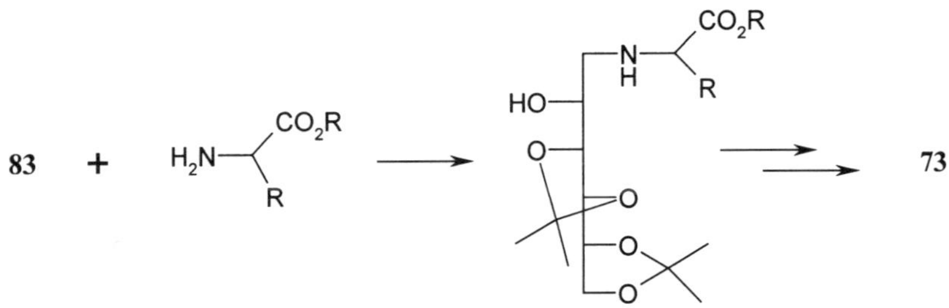

Scheme 27

who successfully performed the crystallization, obtaining D-glucosamine in crystalline form from the corresponding hydrochloride. This allowed first initiatives to investigate this paradigmatic example of a reaction, later coined the Heyns rearrangement.

During their studies of the reaction of free sugars with amines, *N*-aryl-D-fructosylamines were synthesized by Sorokin [14], Kuhn and Birkofer [94], as well as Barry and Honeymoon [95] but, in contrast to *N*-arylaldosylamines, these compounds could not be induced, under the conditions applied, to rearrange into the desired α-aminocarbonyl compounds.

Heyns and Koch succeeded [96] in confirming the structure of the rearrangement product from the reaction of D-fructose and ammonium chloride in the presence of phosphate and proposed a mechanism (Scheme 28). They pointed out that both epimers at position C-2, D-glucosamine (**4**) and D-mannosamine (**5**), were formed with the thermodynamically more stable compound **4** preponderating and recognized that this reaction is related to the Lobry de Bruyn –

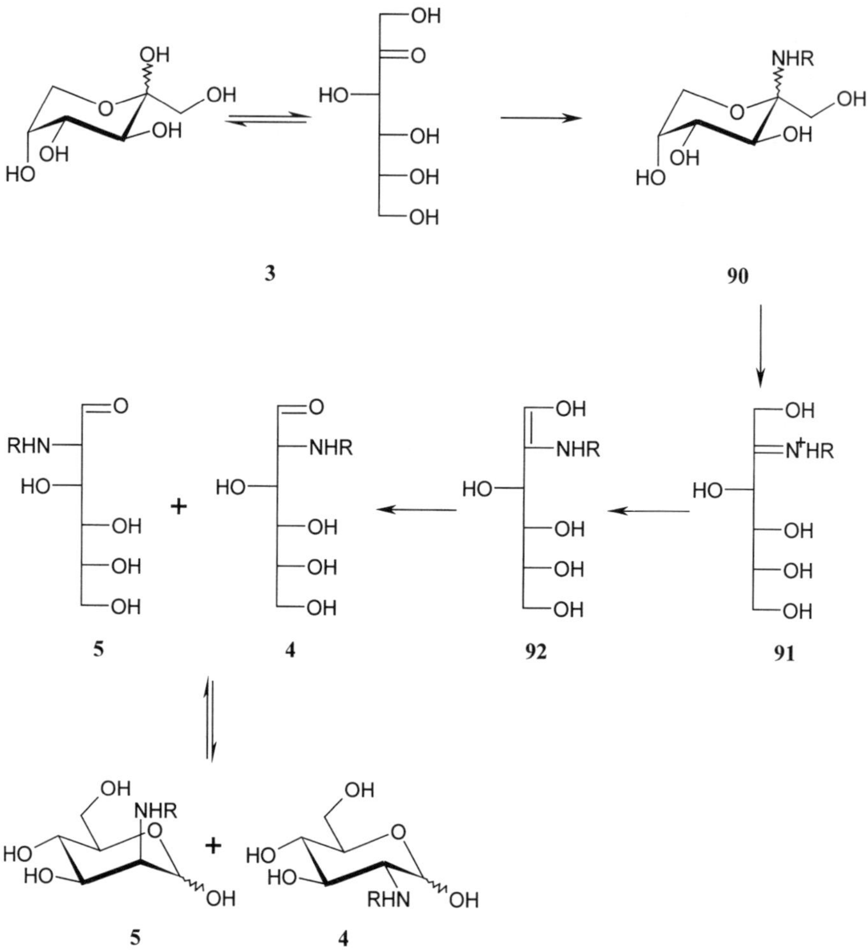

Scheme 28

Alberda van Ekenstein rearrangement; i.e. dependent on the basicity and nucleophilic properties of the amine employed. The proposed mechanism is closely related to the one of the Amadori rearrangement. Ketose **3** reacts with the amine to produce ketosylamine **90** which forms Schiff base **91.** This undergoes isomerization to 1,2-eneamine-ol **92** which subsequently produces epimeric 2-amino-2-deoxyaldose derivatives **4** and **5.**

Optimized conditions to improve yields, such as the range of temperature, pressure and the quantity of liquid ammonia as well as the influence of the addition of ammonium chloride, were investigated by Heyns and Meinecke [97]. Despite all efforts, yields usually did not exceed 35% when the basic component was limited to ammonia.

Around the same time, Erickson investigated the reaction of long-chain aliphatic amines with free sugars and discovered that ketoses were more reactive than aldoses [98a] with this type of nucleophilic component, parallelling the chemistry of the Amadori rearrangement. He also noted that the products, being α-amino aldehydes, were more reactive towards amines then the ketoses he had started from [98b]. A subsequent reaction with a second equivalent of amine resembled the osazone formation. This finding confirmed the relationships between this reaction, the osazone formation, the Amadori rearrangement and the Lobry de Bruyn – Alberda van Ekenstein reaction, as well as the Maillard reaction cascade.

Carson [99] investigated the behavior of different aliphatic amines towards D-fructose, and encountered problems of undesired disubstitution, epimerization, as well as lack of stability of the 2-amino-2-deoxyglucose derivatives during his studies. He performed the reaction in a two-step sequence by initial formation of the ketosylamine, which was isolated as a crystalline intermediate, followed by the rearrangement reaction in methanolic solutions containing glacial acetic acid. These conditions allowed access to increased yields that could be further improved employing benzylamine as the nucleophilic component.

Heyns and co-workers made significant contributions to this field that he called "Amadori like rearrangement", "retro Amadori ketosylamine rearrangement" and "fructosylamine rearrangement". The reaction was soon termed the "Carson-Heyns rearrangement" and is now known as the Heyns rearrangement.

Interesting in this context is the fact that Fischer had already included D-fructose in his studies on the osazone formation [16, 17] and, consequently, would have conducted the Heyns rearrangement quite early in the history of carbohydrate chemistry (see Sect. 1.1).

3.2 Range of Components

Heyns investigated the reactions of different amines with a range of ketoses. Whereas the reaction of D-fructose with liquid ammonia led to D-glucosamine, no such reaction took place with L-sorbose. Either the starting material remained unchanged or, at elevated temperatures, decomposition was observed. Contrasting this result, reaction of L-sorbose with *N*-benzyl-, *N*-butyl-, *N*-hexyl-, and *N*-cyclohexyl-L-sorbosylamine gave the corresponding crystalline rearrangement products in yields ranging from 20 to 30% [100].

Consequently, amino acids were investigated and the products were coined "2-*N*-amino acid-2-deoxyaldoses" [101]. In analogy to the Amadori rearrangement, the acid catalysis of the carboxyl function in the molecule is not always sufficient to induce the rearrangement, so that an additional catalyst, such as ammonium chloride, oxalic acid or glacial acid, is required in some cases. From di- and tripeptides, like gly-gly, gly-leu, gly-tyr, leu-gly, gly-gly-leu and D-fructose, Heyns and Rolle obtained the corresponding "glucose-peptides" [102]. Structural analysis of the products stemming from the reaction of D-fructose with gelatine [103] showed that in the main component the sugar was attached to the ε-amino group of lysine.

Reaction conditions such as the catalytic effect of different organic acids on the rearrangement, the influence of the basicity of the amino component, the effects of different solvent systems, as well as the distribution of epimers, were thoroughly investigated. The reversibility of the rearrangement was also probed by the Heyns group. In addition, the influence of the sugars' configurations on the outcome was compared employing D-fructose, L-sorbose, D-tagatose, as well as L-psicose [104].

Subsequent experiments showed that piperidine with D-fructose gave D-glucose, and that morpholine, dicyclohexylamine as well as tertiary amines converted D-fructose into D-psicose without detectable formation of Heyns rearrangement products. Employing secondary amines such as pyrrolidine, the corresponding Heyns rearrangement products could be obtained very easily, contrasting results with other cyclic amines, for example, piperidine, hexamethylene imine (azepane) and other open-chain amines such as dimethyl-, diethyl-, or methylbenzylamine, which did not react [105].

The study of rearrangement products from D-xylulose (**93**) with amines and amino acids (Scheme 29) was very important to rationalize the general aspects concerning the tendency of rearrangement of different ketoses [106]. D-Xylulose underwent the rearrangement easily followed by D-fructose and L-sorbose. The same sequence was found for the tendency to form furanoid tautomers. This means that the open-chain conformation, which is required for the first step, the condensation between the carbonyl group and the amine, is more readily available than in the pyranose series, because of the higher ring strain of the five-membered ring. The proportion of epimer formation at C-2 was also investigated. In the reaction with glycine, the *lyxo*-configured product **96** was formed preferentially over the corresponding xylose **95** and was also degraded more rapidly. Consequently, with increased reaction times, the proportion of the D-*xylo* compound increased.

Substitution at positions O-5 and O-6 of various ketoses confirmed the suggestion that the open-chain tautomer is the substrate of the ketosylamine formation [107]. Not unexpectedly, NMR spectrometric studies on the configurations and conformations of a range of different rearrangement products demonstrated that the main tautomers were α- and β-pyranose forms in the $^4C_1$ conformation, the configurations at C-2 depending on the starting material, but predominantly being the thermodynamically favored ones [65].

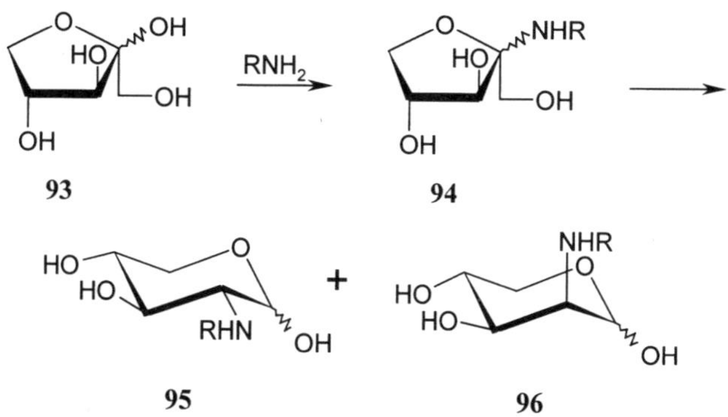

Scheme 29

3.3
Biological Aspects

Early chemistry and first recognition of the biological importance of aminosugars date back to 1876, when Ledderhose discovered D-glucosamine (**4**) in the hydrolysis mixture of crayfish shell chitin [108]. More then 25 years later, Fischer and Leuchs succeeded in the chemical synthesis of this aminosugar (Scheme 30) exploiting the Kiliani – Fischer cyanohydrin procedure [109]. They employed D-arabinose (**97**) as the starting material, which reacted with ammonia to yield D-arabinosylamine (**98**). Addition of hydrocyanic acid led to the corresponding glucononitrile **99**. Subsequent hydrolysis of the nitrile group to the corresponding amino acid **100**, followed by lactone formation to **101** and reduction of the latter, furnished D-glucosamine (**4**) in very low yields. Subsequently, Kuhn and Kirschenlohr modified this synthesis reducing the nitrile group in **99** to the corresponding aldehyde, thus opening up a direct route to D-glucosamine (**4**) from the aminonitrile [110].

Aminosugars are of extraordinary biological importance in nature. Usually, they are found in the *N*-acetylated form or as sulfates and not as free bases. The corresponding aminopolysaccharides serve as scaffold substances and intercellular adhesives. Furthermore, they play an important role in the metabolism and are connected with the pathology of many diseases. Hexosamines are constituents of blood group determinants, partial structures of antibiotics, components

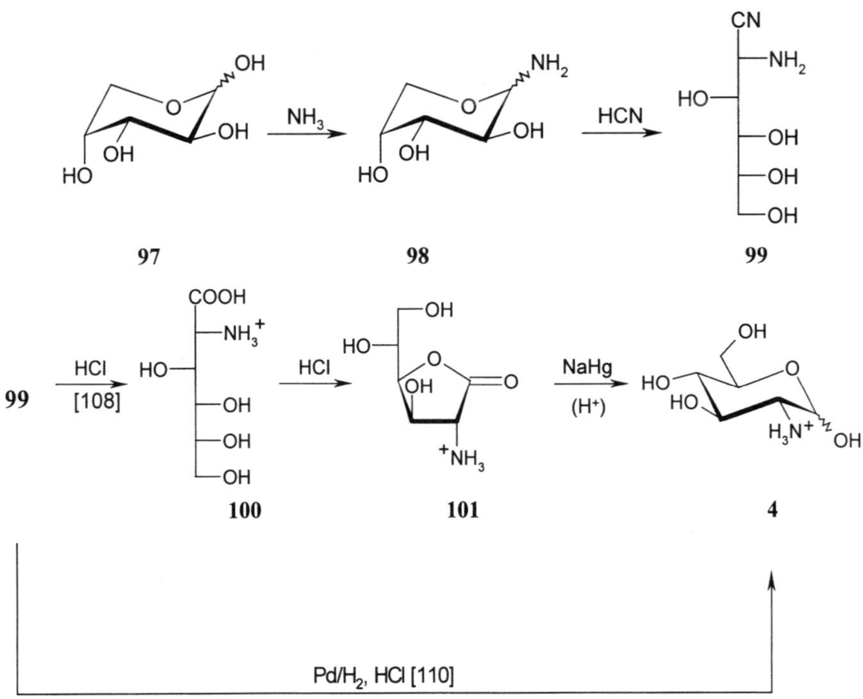

Scheme 30

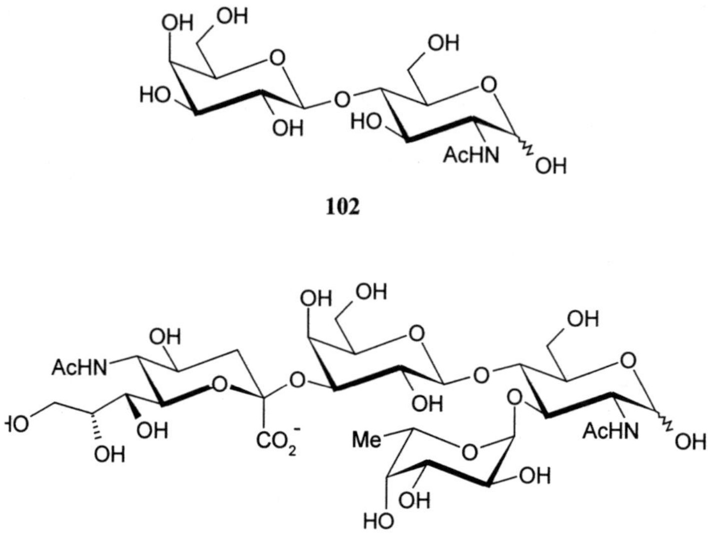

102

Scheme 31

103

of bacterial capsules as well as of immunospecific antibodies. For example, *N*-acetyllactosamine (**102**) is a very important subunit of the Lewis^x (**103**) (Scheme 31) epitope, which plays key roles in pathological processes such as cell adhesion, inflammatory reactions and metastasis. Inhibition of such intercellular interactions is considered a new strategy for the treatment of these diseases [111]. Consequently, synthetic approaches as well as the search for analogs and mimics have emerged as an important research field during the last decade.

Heyns rearrangement products are also intermediates in the non-enzymatic browning reaction of food and, similar to the Amadori rearrangement products, can enter the Maillard reaction cascade as mentioned above (Sect. 2.3). Heyns and co-workers have conducted investigations in this context [112].

Another example of biologically relevant rearrangement products is D-galactosaminuronic acid (2-amino-2-deoxy-D-galacturonic acid), which was found to be a constituent of the Vi-antigen [113]. Compounds of this type appear to be widespread amongst bacterial cell wall constituents.

3.4 Limitations of the Heyns Rearrangement

The Heyns rearrangement suffers from the same shortcomings as the Amadori reaction (Sect. 2.6). One of the main problems is the competition between the rearrangement reaction and the hydrolysis of the initially formed ketosylamine, the latter leading back to the starting material.

Depending on the reaction conditions, each step of the Heyns rearrangement was found to be reversible [104], so that in most cases a mixture of ketosylamines, the desired product and a wide range of side products is obtained. The separation of these constitutional isomers usually proves difficult.

Much attention has been devoted to the structural analysis of these side products. An example of the complexity of the mixture of possible products is the reaction of 2-amino-2-deoxy-D-glucose hydrochloride with ammonia, as conducted by Taha [114] (Scheme 32). He found five main products, which were the unreacted starting material, the cyclization products, 2-methyl-6-D-arabinosylpyrazine (**104**), 2-methyl-5-D-arabinosyl-3-D-erythrosylpyrazine (**105**), 2,5-bis-(D-arabinosyl)pyrazine (**106**), and a dark-brown amorphous substance which consisted of products of further decomposition. From this it can be concluded that careful control of the reaction conditions is very important for the formation of the products.

Another side product which was found in the reaction mixture was the corresponding Amadori rearrangement product. Different explanations for its formation have been suggested. A Lobry de Bruyn – Alberda van Ekenstein rearrangement of the starting ketose to the corresponding aldose, which, in turn, can undergo the Amadori rearrangement is possible, depending on the basicity of the amine employed. On the other hand, in case of D-fructose, the corresponding aldose, D-glucose, could not be detected in the reaction mixture with amino acids [101c], so that the formation of the Amadori side product did not occur via the Lobry de Bruyn – Alberda van Ekenstein isomerization in this case. Another reason (Scheme 33) could be that the 2-aminoaldoses such as **4** and **5** formed in the Heyns rearrangement bear an aldehyde group that is more reactive than the ketone moiety of the starting material. With excess of amine a reaction between these can occur to the corresponding glycosylamine **107** with concomitant isomerization to **108** and subsequent hydrolysis of the amine at position C-2 leading to the Amadori compound **2**.

The stability of the Heyns rearrangement products is considered a problem as they are free amino aldoses which are known to be more or less unstable compounds. Under the conditions for the Heyns rearrangement subsequent reactions along the Maillard reaction cascade can take place leading to a series

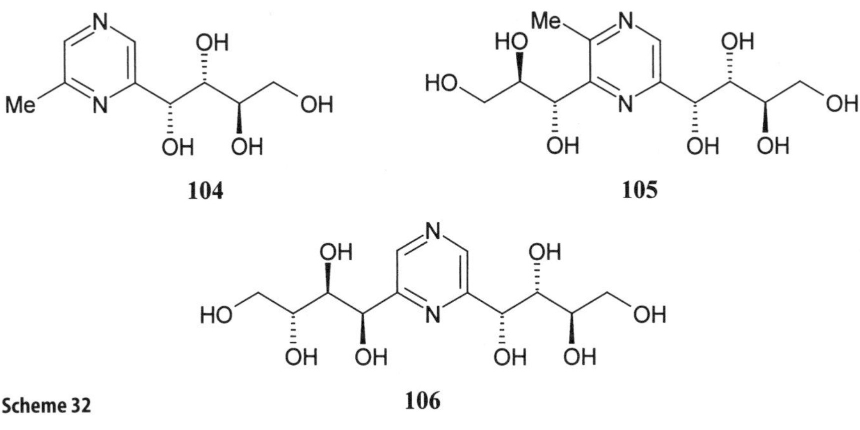

Scheme 32

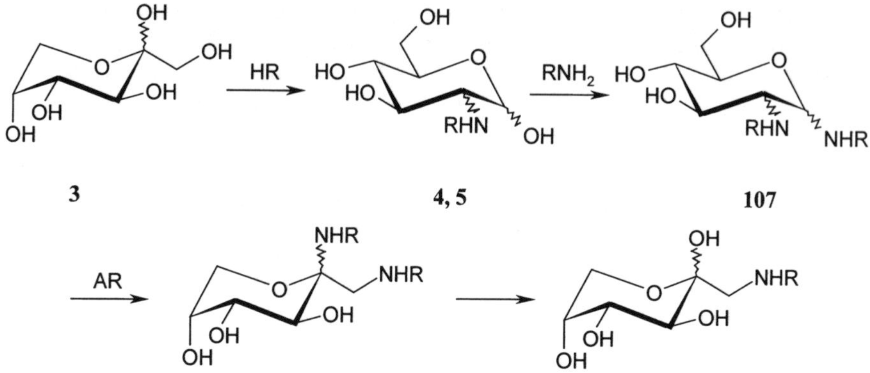

3 4, 5 107

108 2

Scheme 33

of degradation compounds in essentially the same manner as for the Amadori rearrangement.

The problem of epimer formation depends on the reaction time and reaction conditions as well as the components employed. Heyns investigated the ratio of the two possible epimers as well as the fraction of the corresponding Amadori product in the reaction of D-fructose with glycine. In this particular case, the product of kinetic control was the D-*manno*-configured 2-aminosugar, but the proportion of the more stable D-*gluco* epimer increased upon extended reaction times. The analysis of the products from the reaction of D-fructose with alanine and valine provided a similar picture [115].

Considering these facts, the Heyns rearrangement reaction is usually more difficult to perform than the Amadori rearrangement.

3.5 Applications of the Heyns Rearrangement

Because of the low yields frequently observed in the Heyns rearrangement, only a limited number of synthetic applications can be found in the literature.

Recently, various *O*-glycosylated ketoses were recognized to be suitable starting materials in the Heyns rearrangement, leading to the corresponding *O*-glycosylated derivatives of D-glucosamine in good yields (Scheme 34). Lactulose (4-*O*-galactopyranosyl-D-fructose, **51**) led via the Heyns rearrangement to *N*-acetyl-D-lactosamine (**102**), a constituent of a range of natural products, for example, the Lewis^x epitope [116]. Such disaccharides have been available either by multistep chemical syntheses [117] with the inherent need for protecting group manipulations or by enzymatic methods [118]. In this particular example, the Heyns rearrangement proceeded – presumably because of destabilizing interactions of the glycosyl moiety at O-4 with the other polar groups – considerably more efficiently than with unsubstituted D-fructose allowing for yields of 60 to 70 % of the Heyns rearrangement product.

Lactulose (**51**) was reacted with commercial grade benzylamine to give a crystalline mixture of the ketosylamine and unreacted starting material. This

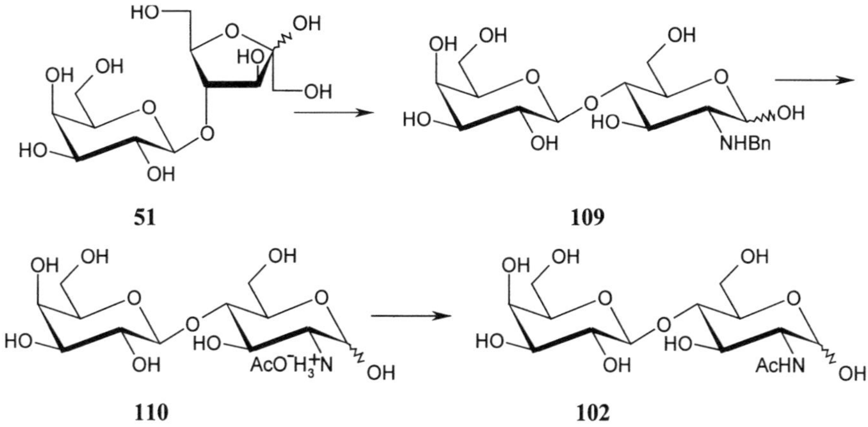

Scheme 34

mixture underwent rearrangement in methanolic solution containing 10% glacial acetic acid as the catalyst to furnish crystalline *N*-benzyllactosamine (**109**) together with small amounts of side products. The corresponding amine **110** was obtained as the acetate by catalytic hydrogenolysis of the *N*-benzyl group.

Subsequent acylation of the free amino group gave *N*-acetyllactosamine (**102**) and other conveniently protected derivatives such as the *N*-tetrachlorophthaloyl- (*N*-TCP) or *N*-(2,2,2-trichloroethyl)oxycarbonyl (*N*-Troc, former *N*-Teoc) derivatives **111** and **113** as the per-*O*-acetates or, in case of the *N*-(2,2,2-trichloroethyl)oxycarbonyl compound, the free sugar **112** is also accessible (Scheme 35).

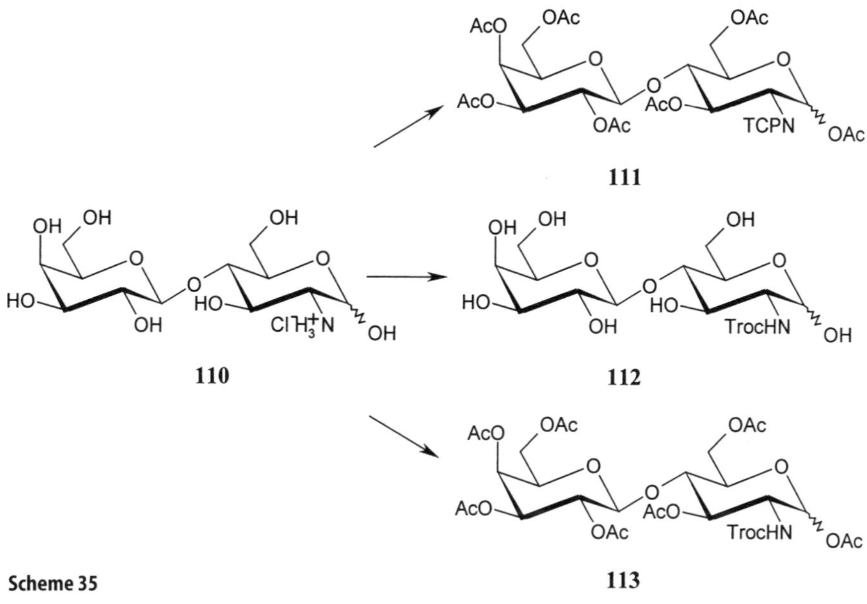

Scheme 35

The sequence can be performed in a three-step/one-pot procedure furnishing overall yields of **102** ranging between 40 and 45% and does not require any protecting group manipulations. Other *O*-glycosylated ketoses (Scheme 36), such as maltulose (**114**) and isomaltulose (**116**), reacted accordingly [119] to give the corresponding glycosylated glucosamine derivatives **115** and **117**, respectively.

In the more general context of hydroxy ketone isomerization with concomitant introduction of an amine group, the Heyns rearrangement was exploited in the area of steroid synthesis. Vicinal hydroxy ketone moieties present in such structures can react with the ε-amino group of lysine residues in vitro and in vivo in a Heyns rearrangement type of reaction to form covalent addition products [120]. One example is the synthesis of Nε-(3-hydroxy-16-oxoestra-1,3,5(10)-trien-17β-yl)-L-lysine (**120**), representing a general case of this reaction from α-hydroxy ketone **118** via the corresponding Schiff base adduct **119** (Scheme 37). The product was planned to serve studies on the relationship between estrogen and breast cancer [121].

The effect of cationic metal ions on the stable adduct formation of 16α-hydroxyesterone (**118**) with primary amines via the Heyns rearrangement has been investigated. Around 30 different metal chlorides were probed, and it was found that Pt^{4+}, Cu^{2+}, Ni^{2+}, Co^{2+} and Mn^{2+} suppressed the formation of the Heyns product significantly, whereas Fe^{2+}, Y^{3+}, Gd^{3+} and Er^{3+} caused a slight increase as compared with reactions in the absence of these cations. These results suggest that the formation of Heyns rearrangement products in vivo can be reduced by several metal ions, which were implied to be useful tools to clarify the significance of the stable adduct of **118** with proteins [122].

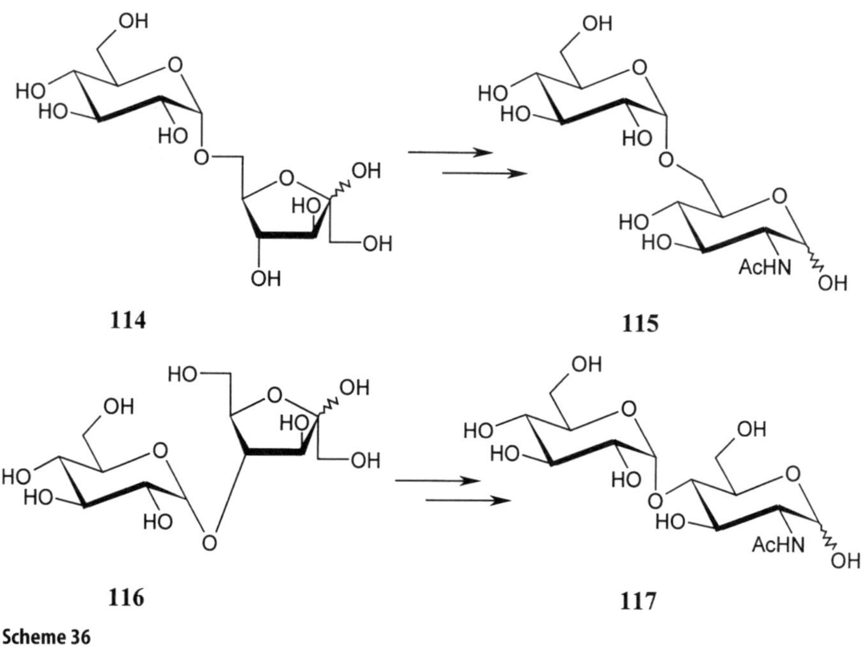

Scheme 36

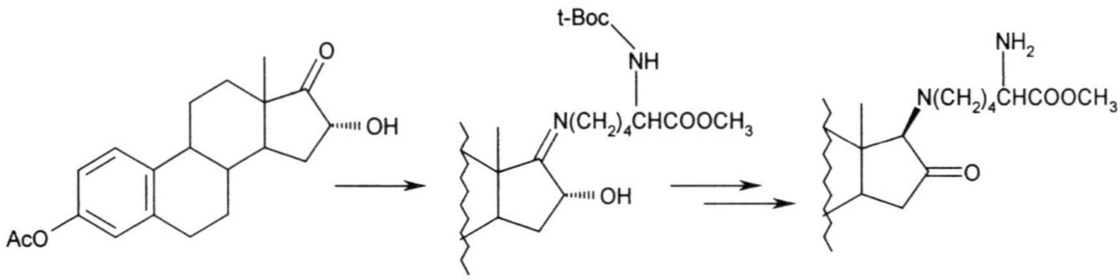

Scheme 37 **118** **119** **120**

Similar intramolecular Heyns-type reactions applied to hydroxy ketones **121** and **123** were employed by Guzi and Mcdonald [123] to synthesize piperidin-3-ones **122** and **124**, respectively (Scheme 38).

An alternative preparative access to selected Heyns rearrangement products was recently reported by reaction of various hexosamine hydrochlorides, for example, glucosamine hydrochloride (**4**), which were reacted (Scheme 39)

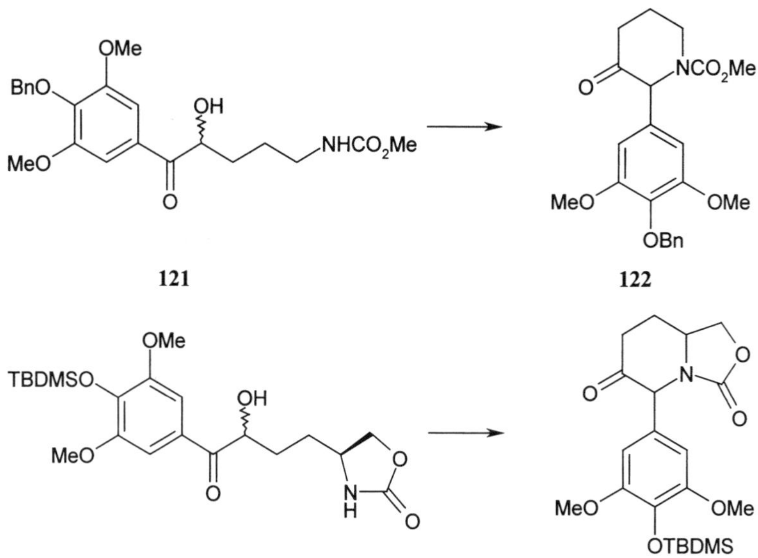

121 **122**

Scheme 38 **123** **124**

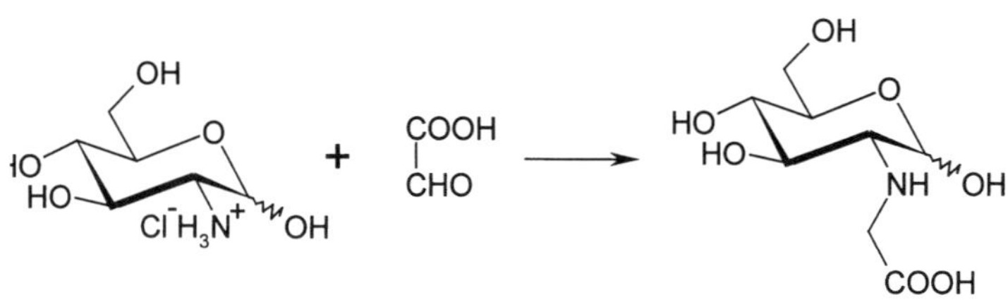

Scheme 39 **4** **125** **126**

with glyoxylic acid (**125**) in formic acid to yield the corresponding rearrangement product, such as **126**, featuring an *N*-glycinyl residue at C-2 of the sugar [124].

4 Summary and Conclusions

General aspects concerning the behavior of sugars in both the Amadori and Heyns rearrangement reactions are as follows. Aldoses give defined glycosylamines with aromatic amines that very easily undergo the Amadori rearrangement reaction. Aliphatic glycosylamines are very sensitive to hydrolysis and give the rearrangement products only under special conditions, for example in methanolic solution with oxalic acid as the catalyst. With *N*-unsubstituted glycosylamines, the rearrangement has not been observed.

Concerning their reaction behavior with amines, ketoses react in the opposite manner. The reaction between aromatic amines and ketoses gives the ketosylamine in low yields and the rearrangement products in poor yields. *N*-Alkylketosylamines, on the other hand, undergo the rearrangement to the corresponding 2-deoxy-2-aminoaldoses very easily. Ammonia yields 2-aminoaldoses. Formation of epimers at C-2 may pose a problem in terms of purification and, consequently, yields.

Despite their potential for very useful synthetic applications, both reactions appear to be highly underrated by synthetic organic chemists. A large variety of products are available by choosing suitable carbohydrates and amino compounds. Protecting group manipulations are not required offering the additional benefit of straightforward synthetic use. Exploiting additional driving forces, such as release of ring strain (by conversion of furanoid starting materials into pyranoid products) and/or destabilizing steric as well as electronic effects of crowding substituents in the starting material, novel and preparatively useful variations of the theme are clearly feasible.

Recently, both rearrangements have also been reported not to be limited to sugars, but could generally be applied to other suitable α-hydroxy aldehydes and ketones.

Acknowledgements. T. M. W. thanks the Austrian Fonds zur Förderung der wissenschaftlichen Forschung (FWF), Vienna, for a Hertha-Firnberg-Fellowship (T-18 CHE).

5 References

1. (a) Hodge JE (1955) Adv Carbohydr Chem 10:169; (b) Hodge JE, Fisher BE (1963) Methods Carbohydr Chem II 99; (c) Stanìk J, Èerny M, Kocourek J, Pacák J (1963) Amadori rearrangement. In: The monosaccharides. Publishing House of the Czechoslovak Academy of Science, Prague, p 456; (d) Verstraeten LMJ (1967) Adv Carbohydr Chem 22:229
2. (a) Heyns K, Bleifuss W (1973) Chem Ber 106:2680; (b) Stanìk J, Èerny M, Kocourek J, Pacák J (1963) Heyns rearrangement. In: The monosaccharides. Publishing House of the Czechoslovak Academy of Science, Praque, p 458

3. (a) Maillard LC (1912) C R Acad Sci Ser 2 154:66; (b) (1912) ibid 155:1554; (c) Ledl F, Schleicher E (1990) Angew Chem 102:597 and references cited therein
4. Van den Ouweland GAM, Peer HG, Tjan SB (1978) Flavor, foods beverages. In: Charalambous G, Inglett GE (eds) Chemical technology. Academic Press, New York, p 131 and references cited therein
5. Mauron J (1981) Prog Food Natur Sci 5:5
6. (a) Amadori M (1925) Atti reale Accad Nazl dei Lincei [6] 2:337; (b) (1929) ibid [6] 9:68; (c) (1929) ibid [6] 9:226; (d) (1931) ibid [6] 13:72
7. Irvine JC, Gilmour R (1909) J Chem Soc 95:1545
8. (a) Claus WH, Ree A (1898) C B 2:695; (b) Claus WH, Ree A (1898) Chem Zeitg 22: 545
9. Kuhn R, Dansi A (1936) Ber 69:1745
10. Kuhn R, Weygand F (1937) Ber 70:769
11. Schiff H (1866) Ann Chem Pharm 140:92
12. Schiff H (1870) Ann Chem Pharm 154:2
13. Sachsse R (1871) Ber 4:834
14. (a) Sorokin B (1886) Ber 19:513; (b) (1887) ibid 20:783
15. Sorokin B (1888) J f prakt Chem 37:291
16. Fischer E (1884) Ber 17:579
17. Fischer E (1886) Ber 19:1920
18. (a) Weygand F (1940) Ber 73:1284; (b) Percival EGV (1948) Adv Carbohydr Chem 3:23; (c) Palm D, Simon H (1963) Z Naturforschung 18b:419; (d) Micheel F, Dijong I (1963) Liebigs Ann 669:137
19. (a) Weygand F, Simon H, Klebe JF (1958) Chem Ber 91:1567; (b) Simon H, Keil K-D, Weygand F (1962) ibid 95:17
20. Weygand F (1940) Ber 73:1259
21. Rosen L, Woods JW, Pigman W (1957) Chem Ber 90:1038
22. Micheel F, Schleppinghoff B (1956) Chem Ber 89:1702
23. Mitts E, Hixon RM (1944) J Am Chem Soc 66:483
24. Hodge JE, Rist CE (1953) J Am Chem Soc 75:316
25. Micheel F, Hagemann G (1959) Chem Ber 92:2836
26. (a) Micheel F, Dijong I (1962) Tetrahedron Lett 1:21; (b) Micheel F, Dijong I (1962) Liebigs Ann 658:120
27. Palm D, Simon H (1965) Z Naturforschung 20b:32
28. Yoshimura J, Funabashi M, Simon H (1969) Carbohydr Res 11:276
29. (a) Gottschalk A, Partrige SM (1950) Nature 165:684; (b) Gottschalk A (1952) Biochem J 52:455 and references cited therein
30. (a) Abrams A, Lowy PH, Borsook H (1955) J Am Chem Soc 77:4794; (b) Lowy PH, Borsook H (1956) ibid 78:3175; (c) Klemer A, Micheel F (1956) Chem Ber 89:1242; (d) Micheel F, Frowein A (1959) ibid 92:304; (e) Heyns K, Paulsen H (1959) Liebigs Ann 622:160; (f) Anet EFLJ (1957) Austr J Chem 10:193
31. (a) Heyns K, Schulz W (1960) Chem Ber 93:128; (b) Heyns K, Baltes W (1960) ibid 93: 1616
32. Heyns K, Behre H, Paulsen H (1967) Carbohydr Res 5:225
33. Mossine VV, Glinsky GV, Feather MS (1994) Carbohydr Res 262:257
34. Büttner U, Ochs S, Severin T (1996) Carbohydr Res 291:175
35. Dixon HBF (1972) Biochem J 129:203
36. Kennedy L, Baynes JW (1984) Diabetologia 26:92
37. Micheel F, Heinemann KH, Schwieger KH, Frowein A (1965) Tetrahedron Lett 42:3769
38. Klemer A, Funcke W (1979) Liebigs Ann Chem 1682
39. Micheel F, Büning R (1957) Chem Ber 90:1606
40. (a) Gottschalk A, Lind PE (1949) Nature 164:232; (b) Gottschalk A (1951) ibid 167: 845
41. Lingens F, Schraven E (1962) Liebigs Ann 655:167
42. Weygand F, Simon H, Keil KD, Millauer H (1964) Chem Ber 97:1002

43. (a) Wygand F, Simon H, Dahms G, Waldschmidt M, Schliep J, Wacker H (1961) Angew Chem 73:402; (b) Simon H, Weygand F, Walter J, Wacker H, Schmidt K (1963) Z Naturforschung 18b:757
44. (a) Kennedy L, Baynes JW (1984) Diabetologia 26:92; (b) Peterson C (ed) (1981) Proceedings of a conference on nonenzymatic glycosylation and browning reactions – their relevance to diabetes mellitus. Diabetes 31 (suppl 3) p 1
45. (a) Bunn HF (1981) Am J Med 70:325; (b) Brownlee M, Vlassara H, Cerami A (1984) Ann Intern Med 101:527
46. (a) Vitek MP, Bhattacharya K, Glendening JM, Stopa E, Vlassara H, Bucala R, Manogue K, Cerami A (1994) Proc Natl Acad Sci USA 91:4766; (b) Smith MA, Taneda S, Richey PL, Miyata S, Yan S-D, Stern D, Sayre LM, Monnier VM, Perry G (1994) ibid 91:5710
47. Stevens VJ, Rouzer CA, Monnier VM, Cerami A (1978) Proc Natl Acad Sci USA 75:2918
48. Monnier VM, Cerami A (1981) Science 211:491
49. Brownlee M (1994) Diabetes 43:836
50. (a) Lo TWC, Westwood ME, McLellan AC, Selwood T, Thornalley PJ (1994) J Biol Chem 269:32299; (b) Wells-Knecht KJ, Zyzak DV, Litchfield JE, Thorpe SR, Baynes JW (1995) Biochem 34:3702; (c) Wetswood ME, Thornalley PJ (1995) J Protein Chem 14:359; (d) Büttner U, Gerum F, Severin T (1997) Carbohydr Res 300:265; (e) Glomb MA, Monnier VM (1995) J Biol Chem 270:10017; (f) Wells-Knecht KJ, Brinkmann E, Bayner JW (1995) J Org Chem 60:6246; (g) Brinkmann E, Wells-Knecht KJ, Thorpe SR, Baynes JW (1995) J Chem Soc Perkin Trans 1 2817; (h) Vasan S, Zhang X, Zhang X, Kapurniotu A, Bernhagen J, Teichberg S, Basgen J, Wagle D, Shih D, Terlecky I, Bucala R, Cerami A, Egan J, Ulrich P (1996) Nature 382:275; (i) Nissl J, Pischetsrieder M, Klein E, Severin T (1995) Carbohydr Res 270:C1; (j) Nagaraj RH, Portero-Otin M, Monnier VM (1996) Arch Biochem Biophys 325:152; (k) Kato H, Shin DB, Hayase F (1987) Agric Biol Chem 51:2009
51. (a) Lederer MO, Gerum F, Severin T (1998) Bioorg Med Chem 6:993; (b) Lederer OM, Bühler HP (1999) ibid 7:1081; (c) Lederer MO, Klaiber RG (1999) ibid 7:2499
52. (a) Röper H, Röper S, Heyns K (1982) In: Bartsch H, O'Neill IK, Castegnaro M, Okada M (eds) N-Nitroso compounds: occurrence and biological effect. International Agency for Research on Cancer, Lyon, p 87; (b) Röper H, Röper S, Meyer B (1984) In: O'Neill IK, von Borstel RC, Miller CT, Long J, Bartsch H (eds) N-Nitroso compounds: occurrence and biological effect and relevance to human cancer. International Agency for Research on Cancer, Lyon, p 101; (c) Pool BL, Röper H, Röper S, Romruen K (1984) Food Chem Toxicol 22:797; (d) Pignatelli B, Malaveille C, Friesen M, Hautefeille A, Bartsch H, Piskorska D, Descotes G (1987) In: Bartsch H, O'Neill IK, Schulte-Hermann R (eds) Relevance of N-nitroso compounds to human cancer: exposures and mechanism. International Agency for Research on Cancer, Lyon, p 277
53. (a) Lynch SC, Gruenwedel DW, Russell GF (1983) Food Chem Toxicol 21:551; (b) Gruenwedel DW, Lynch SC, Russell GF (1984) In: Friedmann M (ed) Nutritional and toxicological aspects of food safety, advances in experimental medicine and biology. Plenum Press, New York, vol 177, p 269; (c) Lynch SC, Eckert PJ, Gruenwedel DW (1987) In: Bartsch H, O'Neill IK, Schulte-Hermann R (eds) Relevance of N-nitroso compounds to human cancer: exposure and mechanism. International Agency for Research on Cancer, Lyon, p 274
54. (a) Glinsky GV (1993) Crit Rev Oncology/Hematology 14:229; (b) (1992) ibid 12:151; (c) Glinsky GV (1990) In: Sell S (ed) Serological cancer markers. Humana Press, Totowa, p 233; (d) Glinsky GV (1989) J Tumor Marker Oncology 4:193; (e) Glinsky GV, Linetsky MD, Livenstov VV, Surmilo NI, Ivanova AB, Sidorenko MV, Vinnitsky VB (1990) ibid 5:119
55. Baynes JW, Watkins NG, Fisher CI (1989) In: Baynes JW, Monnier VM (eds) The Maillard reaction in aging, diabetes and nutrition. Alan L Riss, New York, p 43
56. Hommel U, Eberhard M, Kirschner K (1995) Biochemistry 34:5429
57. Yaylayan VA, Huyghues-Despointes A (1996) Carbohydr Res 286:179
58. (a) Lederer MO, Dreisbusch CM, Bundschuh RM (1997) Carbohydr Res 301:111; (b) Lederer MO, Baumann M (2000) Bioorg Med Chem 8:115

59. Utzmann CM, Lederer MO (2000) Carbohydr Res 325:157
60. Azema L, Bringaud F, Blonski C, Périé J (2000) Bioorg Med Chem 8:717 and references cited therein
61. (a) Ohgaki H, Kusama K, Matsukura N, Marino K, Hasegawa H, Sato S, Takayama S, Sugimura T (1984) Carcinogenesis (London) 5:921; (b) Yamaizumi Z, Shiomi T, Kasai H, Nishimura S, Takahashi Y, Nagao M, Sugimura T (1980) Cancer Lett 9:75
62. Mihara S, Shibamoto T (1980) J Agric Food Chem 28:62
63. Kuhn R, Krüger G (1958) Ann Chem 618:82
64. Funcke W (1978) Liebigs Ann Chem 2099 and references cited therein
65. Altena JH, van den Ouweland GAM, Teunis CJ, Tjan SB (1981) Carbohydr Res 92:37
66. Röper H, Röper S, Heyns K (1983) Carbohydr Res 116:183
67. Neglia CI, Cohen HJ, Garber AR, Thorpe SR, Baynes JW (1985) J Biol Chem 260:5406
68. Kuhn R, Krüger G, Seelinger A (1959) Ann Chem 628:240
69. (a) Montgomery EM, Hudson CS (1930) J Am Chem Soc 52:2101; (b) Bots JPL (1957) Recueil Trav Chim Pays-Bas 76:515
70. Grünnagel R, Haas HJ (1969) Liebigs Ann Chem 721:234
71. (a) Paulsen H (1965) Liebigs Ann 683:187; (b) Paulsen H, Todt K (1968) Adv Carbohydr Chem 23:115
72. Stütz AE (ed) (1999) Iminosugars as glycosidase inhibitors. Wiley-VCH, Weinheim
73. Horvat Š, Roščić M, Varga-Defterdarović L, Horvat J (1998) J Chem Soc Perkin Trans 1 909
74. Vértesy L, Fehlhaber H-W, Kogler H, Schindler PW (1996) Liebigs Ann 121
75. Kosterlitz HW, Smith TW, Hughes J, Forthergill LA, Morgan BA, Morris HR (1975) Nature 258:577
76. Legler G, Korth A, Berger A, Ekhart C, Gradnig G, Stütz AE (1993) Carbohydr Res 250:67
77. Berger A, de Raadt A, Gradnig G, Grasser M, Löw H, Stütz AE (1992) Tetrahedron Lett 33:7125
78. (a) Wrodnigg TM, Stütz AE, Withers SG (1997) Tetrahedron Lett 38:5463; (b) Wrodnigg TM, Gaderbauer W, Greimel P, Häusler H, Sprenger KF, Stütz AE, Virgona C, Withers SG (2000) J Carbohydr Chem 19:975
79. Kadokaw J-i, Hino D, Karasu M, Tagaya H, Chiba K (1998) Chem Lett 383
80. Kort MJ (1970) Adv Carbohydr Chem Biochem 25:311
81. Mills FD, Hodge E (1976) Carbohydr Res 51:9
82. Olsson K, Pernemalm P-A, Popoff T, Theander O (1977) Acta Chem Scand Ser B 31: 469
83. Olsson K, Pernemalm P-A, Theander O (1978) Acta Chem Scand Ser B 32:249
84. Xenakis D, Moll N, Gross B (1983) Synthesis 541
85. Lopez MG, Gruenwedel DW (1991) Carbohydr Res 212:37
86. Noomen SN, Breel GJ, Winkel C (1995) Recl Trav Chim Pays-Bas 114:321
87. Walton DJ, McPherson JD (1987) Carbohydr Res 167:123
88. Kojić-Prodić B, Milinković V, Kidriè J, Pristovšek P, Horvat Š, Jakas A (1995) Carbohydr Res 279:21
89. Iwamoto T, Kan T, Katsumura S, Ohfune Y (1996) Synlett 169
90. Turner JJ, Wilschut N, Overkleeft HS, Klaffke W, van der Marel GA, van Boom JH (1999) Tetrahedron Lett 40:7039
91. Tauss A (1997) Diploma thesis, Technical University of Graz, Austria
92. de Bruyn CAL (1898) Ber 31:2476
93. Breuer R (1898) Ber 31:2193
94. Kuhn R, Birkofer L (1938) Ber 71:621
95. Barry CP, Honeymoon J (1952) J Chem Soc 97:4147
96. Heyns K, Koch W (1952) Z Naturforschung 7b:486
97. Heyns K, Meinecke K-H (1953) Chem Ber 86:1453
98. (a) Erickson JG (1953) J Am Chem Soc 75:2784; (b) (1955) ibid 77:2839
99. (a) Carson JF (1955) J Am Chem Soc 77:1881; (b) (1955) ibid 77:5957; (c) (1956) ibid 78: 3728
100. Heyns K, Eichstedt R, Meinecke K-H (1955) Ber 88:1551

101. (a) Heyns K, Paulsen H, Breuer H (1956) Angew Chem 68:334; (b) (1957) Chem Ber 90:1374; (c) (1958) ibid 91:2750; (d) Heyns K, Noack H (1964) Chem Ber 97:415
102. Heyns K, Rolle M (1959) Chem Ber 92:2439
103. Heyns K, Rolle M (1959) Chem Ber 92:2451
104. Heyns K, Paulsen H, Eichstedt R, Rolle M (1957) Chem Ber 90:2039
105. Heyns K, Pflughaupt K-W, Müller D (1968) Chem Ber 101:2807
106. (a) Heyns K, Pflughaupt K-W, Paulsen H (1968) Chem Ber 101:2800; (b) Heyns K, Bleifuß W (1970) ibid 103:2873
107. Heyns K, Reinhold R (1981) Liebigs Ann Chem 138
108. Baer HH (1958) Fortschr chem Forsch 3:822 and references cited therein
109. Fischer E, Leuchs H (1903) Ber 36:24
110. Kuhn R, Kirschenlohr W (1955) Angew Chem 67:786
111. Simanek EE, McGarvey GJ, Jablonowski JA, Wong C-H (1998) Chem Rev 98:833 and references cited therein
112. Heyns K, Noack H (1962) Chem Ber 95:720
113. Heyns K, Kiessling G, Lindenberg W, Paulsen H, Webster ME (1959) Chem Ber 92:2435
114. Taha MI (1961) J Chem Soc 2468
115. Heyns K, Müller G, Paulsen H (1967) Liebigs Ann Chem 703:202
116. Fukuda M (1995) Bioorg Med Chem 3:207 and references cited therein
117. (a) Paulsen H (1990) Angew Chem 102:851; (b) Toshima K, Tatsuat K (1993) Chem Rev 93:1503; (c) Schmidt RR, Kinzy W (1994) Adv Carbohydr Chem Biochem 50:21; (d) Garegg PJ (1997) ibid 52:179; (e) Deshpande PP, Kim HM, Zatorski A, Park T-K, Ragupathi G, Livingston PO, Live D, Danishefsky SJ (1998) J Am Chem Soc 120:1600
118. (a) Wong C-H, Halcomb RL, Ichikawa Y, Kajimoto T (1995) Angew Chem 107:569 and references cited therein; (b) Halcomb RL (1995) In: Drauz K, Waldmann H (eds) Enzyme catalysis in organic synthesis. VCH, Weinheim, p 279
119. Wrodnigg TM, Stütz AE (1999) Angew Chem 111:854
120. Bucala R, Ulrich PC, Chait BT, Bencsath FA, Cerami A (1986) J Steroid Biochem 25:127 and references cited therein
121. Miyairi S, Ichikawa T, Nambara T (1991) Tetrahedron Lett 32:1213
122. Miyairi S, Maeda K, Oe T, Kato T, Naganuma A (1999) Steroids 64:252
123. Guzi TJ, Macdonald TL (1996) Tetrahedron Lett 37:2939
124. Mossine VV, Barnes CL, Glinsky GV, Feather MS (1996) Carbohydr Res 284:11

Substitution-with-Allylic-Rearrangement Reactions of Glycal Derivatives

Robert J. Ferrier

Industrial Research, PO Box 31-310, Lower Hutt, New Zealand
E-mail: r.ferrier@irl.cri.nz

Glycals (or usually their O-substituted derivatives) are readily converted into 2,3-unsaturated glycosyl compounds with O-, C-, N-, S- or otherwise linked substituents at the anomeric position. These products have been found to be useful for a range of synthetic purposes. In particular, the C-glycosidic compounds have served as readily available starting materials for the preparation of useful non-carbohydrate compounds. While these allylic rearrangement processes are usually conducted under the influence of Lewis acid catalysts, adaptations that involve activation of the allylic substituents of the starting glycals as leaving groups under neutral conditions have been developed. General features of the reactions are described as well as applications in synthesis and extensions of the basic processes.

Keywords: Carbohydrates, Glycals, *O*-Glycosides, *C*-Glycosides, *N*-Glycosides, *S*-Glycosides, Nucleosides, Allylic rearrangements

1
Introduction

Tri-*O*-acetal-D-glucal (**1**) is the best known and most useful member of the family of cyclic 1,2-unsaturated aldose derivatives called "glycals". These compounds are of major significance in synthesis – to a large extent because of the wide range of biologically important compounds that can be obtained from them by addition processes [1]. Of special significance are the additions that lead to glycosides of different kinds – notably 2-amino-2-deoxy- and 2-deoxy-glycosides. On the other hand, the allylic ester feature within the structure of compound **1** makes it susceptible to displacement processes that can also be used to yield modified products – again, notably, a wide range of glycosidically bonded compounds.

The subject of the present review is the displacement reaction of glycal derivatives like **1** that gives rise to 2,3-unsaturated glycosyl products, e.g. **2** (X = OR, SR, NR^1R^2, $CR^1R^2R^3$, Scheme 1). This reaction can also be used for synthetic purposes, the *O*- and *C*-glycosides being of particular significance. The general features of the reactions exemplified by **1** → **2** will be discussed as well as specific examples, uses and modifications. A more extensive survey of this topic has been prepared [2].

First evidence of the allylic rearrangement reaction was provided by Emil Fischer in the land-mark 1914 paper which described the preparation of compound **1** by the reaction of tetra-*O*-acetyl-α-D-glucopyranosyl bromide with zinc and acetic acid [3]. Heating with water caused it to dissolve and apparently to lose one of the acetyl groups by hydrolysis, the main product later being shown to be the allylically rearranged hemiacetal **2** (X = OH) [4]. The possible presence of compound **2** (X = OH) and its *E*-enal isomer (Sect. 4) in his samples of tri-*O*-acetyl-D-glucal is deemed to be the cause of Fischer erroneously applying the "al" suffix in the name of the glycals [5].

Following the early recognition of the allylic rearrangement reaction undergone by compound **1** with hot water, the synthetic potential of this type of reaction was not recognized until the early 1960s when, within a short period of time, *p*-nitrophenol [6], thioacetic acid [7] and theophylline [8] were all shown to take part to give the *O*-, *S*- and *N*-linked unsaturated glycosyl compounds **2** (X = O-C_6H_4-NO_2-(*p*), SAc and theophyllyl, respectively). Since then the general reaction has been developed appreciably as a procedure for making synthetically useful derivatives of 2,3-unsaturated sugars, the relevant *C*-linked compounds being particularly useful in the synthesis of non-carbohydrate natural products. General features of the reactions, ways of promoting them and more recent applications will now be described.

2 General Features

Enol ethers with potential leaving groups at the allylic position (e.g. compounds of general structure **3**) can be subject to acid-catalyzed additions or nucleophilic displacements, protonic acids commonly giving products (**5**) of the former reactions while Lewis acids favour the formation of unsaturated products (**7**, **8**)

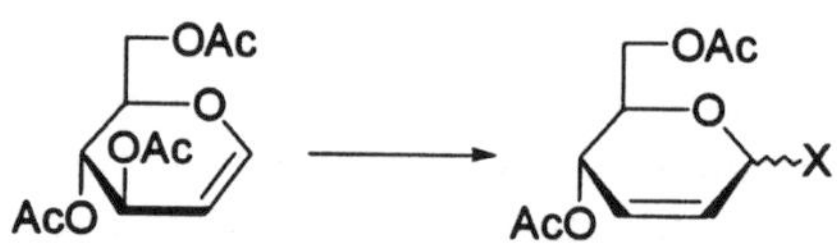

Scheme 1 **1** **2**

of the latter. Ionic intermediates **4** and **6** are involved in the respective processes, as indicated in Scheme 2.

Furthermore, however, some reactions which yield unsaturated compounds proceed by addition/elimination processes – for example, palladium-promoted reactions (see later in this section). With glycal derivatives the great majority of substitution reactions give allylically rearranged products of type **7**, but some afford isomers of type **8** seemingly formed by direct substitution. In some cases, as for example with azides, ambident nucleophiles attack at one centre under kinetic control to give products that may isomerize either by reversal processes or by sigmatropic rearrangement to products with the alternative allylic structures. Almost always, reactions with *O*- and *C*- nucleophiles result in 2,3-unsaturated *O*- and *C*-glycosides of type **7** (Schemes 4 and 5, respectively), but 3-*N*- and 3-*S*-substituted glycals (type **8**) are encountered when *N*- and *S*-nucleophilic compounds are involved (Sect. 6).

For the preparation of unsaturated substitution products, most commonly *O*-acylated glycals (some of which are commercially available) are used as starting materials. With these, the allylic acyloxy groups have sufficiently good leaving properties to allow reactions with nucleophiles such as alcohols or phenols to occur without the need for added catalyst – provided elevated temperatures are used. Thus, for example, reaction of tri-*O*-acetyl-D-glucal (**1**) with ethanol [9] and phenol [10] at 180 and 140 °C, respectively, gives compounds **2** (X = OEt and X = OPh) in good yield. Reaction is, as expected, facilitated by the use of glycals with allylic groups having better leaving properties than do simple carboxylates, and bis-*O*-(*p*-nitrobenzoyl)-D-xylal and 4,6-*O*-benzylidene-3-*O*-methanesulfonyl-D-allal, treated respectively with purine or pyrimidine bases [11] and a Grignard reagent [12], lead to the corresponding *N*- and *C*- linked 2,3-unsaturated glycosidic products without the need for catalysts. These cases are however exceptional, normally acid promoters are employed and, while there are examples of the use of protonic catalysts successfully employed together with heterocyclic bases [13], these are normally avoided since, as previously noted, they preferentially favour addition processes leading to 2-deoxyglycosyl products with general structure **5**.

Lewis acids are very effective catalysts for promoting reactions of the type **1** → **2** with $BF_3.Et_2O$ being most frequently used; however, it is ineffective in cases involving the use of glycal esters with *cis*-related ester groups at C-3 and C-4, and $SnCl_4$ is commonly used instead (see below). Many others have been employed, with iodine having the advantages that it is both relatively mild and can be used with some nucleophiles that are decomposed by common Lewis acid catalysts [14]. Trimethylsilyl triflate is often used to catalyze the production of

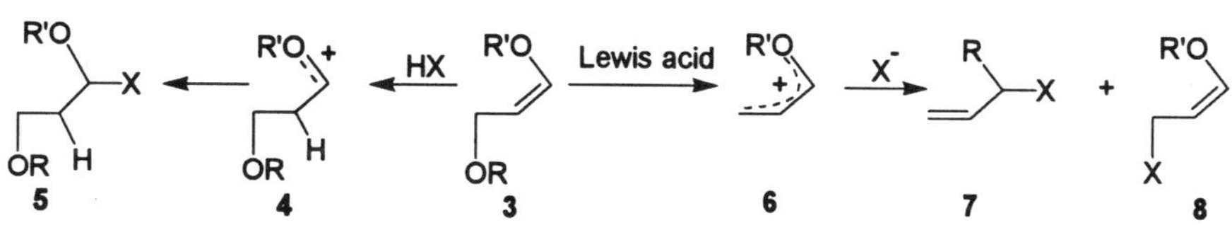

Scheme 2

2,3-unsaturated *C*-glycosides and can promote reaction, even of *O*-unsubstituted glycals, at very low temperatures [15].

Important developments have seen the introduction of means of activating the allylic groups of the glycals without resort to acid catalysts. Thus, C-3 hydroxyl groups can be activated by the Mitsunobu procedure to yield phenyl *O*-glycosides (e.g. Scheme 3) [16].

A similar approach involves the use of glycal derivatives, e.g. **9** (R = Me_2C), with pent-4-enyloxy or pent-4-enoate functions as allylic groups which are activated to leave by use of electrophilic species such as iodonium dicollidine perchlorate (Scheme 4). This approach led to the successful synthesis of the disaccharide **11** (R = Me_2C, α-anomer, 65%) when the orthodox method involving Lewis acids failed [17]. In a related manner, the 3-(phenylthio)-D-allal derivative **10**, activated with the thiophilic *N*-iodosuccinimide, gave disaccharide **11** (R = Ac, α-anomer) in 70% yield [17].

An alternative procedure relies on DDQ to act as a Lewis acid under neutral conditions to catalyze the departure of allylic hydroxy [18] or acyloxy [19] groups, and give *O*-[18, 19] or *C*-[20] 2,3-unsaturated glycosides.

Glycals with unsubstituted hydroxyl groups at the allylic centres can take part in acid-catalyzed allylic rearrangements; D-glucal (**13**) itself giving methyl 4,6-*O*-benzylidene-2,3-dideoxy-α-D-*erythro*-hex-2-enopyranoside (**12**) in high yield on treatment with benzaldehyde dimethylacetal in the presence of *p*-toluenesulfonic acid (Scheme 5) [21]. Under these conditions, proton-catalyzed addition is therefore unfavoured relative to the rearrangement reaction. Very unexpected, but usefully, unprotected glycals give excellent yields of allyl *C*-glycosides almost stereospecifically when treated at very low temperatures with allylsilanes in dichloromethane/acetonitrile with trimethylsilyl triflate as catalyst [15], D-glucal (**13**) now affording 99% of the allyl 2,3-unsaturated α-compound (**14**) under these conditions (Scheme 5).

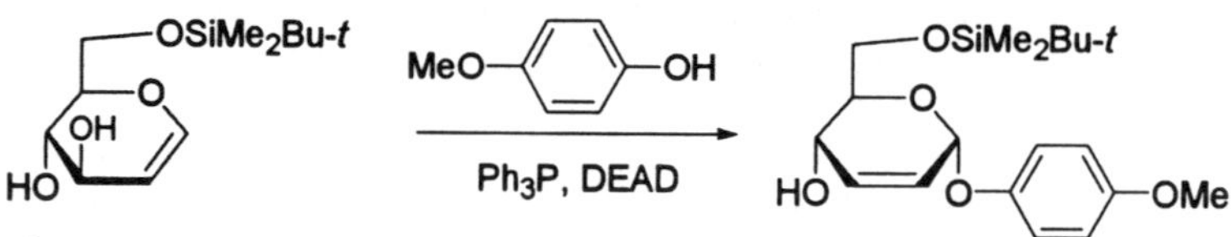

Scheme 3

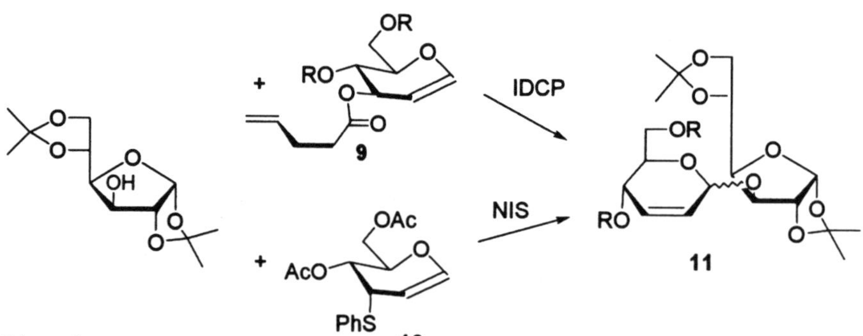

Scheme 4

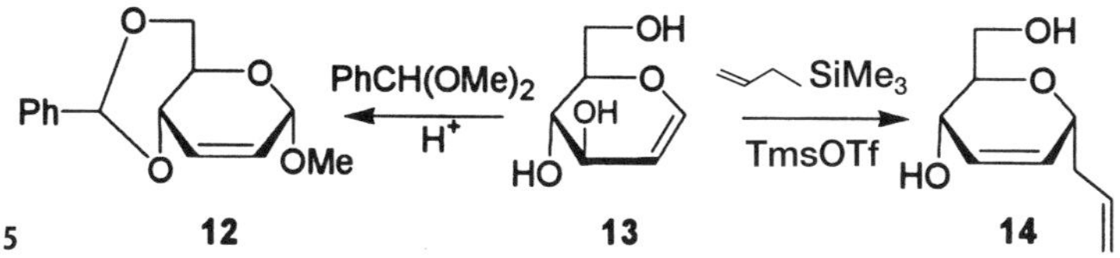

Scheme 5 **12** **13** **14**

Glycal derivatives with ether allylic groups can be induced to undergo the rearrangement reaction but yields can be poor [22]. Nevertheless, tri-*O*-methyl-D-glucal reacts with 2-[(trimethylsilyl)thio]pyridine in benzene in the presence of boron trifluoride etherate to give the relevant 2,3-unsaturated *S*-pyridyl thioglycosides in good yield [23], and tri-*O*-benzyl-D-glucal (**16**) in ether with the same catalyst isomerizes in unexpected fashion to afford benzyl 4,6-di-*O*-benzyl-2,3-dideoxy-D-*erythro*-hex-2-enopyranoside (**15**) (Scheme 6) [24]. While this reaction seems related to the isomerization undergone by tri-*O*-acetyl-D-glucal (Sect. 4), the means of the intramolecular migration of the C-3 acetoxy group in the latter compound is much easier to comprehend since a [3.3]-sigmatropic process is possible. Whether **15** is an intermediate in the formation in 45% yield from tri-*O*-benzyl-D-glucal (**16**) of the more unusual further isomer **17** on treatment with acetyl perchlorate (Scheme 6) is uncertain, a more probable intermediate being the derived delocalized carbenium ion [25].

An alternative means of making 2,3-unsaturated *O*- [26] and especially *C*-[27] glycosides from glycal derivatives depends on the use of palladium mediation. It is mechanistically quite different from the reactions otherwise referred to in this survey, but should be recognized as a complementary means of carrying out the glycal to 2,3-unsaturated glycoside conversion. In outline the palladium-dependent reactions involve additions to give intermediates containing the glycal, the palladium species and the nucleophile, which then often undergo elimination of a palladium-containing moiety together with the oxygen-bonded substituent at C-3, with the nucleophile bonding to C-1. In the case of the *C*-glycosides, various factors (some extremely subtle) operating during the formation and break down of the intermediate adducts determine whether α- or β-glycosides result, and also whether H-3 replaces the C-3 oxygenated substituent as the eliminating partner of the palladium species. If so, the products retain their *O*-substituent on their double bonds. For example, when 3,4-di-*O*-acetylglycals with H, Me or CN groups bonded at C-5 are treated under argon with alcohols and palladium chloride, alkoxypalladation occurs and the adducts, on reduction with sodium cyanoborohydride, undergo elimination to give 2,3-*O*-unsaturated glycosides with α configuration [26]. When, however, the glycals have oxygen-carrying carbon groups at C-5 the final products are 3,4-unsaturated glycosides with acetoxy

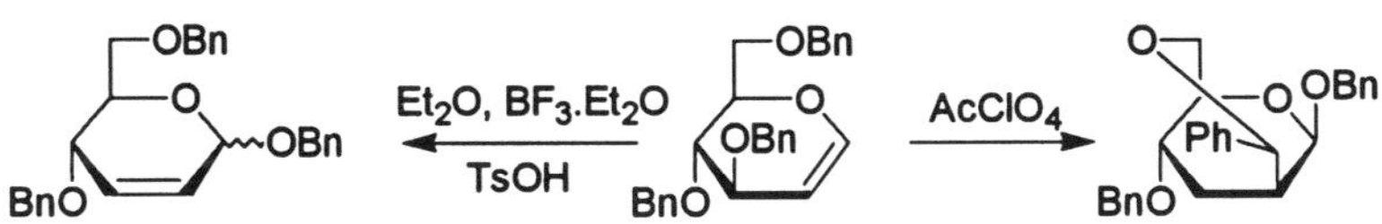

15 **16** **17**

Scheme 6

groups at C-2, which illustrates the propensity for the palladium-containing intermediates to undergo rearrangement processes prior to dissociations. A further restriction on this approach to *O*-glycosides is the need for excess of the alcohols, but means of overcoming this limitation have been demonstrated [28], and this way of making 2,3-unsaturated glycosides of simple alcohols has on occasion been preferred to the use of Lewis acid catalysts [29].

To make *C*-glycosides, organopalladium(II) reagents produced from the corresponding organomercurials or -stannanes are used or iodoaromatics can be employed with catalytic proportions of palladium salts. Compound **19**, made by treatment of the adduct **18** with triphenylphosphine, gives **20** (R = H) in near quantitative yield on reaction with aqueous sodium bicarbonate, but when it is simply heated briefly in toluene a different elimination occurs to give the acetoxy compound **20** (R = OAc, Scheme 7) [30].

In the furanoid glycal series the *O*-substitution patterns of the starting materials can determine the nature of the products obtained: when the C-3 and C-5 hydroxyl groups of compound **21** ($R^1 = R^2 = H$) are substituted, β-nucleoside derivatives **22** with a retained substituent at C-3 predominate; removal of the O-3 substituent of the starting material results in α-compound **23** with no substituent at C-3 (Scheme 8) [27a].

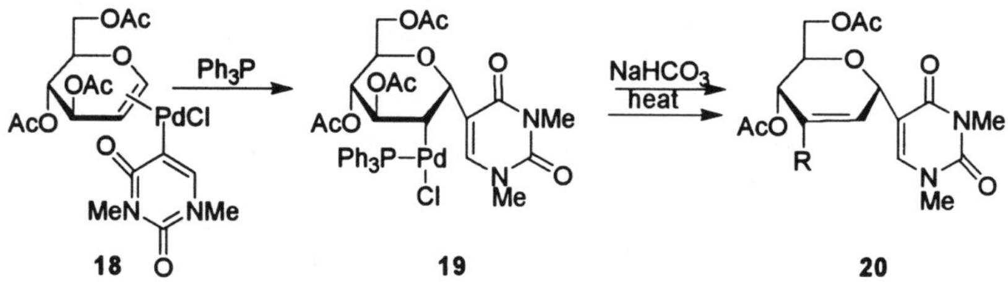

Scheme 7

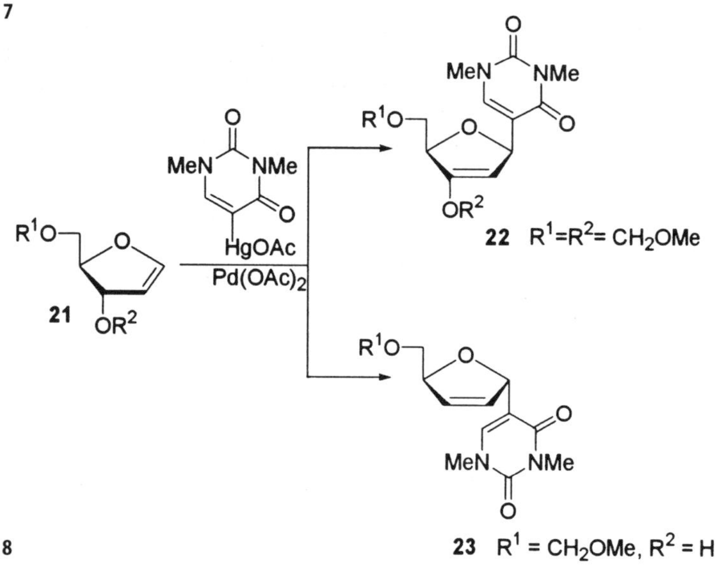

Scheme 8

Likewise, in the production of 2,3-unsaturated aryl *C*-glycosides, mixed products with H or OAc at C-3 are obtained on treatment of tri-*O*-acetyl-D-glucal with aromatic hydrocarbons in acetic anhydride containing palladium acetate [31]. When benzene itself is used the latter type predominates (54%), and the former is formed in small proportions (10%), both compounds having the α configuration.

Although it has been stressed that the palladium-mediated method of making 2,3-unsaturated glycosides from glycals differs fundamentally from the Lewis acid catalyzed procedure, the results of comparative experiments involving coupling of tri-*O*-acetyl-D-glucal with pentane-2,4-dione separately with $Pd(PhCN)_2Cl_2$ and $BF_3.Et_2O$ as catalysts show surprising similarity. The 3′-linked 2,3-unsaturated *C*-glycosides are produced in 83 and 73% yield, respectively, with an α/β ratio of 5:1 in both cases. With tri-*O*-acetyl-D-galactal the figures are 59 and 72%, with the α-anomer being formed exclusively [28].

3
Stereochemical Features

Stereochemical factors pertaining to the starting glycal derivatives and to the products are of appreciable relevance in the acid-catalyzed allylic rearrangement reaction. Epimers of glycals at the allylic centres can react similarly or very differently. Thus the D-allal derivative **10**, activated under neutral conditions (see above), gives the same 2,3-unsaturated glycoside products as does the D-glucal C-3 epimer, but with a quasi-axial leaving group, the rate of reaction of the former is enhanced [32] in accordance with the influence of Curran's "vinylogous anomeric effect" [33]. In contrast, however, in some circumstances, several D-allal derivatives react to give non-specific products in the presence of Lewis acids, conceivably because *trans*-diaxial elimination at C-3–C-4 to give unstable pyrans is possible for these compounds, whereas the D-glucal epimers are not vulnerable in this way [34].

A further stereochemical factor that can be of major significance in determining the effectiveness of the allylic rearrangement reaction is the relative orientation of ester groups at C-3 and C-4 of acylated glycals. For example, tri-*O*-acetyl-D-galactal (**24**), the C-4 epimer of compound **1**, does not react satisfactorily with alcohols in the presence of boron trifluoride as catalyst, whereas the latter glycal ester gives 2,3-unsaturated *O*-glycosides very efficiently apparently as a consequence of the cleavage of the C-3–O bond being assisted by the *trans*-related ester group [35]. Similarly, while tri-*O*-acetyl-D-glucal reacts with hot acetic acid to give the expected tri-*O*-acetyl-2,3-dideoxy-D-*erythro*-hex-2-enoses (and some tri-*O*-acetyl-D-allal, Sect. 4), tri-*O*-acetyl-D-galactal undergoes addition and elimination reactions as well as these isomerizations under these conditions [36].

Tin tetrachloride appears to be more effective than boron trifluoride in coordinating with the allylic ester groups to enhance their leaving properties. It catalyzes the rearrangement reaction of tri-*O*-acetyl-D-glucal with alcohols more efficiently [37], and it allows the analogous reaction of tri-*O*-acetyl-D-galactal such that the α-D-*threo*-2,3-unsaturated glycosides are obtainable in high yield [35].

A generalization regarding the stereochemistry at the anomeric centre of the products formed during the allylic rearrangements of glycals is that stereoselectivity is usually pronounced, and occasionally almost complete. Thus, for example, normal Lewis acid catalyzed reactions of tri-*O*-acetyl-D-glucal with alcohols lead to the 2,3-unsaturated *O*-glycosidic products with α/β ratios in the range $7 \pm 2{:}1$ which, since it can be accepted that the anomers are interconvertible under the conditions of their preparation, are equilibrium values. With the products derived from tri-*O*-acetyl-D-galactal the ratio is appreciably higher [35], which suggests that there is a stereoelectronic factor favouring the dihydro-[2*H*]-pyran structure with the electronegative allylic groups both *quasi*-axial.

Mixed anomers of 2,3-unsaturated compounds formed by use of *N*- or *S*-nucleophilic species normally also contain α-isomers predominantly, and the same can be said of the products derived by use of *C*-nucleophiles. This last group has however to be considered specially because, having ether rather than acetal character at the anomeric centre, they are less prone to acid-catalyzed anomerization than are, for example, the *O*-glycosidic analogues and the products are normally those of kinetically controlled processes. Reaction of tri-*O*-acetyl-D-glucal with allyltrimethylsilane in dichloromethane at −78 °C with titanium tetrachloride as catalyst gives the *C*-allyl products in 85% yield and with an α/β ratio of 16:1 (Scheme 9) [38]. Again this ratio is higher (30:1) in the case of the analogous glycosides derived from tri-*O*-acetyl-D-galactal. In related work, also conducted at −78 °C, high yields and ratios as high as >99:1 have been recorded by use, not just of *O*-acetylated glycals but, remarkably, of *O*-unprotected compounds themselves [15].

Alkenes which can undergo additions of electrophilic ions to give tertiary carbocations also react with the delocalized ions derivable from glycals and yield α-compounds with high selectivity (e.g. Scheme 10) [39].

The 2,3-unsaturated glycosyl cyanides obtainable from glycal derivatives are of appreciable interest as synthetic intermediates and have been produced by use of trimethylsilyl cyanide in nitromethane with boron trifluoride etherate as catalyst [40]. While under these conditions the anomeric cyanides are obtained from tri-*O*-acetyl-D-glucal in an α/β ratio of 1.4:1, the product formed with dichloromethane as solvent contains only the α-isomer [41]. Similar results are obtained by use of the Lewis acid diethylaluminium cyanide which, in benzene at room temperature, affords the 2,3-unsaturated glycosyl cyanides from tri-*O*-acetyl-D-

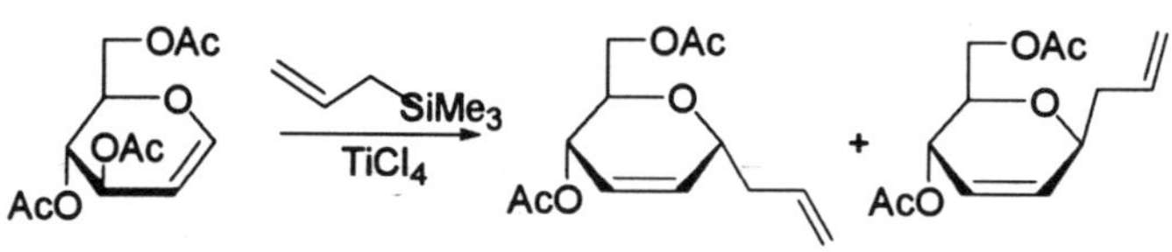

Scheme 9

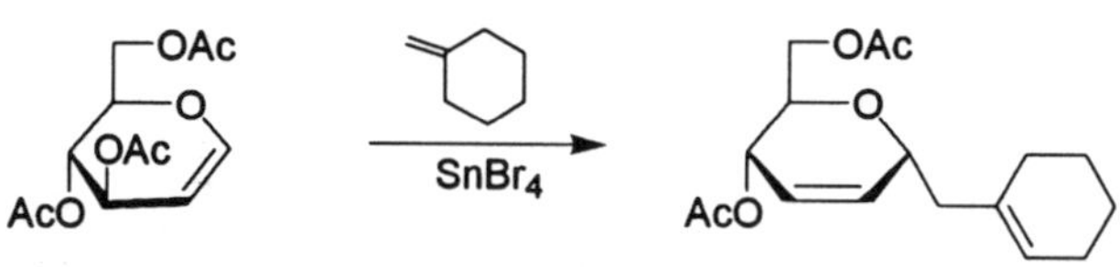

Scheme 10

glucal in the ratio 3:2, this becoming 9:1 when the reaction is conducted in refluxing solvent, which corroborates earlier evidence that the β-isomer (**25**) can anomerize [42]. These results suggest that α-cyanides are kinetically somewhat favoured and that they are strongly thermodynamically preferred. The same does not apply to all such *C*-glycosides, however: compounds **26** equilibrate under basic conditions to give mainly the β-D-*erythro*-*C*-hexosides, presumably via derived stabilized carbanions which can ring open to allow anomerization [43].

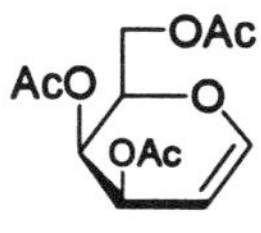

24

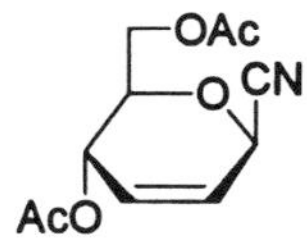

25

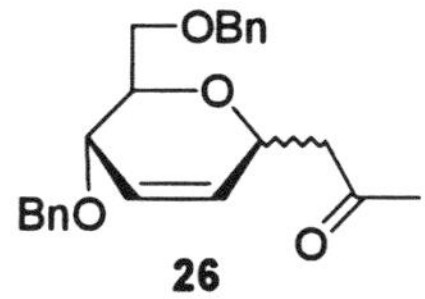

26

Likewise, activated aromatic compounds give mainly β-linked aryl *C*-glycosides from tri-*O*-acetyl-D-glucal (**1**) in the presence of Lewis acids, presumably because the bulky aryl groups can attain the favoured ψ-equatorial orientation by isomerization via acyclic C-1 carbenium ions which are stabilized by both the allylic and aryl π-systems [44].

4 2,3-Unsaturated *O*-Glycosides and Other 1-*O*-Linked Products

Recent studies have revealed that the seminal reaction whereby tri-*O*-acetyl-D-glucal is converted to the rearranged free sugar **2** (X = OH) by heating in water (Scheme 1) [3, 4] is part of a complex process with light causing the formation of the acyclic *E*-isomer **28** of the aldehydic isomer **27** of the first product (Scheme 11) [5]. Exclusion of light inhibits its formation, but continuation of the reaction without such exclusion results in the *E*-enal becoming the main reaction product with several by-products being formed [45].

Otherwise acetylated glycals treated in aqueous dioxane containing sulfuric acid and catalytic amounts of mercury(II) sulfate are efficiently converted directly to *E*-enals (e.g. **28**) [46] by reactions that probably involve addition/elimination processes rather than allylic rearrangements.

Both in dioxane containing sulfuric acid [47] or in water with molybdenum trioxide [48] hydrogen peroxide reacts with tri-*O*-acetyl-D-glucal to give 50–70% yields of the crystalline hydroperoxide **29**. On the other hand, *m*-chloroperbenzoic acid in dichloromethane containing boron trifluoride gives peroxide **30**, which spontaneously loses the benzoic acid to afford the unsaturated lactone **31** [49].

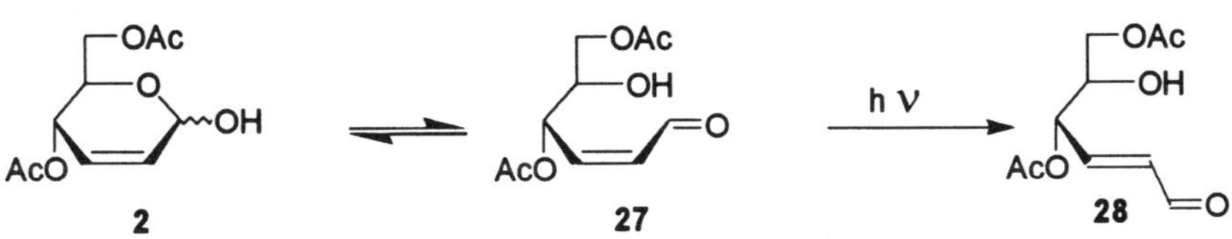

Scheme 11

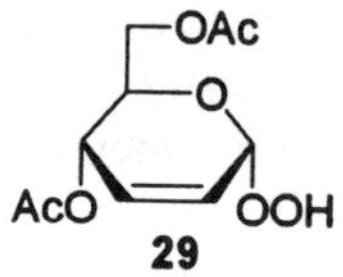

29

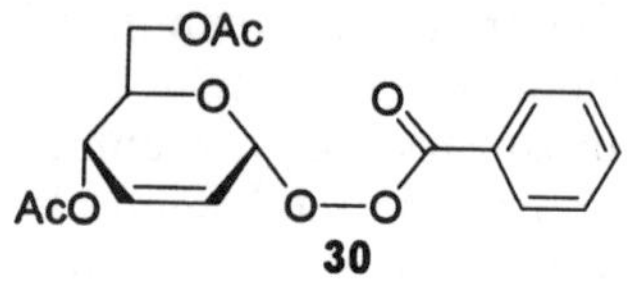

30

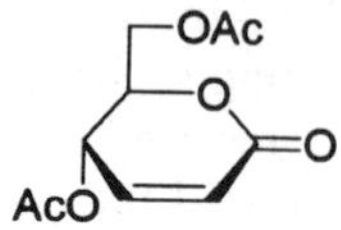

31

Simple thermal rearrangement of tri-*O*-acetyl-D-glucal, or treatment in benzene containing boron trifluoride [50], or heating in acetic anhydride containing nickel(II) chloride [51], all result at equilibrium not just in the expected 2,3-unsaturated glycosyl acetates **32** (65%) but in some of the starting material (15%) and its C-3 epimer tri-*O*-acetyl-D-allal **33** (20%) (Scheme 12).

While the thermal rearrangement probably proceeds by a sigmatropic process as illustrated, followed by anomerization of the derived 2,3-unsaturated β-acetate to the α-compound, and in turn its sigmatropic rearrangement, the other reactions may involve the stabilized carbocation derived from the starting material by catalyzed loss of the allylic acetoxy group (cf. Scheme 2). A further means of making 2,3-unsaturated glycosyl esters involves the use of the Mitsunobu reaction, as illustrated in Scheme 13, but when the C-3 epimer of the starting material is used the reaction is not specific yielding some of the 3-*O*-benzoyl-glucal compound derived by direct displacement of the hydroxyl group [52].

To a large extent the rearrangement reaction of glycals has been used to give access to a wide range of *O*-glycosides with boron trifluoride being the favoured catalyst, but it is ineffective when tri-*O*-acetyl-D-galactal is involved and tin(IV) chloride is used instead (Sect. 3). Yields are frequently >80% for reactions with simple primary alcohols and rather less for secondary and tertiary compounds, *t*-butyl glycosides were produced from tri-*O*-acetyl-D-glucal and di-*O*-acetyl-L-rhamnal in 73 and 50% yield, respectively (catalysts: I_2 [14] and $SnCl_4$ [22]).

Phenols can take part in the rearrangement process without the need for catalysts but elevated temperatures are required, phenol reacting with tri-*O*-acetyl-D-glucal in 3 h at 140 °C to give 80% yield of the α- and β-glycosides in the ratio 6:1 [10]. Pure α-anomers can be isolated directly in about 40% yield after reaction of this glycal ester and di-*O*-acetyl-L-rhamnal with various substituted phenols [53]. Again, Lewis acids promote the reaction, and again tin(IV) chloride is required for the efficient reaction of tri-*O*-acetyl-D-galactal. A special feature

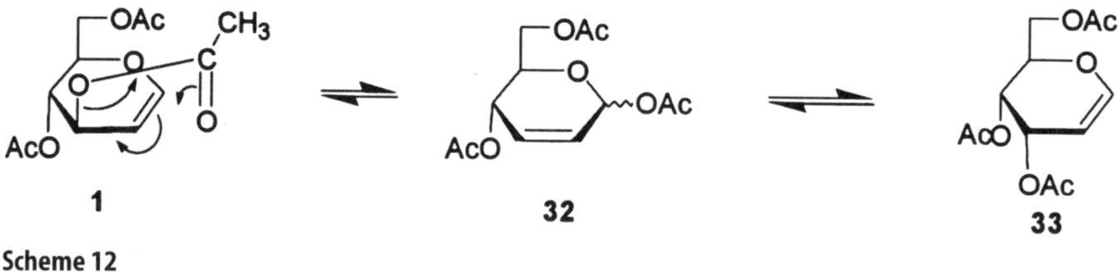

Scheme 12

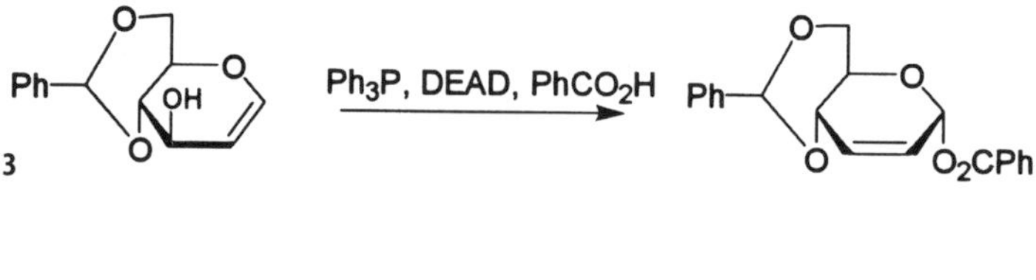

Scheme 13

of the acid-catalyzed reactions that involve the use of phenols containing activating substituents is the further rearrangement to *C*-glycosidic isomers that the initial products may undergo. Thus, for example, treatment of tri-*O*-acetyl-D-glucal with *p*-methoxyphenol in toluene with small amounts of boron trifluoride etherate results in 92% of the expected *O*-linked kinetic products **34** (Scheme 14) which, however, can be converted to the thermodynamically favoured **35** (72%). Otherwise, the latter can be derived directly from the starting materials by use of dichloromethane as solvent and higher proportions of the Lewis acid catalyst [54].

Aryl 2,3-unsaturated *O*-glycosides also result from the reaction of unsubstituted glycals with phenols in the presence of diethyl azodicarboxylate and triphenylphosphine, some reactions affording the products in high yields and with stereospecific introduction of the aglycone. In this way, the *p*-methoxyphenyl 2,3-unsaturated α-glycoside is obtained in 80% yield from L-rhamnal using dichloromethane as solvent [55].

Coupling of glycals with carbohydrate alcohols has obvious potential for the synthesis of di- and higher saccharides, and compounds **36** (R^1 = AcO; R^2 = H) [35] and **37** (R^1 = H; R^2 = AcO) [56] exemplify the types of products that can be made, the yields each being 56% with tin(IV) chloride being used in the former case since triacetyl-D-galactal was involved. These Lewis acid catalyzed couplings can be conducted with glycosyl acceptors containing the following groups: acyl and sulfonyl esters; alkyl, benzyl and silyl ethers; glycosidic, acetal, epoxide, halide. Occasionally, monosaccharide derivatives do not take part satisfactorily in the normal type of coupling reactions, for example, 1,2:5,6-di-*O*-isopropylidene-D-glucose (perhaps because the alcohol group is sterically hindered) or 1,3,4,6-tetra-*O*-acetyl-D-fructose (because of its acid sensitivity). In these cases glycals bearing the pent-4-enoyl group at C-3 have been used successfully to produce the respective disaccharide derivatives (Scheme 4; R = Ac) [17]. Otherwise iodine can be used to provide mild coupling conditions [14].

Applications of the rearrangement reaction with dihydroxy acceptors are so far rare, but diols **38** [57], **39** [58] and **40** (chloramphenicol) [14] are examples of compounds that have been successfully diglycosylated, and compound **41**,

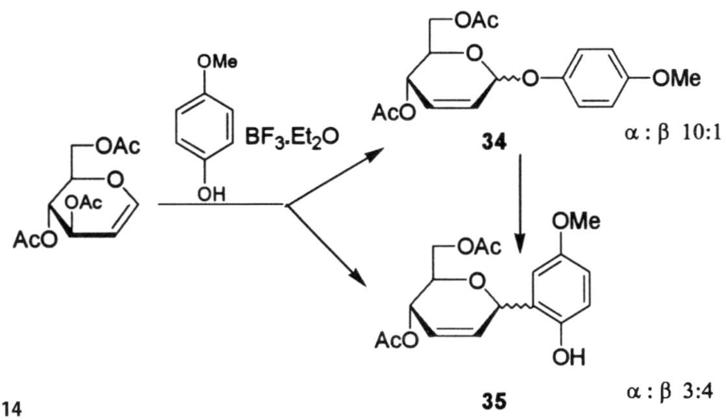

Scheme 14

produced by two sequential applications of the reaction, has been converted into saturated compounds of relevance in aminoglycoside antibiotic chemistry [59].

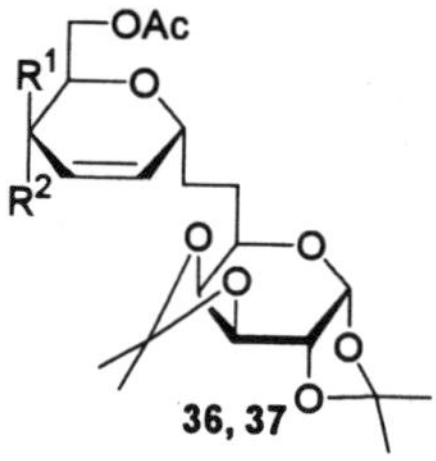

36, 37

38

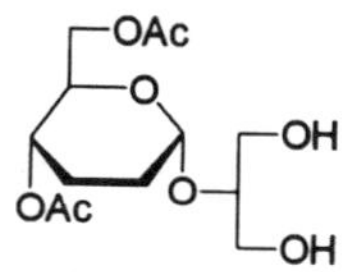

39

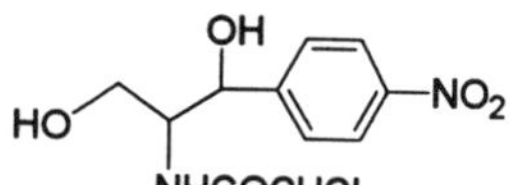

40

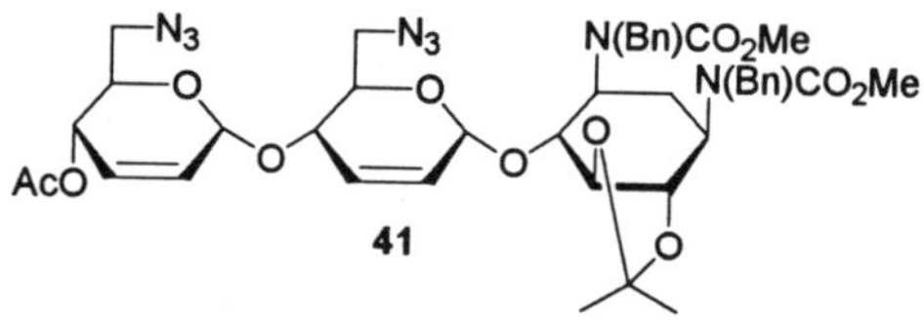

41

Other glycosides of significance in medicinal chemistry to have been obtained by application of the allylic rearrangement reaction to alcohols are **42** (R^1 = OAc, R^2 = H) [60] and **43** (R^1 = H, R^2 = OAc) [61]. In the former case reaction occurs selectively at the secondary alcohol site, and in the latter the glycosylation has been used to resolve racemic 3-hydroxy-2-azetidinones, the illustrated enantiomer being required for synthesis studies on taxol.

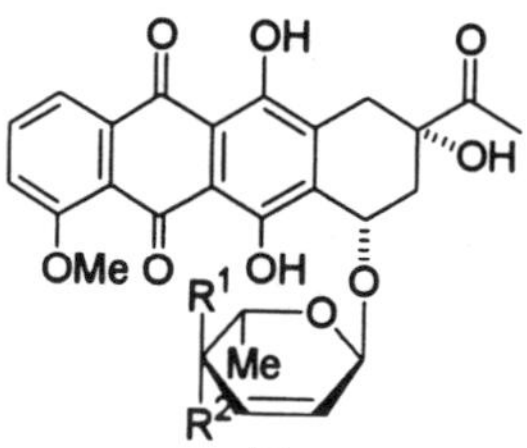

42

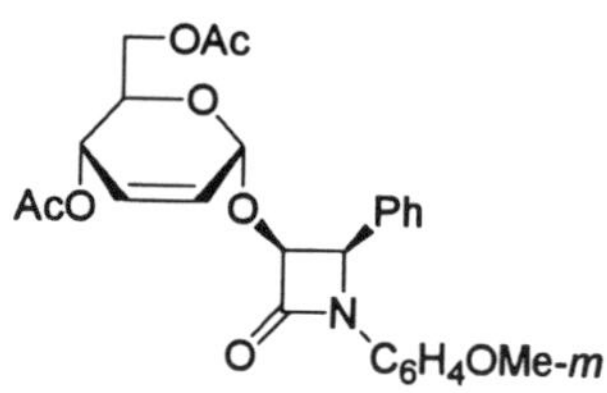

43

5 2,3-Unsaturated *C*-Glycosides

For the "normal" allylic rearrangement applied in the synthesis of *C*-glycosides, reagents are required that afford *C*-nucleophilic species under the (usually acidic) reaction conditions, and the following types of compounds have proved suitable: organometallics, cyanides, *C*-silylated compounds, alkenes (especially those that react with electrophiles to give stable carbocationic intermediates; enol ethers and esters being notable examples), activated aromatic and β-dicarbonyl compounds. Reagents used in palladium-mediated reactions are special cases (Sect. 2). With few exceptions all these *C*-nucleophile sources react with glycals at the anomeric centre, but very occasional instances of the formation of C-3 branched-chain glycals have been reported.

By use of trialkylaluminums a range of alkyl and alkenyl groups can be introduced at C-1; compound **44** can be obtained in 72% yield by treatment of di-*O*-acetyl-6-deoxy-D-galactal in dichloromethane at −78 °C with trimethylaluminum in the presence of titanium tetrachloride (Scheme 15) [62], and the doubly C-1-substituted **45**, required for natural product synthesis, is produced stereospecifically from tri-*O*-acetyl-1-*C*-allyl-D-glucal by a similar methylation procedure [63].

To introduce the cyano group at C-1, diethylaluminum cyanide can be used in benzene solution without a catalyst and gives anomeric mixtures of the α,β anomers, the ratios of which depend on the temperature used [42] (Sect. 3). Otherwise trimethylsilyl cyanide together with boron trifluoride etherate can be employed to introduce the cyano group [40, 41]. Organozinc derivatives of the Reformatski type, e.g. (cyanoethyl)zinc iodide, can be used with Lewis acid catalysis [64]; exceptionally, however, zinc/copper analogues favour the formation of C-3-substituted glycals [65].

Methylenecyclohexane is an example of an alkene that adds an electrophile to give a stable cationic intermediate, and with tri-*O*-acetyl-D-glucal and tin(IV) bromide as catalyst it gives the diene indicated in Scheme 10 stereospecifically in 94% yield. On the other hand, reaction of methylenecyclobutane ($EtAlCl_2$ as catalyst) results in the chlorinated adduct **46,** since the ionic reaction intermediate favours addition rather than elimination [39].

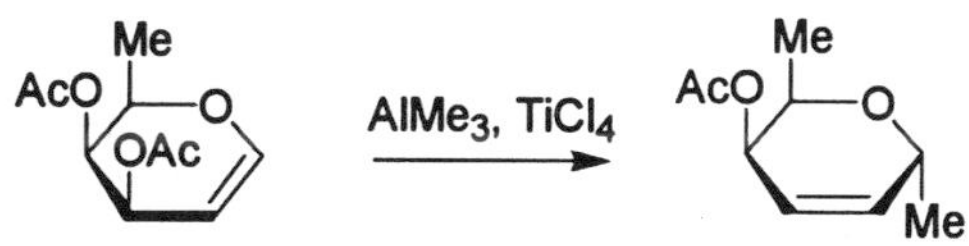

Scheme 15 **44**

An alternative way of generating *C*-nucleophilic species depends on the use of allylsilanes together with Lewis acids; allyltrimethylsilane having been frequently employed. Consequently, compound **47** has been described on several occasions [38, 66] and this compound can also be derived by the use of DDQ as promoter [20]. Most strikingly, allyl 2,3-unsaturated *C*-glycosides can be obtained from the *O*-unprotected glycals using trimethylsilyl triflate in dichloromethane/acetonitrile at low temperatures; the reactions are both highly efficient and highly stereoselective, with α/β ratios of >99:1 being recorded for the most common glycals [15]. Scheme 16 illustrates the ingenious use of a substituted allylsilane in the production of a *C*-linked disaccharide derivative [67].

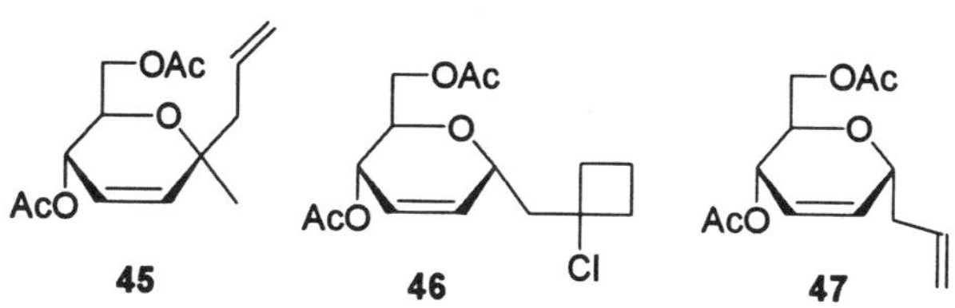

45 **46** **47**

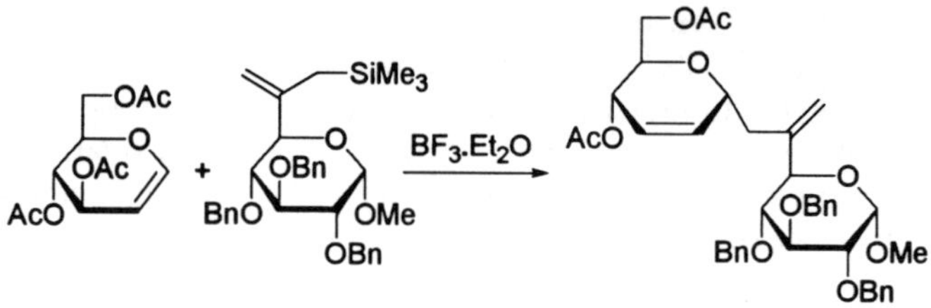

Scheme 16

Further examples of reagents that react with electrophilic species to form C–C bonds are alkynylsilanes: with the assistance of Lewis acid catalysts a set of compounds **48** (e.g. R = $SiMe_3$, C ≡ $CSiMe_3$, CH = CHCl, CH = CH-C ≡ $CSiMe_3$) have been made [68], and the parent compound **48** (R = H) is available from the trimethylsilyl analogue [69].

As is to be expected from the above, vinyl ethers or esters react with glycals in standard rearrangement processes to yield 2,3-unsaturated *C*-glycosides, tri-*O*-acetyl-D-glucal with enol acetates [70] or enol silyl ethers [71] of acetaldehyde, acetone and acetophenone giving access to the carbonyl compounds **49** (R = H, Me, Ph), respectively.

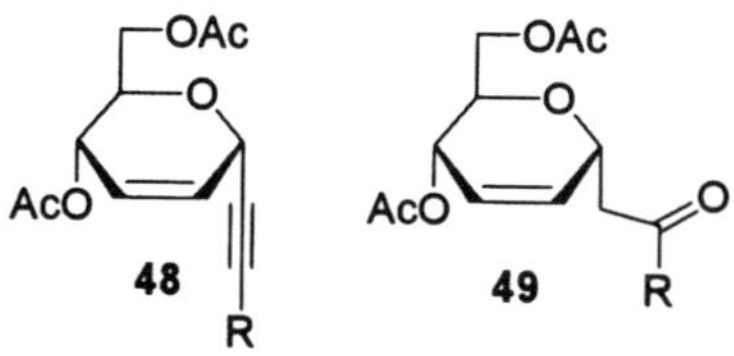

48 **49**

It follows that glycals themselves, being vinyl ethers, could react in intermolecular fashion in the presence of Lewis acids, and in a control experiment conducted in early work on the reaction of tri-*O*-acetyl-D-glucal with alcohols it was found that dimerization does occur, as indicated in Scheme 17 [50].

Initially, the product **50** was isolated in 10% yield, but acetyl perchlorate in dichloromethane at –78 °C has given a yield of 61% (α/β ratio 1.5:1, at the C-1–OAc centre) from tri-*O*-acetyl-D-glucal; 77% of the pure α,α-analogue is obtainable from tri-*O*-acetyl-D-galactal [25a]. Not surprisingly, dimers of this kind have been encountered following several reactions of glycal esters under acidic conditions [2].

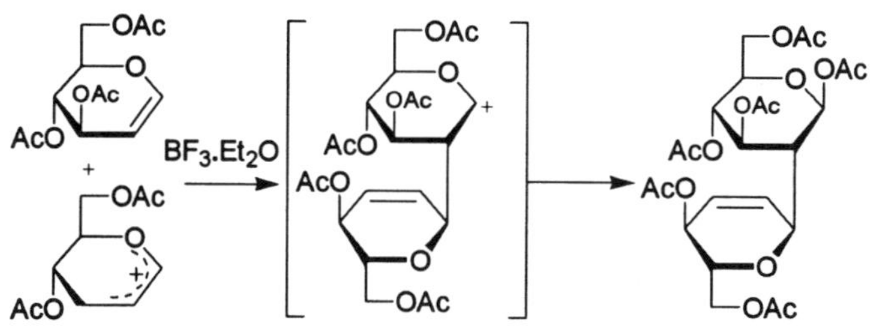

Scheme 17

50

As already indicated in Sect. 4, heating acetylated glycals with activated phenols can result in the isolation of 2,3-unsaturated *C*-aryl glycosides [54], formed via *O*-glycosides (Scheme 14), with the yields correlating with the acidity of the phenols used [72]. Otherwise, reaction of acetylated glycals with bromomagnesium phenates at room temperature under ultrasonication gives 2,3-unsaturated *C*-aryl glycosides in ca. 70% yield with α/β ratios >20:1 for tri-*O*-acetyl-D-glucal-derived products [73].

For the synthesis of 2,3-unsaturated *C*-glycosides with carbonyl groups at C-3 of the C-1 substituents the use of allyl silyl ethers together with a Lewis acid catalyst is recommended [74].

6 Other 2,3-Unsaturated Compounds

The allylic rearrangement of glycals also offers a means of introducing *S*-, *N*-, and *P*-linked groups and hydride (and to a limited extent halide ions) into glycals at C-1 at the expense of the allylic substituents. With alcohols and phenols taking part in the reaction to give 2,3-unsaturated products exclusively, it might be expected that thiols would react similarly, and tri-*O*-acetyl-D-glucal and -galactal with simple thiols, under mild conditions, have given largely the expected glycosides, but with smaller proportions of 3-thioglycal isomers. This, however, represents the kinetic situation and, when equilibrium is reached, the 3-thioglycals are favoured by more than 10:1, establishing that the alcohol and thiol reactions are dissimilar – a difference that has been attributed to the relative hard/soft natures of these nucleophiles and the C-1, C-3 centres of glycals [75]. Notwithstanding the above the most readily isolated products from tri-*O*-acetyl-D-glucal and -galactal treated with thiophenol and appropriate Lewis acids are the phenylthio 2,3-unsaturated-α-glycosides [76], and thioglycosides are the sole products when (trimethylsilyl)thiols are used as nucleophiles [23]. Reaction of tri-*O*-acetyl-D-glucal with benzenesulfinic acid at –78 °C in the presence of boron trifluoride provides a novel way of producing the 2,3-unsaturated glycosyl sulfones [77]. In this reaction the carbohydrate is pretreated with the catalyst prior to the addition of the sulfinic acid, to generate the intermediate carbocation; this important experimental detail could possibly ensure success in applications other than this particular rearrangement reaction.

Heating of various glycal derivatives with the soft nucleophile sodium azide in the presence of boron trifluoride initially gives the 2,3-unsaturated glycosyl azides – but accompanied by larger proportions of 3-azido-3-deoxyglycal isomers. Apparently, like the thio compounds, the former are kinetically favoured and they rearrange to the latter, but it is not known whether reionization or [3.3]-sigmatropic thermal rearrangement is involved. The equilibrium products derived from tri-*O*-acetyl-D-glucal are shown in Scheme 18, the clearly dominant one being the 3-azidoallal isomer [78].

With *N*-nucleophilic species the purine and pyrimidine bases have received the most attention because of their potential to yield nucleoside analogues of interest in medicinal chemistry. In Scheme 19 the consequences of heating tri-

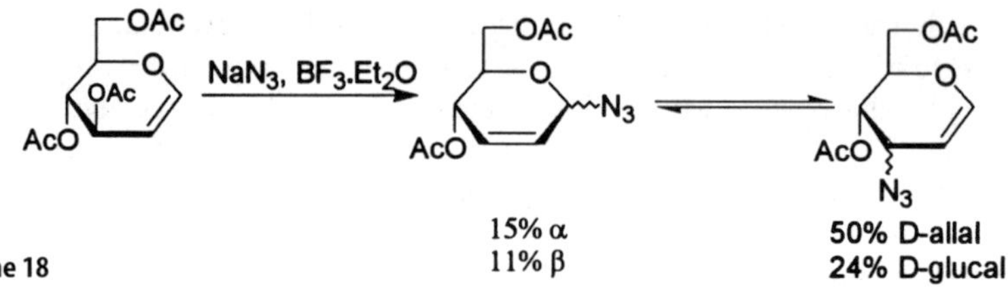

Scheme 18

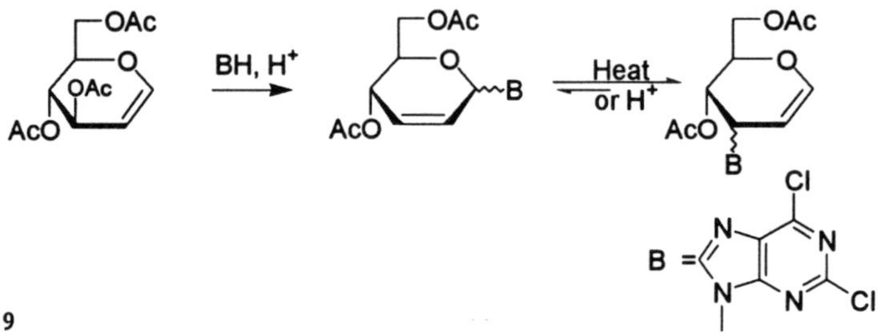

Scheme 19

O-acetyl-D-glucal with 2,6-dichloropurine in boiling nitromethane containing catalytic proportions of *p*-toluenesulfonic acid are illustrated. Again, the kinetic products are the 2,3-unsaturated glycosyl compounds, which convert to the more stable 3-substituted glycal isomers [79], the latter step apparently requiring acid [80], which indicates that the equilibration involves the regeneration of a stabilized carbocation.

To inhibit the formation of the 3-substituted glycals, starting materials with allylic groups with enhanced leaving properties are recommended [81] or, otherwise, *N*-trimethylsilylated bases can be used with catalysts such as trityl on lithium perchlorate [82].

Pyrimidine bases can be less reactive than purines in these processes, but the tri-*O*-acetyl-D-galactal-derived compounds **51** have been obtained (α, 40%; β, 24%) by use of bis-*O*-(trimethylsilyl)uracil in ethyl acetate with antimony pentachloride as catalyst, the β-anomer being required for the synthesis of an antibiotic [83]. Although trace proportions of 3-substituted glycals are present in the products of this reaction, their formation from the initial 2,3-unsaturated compounds is apparently less easy than is the case with the purine analogues.

While 2,3-unsaturated glycosyl chlorides and bromides are effectively unknown (but such chlorides with acyloxy groups at C-2 have been reported [84]), analogous fluorides, e.g. compound **52**, have been made as reactive syrups by use of hydrogen fluoride in benzene [85] or pyridinium poly(hydrogen fluoride) [86]. Analogous phosphonates, such as compound **53** [87], can be obtained by use of dialkyl [87] or trialkyl [88] phosphites. Mixed anomers, with the β-compounds predominating, are formed, and the compounds are characterized by undergoing prototropic shifts under basic conditions to give 3-deoxyglycal 1-phosphonates with the π bonds now in conjugation.

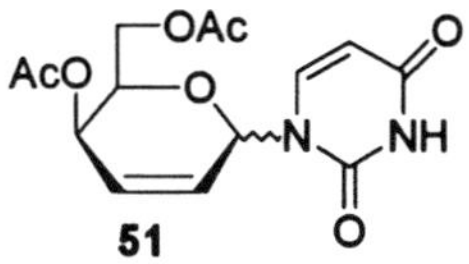

51

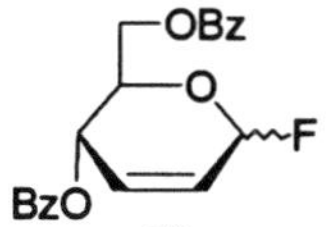

52

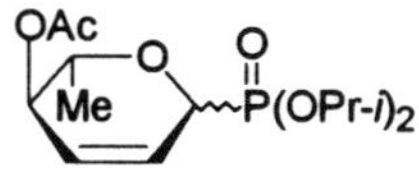

53

On treatment of *O*-acetylated glycals with triethylsilane and boron trifluoride in inert solvents, hydride enters at C-1 to displace the allylic groups and give 2,3-unsaturated 1,5-anhydroalditols (Scheme 20) [89].

7 Extensions of the Reaction

While the great majority of the substitution reactions of glycals and their derivatives which occur with allylic rearrangement have involved simple pyranoid compounds in intermolecular processes, there are some variations that usefully extend the scope of the reaction. A well-known set of glycal derivatives that carry acyloxy groups at C-2 are readily produced by removal of hydrogen halide from acylated glycosyl halides, and they undergo most of the allylic rearrangement reactions already described for the glycals. There are, however, characteristics that are peculiar to these 2-substituted compounds. Thus, treatment of the best-known member of the set, compound **54**, in inert solvents with boron trifluoride causes both intramolecular allylic rearrangement and anomerization, and compound **55** (R = Ac, X = H, α-anomer) is obtained as the main product (60%) [90]. However, if the reaction time is extended, loss of acetic anhydride occurs with further allylic rearrangement, and the enone **56** and its β-anomer are thereby available in 70% yield (α/β ratio 3:1). Likewise, the same products are obtained in 83% yield (α/β 6:1) from the 4-epimer of **54** [91]. On the other hand, simple thermal rearrangement of compound **54** gives the acetate **55** (R = Ac, X = H, β-anomer) specifically (75% isolated) [90]. The C-4 epimer of **54**, derived from tetra-*O*-acetyl-α-D-galactopyranosyl bromide, does not react under these thermal conditions, which indicates that the reactivity of compound **54** is dependent on the ester group at C-4 assisting the departure of the allylic group (cf. the analogous behaviour of epimers **1** and **24**, Sect. 3).

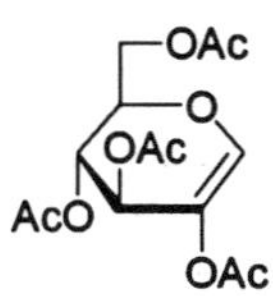

54

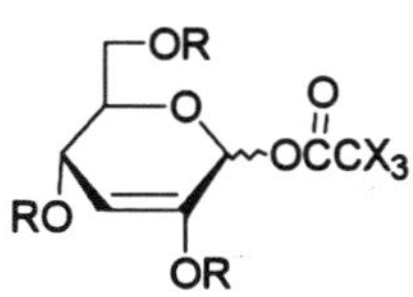

55

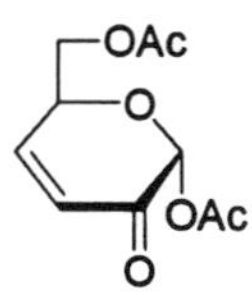

56

Scheme 20

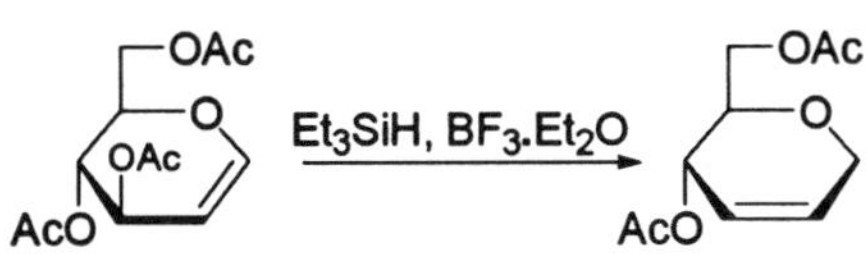

Compounds such as **54** take part in coupling reactions with alcohols in the presence of Lewis acids to give 2,3-unsaturated glycosides; disaccharide **57**, for example, can be obtained in 59% yield (α/β ratio 4:1) [92]. β-Glycosides of the series can be made by uncatalyzed displacements at C-1 applied to the α-trichloroacetate **55** (R = Bz, X = Cl) made from the benzoyl analogue of tetraacetate **54** [93]. Like the glycal esters, their 2-acyloxy analogues also react with heterocyclic bases [94] and with a range of *C*-nucleophilic species to give access to products such as **58** [70], and analogous glycosyl chlorides and fluorides are known [84, 95].

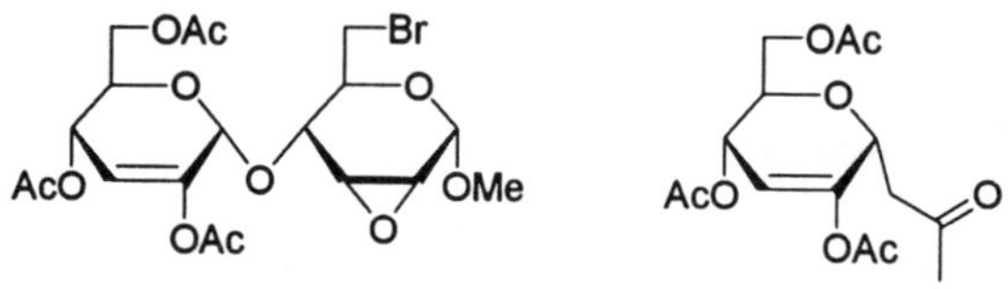

57 **58**

An unusual reaction appears to take place when the 2-*C*-formyl glycal **59** reacts in the presence of boron trifluoride with *p*-cresol to give the C-3-substituted **60** apparently by direct displacement [96]. More probably, however, the reaction proceeds in the normal way to give the 2,3-unsaturated α-*O*-glycoside which undergoes Claisen rearrangement (cf. Scheme 14).

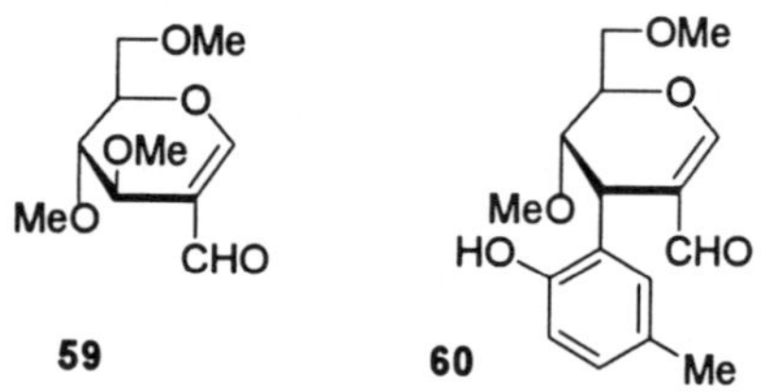

59 **60**

Furanoid glycal and 2-acyloxyglycal derivatives and the products formed as a result of their substitution with allylic rearrangement are all much less robust than their pyranoid analogues because of their vulnerability to eliminations to give furans. Nevertheless, they have potential in synthesis, as illustrated by the reaction undergone by the sodio derivative **61** on treatment with 2-chloropyrimidine, the intermediate 3-*O*-(2-pyrimidinyl)glycal **62** rearranging spontaneously to give the *N*-(2,3-unsaturated glycofuranoyl)pyrimidinone **63** in 61% yield (Scheme 21) [97].

For the production of 2,3-unsaturated furanoid *C*-glycosides, palladium-promoted addition/elimination reactions can be used to good advantage (Scheme 8).

While the majority of relevant reactions of glycals discussed so far involve intermolecular processes, intramolecular examples are well established and very useful, the simplest involving attack of the O-6 at C-1 so that when D-glucal itself is dehydrated, or when 3,4-di-*O*-substituted derivatives or compounds with silyl protection at O-6 are treated with Lewis acids, they give 2,3-unsaturated 1,6-anhydrides of type **64** usually in high yield [98]. A related example involves the formation of the *spiro* compounds **65** which are produced on activation with *N*-iodosuccinimide of the appropriate glycal derivative possessing a

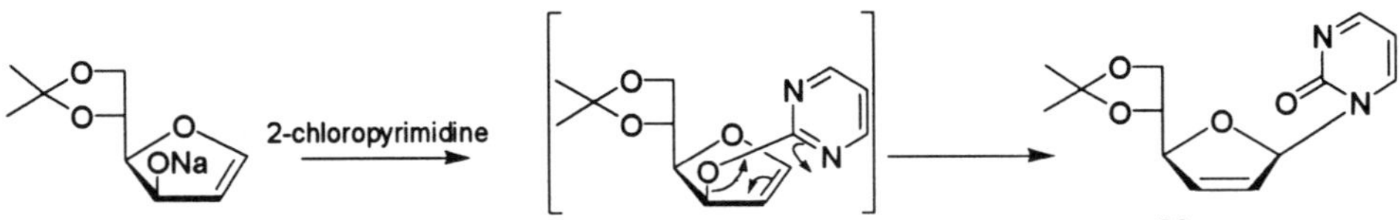

61 62 63

Scheme 21

phenylthio group at C-3 and a C_6 hydroxyl-containing *C*-bonded substituent at C-1 [99].

Most cases, however, of intramolecular examples of the allylic rearrangement reaction of glycals involve the participation of nucleophiles bonded to C-3 and instances have already been given in the rearrangements of C-3 esters and of an unstable pyrimidinoxy intermediate (Scheme 21). Glycals that possess 3-azido groups and 3-vinyl ethers take part in related [3.3]-sigmatropic rearrangements to give 2,3-unsaturated glycosyl azides [100] and *C*-formylmethyl glycosides (e.g. **66**) [42], respectively, with retention of configuration. Extensions of the latter reaction have proved useful for the production of *C*-glycosides of relevance in natural product synthesis, the propanoate and silyl ether **67**, for example, thermally giving mainly compound **68** from which the Prelog – Djerassi lactone was made [101].

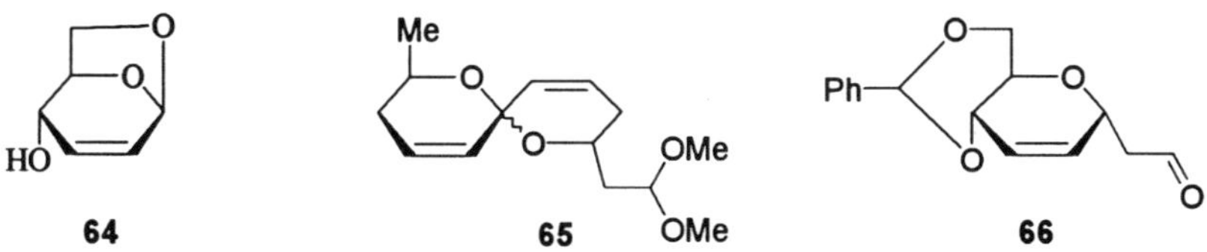

64 65 66

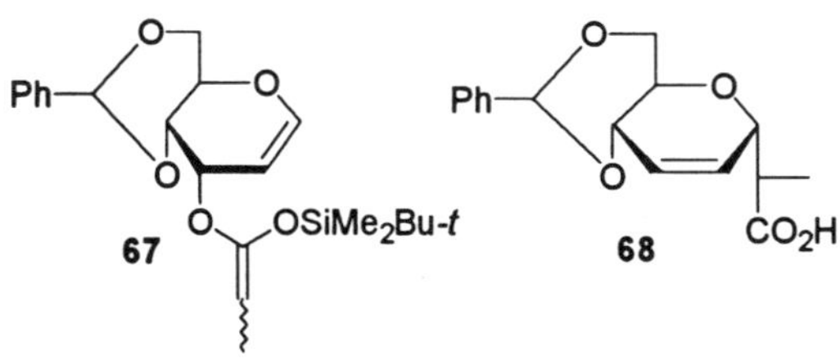

67 68

On the other hand, spontaneous [2,3]-Wittig rearrangement of the anion **69** affords the epimers **70** (Scheme 22) [102], illustrating that useful 2,5-dihydrofurans with *C*-substituents on the carbon atoms on either side of the ring oxygen atom can be obtained by this and related approaches.

Rearrangement of compound **71** under acidic conditions illustrates the use of the reaction in the formation of *spiro* compounds, the product **72** being obtained in 53% yield (Scheme 23) [103].

The allylic rearrangement reaction has been extended further in applications with the allylic leaving group at a branch-point carbon atom, so that compound

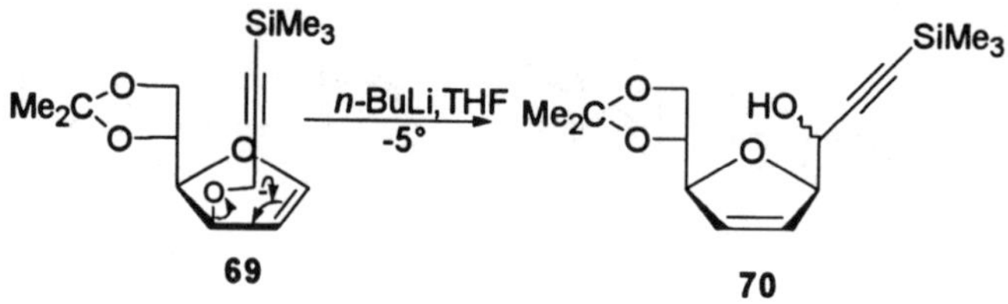

Scheme 22

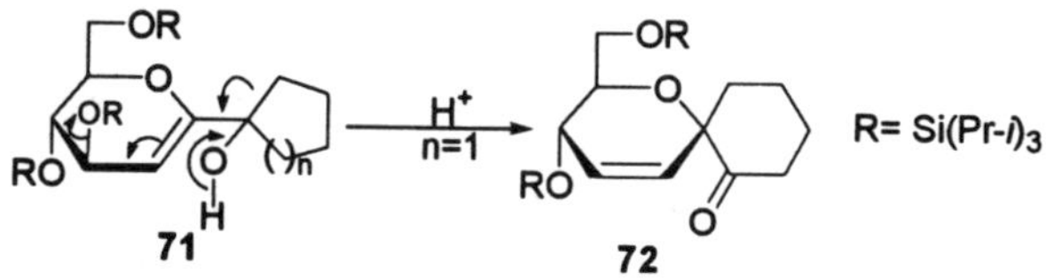

Scheme 23

73 affords the reactive *exo*-methylene glycoside **74** under the usual conditions and hence gives pyranobenzopyrans **75** (Scheme 24) [104].

Otherwise, the leaving group may be in the homoallylic situation either in the glycal itself or on a branch carbon atom; in both situations the double-bond electrons of the glycal bond to the homoallylic carbon atom to give products with cyclopropane rings fused to furanoid or pyranoid rings, respectively. In this way compounds **76** [105] and **77** [106] are converted to cyclopropa-furanoid and -pyranoid nucleosides (Schemes 25 and 26, respectively).

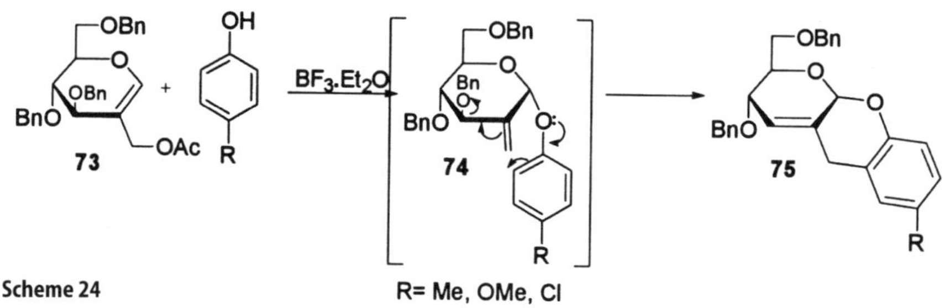

Scheme 24

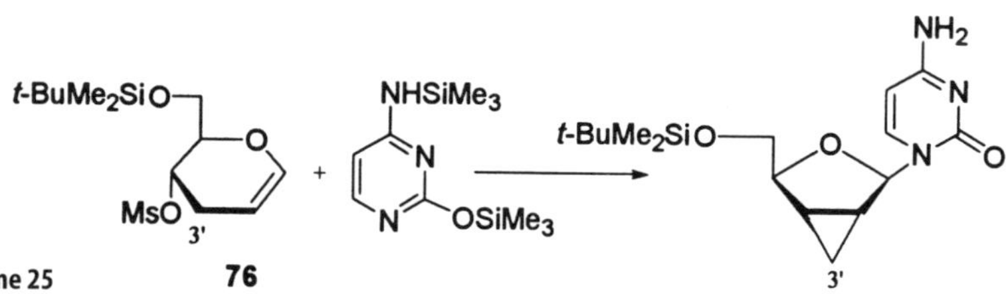

Scheme 25

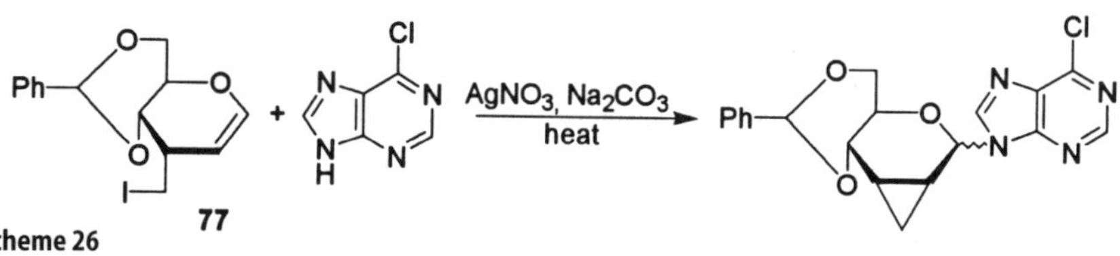

Scheme 26

8 References

1. (a) Danishefsky SJ, Bilodeau MK (1996) Angew Chem Int Ed Engl 35:289; (b) Bhattacharya SK, Danishefsky SJ (2000) J Org Chem 65:144
2. Ferrier RJ, Zubkov OA (2000) Org React submitted
3. Fischer E (1914) Chem Ber 47:196
4. Bergmann M (1925) Liebigs Ann Chem 443:223
5. Fraser-Reid B, Radatus B (1970) J Am Chem Soc 92:5288
6. Ferrier RJ, Overend WG, Ryan AE (1962) J Chem Soc 3667
7. Tejima S, Haga M, Nakamura H, Maki T, Sakata M, Agaki M (1963) Abstr Papers Am Chem Soc Meeting 145:7D
8. Bowles WA, Robins RA (1964) J Am Chem Soc 86:1252
9. Ferrier RJ (1964) J Chem Soc 5443
10. Brakta M, Lhoste P, Sinou D (1989) J Org Chem 54:1890
11. Doboszewski B, Blaton N, Herdewijn P (1995) J Org Chem 60:7909
12. Ogihara T, Mitsunobu O (1983) Tetrahedron Lett 24:3505
13. Leutzinger EE, Robins RK, Townsend LB (1970) Tetrahedron Lett 3751
14. Koreeda M, Houston TA, Shull BK, Klemke E, Tuinman RJ (1995) Synlett 90
15. Toshima K, Matsua G, Ishizuka T, Ushiki Y, Nakata M, Matsumura S (1998) J Org Chem 63:2307
16. Sobti A, Sulikowski GA (1994) Tetrahedron Lett 35:3661
17. López JC, Gómez AM, Valverde S, Fraser-Reid B (1995) J Org Chem 60:3851
18. Kjølberg O, Neumann K (1993) Acta Chem Scand 47:843
19. Toshima K, Ishizuka T, Matsuo G, Nakata M, Kinoshita M (1993) J Chem Soc Chem Commun 704
20. Toshima K, Ishizuka T, Matsuo G, Nakata M (1993) Chem Lett 2013
21. Florent J-C, Monneret C (1982) Synthesis 29
22. Bhaté P, Horton D, Priebe W (1985) Carbohydr Res 144:331
23. Dunkerton LV, Adair NK, Euske JM, Brady KT, Robinson PD (1988) J Org Chem 53:845
24. Descotes G, Martin J-C (1977) Carbohydr Res 56:168
25. (a) Byerley ALJ, Kenwright AM, Steel PG (1996) Tetrahedron Lett 37:9093; (b) Byerley ALJ, Kenwright AM, Steel PG (1997) ibid 38:2195; (c) Byerley ALJ, Kenwright AM, Lehmann CW, MacBride JAH, Steel PG (1998) J Org Chem 63:193
26. Dunkerton LV, Brady KT, Mohamed F, McKillican BP (1998) J Carbohydr Chem 7:49
27. (a) Daves GD (1990) Acc Chem Res 23:201; (b) Daves GD, Hallberg A (1989) Chem Rev 89:1433
28. Yougai S, Miwa T (1983) J Chem Soc Chem Commun 68
29. Basson MM, Holzapfel CW, Verdoorn GH (1998) Heterocycles 29:2261
30. Arai I, Daves GD (1981) J Am Chem Soc 103:7683
31. (a) Czernecki S, Dechavanne V (1983) Can J Chem 61:533; (b) Bellosta V, Czernecki S, Avenel D, Bahij SE, Gillier-Pandraud H (1990) ibid 68:1364
32. Takeda K, Nakamura H, Ayabe A, Akiyama A, Harigaya Y, Mizuno Y (1994) Tetrahedron Lett 35:125
33. Curran DP, Suh Y-G (1987) Carbohydr Res 171:161
34. (a) Hicks DR, Fraser-Reid B (1975) Can J Chem 53:2017; (b) Rajanbabu TV (1985) J Org Chem 50:3642
35. Grynkiewicz G, Priebe W, Zamojski A (1979) Carbohydr Res 68:33
36. Ciment DM, Ferrier RJ (1966) J Chem Soc (C) 441
37. Ferrier RJ, Ilanko G unpublished results
38. Danishefsky SJ, Kerwin JF (1982) J Org Chem 47:3803
39. Herscovici J, Muleka K, Boumaîza L, Antonakis K (1990) J Chem Soc Perkin Trans 1 1995
40. De las Heras FG, San Felix A, Fernández-Resa P (1983) Tetrahedron 39:1617
41. Grynkiewicz G, BeMiller JN (1982) Carbohydr Res 108:229
42. Tulshian DB, Fraser-Reid B (1984) J Org Chem 49:518

43. Dawe RD, Fraser-Reid B (1981) J Chem Soc Chem Commun 1180
44. Casiraghi G, Cornia M, Colombo L, Rassu G, Fava GG, Belicchi MF, Zetta L (1988) Tetrahedron Lett 29:5549
45. (a) Tam SY-K, Fraser-Reid B (1975) Carbohydr Res 45:29; (b) Madaj J, Rak J, Sokolowski J, Wisniewski A (1996) J Org Chem 61:2988
46. Gonzalez F, Lesage S, Perlin AS (1975) Carbohydr Res 42:267
47. Fehlhaber H-W, Snatzke G, Vlahov I (1987) Liebigs Ann Chem 637
48. Mostowicz D, Jurczak M, Hamann H-J, Hoft E, Chmielewski M (1998) Eur J Org Chem 2617
49. Jarglis P, Lichtenthaler FW (1982) Tetrahedron Lett 23:3781
50. Ferrier RJ, Prasad N (1969) J Chem Soc (C) 581
51. Inaba K, Matsumura S, Yoshikawa S (1991) Chem Lett 485
52. Guthrie RD, Irvine RW, Davison BE, Henrick K, Trotter J (1981) J Chem Soc Perkin Trans 2 468
53. Frappa I, Sinou D (1995) Synth Commun 25 2941
54. Noshita T, Sugiyama T, Kitazumi Y, Oritani T (1995) Biosci Biotech Biochem 59:2052
55. Sobti A, Sulikowski GA (1994) Tetrahedron Lett 35:3661
56. Ferrier RJ, Prasad N (1969) J Chem Soc (C) 570
57. Canas-Rodriguez A, Martinez-Tobed A (1979) Carbohydr Res 68:43
58. Wieczorek E, Thiem J (1998) Carbohydr Res 307:263
59. Canas-Rodriguez A, Ruiz-Poveda SG, Coronel-Borges LA (1987) Carbohydr Res 159:217
60. Klaffke W, Pudlo P, Springer D, Thiem J (1991) Liebigs Ann Chem 509
61. Banik BK, Manhas MS, Bose AK (1994) J Org Chem 59:4714
62. Deshpande PP, Price KN, Baker DC (1996) J Org Chem 61:455
63. Nicolaou KC, Hwang C-K, Duggan ME (1986) J Chem Soc Chem Commun 925
64. (a) Dorgan BJ, Jackson RFW (1996) Synlett: 859; (b) Orsini F, Pelizzoni F (1993) Carbohydr Res 243:183; (c) Thorn SN, Gallagher T (1996) Synlett 185
65. Thorn SN, Gallagher T (1996) Synlett 856
66. (a) Ichikawa Y, Isobe M, Goto T (1984) Tetrahedron Lett 25:5049; (b) Wincott FE, Danishefsky SJ, Schulte G (1987) ibid 28:4951; (c) Nicolaou KC, Duggan ME, Hwang C-K (1989) J Am Chem Soc 111:6666
67. de Raadt A, Stütz AE (1991) Carbohydr Res 220:101
68. Tsukiyama T, Isobe M (1992) Tetrahedron Lett 33:7911
69. Tsukiyama T, Peters SC, Isobe M (1993) Synlett 413
70. Grynkiewicz G, BeMiller JN (1982) J Carbohydr Chem 1:121
71. Dawe RD, Fraser-Reid B (1984) J Org Chem 49:522
72. Ramesh NG, Balasubramanian KK (1992) Tetrahedron Lett 33:3061
73. (a) Casiraghi G, Cornia M, Rassu G, Zetta L, Fava GG, Belicchi MF (1988) Tetrahedron Lett 29:3323; (b) Casiraghi G, Cornia M, Rassu G, Zetta L, Fava GG, Belicchi MF (1989) Carbohydr Res 191:243
74. Herscovici J, Delatre S, Antonakis K (1987) J Org Chem 52:5691
75. Priebe W, Zamojski A (1980) Tetrahedron 36:287
76. (a) Wittman MD, Halcomb RL, Danishefsky SJ, Golik J, Vyas D (1990) J Org Chem 55:1979; (b) de Raadt A, Ferrier RJ (1991) Carbohydr Res 216:93
77. Brown DS, Bruno M, Davenport RJ, Ley SV (1989) Tetrahedron 45:4293
78. Guthrie RD, Irvine RW (1980) Carbohydr Res 82:207
79. Ferrier RJ, Ponpipom MM (1971) J Chem Soc (C) 553
80. De las Heras FG, Stud M (1977) Tetrahedron 33:1513
81. Doboszewski B, Blaton R, Herdewijn P (1995) Tetrahedron Lett 36:1321
82. Bessodes M, Egron M-J, Filippi J, Antonakis K (1990) J Chem Soc Perkin Trans 1 3035
83. Kondo T, Nakai H, Goto T (1973) Tetrahedron 29:1801
84. Bock K, Pedersen C (1970) Acta Chem Scand 24:2465
85. Bock K, Pedersen C (1969) Tetrahedron Lett: 2983
86. Macdonald SJF, McKenzie TC (1988) Tetrahedron Lett 29:1363
87. Paulsen H, Thiem J (1973) Chem Ber 106:3850

88. Alexander P, Krishnamurthy VV, Prisbe EJ (1996) J Med Chem 39:1321
89. Grynkiewicz G (1984) Carbohydr Res 128:C9
90. Ferrier RJ, Prasad N, Sankey G H (1968) J Chem Soc (C) 974
91. Köll P, Klenke K, Eiserman D (1984) J Carbohydr Chem 3:403
92. Thiem J, Schwentner J, Schüttpelz E, Kopf J (1979) Chem Ber 112:1023
93. Ferrier RJ, Prasad N, Sankey GH (1969) J Chem Soc (C) 587
94. (a) Onodera K, Yajima T (1970) Carbohydr Res 13:97; (b) Ferrier RJ, Ponpipom MM (1971) J Chem Soc (C) 560
95. Bock K, Pedersen C (1971) Acta Chem Scand 25:1021
96. Booma C, Balasubramanian KK (1992) Tetrahedron Lett 33:3049
97. Armstrong PL, Coull IC, Hewson AT, Slater MJ (1995) Tetrahedron Lett 36:4311
98. (a) Mereyala HB, Venkataramanaiah KC, Dalvoy VS (1992) Carbohydr Res 225:151; (b) Lauer G, Oberdorfer F (1993) Angew Chem Int Ed Engl 32:272; (c) Sharma GVM, Ramanaiah KCV, Krishnudu K (1994) Tetrahedron Asymmetry 5:1905
99. Gomez AM, Valverde S, Fraser-Reid B (1991) J Chem Soc Chem Commun 1207
100. Guthrie RD, Irvine RW (1980) Carbohydr Res 82:225
101. Ireland RE, Daub JP (1981) J Org Chem 46:479
102. Bertrand P, Gesson J-P, Renoux B, Tranoy I (1995) Tetrahedron Lett 36:4073
103. Paquette LA, Kinney MJ, Dullweber U (1997) J Org Chem 62:1713
104. Booma C, Balasubramanian KK (1993) Tetrahedron Lett 34:6757
105. Okabe M, Sun R-C (1989) Tetrahedron Lett 30:2203
106. Tam SY-K, Fraser-Reid B (1977) Can J Chem 55:3996

Synthetically Useful Base-Induced Rearrangements of Aldonolactones

Inge Lundt, Robert Madsen

Department of Organic Chemistry, Technical University of Denmark, Building 201, 2800 Lyngby, Denmark
E-mail: okil@pop.dtu.dk

Aldonolactones can be activated at the α and ω positions by selective bromination or tosylation. The activated aldonolactones can be transformed into epoxyaldonolactones by treatment with base under non-aqueous conditions. Treatment of epoxy- or bromodeoxyaldonolactones with aqueous base gives epoxyaldonates in which the epoxide can undergo Payne rearrangement to more stable epoxyaldonates. These can subsequently be opened by the carboxylate group with inversion of the configuration at the attacked carbon. Using this method a number of less available aldonolactones/acids have been prepared, in a reaction sequence where the configuration at one, two or three carbon centers has been stereospecifically interconverted. An attractive synthesis of L-gluconic acid from D-gluconolactone is presented.

Keywords: Aldonolactones, Epoxides, Rearrangements

1 Introduction

1.1 Aldonolactones: Preparation and Chemical Reactivity

In the past 20 years aldonolactones have found widespread application as cheap, chiral synthons for the synthesis of many biologically important compounds and natural products [1]. The area of aza- and carbasugar synthesis, in particular, has seen aldonolactones emerge as versatile starting materials [1, 2]. Aldonolactones constitute a more diverse chiral pool of compounds than aldoses. From each aldose several aldonolactones/aldonic acids are available in just one step, i.e. by anomeric oxidation of the aldose [3], by one-carbon Kiliani chain elongation [4] or by one-carbon oxidative degradation [5]. In addition, a number of aldonolactones are available via some more special reactions including reduction, or oxidative cleavage of the double bond in vitamin C [6, 7]. As a result, the aldonolactones are in many cases more easily available than the corresponding aldoses, especially in the L-series.

The chemical reactivity of aldonolactones also differs remarkably from the reactivity of aldoses. For synthetic manipulation of aldoses several steps are usually required to protect and define the stereochemistry at the C-1 hemiacetal function, e.g. via a glycoside. The lactone group in aldonolactones can be, however, maintained through a number of transformations [1, 2]. The reactivity of the hydroxy groups in aldonolactones is also different from that normally observed in aldoses. The hydroxy group α-positioned to the lactone moiety, in particular, shows similar or enhanced reactivity as compared with the primary hydroxy group. This gives rise to a number of regioselective reactions in aldonolactones and diminishes the need for many different protecting groups [1, 2]. Furthermore, it should be noted that aldonolactones usually prefer the five-membered 1,4-lactone over the six-membered pyranoid form predominant in aldoses.

1.2 Isomerisation at C-2

Due to the electronegativity of the carboxylic function, it is possible to isomerise aldonolactones at C-2 in the presence of a base. The reaction, which has been known for more than a century, can be carried out in water at 100–140 °C with calcium or barium hydroxide or with pyridine as the base [8]. With hydroxide it is possible to isomerise the aldonate while the weaker base pyridine can isomerise the lactone form. The reaction is slow to reach equilibrium and then typically gives an epimeric ratio of 3:1 in favour of the 2,3-*threo*-epimer [8]. After separation of the two epimers by crystallisation, the starting material can be recycled. However, due to the harsh conditions, the isolated yields of the two epimers are often moderate to low. As a result, the base-catalysed C-2 isomerisation is most useful in cases where the starting aldonolactone is very cheap and the C-2 epimer very expensive, as is the case for the isomerisation of galactonolactone to talonolactone and arabinonolactone to ribonolactone [8]. The latter isomerisation is in use industrially for the manufacture of riboflavine [9]. Although useful, these isomerisation procedures have not seen much additional development for several decades and therefore the methods will not be further discussed here. This chapter will focus on base-mediated isomerisations of aldonolactones via epoxides.

2 Epoxides of Aldonolactones

2.1 Preparation

Aldonolactones can be converted into epoxylactones with base if they have been regioselectively activated by introduction of leaving groups. The regioselective functionalisation is possible at the position α to the lactone and at the terminal primary (ω) position. The most efficient method for this functionalisation is treatment of the aldonolactone/aldonic acid with hydrogen bromide in acetic acid (Scheme 1). In this strongly acidic medium the lactone is partly acetylated followed by formation of acetoxonium ions. These then undergo opening with bromide ions to give acetylated bromodeoxyaldonolactones [10]. The formation of acetoxonium ions controls the regio- and stereoselectivity of introduction of the bromine: bromine will always open the acetoxonium ion at the primary position, as well as at C-2 of a 2,3-acetoxonium ion, with inversion of configuration.

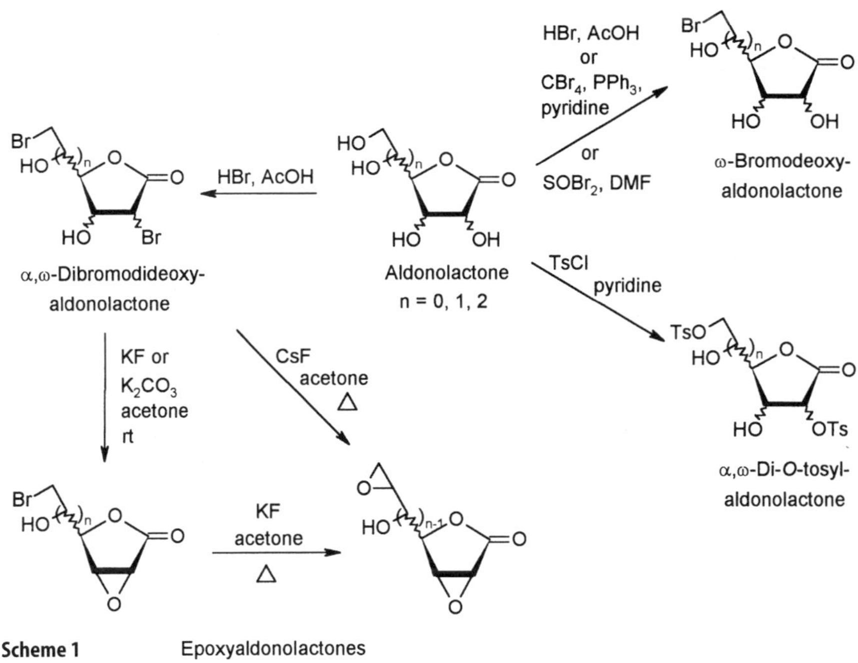

Scheme 1

2,3-Acetoxonium ions are formed in aldono-1,4-lactones that have the OH-2 and OH-3 in a *cis* orientation.

In two cases, however, bromine is also introduced at C-2 with inversion of configuration, although the OH-2 and OH-3 are in a *trans* orientation. Aldonic acids do in general adopt the 1,4-lactone form in strongly acidic media, but gluconic and xylonic acid exist to a certain degree as the open-chain acid form, and thus, because the OH-2 and OH-3 groups are *trans*-oriented in the 1,4-lactone, are able to form acetoxonium ions between two *trans* hydroxy groups from the open aldonic acids. Thus, D-glucono-1,5-lactone yields 2,6-dibromo-2,6-dideoxy-D-mannono-1,4-lactone, while 2,5-dibromo-2,5-dideoxy-D-lyxono-1,4-lactone is formed from D-xylonic acid by treatment with hydrogen bromide in acetic acid. In general, bromodeoxyaldonolactones adopt the 1,4-lactone form [1b].

Regioselective bromination of the primary position can also be achieved with tetrabromomethane (carbon tetrabromide) and triphenylphosphine in pyridine [11] or with thionyl bromide in dimethylformamide (DMF) [12]. Regioselective mesylation or tosylation of the primary hydroxy group, however, is less efficient due to competing sulfonylation of the α-hydroxy group. Instead, α,ω-di-*O*-tosylated aldonolactones can in some cases be formed in good yields by using 2.0–2.3 equiv of tosyl chloride in pyridine [13]. The 2-*O*-tosylated aldonolactones thus formed have *the same configuration as the starting lactones and thus are C-2 epimers to the bromodeoxy aldonolactones.* Access to new activated lactones has thus been achieved.

Epoxylactones can be prepared from the α,ω-dibromodideoxyaldonolactones by treatment with base under non-aqueous conditions. The bromine α to the lactone is more reactive than the primary one. As a result a 2,3-epoxide can be formed selectively by treatment with potassium fluoride or carbonate in acetone or acetonitrile at ambient temperature [14]. A bis-epoxide can be formed by prolonged treatment or by using the stronger base cesium fluoride [14] (Scheme 1).

2.2
Reactions in Aqueous Base

When epoxylactones or bromodeoxyaldonolactones are treated with excess aqueous sodium or potassium hydroxide, ring opening of the lactone will take place. In addition, the resulting epoxyaldonic acid can undergo a Payne rearrangement under these basic conditions. The Payne rearrangement is the isomerisation of a 2,3-epoxy alcohol with inversion of configuration at carbon number 2 (Scheme 2) [15]. During the isomerisation less stable epoxides will isomerise to more stable ones. In general, 1,2-disubstituted (secondary) epoxides are more stable than monosubstituted (primary) epoxides and, among the disubstituted epoxides, the *trans*-epoxides are more stable than the corresponding epoxides that possess *cis* configuration.

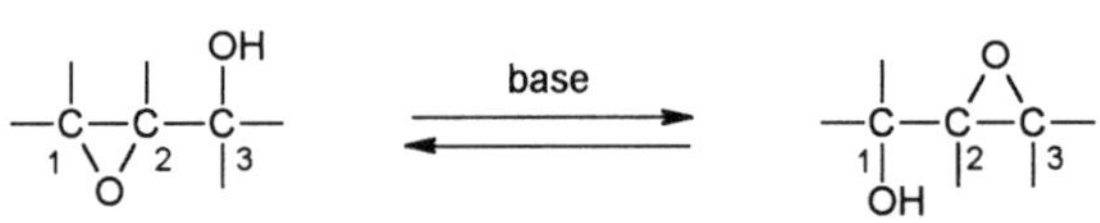

Scheme 2

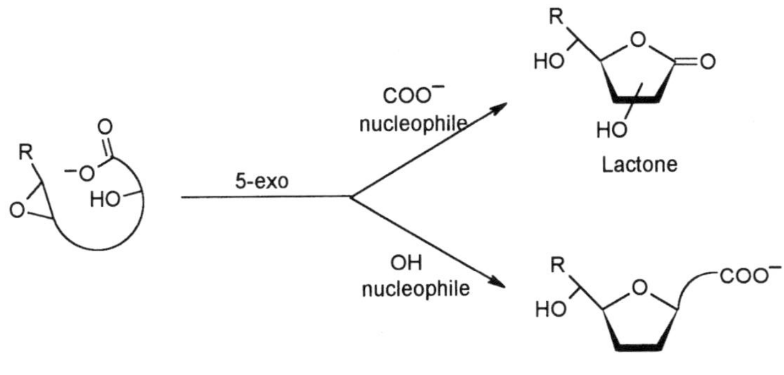

Scheme 3

However, the Payne rearrangement tends to produce mixtures of epoxides and is therefore of limited preparative value. The most important application of the Payne rearrangement occurs when one of the epoxides ring opens preferentially with a nucleophile under the reaction conditions and often intramolecular substitutions are energetically favoured compared with reactions with external nucleophiles. In the epoxyaldonic acids two different internal nucleophiles are present: the carboxylate group and the hydroxy groups. Ring opening of the epoxide with the carboxylate gives inversion of configuration at the attacked carbon, thereby producing a lactone, which opens easily in the basic medium to give an aldonic acid, isomeric to the starting aldonic acid (Scheme 3). Ring opening with a remote hydroxy group will produce an oxolane that, in this context, is considered an unwanted side reaction (Scheme 3). The ring opening of the epoxide will in both cases occur in a 5-*exo* mode according to Baldwin's rules [16]. In the following sections several methods for isomerisation of aldonolactones will be presented by formation and ring opening of epoxyaldonic acids.

3 Base-Induced Rearrangements of ω-Activated Aldonolactones via Primary Epoxides Formed in Situ

3.1 Stereoselective Inversion at C-4 in Pentonolactones

Aldonolactones activated at the primary position very rapidly form a primary epoxide of the open aldonic acid when dissolved in aqueous base. Usually, 3 molar equivalents of potassium hydroxide are used to keep the pH above 14. This is a prerequisite for a Payne rearrangement and/or for opening of the epoxide by an intramolecular nucleophile. Using the weaker base potassium carbonate, no rearrangement will take place and hence the primary epoxide will be opened at the primary carbon by hydroxide. Pentonolactones activated at C-5 will isomerise stereospecifically at C-4 when treated with strong potassium hydroxide. This is due to the opening of the primary epoxide by the carboxylate group, since this is the only possible intramolecular ring opening with a five-membered transition state with the epoxide in an *exo* mode.

When 5-bromo-2,5-dideoxy-D-*threo*-pentonolactone (**1**) (Scheme 4) was treated with aqueous potassium hydroxide and subsequently acidified, 2-deoxy-L-*erythro*-pentono-1,4-lactone (2-deoxy-L-ribonolactone) (**2**) was formed and isolated as the crystalline benzoate [17]. A stereospecific inversion at C-4 had

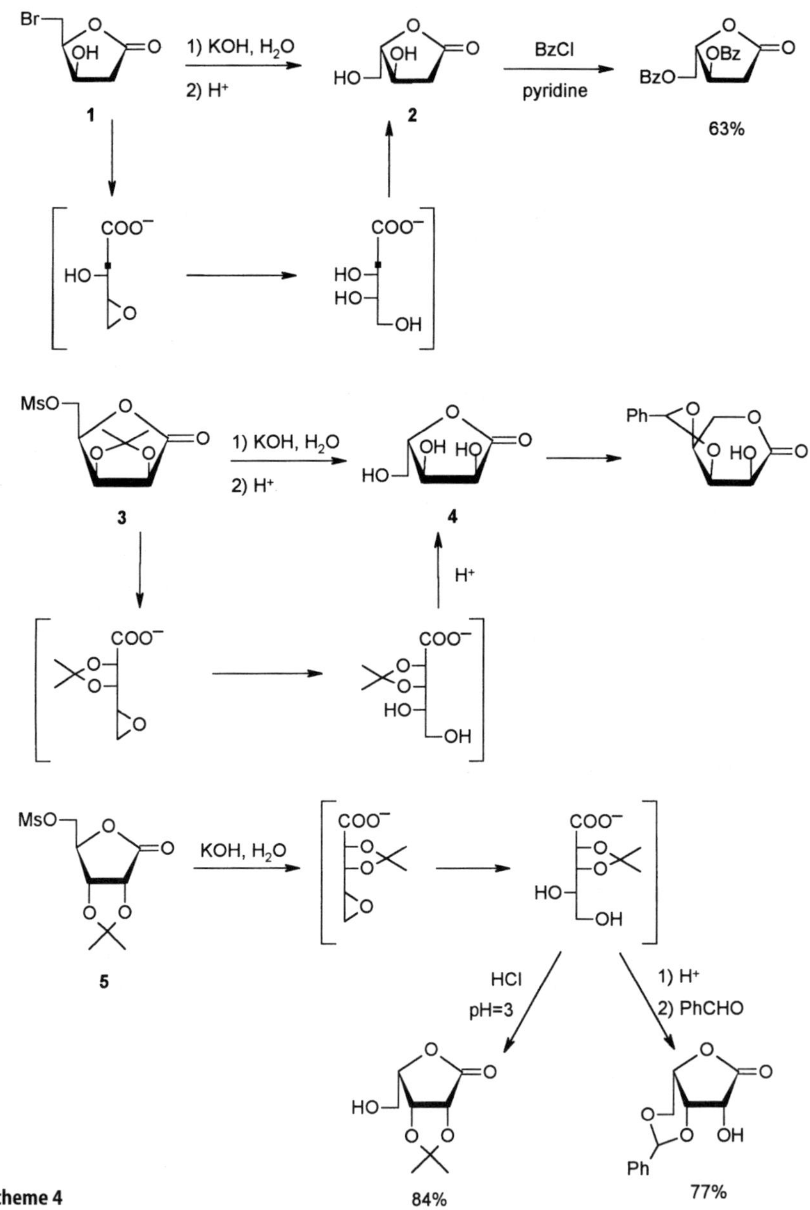

Scheme 4

taken place. Monitoring the reaction by ^{13}C NMR spectroscopy revealed immediate formation of the open 4,5-epoxycarboxylate. No other intermediates were observed before the appearance of the 2-deoxy-L-*erythro*-aldonate. The inversion at C-4 is a result of the opening of the epoxide by the carboxylate group leading to an intermediate lactone that is opened immediately by the aqueous base.

Similarly, 5-*O*-mesyl-2,3-*O*-isopropylidene-D-lyxonolactone (**3**) [18] was rearranged to L-ribonolactone (**4**) when treated with 3 molar equivalents of potassium hydroxide followed by acidification [19] (Scheme 4). The compound was isolated as the highly crystalline 3,4-*O*-benzylidene-L-ribono-1,5-lactone. The method was also used to prepare L-lyxonolactone from 5-*O*-mesyl-2,3-*O*-isopropylidene-D-ribonolactone (**5**) [19] but when during the workup the pH was adjusted to 3, the 2,3-*O*-isopropylidene-L-lyxonolactone was isolated in 84% yield [20] (Scheme 4). Again the reactions were monitored by ^{13}C NMR spectroscopy to observe the intermediate epoxides.

Using these simple procedures, which are easy to perform, a method for the synthesis of L-ribono- and L-lyxonolactone from D-lyxono- and D-ribonolactone, respectively, has been developed.

As pointed out above, aldonolactones/acids may isomerise at C-2 in the presence of strong base. No isomerisations were detected in the reaction of the 2,3-*O*-isopropylidene-protected lactones **3** and **5** with strong aqueous potassium hydroxide. Likewise, unprotected 5-bromo-5-deoxy-D-lyxono- and -D-ribonolactones give rearrangements to L-ribonic and L-lyxonic acids, respectively [19]. In these cases isomerisation at C-2 was not observed either. It has thus been confirmed that aqueous base requires higher temperatures in order to epimerise at the carbon center α to the carboxylate group.

3.2
Stereoselective Inversion of Configuration at More Than One Carbon Atom in Aldonolactones by Payne Rearrangements

When ω-activated aldonolactones with six or more carbon atoms are treated with strong aqueous base, the primary epoxides of the polyhydroxy carboxylic acids first formed may rearrange to secondary epoxides, which can finally be opened at C-4 by the carboxylate in a five-membered transition state. *The driving force, and thus the success of the stereoselective inversion at more carbon atoms in the polyhydroxy acids, is the final opening of an epoxide at C-4.* The method has been used to prepare a number of less available aldonic acids/lactones, as illustrated in the following examples.

6-Bromo-2,6-dideoxy-D-*arabino*-hexono-1,4-lactone (**6**) was cleanly transformed into 2-deoxy-L-*ribo*-hexono-1,4-lactone (**7**) [21] (Scheme 5). Interestingly, the product **7** can be converted into the starting lactone **6** by a three-step sequence: bromination at C-6 with HBr/HOAc, base-catalysed epimerisation to the *arabino*-lactone, followed by another bromination [21] (Scheme 5). Monitoring the reaction by ^{13}C NMR spectroscopy revealed immediate formation of the aldonate with a primary epoxy group, which rearranged to the secondary *trans*-4,5-epoxide, thereby inverting the configuration at C-5. The L-*ribo*-aldonate appeared within 5 min, as a result of opening of the epoxide by the carboxylate

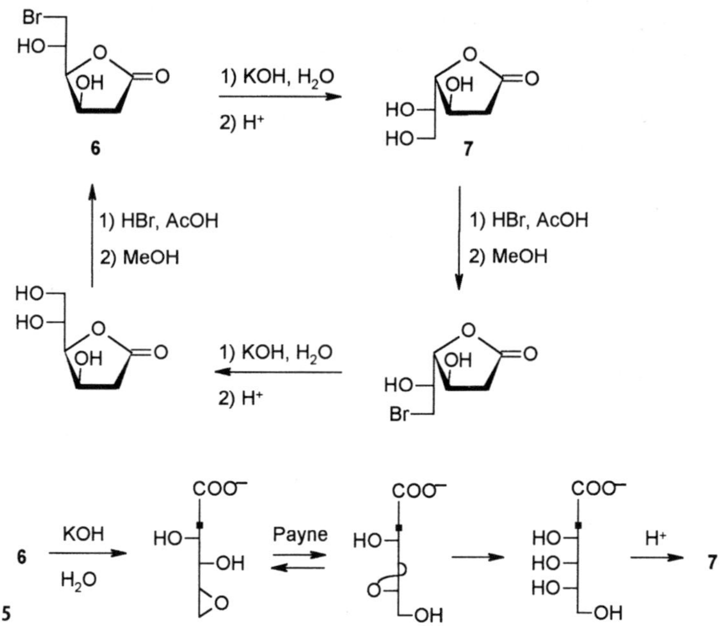

Scheme 5

group, thereby inverting the configuration at C-4 (Scheme 5). After treatment of **6** with 4 molar equivalents of aqueous potassium hydroxide, the reaction was complete within 30 min.

When the 3-deoxy-D-*arabino*- (**8a**) [22] or the 2,3-dideoxy-D-*erythro* analogue (**8b**) [23] was treated similarly, 3-deoxy-L-*ribo*- (**9a**) or 2,3-dideoxy-L-*erythro*-hexono-1,4-lactone (**9b**) was formed, respectively (Scheme 6). In all three cases a clean inversion at both C-5 and C-4 had taken place. As outlined above, this is explained by a mechanism involving rearrangement of the primary epoxide to a *trans*-secondary epoxide (Scheme 6). Both epoxides were observed as intermediates when monitoring the reactions by ^{13}C NMR spectroscopy [21, 22]. In the reaction of the 6-bromo-3,6-dideoxyhexonolactone with strong base, however, a minor amount of a five-membered tetrahydrofuran derivative **10** was also observed (17%). This might be formed from the 3-deoxy-5,6-epoxide by attack at C-5 by the hydroxy group at C-2 in competition with attack of the C-4 hydroxy group (Scheme 6) [22]. This is further discussed in Sect. 3.3. In all cases synthetic procedures for the preparation of the lactones **7**, **9a** and **9b** in good yields (Schemes 5 and 6) are outlined.

An interesting example of the rearrangements discussed above is the formation of the enantiomeric lactone from a 7-bromo-7-deoxy-heptonolactone, which means that the configurations at all chiral centers were inverted. For example, when 7-bromo-2,3,7-trideoxy-D-*arabino*-heptono-1,4-lactone (**11**) was treated with strong potassium hydroxide, 2,3-dideoxy-L-*arabino*-heptono-1,4-lactone (**12**) was formed and isolated as the crystalline peracetate in 47% yield [24]. In contrast, when **11** was treated with aqueous potassium carbonate, work-up gave 2,3-dideoxy-D-*arabino*-heptono-1,4-lactone (**13**), i.e. retention of con-

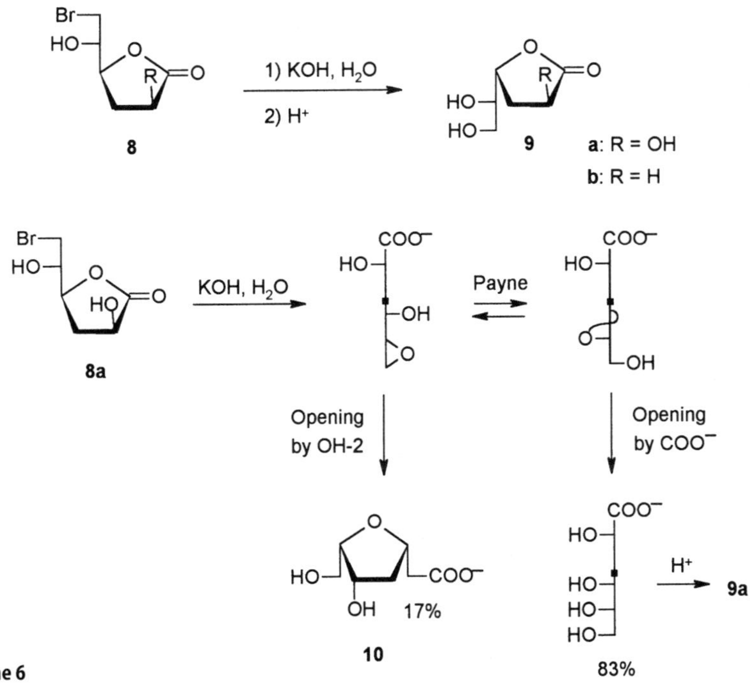

Scheme 6

figuration at all stereocenters. Accordingly, depending on the strength of the base used, either of the enantiomeric 2,3-dideoxy-*arabino*-1,4-heptonolactones can be isolated [24]. The mechanism, which is outlined in Scheme 7, explains the difference. The initially formed 6,7-epoxyaldonate **11a** will not undergo a rearrangement in the weaker base, potassium carbonate (pH ~12), as discussed above, and is thus opened at the less-hindered primary carbon by the external nucleophile hydroxide to give the aldonate **11b**. Acidification gives the D-*arabino*-lactone **13**. In the more strongly basic potassium hydroxide (pH >14) the epoxide **11a** rearranges and in the ^{13}C NMR spectra the 5,6-*trans*-epoxide **11c** was observed. No other epoxides were observed before the final aldonate **11e** appeared. Acidification in this case gave the L-*arabino*-lactone **12**. The L-*arabino*-aldonate **11e** is, however, not obtained by opening of the *trans*-epoxide **11c**, but from the *cis*-epoxide **11d**. The equilibrium between the two epoxides **11c** and **11d** is shifted towards the more stable *trans*-epoxide, but the requirements for an internal nucleophilic opening by the carboxylate is only fulfilled by **11d**.

3.3 Formation of Five-Membered Cyclic Ethers (Oxolanes) in Competition with Rearrangements of Epoxides

The general requirement for intramolecular nucleophilic opening of an epoxide is, as discussed in the introduction, formation of a five-membered transition

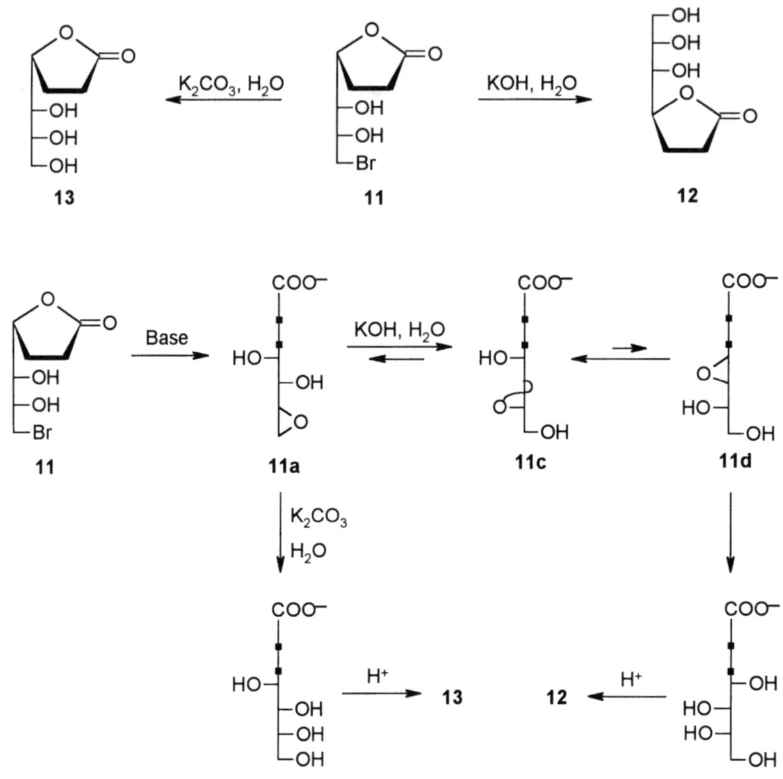

Scheme 7

state that has the epoxy group in an *exo* mode. When a hydroxy group is in such a position this will open the epoxide to give an oxolane, as illustrated in Schemes 3 and 6. In the treatment of 6-bromo-3,6-dideoxy-D-*arabino*-hexono-1,4-lactone (**8a**) with a base (Scheme 6) formation of a minor amount of the oxolane **10** was an unwanted side reaction. In other cases the oxolane formation was observed to be the main product, as illustrated in Scheme 8. Thus, when 7-bromo-2,7-dideoxy-D-*gluco*-heptono-1,4-lactone (**14**) was treated with potassium hydroxide, a ^{13}C NMR spectrum of the reaction mixture showed after 5 min the oxolane **15** as the main product, together with the primary epoxide **14a**.

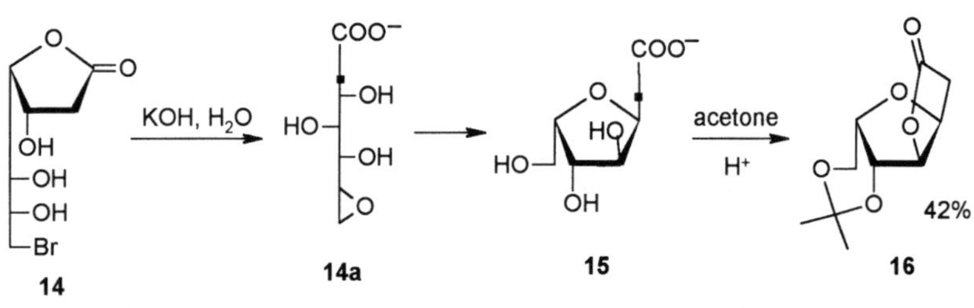

Scheme 8

The reaction was complete within 2–3 h when the ^{13}C NMR spectrum showed **15** and a minor unidentified product in a ratio of 4:1. After acidification and treatment with acetone, the isopropylidene-protected lactone **16** was isolated in 42% yield based on **14** [24].

These examples indicate that ω-bromodeoxyhexono- and higher aldonolactones on treatment with base will give primary epoxides which directly, or after epoxide rearrangements, can be opened by a remote hydroxy group, rather than the carboxylate group, to yield oxolanes. This also proved to be the case when 6-bromo-6-deoxy-D-galactono-, -D-altrono- or -D-mannono-1,4-lactone was treated with strong potassium hydroxide. They all gave mixtures of aldonates with inverted configurations at C-4 and C-5, together with oxolanes [25]. Thus, a competition exists between attack of the neighbouring hydroxy group to give a three-membered ether (secondary epoxide) and a remote hydroxy group to give a five-membered ether (oxolane) according to Baldwin's rules. The preferred pathway will depend on the relative energy in the respective transition states leading to the two types of products and will be dependent on the relative stereoisomeric arrangements of the hydroxy groups in the aldonates.

3.4 Base-Induced Rearrangement of α,β-Epoxyaldonolactones. Synthesis of L-Gluconic Acid from D-Gluconolactone

In the base-induced rearrangements of epoxyaldonolactones discussed so far, the starting epoxide was the primary one, which rearranged to a secondary, more stable 1,2-disubstituted epoxide. We then asked the question whether α,β-epoxyaldonolactones, which can be prepared from α-bromodeoxylactones [14] (Scheme 1), might be able to participate in similar base-induced rearrangements, although secondary epoxides "conjugated" to a carboxylic function are rather stable [26]. We have now found that, by treatment with strong potassium hydroxide, 2,3-anhydro-6-bromo-6-deoxy-D-mannono-1,4-lactone (**18**) can be transformed cleanly into the L-gluconic acid/lactone **19** (Scheme 9) [27]. By monitoring the reaction by ^{13}C NMR spectroscopy the diepoxide **18a** was observed, immediately followed by rearrangement of the primary epoxide to a secondary one, to give a secondary di-epoxide, probably **18b**. The disappearence of **18b** and appearance of an epoxyaldonate, probably **18c**, could be explained by the nucleophilic opening of the 4,5-epoxide at C-4 in **18b** by the carboxylate group. By these rearrangements inversions at C-5 and C-4 had taken place. After 3 days the L-gluconate **19** was the final and only product present. No other intermediates were observed before the final aldonate **19** appeared in the spectra. The structure has been confirmed by isolation and identification as the ethyl L-gluconate and the L-gluconamide.

The mechanism, which explains the configuration of the final product, involves a rearrangement of the epoxide **18c** *away* from the carboxylate group through the *trans*-epoxides **18d** and **18e**, where the epoxy group is in a position to be opened at C-4 by the carboxylate (Scheme 9). It is worth noting that none of these intermediates were present in detectable amounts during the reaction, showing both the stability of the 2,3-epoxide and the favoured opening at C-4 in

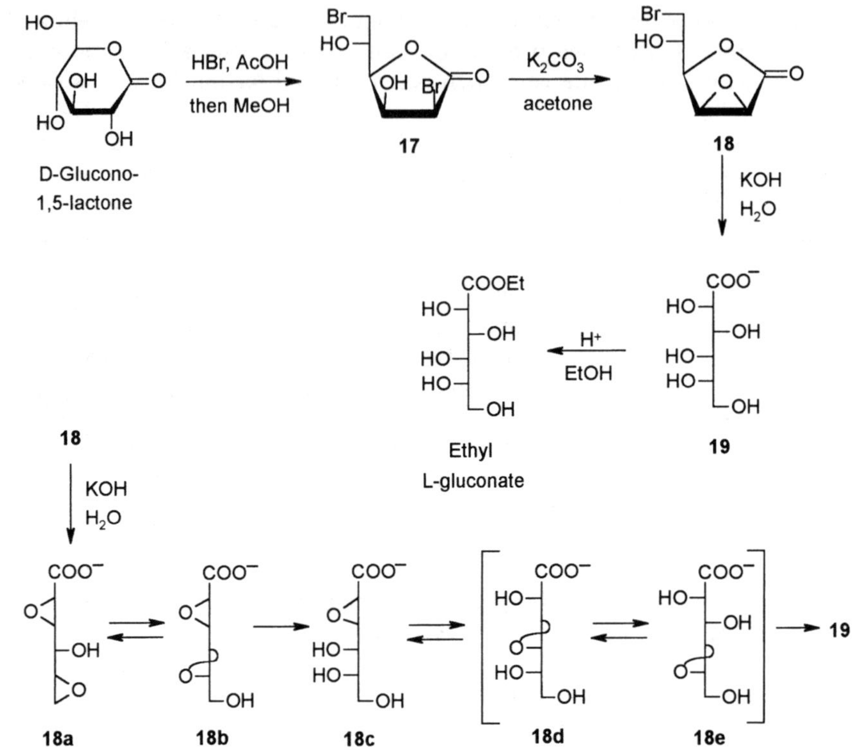

Scheme 9

an *exo* mode in accordance with Baldwin's rules, as discussed above. Compared with the epoxide **18c** the configuration at C-3 in **19** has been inverted [27]. The bromoepoxylactone **18** can be synthesised in two steps from D-gluconolactone [14], and thus a synthetic procedure has been developed to obtain the enantiomeric L-gluconic acid/lactone by a range of stereospecific inversions at all four chiral centers in D-gluconolactone. A simple method for preparation of L-glucose by reduction of the ethyl L-gluconate with sodium borohydride has also been demonstrated [27].

Similarly, we have synthesised 6-deoxy-L-altrono-1,4-lactone (**22**) from 6-deoxy-L-mannono-1,4-lactone (L-rhamnonolactone) [27] (Scheme 10). In this synthesis, the L-rhamnonolactone was converted into the 2-bromodeoxylactone **20** and further to the 2,3-epoxylactone **21** [14]. Treatment with strong base gave a final aldonate that, after acidification, gave the 1,4-lactone **22**, identified as the peracetate. The mechanism was again deduced from ^{13}C NMR spectra of the reaction mixture, as outlined in Scheme 10.

An interesting feature of these two examples is the strong driving force for an intramolecular opening of an epoxide according to Baldwin's rules. In these cases the Payne rearrangement causes equilibration between several epoxides, whereby one of those will finally be opened by the first nucleophilic group that is in the right position. In the examples starting from a 2,3-epoxide, the nucleo-

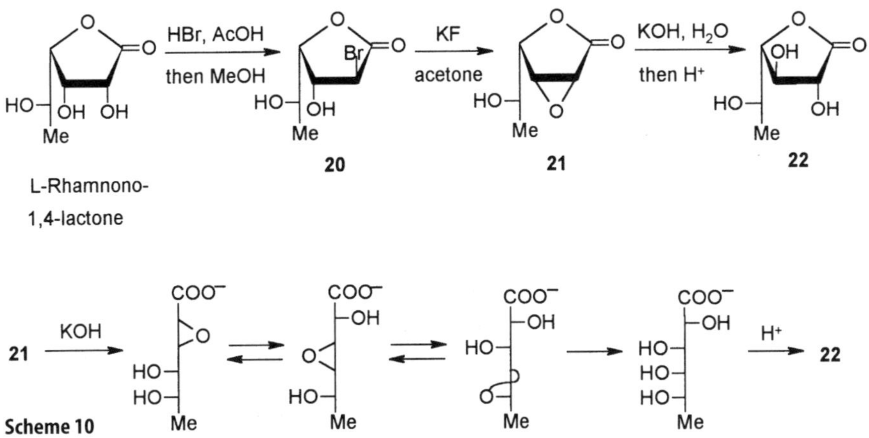

Scheme 10

phile will always be the carboxylate group, the first internal nucleophile to fulfill Baldwin's rules.

At this point a comment is required about the formation of the 2,3-epoxy-D-mannonolactone **18** from the bromodeoxy-D-mannono-1,4-lactone **17** that has the bromine and the hydroxy groups in a *cis* relationship (Scheme 9). Obviously, a *trans* relationship between the bromine and the hydroxy group is necessary to form the epoxide within a five-membered ring.

As discussed in the introduction aldonolactones may epimerise at C-2 when treated with strong base due to the acidity of the H-2 proton. This acidity is enhanced by substituents at C-2 that are more electronegative than a hydroxy group. We have shown that both bromine and azido substituents at C-2 are configurationally unstable in the presence of base, also as weak bases as azide ions in acetonitrile or DMF [28]. We have also shown that 2-bromo-2-deoxy-lactones isomerise at C-2 in a, preferably polar, organic medium in the presence of bromide ions [29]. Hence, treatment of the *manno*-configured bromodeoxylactone **17** in acetone or acetonitrile with solid anhydrous potassium carbonate gives the *manno*-configured epoxylactone **18** quantitatively within 30 min [14]. The epoxylactone **18** was also prepared from the C-2-epimeric bromodeoxylactone with the *gluco* configuration [14]. This is prepared from D-mannonolactone, while **17** is prepared from the inexpensive D-gluconolactone [14].

The advantage of using the prepared 2,3-epoxylactones for the study of the base-induced rearrangements discussed above is now clear. Direct treatment of the 2-bromolactones with *aqueous* base will cause rapid epimerisation at that center, and thus in aqueous base the bromodeoxylactone **17** will give both the 2,3-*cis*- as well as the 2,3-*trans*-epoxide of the open aldonates. In the lactone form of course only *cis*-epoxides can be formed. The procedure for preparation of L-gluconic acid is thus performed by stirring **17** in acetone with excess of solid anhydrous potassium carbonate for about 30 min, followed by filtration, concentration, addition of water, and 3 molar equivalents of potassium hydroxide. After 3 days at room temperature the mixture is acidified and the product isolated as the ethyl ester [27].

4 Conclusions

Some general features can be drawn from these investigations. On treatment of a ω-bromo-ω-deoxyaldonolactone with aqueous base a primary epoxide of an aldonate was formed immediately. Using a weak base (potassium carbonate, pH <14) this epoxide was hydrolysed by the solvent. In a stronger base (potassium hydroxide, pH>14), Payne rearrangement took place yielding 1,2-disubstituted (secondary) epoxides preferentially. *trans*-1,2-Disubstituted epoxides were formed faster and are thermodynamically more stable than their *cis*-disubstituted counterparts. Epoxide migration to give *cis*-1,2-disubstituted epoxides was only detected when these subsequently underwent intramolecular ring opening. Opening of a final epoxide, formed by epoxide rearrangements, was favoured by a five-membered transition state that possessed the epoxy group in the *exo* mode. The carboxylate or a remote hydroxy group can act as a nucleophile, resulting either in an aldonate with inverted configuration or in an oxolane.

Synthetically useful procedures for the preparation of less readily available aldonic acids from bromodeoxylactones by treatment with strong bases have been demonstrated. These very simple procedures, with no or very little use of protecting groups, are attractive compared with other synthetic methods for the same compounds.

5 References

1. (a) Rassu G, Zanardi F, Battistini L, Casiraghi G (1999) Synlett 1333; (b) Lundt I (1997) Top Curr Chem 187:117; (c) de Lederkremer RM, Varella O (1994) Adv Carbohydr Chem Biochem 50:125; (d) Fleet GWJ (1993) Sugar lactones as useful starting materials. In: Krohn K, Kirst HA, Maas H (eds) Antibiotics and antiviral compounds: chemical synthesis and modifications. VCH, Weinheim, p 333
2. Lundt I, Madsen R (1999) Iminosugars as powerful glycosidase inhibitors – synthetic approaches from aldonolactones. In: Stütz AE (ed) Iminosugars as glycosidase inhibitors – nojirimycin and beyond. Wiley-VCH, Weinheim, p 93
3. Lundt I (in press) Reaction on C: Oxidation, reduction and deoxygenation at the anomeric position. In: Fraser-Reid B, Tatsuta K, Thiem J (eds) Glycoscience: chemistry and chemical biology. Springer, Heidelberg
4. Györgydeák Z, Pelyvás IF (1998) Monosaccharide sugars – chemical synthesis by chain elongation, degradation and epimerisation. Academic Press, San Diego, p 18
5. Györgydeák Z, Pelyvás IF (1998) Monosaccharide sugars – chemical synthesis by chain elongation, degradation and epimerisation. Academic Press, San Diego, p 450
6. (a) Isbell HS, Frush HL (1979) Carbohydr Res 72:30; (b) Czarnocki Z, Mieczkowski JB, Ziólkowski M (1996) Tetrahedron Asymmetry 7:2711
7. Recently, a two-step two-carbon chain elongation of aldoses by a Wittig/dihydroxylation protocol has been developed: Jørgensen M, Iversen EH, Madsen R (2000) 20th Int Carbohydrate Symposium, 27.08–01.09, Hamburg, Germany, poster B-158
8. (a) Lundt I, Madsen R (1993) Synthesis 720; (b) Steiger M (1936) Helv Chim Acta 19:189; (c) Bonnett HT, Upson FW (1933) J Am Chem Soc 55:1245 and references cited therein
9. Kalman M, Sokolowski J, Szafranek J (1987) J Carbohydr Chem 587 and references cited therein
10. For an overview of the available bromodeoxylactones by this method, see Lundt I (1997) Top Curr Chem 187:117

11. (a) Lundt I, Frank H (1994) Tetrahedron 50:13285; (b) Kold H, Lundt I, Pedersen C (1994) Acta Chem Scand 48:675
12. Bouchez V, Stasik I, Beaupère D, Uzan R (1997) Carbohydr Res 300:139
13. Lundt I, Madsen R (1992) Synthesis 1129
14. Lundt I, Pedersen C (1992) Synthesis 669
15. Payne GB (1962) J Org Chem 27:3819; Behrens CH, Ko SY, Sharpless KB, Walker FJ (1985) J Org Chem 50:5687
16. Baldwin JE (1976) J Chem Soc Chem Commun 734
17. Bock K, Lundt I, Pedersen C (1981) Carbohydr Res 90:17
18. Godskesen M, Lundt I, Søtofte I (2000) Tetrahedron Asymmetry 11:567
19. Kold H, Lundt I, Pedersen C (1994) Acta Chem Scand 48:675
20. Godskesen M, Lundt I, Madsen R, Winchester B (1996) Bioorg Med Chem 4:1857
21. Bock K, Lundt I, Pedersen C (1984) Acta Chem Scand B 38:555
22. Bock K, Lundt I, Pedersen C (1986) Acta Chem Scand B 40:163
23. Pedersen C, Schubert L (1993) Acta Chem Scand 47:885
24. Bock K, Lundt I, Pedersen C (1988) Carbohydr Res 179:87
25. Frank H, Lundt I unpublished results
26. Sorriso S, Stefani F, Semprini E, Flamini A (1976) J Chem Soc Perkin 2 374
27. Lundt I (2000) 20th Int Carbohydrate Symposium, 27.08–01.09, Hamburg, Germany, poster B-143
28. Bols M, Lundt I (1988) Acta Chem Scand B 42:67
29. Bock K, Lundt I, Pedersen C (1987) Acta Chem Scand B 41:435

Rearrangements in the Course of Nucleophilic Substitution Reactions

Karl Dax, Martin Albert

Institute of Organic Chemistry, Technical University of Graz, Stremayrgasse 16, 8010 Graz, Austria
E-mail: dax@orgc.tu-graz.ac.at

Nucleophilic substitutions in polyfunctional heterocyclic compounds such as carbohydrates are accompanied by a series of side reactions including rearrangements. As the mechanistic interpretation for their different types has disclosed the same steric prerequisites and even common intermediates, in this compilation the whole set of side reactions observed in each single case is presented. The neighbouring group (NG) attacks either in a "simple" way [by the one atom which is directly bound to the sugar chain; this form requires antiperiplanar orientation of the NG to the leaving group (LG)] or in a "complex" type [by its third atom (usually a carbonyl oxygen), for which, in pyranoid systems, a *trans*-diequatorial arrangement of LG and NG is essential]. Possible reactions, besides straight S_N2 by an external nucleophile (including the solvent) or internal nucleophilic atom under formation of a new ring system (**IS**) as well as elimination (**E**), arise from anchimeric assistance by a NG and the ring heteroatom, respectively. The latter either causes substitution with retention of configuration (**AS**) or migration, to the reaction centre, of the participating group (**MS**). Special types result when in the "complex" case the intermediate acyloxonium ion reacts to give derivatives of an orthoester (**ISC**) and, generally, besides the anchimerically assisting atom, also the attacking group is an internal one (**AIS/AISC** and **MIS**), respectively. Other paths are opened by the involvement of carbon atoms from the sugar chain, which gives rise either to ring contraction with formation of a formyl side chain (**MS**) or causes fragmentation (**F**).

Keywords: Anchimeric assistance, Carbohydrates, Deamination, Fragmentation, Group migration, Neighbouring group participation, Nucleophilic substitution, Rearrangement, Ring contraction, Sulfonate displacement

Abbreviations

sym-collidine	2,4,6-trimethylpyridine
DAST	diethylaminosulfur trifluoride
diglyme	diethylene glycol dimethyl ether
DTBMP	2,6-di-*tert*-butyl-4-methylpyridine
HF*i*P	1,1,1,3,3,3-hexafluoro-2-propanol
MeCN	acetonitrile
TASF	tris(dimethylamino)sulfonium difluorotrimethylsilicate
TEA	triethylamine
TMU	*N,N,N′,N′*-tetramethylurea
TPP	triphenylphosphane

Symbols with relation to the reaction mechanism are listed and explained in Sect. 4.1.

1 Introduction

The migration of a single group from one atom to another within the same molecule is called a rearrangement reaction. As the group that migrates in the course of a nucleophilic substitution reaction has to carry the electron pair with it, this transformation is also classified as an intramolecular nucleophilic substitution. When taking mechanistic details into consideration, such as, e.g., the existence of a charged cyclic intermediate or stability criteria for carbenium ions, the close relationship of this type of reaction to other possible pathways like substitution, by an external nucleophile, with retention of configuration and side reactions like elimination or fragmentation becomes evident. Thus, in this chapter, the primary aim is to give a synopsis of the whole reaction pattern observed in selected nucleophilic displacement or solvolysis reactions. Secondly, the potential of some recent developments made in this area for synthesis will be briefly highlighted.

2 The Nucleophilic Substitution and Its Possible Side Reactions

The nucleophilic substitution reaction at alkyl carbons represents one of the most investigated types of reactions in organic chemistry [1]. The plethora of data concerning its dependence on factors like the structure of the educt (neutral or charged; constitution, configuration and possible conformation), the character (neutral or charged; base strength) of the leaving group as well as the attacking nucleophile (nucleofugicity vs. nucleophilicity) and, especially, the solvent (polarity, ionising power, possible nucleophilic assistance) and other reaction parameters (pH-value, catalyst, temperature) represents a unique training facility in chemical education. For a comprehensive mechanistic explanation, despite the generally accepted types of a mono- (S_N1) or bimolecular (S_N2) substitution, their "simultaneous operation" and several other possibilities like the concept of "ion pairs which react at different states of solvation" (a mechanism somewhere "in-between S_N1 and S_N2") or a "single-electron transfer (SET) mechanism" have been proposed to cover even borderline phenomena [1]. In particular, the controversies surrounding a possible discrete intermediate in such reactions, the classical or nonclassical carbenium ion, have given much impact to the general understanding and further development of organic chemistry. With the studies on the scope of nucleophilic substitution reactions, even with monofunctional educts, their close relationship to elimination (E1, E2, E1cA and E1cB) as well as intramolecular, nucleophilic rearrangement reactions (alkyl and hydride shifts) has become evident. For many instances, depending on the governing mechanism, common intermediates (classical/nonclassical carbenium ions or ion-pairs) or identical transition states (E2C/E2H) have been proposed.

In the case of educts which contain other functionalities of nucleophilic character (OH/OR, SH/SR, NH_2/NHR/NR_2 with R = alkyl, acyl, alkyl- or arylsulfonyl, etc.) besides the leaving group (LG), additional reaction pathways are opened up by a possible intramolecular participation of these neighbouring groups (NGs). As exemplified with the participation of oxygen-containing species from a vicinal position (Schemes 1 and 2), the paths may be classified, according to the mode of interaction involved, as "simple" (Scheme 1) and "complex" (Scheme 2) and include the following possibilities:

1. Direct intramolecular nucleophilic substitution (by unsubstituted OH, SH, NH_2 or NHR) with formation of a heterocyclic ring (route **a**, symbol **IS**).
2. "Anchimeric assistance", which results – via formation of positively charged cyclic intermediates (route **b** or **b′**) – either in (an enhanced rate of) direct substitution by the external nucleophile under retention of configuration [route (**b** + **c**) or (**b′** + **c′**), symbol **AS**] or nucleophilic migration of the neighbouring group and entry of the external nucleophile at the vicinal position with configurational inversion at both ends [route (**b** + **d**) or (**b′** + **d′**), symbol **MS**].
3. Additionally, attack of the external nucleophile at the first C-atom of the participating group in the charged intermediate can occur to give an anhydro ring (route **b** + **e**, "simple" NG, symbol **IS**) or a cyclic orthoester derivative [route (**b′** + **e′**) in the case of "complex" type, symbol **ISC**]; the latter product may, when water is used as the nucleophile, lead to (one or both of) the isomeric diol monoesters shown in Scheme 2 (symbol **Hy**) with inverted configuration at the position of the original activation [for clarity it should be added that these latter structures are also prone to *O*,*O*-acyl migration (symbol **AcM**)]. Thiols and their acylated derivatives react analogously.

For NGs consisting of an acylated amino group (as shown in Scheme 3 with amides and urethanes) participation may occur either via the amino nitrogen ("*N*-attack", "simple" case, **IS**) or the carbonyl oxygen ("*O*-attack", "complex" case, **ISC**). Thus oxazolines **1** or aziridines **2** and oxazolones **3**, respectively, are formed.

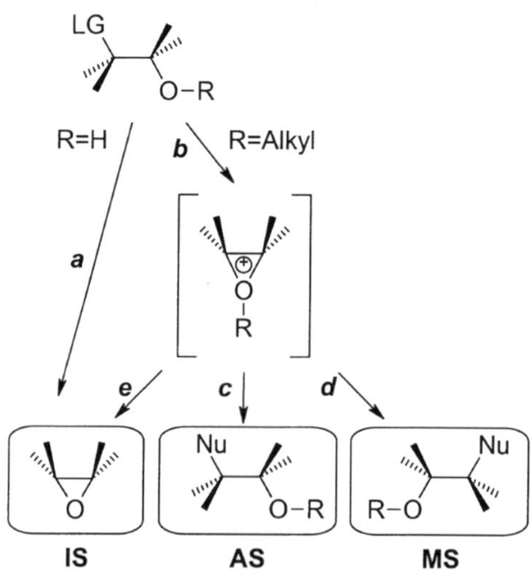

Scheme 1

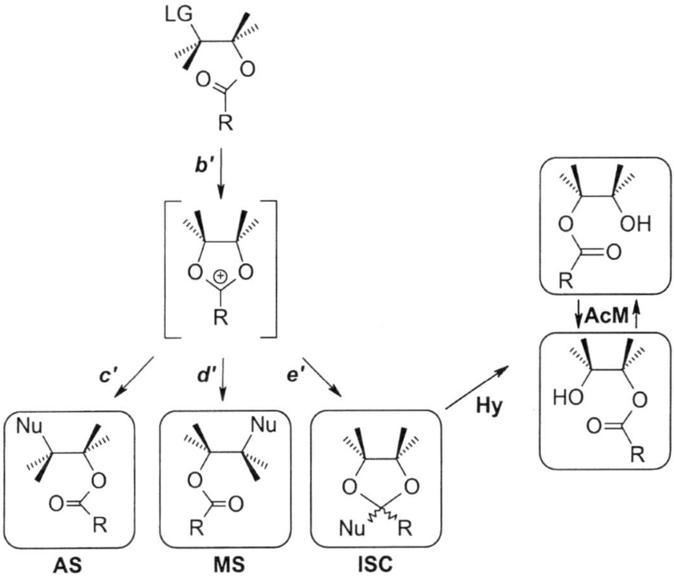

Scheme 2

In cyclic systems, these modes of intramolecular participation from vicinal positions are restricted to a *trans*-orientation between the LG and the participating NG. Strictly speaking, for the attack of the carbonyl oxygen in educts containing an ester, amide or urethane grouping (see Scheme 2 and the cases of "*O*-attack" in Scheme 3), an arrangement close to a *trans*-diequatorial one is essential, whereas for reactions leading to three-membered rings (see **IS**, Schemes 1 and 3) the existence of a conformation containing an antiperiplanar orientation of the LG and the NG (at least under the reaction conditions applied) is required.

An antiperiplanar steric relationship between the LG and the respective participating group is also a general prerequisite for the remaining types of side

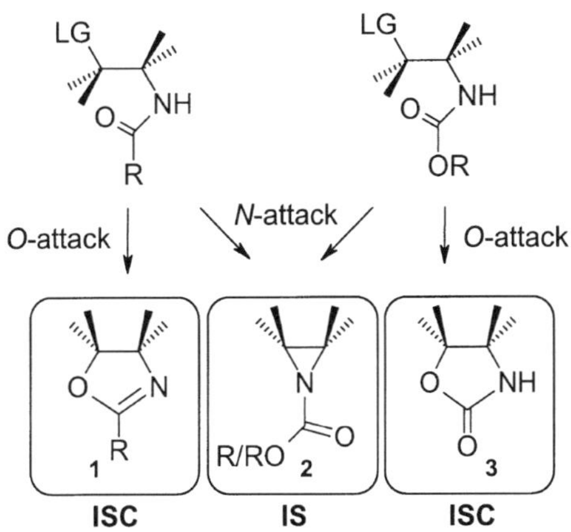

Scheme 3

reactions of nucleophilic displacements. These are two mechanistically closely related pairs of reactions: hydride and alkyl shift on the one hand, elimination (E2) and fragmentation on the other. Whereas within the first pair a hydrogen atom or an alkyl group migrates intramolecularily as a nucleophile (together with the electron pair of the former covalent bond), in the second case, a proton or carbocation, the latter necessarily stabilised by an unshared pair of electrons from an OH/OR or NR_2-substituent, is released (as an electrofuge) from the vicinal position to the LG.

Concerning the mechanisms of these kinds of reactions, currently no general pattern can be given. However, useful hints are to be found in the stereochemical consequences of the reaction inasmuch as concertedness manifests itself by stereochemical homogeneity of the final product, whereas the existence of intermediates (stabilised carbenium ions) will, to some extent, cause racemisation at the centre in question. In Scheme 4 plausible mechanistic versions of the side reactions in nucleophilic displacements as observed in oligofunctional educts are depicted (together with the corresponding symbols and, in brackets, the specification of the participating atom). These include hydride shift (**MS**[H], route **f**), alkyl shift (**MS**[C], routes **g**), elimination (**E**[H], route **h**) and fragmentation (**F**[C], routes **i**), respectively. [Depending on the kind of protecting group present in the educt, some of the products thus formed (especially hemiacetals and enol esters) are prone to rapid hydrolysis (**Hy**).]

The strong connection between some of these different routes and the stereochemical parameters discussed above is illustrated in Table 1 with the results obtained by Sicher and his group [2] from the nitrous acid deamination of the four diastereomeric 2-amino-4-*tert*-butylcyclohexanols, which predominantly

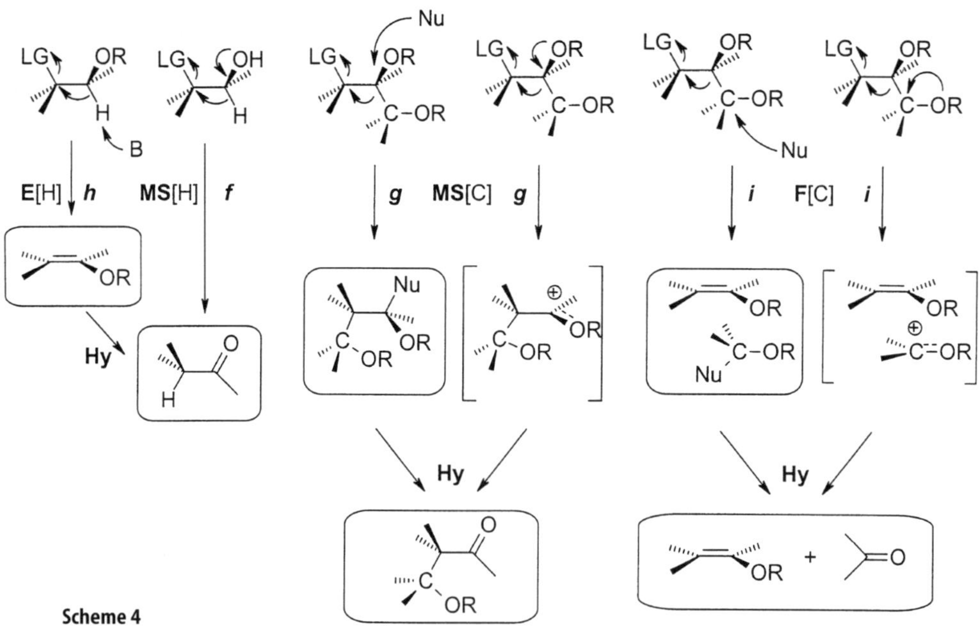

Scheme 4

Table 1

	Intermediate	Main Product	Conditions	yield (%)	Ref.
1	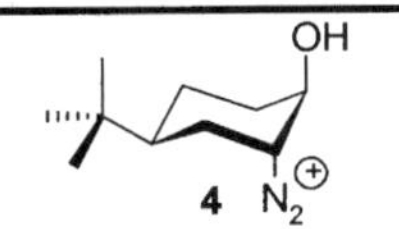 **4**	 **IS**[O] **5**	NaNO$_2$, HClO$_4$, H$_2$O [pH4] pentane, *trans*-decaline 0 °C, 2 h	77	2
2	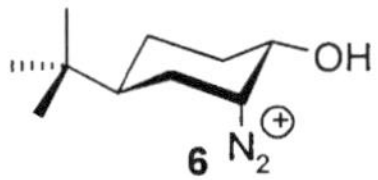 **6**	 **MS**[H] **7**		76	
3	 **8**	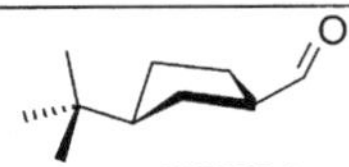 **MS**[C] **9**		90	
4	 **10**	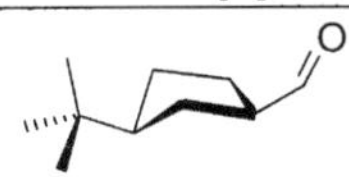 **MS**[C] **9**		98	

assume a chair conformation with the voluminous *tert*-butyl group in equatorial orientation. [Treatment of primary amines with nitrous acid leads to alkyldiazonium compounds (RN_2^+, as shown with structures **4**, **6**, **8** and **10**). These release N_2 (one of the best leaving groups) to generate highly reactive, unsolvated ("hot") carbenium ions, which are believed to allow the formation of the final products determined by the conformation of the ground state (rather than by electronic factors in the transition state).]

In each case, product formation was found to have been dominantly or exclusively determined by the participation of an antiperiplanar arranged neighbouring group. Thus, from the educts containing an axially oriented amino group, the one with *trans*-diaxially arranged substituents (entry 1) gave anhydro compound **5** (by intramolecular S_N2 reaction according to **IS**, route **a**, Scheme 1), whereas in the other case (entry 2), deoxy ketone **7** was formed (by a hydride shift corresponding to **MS**[H], route **f**, Scheme 4). In the reactions of structures **8** and **10**, which each contain an equatorially oriented LG, participation of the antiperiplanar C-1/C-6 bond (according to an alkyl shift **MS**[C], path **g**, Scheme 4) elapses independently of the configuration at C-1 and gives rise to ring contraction under generation of a formyl side chain (product **9**). The formation of the same product **9** (1–3%) in the reactions shown in entries 1 and 2 was rationalised [2] as having been caused by "some stereochemical leakage at the carbenium ion stage". The lower yields of unpolar products generally observed in these latter reactions as compared to those found in entries 3 and 4 were explained as the result of the concurring solvolysis of the intermediate diazonium compounds **4** and **6** to give polar glycols, which were not isolated.

In nucleophilic substitution reactions with educts derived from heterocyclic rather than alicyclic systems, the ring heteroatom (rO, rS or rN) may, under the steric requirement of an antiperiplanar arrangement with the (necessarily equatorial oriented) LG, participate in the reaction mechanism in a similar fashion to neighbouring groups in acyclic structures or side chains (see Scheme 1): **AS** [route (**b** + **c**)], **MS** [route (**b** + **d**)] or **IS** [route (**b** + **e**)]. This is exemplified in Scheme 5 by the solvolysis of (*R*)-3-chloro-1-ethylpiperidine (**11**), which on treatment with aqueous sodium hydroxide, by way of the intermediate aziridinium ion **12**, gives [3] the product of ring contraction **13** (**MS**[rN]; rN denotes the ring nitrogen) as well as of substitution with retention of configuration **14** (**AS**[rN]).

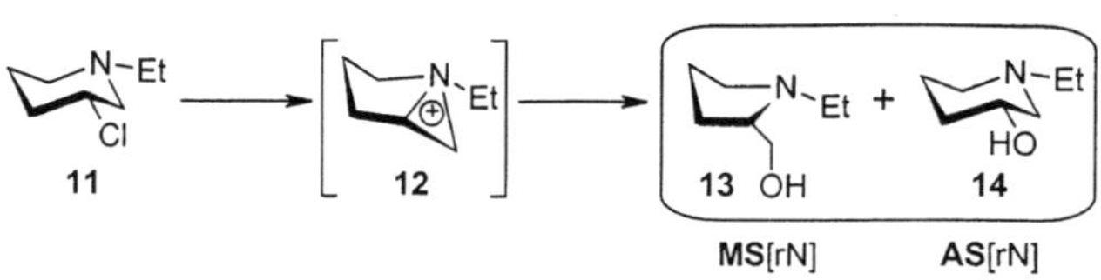

Scheme 5

3
The Nucleophilic Substitution and Its Side Reactions with Carbohydrate-Derived Educts

The higher number of given functionalities and, in particular, the exponential number of possible steric relationships with each other in carbohydrates allow an even more complex reaction pattern to be expected. A further multiplier to these options arises from any feasible conformational flexibility of the educt under the reaction conditions applied. Although the results reported in the literature so far do not cover reactions of all possible constitutional and configurational instances, and systematic investigations under comparable conditions are rare, essential questions have long been rationalised by using conformational principles. This is also evident from important reviews in this field, where either the mechanistic background of these reactions or their preparative potential has been illustrated. Of special interest on the subject of deamination reactions is the excellent contribution of Williams [4] and, on sulfonate displacement reactions, those of Goodman [5], Ball and Parrish [6], as well as Binkley [7]. Furthermore, Defaye [8] comprehensively described the formation of 2,5-anhydrides of sugars by ring-contraction reactions based on intramolecular nucleophilic substitution at C-2 of aldopyranosides by the tetrahydropyran ring oxygen. Hanessian and Pernet [9], Yoshimura [10], de Lederkremer and Varela [11], as well as Redlich [12] have also briefly highlighted the significance of this and other types of ring contractions in natural product synthesis.

In the following outline of the present state of our knowledge concerning side reactions of nucleophilic substitutions, emphasis will primarily be placed on the situation in deamination and "desulfonyloxylation" (sulfonate displacement) reactions with the detailed review of Williams [4] on the one hand and that of Ball and Parrish [6] on the other serving as points of reference. As it is not intended to discuss the actual mechanistic implications (S_N1- vs. S_N2-type) of each reaction presented, we propose the general "working hypothesis" that all steps in a single transformation elapse simultaneously except those parts where "hot" carbenium or stable oxocarbenium ions (including cyclic oxiranium, thiiranium or aziridinium ions, which contain C-1) are involved. The latter species is of importance in ring contractions and fragmentations where a C-atom, especially C-1, takes the role of an electrofuge. The main arguments serving as a basis for this mechanistic simplification consist of the stereochemical consequences deduced from the configuration of the respective products and the dependence of product ratios as well as reaction rates on the nucleofugicity of the LG as well as the nucleophilic strength of partners present.

Thus, beyond the seminal contributions on deaminations, side reactions were found to most frequently occur in attempted sulfonate displacements when poor nucleophiles, especially fluoride, or the conditions of solvolysis were applied. In this way, impetus for the development of better LGs, represented by trifluoromethanesulfonates (triflates) and imidazolesulfonates (imidazylates), as well as more powerful nucleophiles, e.g., tetraalkylammonium salts, was added. In the case of intended fluorinations, the classical "sulfonate/fluoride tandem" [usually triflate/tetrabutylammonium fluoride (TBAF) or triflate/tris(dimethylamino)

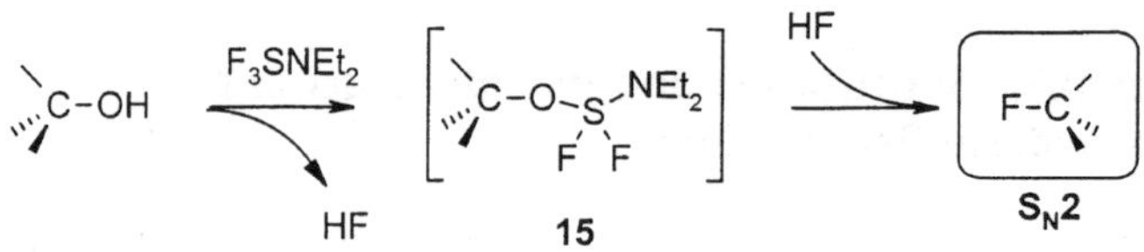

Scheme 6

sulfonium difluorotrimethylsilicate (TASF)] as well as the even more powerful "activation/fluorination route" employing diethylaminosulfur trifluoride (DAST) reached their limits. (In Scheme 6 the general equation for the latter transformation, which involves the activated species **15**, is shown.) Thus, problems in nucleophilic fluorination reactions could not be generally overcome; however, formation of certain side products in nucleophilic substitution reactions (neighbouring group migrations including ring contractions) reached preparative significance and opened new synthetic strategies. The fascinating results obtained in this field until 1989 are collected in the outstanding review of Tsuchiya [13]; an attempt to update this work was recently undertaken by us [14].

The following presentation of a selection of the most significant results is organised in such a way that, first, reactions of educts derived from aldopyranoses are generally treated. To allow a rapid scan through the vast number of individual entries, the cases are separated into those starting from educts containing an axially and equatorially oriented LG, respectively. After characteristic examples obtained with pyranose derivatives of different configuration, single results obtained with educts from the furanose and septanose series are presented. This section is followed by a description of the situation with analogous educts containing nitrogen or sulfur as a ring member. Throughout, intramolecular reactions with participation of exocyclic (side chain) functionalities are discussed in brief only, whereas those observed with acyclic compounds are generally omitted. Another section is devoted to the reactions of the 2-triflates of aldonolactones and the preparative potential thereof. Although not strictly fitting into the topic of nucleophilic substitution (but showing parallels to it), the mode of reaction of the carbonyl group in dialdoses and aldosuloses with the nucleophilic reagent diethylaminosulfur trifluoride (DAST) will also be discussed. Finally, some further facets of side reactions are shown from examples related to those discussed before.

Not embodied in this collection are results which were obtained from reactions of carbohydrate-derived educts activated (or latently activated) for nucleophilic substitution at C-1. These species allow easy cleavage of the activating group to form oxocarbenium ions, which involve participation of vicinal ester or amide groups. Thus, independently of the steric situation in the original educt, attack occurs by the carbonyl oxygen (**ISC**, cf. Schemes 2 and 3) to form a cyclic acyloxonium ion **16** and an oxazoline **17**, respectively. The regioselective opening, at C-1 or at the C-atom of the former carbonyl group, allows the synthesis of glycosides with 1,2-*trans* configuration and of 1,2-cyclic orthoesters **18**, respectively; the latter in turn are also prone to rearrange into glycosides.

Another interesting case is the so-called "intramolecular aglycon delivery", in which, from the centre of a cyclic 2,3-orthoester **19** or of a mixed acetal **20** or **21**,

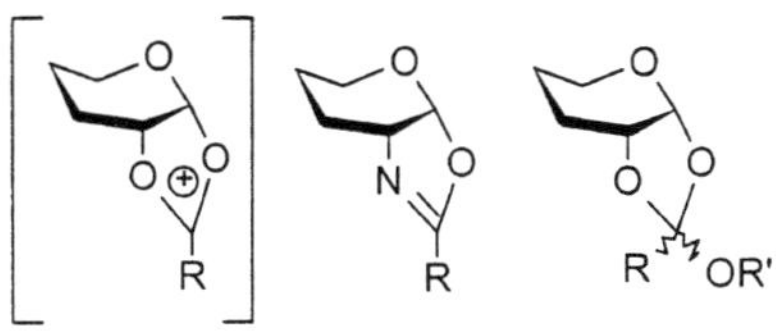

16 **17** **18**

the aglycon (OR′) stereoselectively migrates to the anomeric carbon. This latter reaction is of special value in the synthesis of 1,2-*cis*-configured β-D-mannopyranosides. As excellent reviews covering all facets of interest on this topic are available [15], no further details will be mentioned in this context.

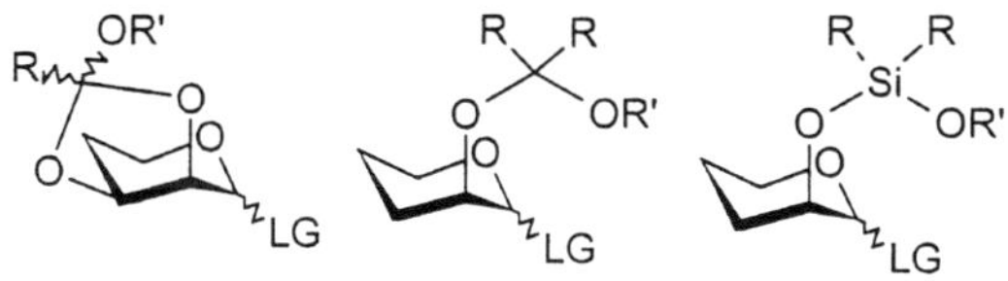

19 **20** **21**

For a correct interpretation of the collected data in the following tables, it should be mentioned that the significance of each single reaction cited finds its measure in the given total yield of isolated products. In only a few cases were all compounds formed isolated and identified, whereas in many other examples formation of a single product in only low yield is reported. This latter situation may not be interpreted as an indication that other possible reaction paths have been excluded under the conditions applied.

No special mention is made concerning the reaction conditions as they are given in the tables. Generally, deamination (amino sugar/$NaNO_2$/HOAc/H_2O), hydrolysis [sulfonates/acid scavenger (pyridine)/H_2O], substitution reactions [sulfonates/poor nucleophiles (especially the triflate/fluoride tandem)] or the DAST route (OH-unprotected sugar/DAST) are the main sources of side reactions.

4 Reactions of Pyranose Derivatives

4.1 Reactions of Educts Containing an Axially Oriented Leaving Group

For reactions starting from educts containing an axially oriented leaving group in pyranose derivatives, according to the general statements made in the introduction, besides the direct nucleophilic substitution with inversion of configuration (including nucleophilic participation of the solvent), intramolecular reactions under participation of an axially oriented NG from a vicinal position only are possible. For this reason, no ring atom can participate and, as a consequence, no change in ring size will be observed. The possible side reactions thus left include intramolecular substitution (**IS**, route **a**, Scheme 1), anchimeric assistance with or without group migration (**MS**, **AS** and **IS**, route **b** together with **c**, **d** or **e**, Scheme 1), hydride shift (**MS**[H], route **f**, Scheme 4) and elimination (**E**[H], route **h**, Scheme 4). As the choice depends [16] on the given position of activation, the stereochemical relationship of the LG to possible participating groups in the educt, their ranking in terms of nucleophilicity, as well as the relative stability of the carbenium ion possibly formed, representative examples for all kinds of reactions, including older but essential results from deamination reactions, are presented. According to the position of the LG (C-2, 3 or 4), the results are collected in Tables 2, 3 and 4.

Within these tables, the mechanisms, as anticipated to have been operative in the formation of individual products, are indicated by the following symbols:

- **IS** [intramolecular nucleophilic substitution with inversion of configuration by a neighbouring substituent under formation of a (small) heterocyclic ring]
- **ISC** [intramolecular substitution by the third atom of a carbonyl-containing NG under inversion of configuration at the position of the original activation and either entry of the nucleophile at the carbon of the original carbonyl group (to form an orthoester derivative) or elimination of another substituent (to give an oxazoline, oxazolone, etc.)]
- **AS** (anchimerically assisted substitution with retention of configuration at the position of the original activation)
- **MS** (substitution by migration of the participating group and entry of the nucleophile at the position where the NG departs with inversion of configuration at both centres involved)
- **E** (elimination, E2)
- **M/E** (in cases where a plausible decision between hydride shift and an elimination cannot be made)
- **F** (fragmentation)

For the very few cases where a second intramolecular grouping of nucleophilic character participates in one single reaction, additional symbols are introduced herewith:

- **AIS** (anchimerically assisted substitution, by another internal nucleophilic group, with retention of configuration at the position of the original activation)
- **MIS** (substitution by migration of the first participating group and entry of another internal nucleophilic atom at the position where the group having lent anchimeric assistance departs, with inversion of configuration at both centres involved)
- **MISC** [substitution by migration of the anchimerically assisting group and intramolecular substitution by a carbonyl oxygen of another NG under inversion of configuration at the position of the original activation as well as that of the departure of the assisting group with entry of the nucleophile at the carbon of the original carbonyl group (to form an orthoester derivative)].

In all instances where a NG is involved, the position of activation, the type ["H" for hydrogen, "NG" for a functional group, "rO", "rN" or "rS" for the heteroatom of the sugar ring] and the original location of the partner is given in brackets; for all cases where anchimeric assistance and intramolecular substitutions are involved, both partners are cited. Additionally, general chemical transformations, such as S_N2 (nucleophilic substitution with inversion of configuration by an external nucleophile including the solvent) or hydrolysis of certain functionalities (**Hy**), acyl migration (**AcM**), transesterification (**TE**), epimerisation in the vicinal position to a newly formed carbonyl group (**Ep**), hemiacetal or acetal formation between a newly formed carbonyl and an alcoholic OH-group (**HAF** or **AF**) or E1cB-elimination (to form an α,β-unsaturated carbonyl compound, **E1cB**), are indicated.

The numerous reports on successful substitutions with inversion of configuration (often in joint occurrence with eliminations) are not cited and their dependence on the type of leaving group, the character of the nucleophile and details of the conditions are not discussed.

4.1.1
Reactions of Educts Containing an Axially Oriented Leaving Group at C-2

In the selected educts with D-*manno* configuration (Table 2, entries 1–6), the potentially participating, axially oriented NGs are: the hydrogen atom at C-3 (in all instances), and either the hydrogen atom at C-1 [in the case of 2-amino-1,5-anhydro-2-deoxy-D-mannitol (entry 1) and the β-anomers (entries 2 and 3)] or, in educts of α configuration, the aglycon (entries 4–6). The participation of axially oriented hydrogen atoms, principally allowing elimination as well as hydride shift reactions (the latter obviously predominating in deamination reactions), under comparable conditions shows preference for the involvement of the more stable carbocations (cf. entries 1 [17] and 2 [18]). 3-Uloses, formed by hydride shift or hydrolysis of enol esters/ethers (resulting from eliminations), are prone to E1cB elimination (**Hy + E1cB**, entries 3 [19] and 4 [20]). Participation of an antiperiplanarly arranged alkoxy substituent at C-1 or C-3 leads, by anchimeric assistance, either to alkoxy migration under inversion of configuration at both centres involved (**MS**, entries 5 [21], 6 [22], 7 [23], 8 [24] and 10 [25]) or to substitution at C-2 with retention of configuration (**AS**, entry 8 [24]). An axial OH-group at C-3 causes oxirane formation with inversion of configuration at C-2 (**IS**, entry 9 [26]). A 1,2-epoxide, formed by the corresponding attack of an axial OH-1, was not isolated, but was proposed [18c] as a possible intermediate. The result of the deamination reaction shown in entry 10 [25] is of special interest as, obviously for steric reasons, a group migration (resulting in an enlargement of the anhydro ring) and not oxirane formation (by attack of OH-3) was observed. In all cases where the NG departs from C-1, both anomers of the respective **MS**-product were isolated, thus giving evidence for the intermediate existence of a stable oxocarbenium ion.

Results of special significance have been reported from attempted fluorinations of benzyl 3-azido-4,6-*O*-benzylidene-3-deoxy-D-altro- and -D-idopyranosides using the triflate/fluoride tandem as well as the DAST route [27]. As is evident from the product pattern in entry 11, participation of both axially oriented NGs (**MS**[2,NG-1] and **MS**[2,NG-3]) occurs; therefore, strict mechanistic

Table 2

	Educt	Product(s)	Conditions	Σ (%)	Ref.
1		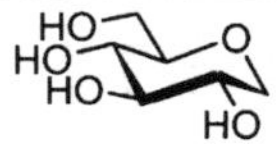 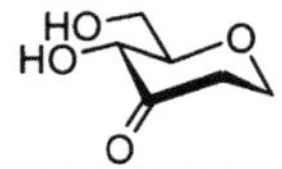 **S_N2** **MS**[2,H-3] (major)	not given	not given	17
2	 R = sugar	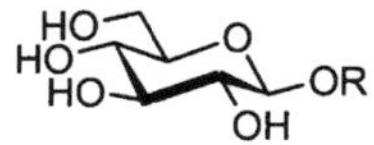  **S_N2** **MS**[2,H-1]	$NaNO_2$, HOAc/H_2O [pH 3.6], rt, 6 h	not given	18
3		 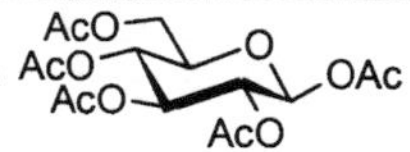 **S_N2** (23%) **S_N2** (11%) 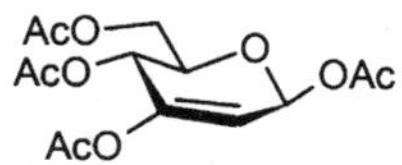 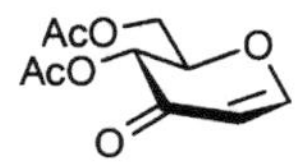 **E**[2,H-3] (rapidly hydrolysed) **+Hy+E1cB** (23%)	TBAF, benzene reflux, 2 h	⩾57	19
4	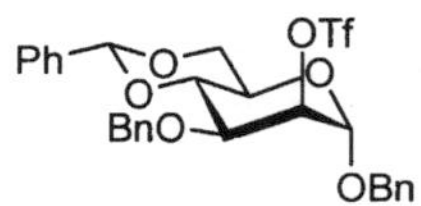	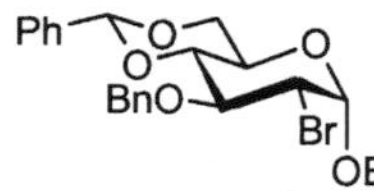 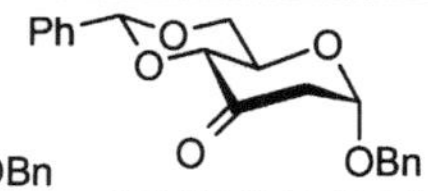 **S_N2** (26%) **M/E** [2,H-3] (21%) 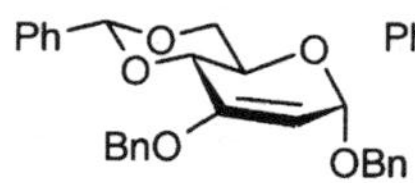 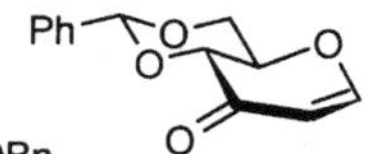 **E**[2,H-3] (24%) **+Hy+E1cB** (24%)	TBABr DMF/HMPA 50 °C, 15 h	95	20

Table 2 (continued)

	Educt	Product(s)	Conditions	Σ (%)	Ref.
5	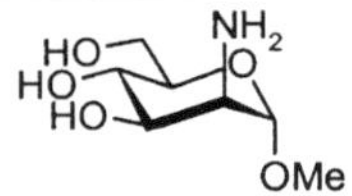	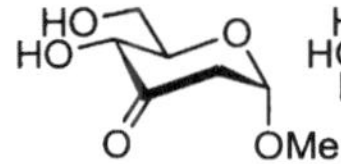 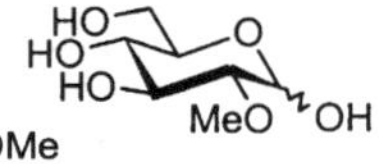 **MS**[2,H-3] (50%) **MS**[2,NG-1] (30%)	$NaNO_2$, $HOAc/H_2O$ rt, 2 h	80	21
6	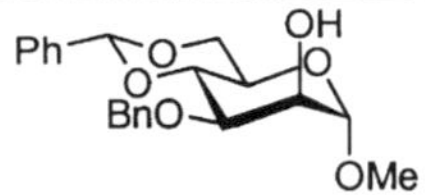	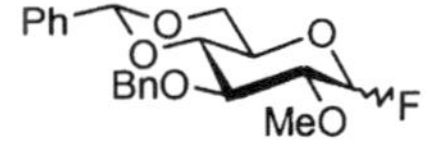 **MS**[2,NG-1] (80%)	DAST, diglyme 100 °C, 30 min	80	22
7	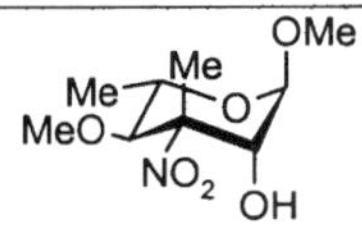	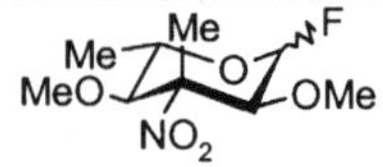 **MS**[2,NG-1] (58%)	DAST, CH_2Cl_2 0 °C→reflux, 15 h	58	23
8	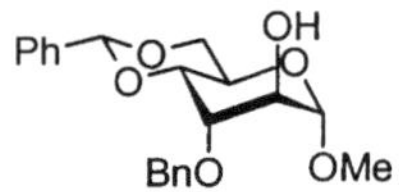	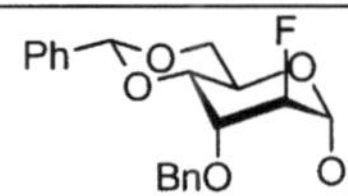 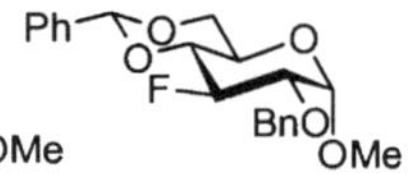 **AS**[2,NG-3 and/or 1] (47%) **MS**[2,NG-3] (5%)	DAST, CH_2Cl_2 rt, 15 h	52	24
9	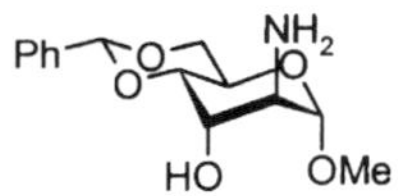	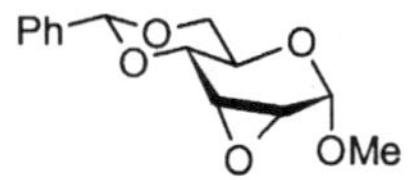 **IS**[2,O-3] (100%)	$NaNO_2$, $HOAc/H_2O$ 0 °C	100	26
10	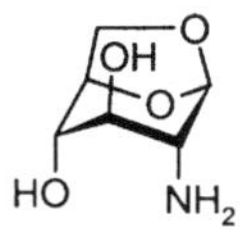	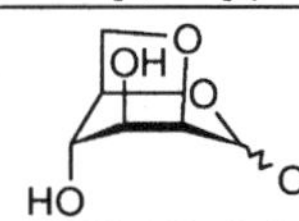 **MS**[2,NG-1] (37%)	$NaNO_2$, $HOAc/H_2O$ rt, 6 h	37	25

Table 2 (continued)

	Educt	Product(s)	Conditions	Σ (%)	Ref.
11	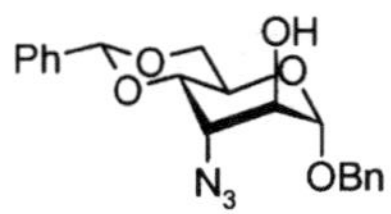	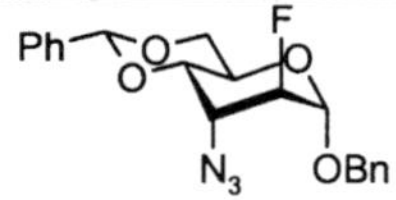 **AS**[2,NG-1 and/or 3] (40%) 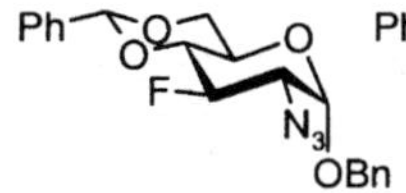 **MS**[2,NG-3] (40%) 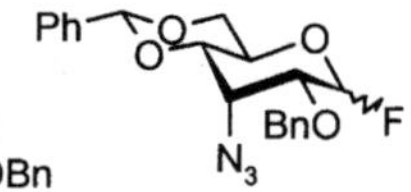 **MS**[2,NG-1] (15%)	DAST, benzene reflux, 2 h	95	27a
12	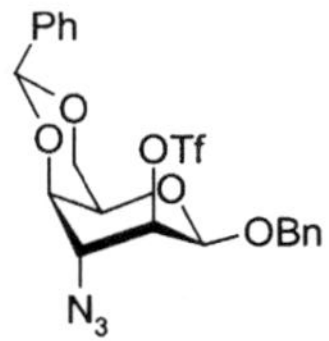	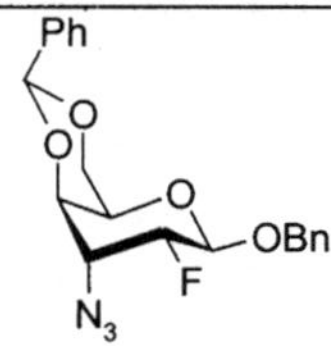 $\mathbf{S_N2}$ (71%)	CsF, DMF 70 °C, 5 h	71	27b
13	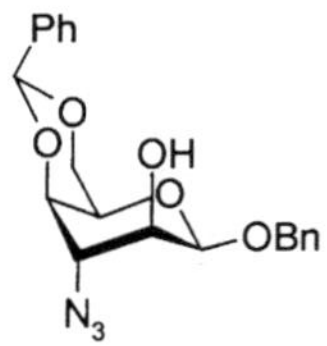	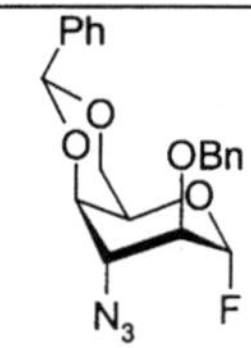 (82%)	DAST, benzene reflux, 1 h	82	27b

discrimination between **AS**[2,NG-3] and **AS**[2,NG-1] cannot be made for the product of fluorination with retention of configuration. For the involvement of the azido group, an unusual intermediate "azidonium (= 1-aziridinediazonium) ion" has been postulated (see the discussion on azide participation in [5]). Whereas on treatment with CsF in DMF the triflate with β-D-*ido* configuration (entry 12 [27b] underwent smooth $\mathbf{S_N2}$ displacement, under the same conditions its C-4 epimeric analog (D-*altro* configuration) remained unchanged. The formation of a product of aglycon migration from a *cis* orientation (entry 13 [27b]) is unexpected as long as the role of the axial azido group at C-3 is not fully understood.

The type of aglycon shift depicted in entry 6 [22] was at the same time also described by Nicolaou's group [28] from the DAST reactions of a series of 2-OH-unprotected α-D-mannopyranose derivatives with an alkoxy, acetoxy, phenylthio or azido group as substituent at C-1, respectively. A general mechanistic explanation given for all migrating groups involves a positively charged three-membered ring as well as the (C-1)-oxocarbenium ion (see also Sect. 4.2.1). Of special interest from these series is the reaction sequence (Scheme 7) that sets off from thioglycoside **22** to initially give a mixture of 2-phenylthio-D-glucopyranosyl fluorides **23** and allows selective access to 2-deoxy-α- or -β-glycosides **25** in excellent yields. In the second step, the stereocontrolled glycoside formation to **24**, use is made of NG and solvent participation, respectively. The synthesis completes with desulfurisation.

Another example of outstanding synthetic importance and based on a 1,2-migration is the so-called "sulfonamidoglycosylation" [29a], developed by Griffith and Danishefsky. The reaction sequence starts with a glycal and leads to glycosides of 2-amino-2-deoxy-hexoses. Scheme 8 shows the respective transformation of tri-*O*-acetyl-D-glucal (**26**) to (1,2-*trans*-diaxially substituted) *N*-benzenesulfonyl-2-deoxy-2-iodo-α-D-mannopyranosyl amine (**27**), which (in the presence of a strong base, silver triflate and another sugar containing an unprotected OH-group) reacts to yield the disaccharidic β-glucosaminide **29**. In the second step, involvement of the intermediate aziridine **28** (compare structure **2** formed by "*N*-attack" in Scheme 3) was anticipated. Recently, this methodology was extended [29b] to the solid-phase synthesis of oligosaccharides.

4.1.2
Reactions of Educts Containing an Axially Oriented Leaving Group at C-3

With educts containing an axially oriented LG at C-3, in addition to an intramolecular S_N2 reaction with an axially oriented hydroxy or acetamido group in a vicinal position (**IS**, Table 3, entries 1 [26] and 2 [30]), elimination and/or hydride shift (**E** or **M/E**, entries 3 [30], 4 [31] and 6 [32]), as well as substitution with retention of configuration (**AS**, entries 7 and 8 [27b]) have been observed. For the transformation outlined in entry 4, 2,6-di-*tert*-butyl-4-methylpyridine was used as acid scavenger, since pyridine per se in a similar reaction had caused S_N2 displacement (entry 5 [33]). Of special interest are the results from the deamination of methyl 3-amino-3-deoxy-β-D-allopyranoside (entry 6), where the main reaction consists of direct $\mathbf{S_N2}$ displacement with formation of methyl

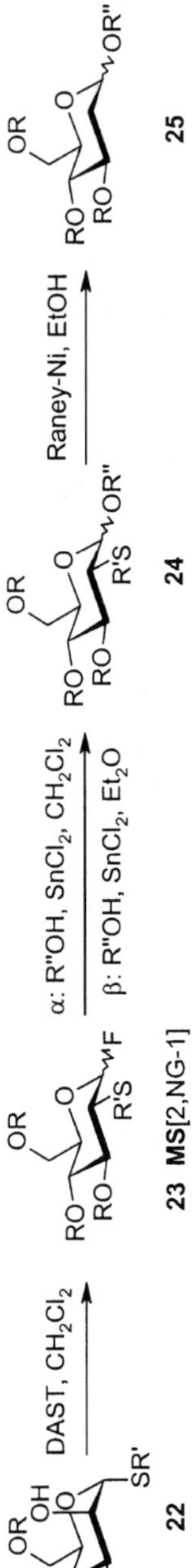

Scheme 7

Scheme 8

β-D-glucopyranoside. Amongst the side products **31**–**34**, a new principle appears, that further multiplies the possible output-number: Epimerisation (**Ep**) in a vicinal position to a carbonyl group. This functionality results from a rearrangement reaction, in this case, hydride shift (**MS**[3,H-4] to yield **31**) and ring contraction (**MS**[3,C-5] to give **33**), respectively. [Compound **33** originates from the participation of the C-4/C-5 bond and thus proves conformational inhomogeneity of the educt, as indicated by the depiction of both chair conformers (although for this particular reaction other conformers may be responsible [32]). The reactions of educts containing an equatorially oriented LG at C-3 are treated in detail in Sect. 4.2.2.] The epimerization at C-3 of product **33** leads to a species containing both side chains on the same ring side, which gives rise to the formation of furanoid hemiacetals **34** (**HAF**).

As seen from the results shown in entries 11–13 [27] in Table 2, an axially oriented azido group is capable of participation and this is confirmed with the substitution under retention of configuration (**AS**) found in entries 7 and 8. The latter case may alternatively or additionally also have been caused by the axial benzyloxy substituent at C-4.

4.1.3
Reactions of Educts Containing an Axially Oriented Leaving Group at C-4

In Table 4 representative results as obtained with educts containing an axially oriented LG at C-4 are compiled. These include direct nucleophilic substitution, by the external nucleophile ($\mathbf{S_N}$, entries 1 [32], 3 [34], 5 [35] and 6 [36]) or the solvent (entries 3 and 4 [35]), hydride shift and/or elimination (**MS**[H], **E**[H], **M/E**[H], entries 1, 2 [37], 4 and 6 [36]). Special attention is called to the following results:

1. In entry 1 [32], formation (2%) of substitution product **35** with retention of configuration. (As anchimeric assistance cannot be visualised, it may be the consequence of the existence of a "hot" carbenium ion; therefore, the mechanism is designated as $\mathbf{S_N}$ only.)
2. In entry 2, the DAST reaction of a kanamycin A derivative [37], the loss of the hydrogen atom at C-3 under cleavage of the cyclohexylidene acetal with concomitant fluorination, under inversion of configuration, at C-5 of the deoxystreptamine (DOS) part (3-AG denotes the 3-amino-3-deoxy-D-glucose moiety). [In reactions of educts also containing *trans*-diaxial orientation of the LG (at C-3) and hydrogen (at C-4) together with a 4,6-benzylidene acetal grouping (Table 3, entries 1–3) this **M/E**-type was not observed; see also Sect. 4.2.2.]
3. In entry 5 [35], the isolation of "glycosyl fluoride" **36** (15% yield). [Its formation was assumed to have involved elimination, stereoselective fluoride attack at C-5 and protonation. An alternative mechanism may consist of hydride shift (**MS**[4,H-5]) and reaction, with fluoride, of the intermediate oxocarbenium ion to form the more stable anomer.]
4. In entry 6 [36], the isolation of either of the possible side products formed by elimination **37** (**E**[4,H-5] and **38** (**E**[4,H-3]).

Table 3

	Educt	Product(s)	Conditions	Σ (%)	Ref.
1	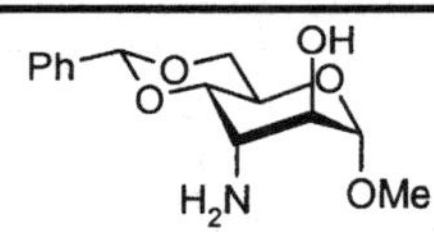	 **IS**[3,O-2] (100%)	not given	100	26
2	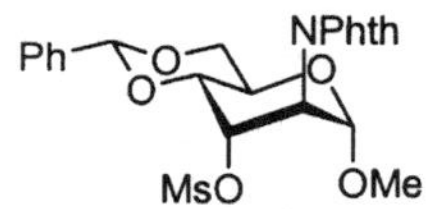	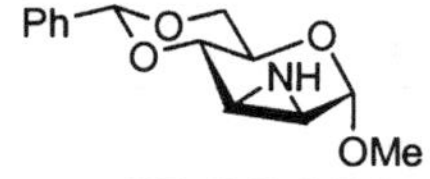 **IS**[3,N-2] (84%)	TBAF, MeCN reflux, 20 h	84	30
3	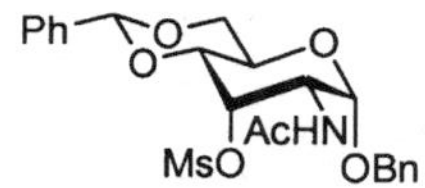	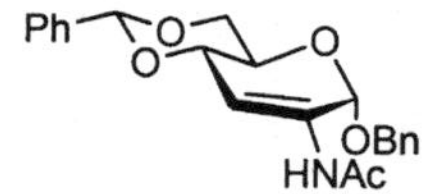 **E**[3,H-2] (75%)	TBAF, MeCN reflux, 4 h	75	30
4	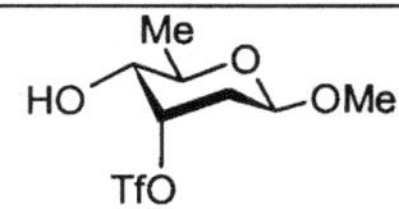	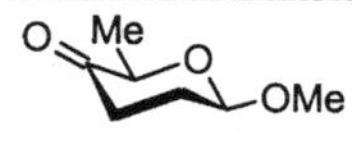 **MS**[3,H-4] (72%)	DTBMP/benzene 80 °C, 30 min	72	31
5	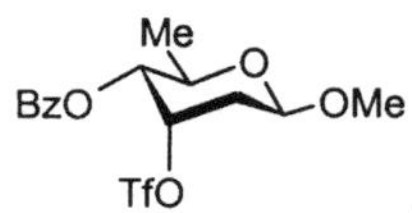	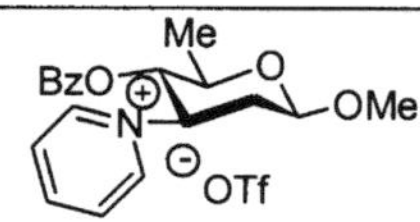 $\mathbf{S_N2}$ (62%) **30**	pyridine 100 °C, 15 min	62	33

Table 3 (continued)

	Educt	Product(s)	Conditions	Σ (%)	Ref.
6	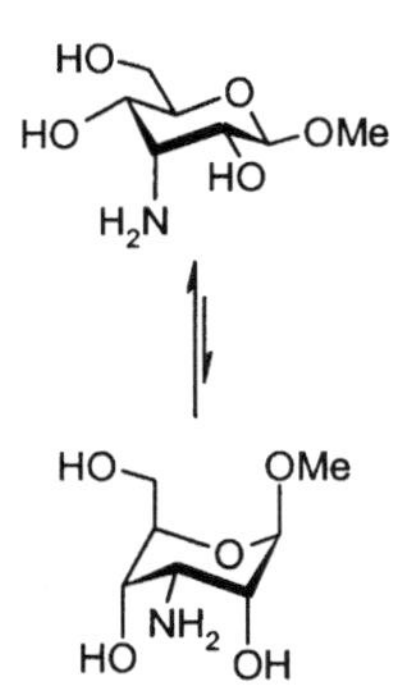	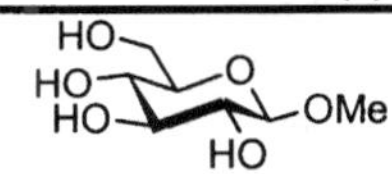 S_N2 (main product) 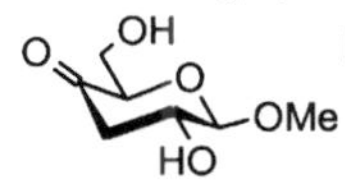 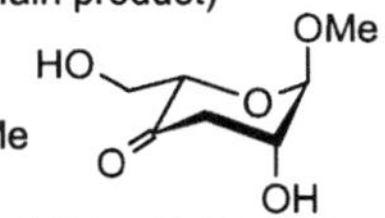 **MS**[3,H-4] **31** (7%) +**Ep**[C-5] **32** 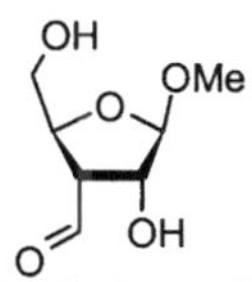 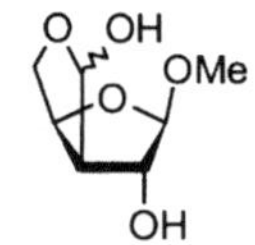 **MS**[3,C-5] **33** (5%) +**Ep**[C-3]+**HAF 34**	not given	not given	32
7	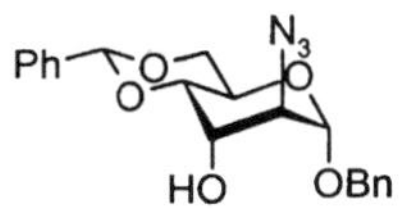	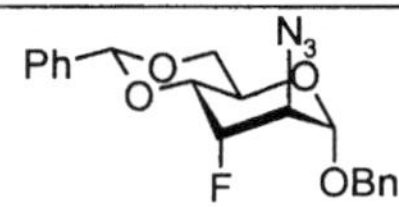 **AS**[3,NG-2]	DAST, CH_2Cl_2 rt, 5 h	75	27b
8	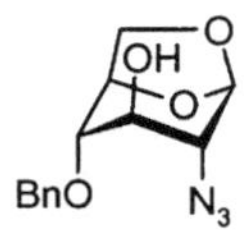	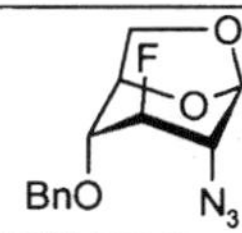 **AS**[3,NG-2 and/or 4]	DAST, benzene reflux, 30 min	85	27b

Table 4

	Educt	Product(s)	Conditions	Σ (%)	Ref.
1	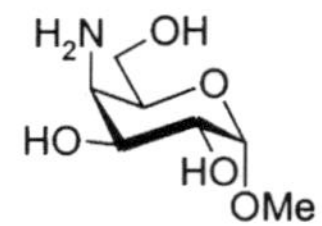	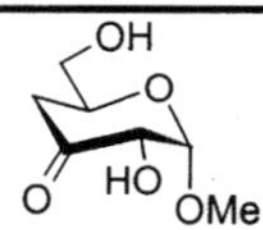 **MS**[4,H-3] (~41%) 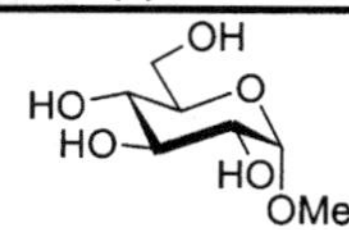 $\mathbf{S_N}$ (9%) 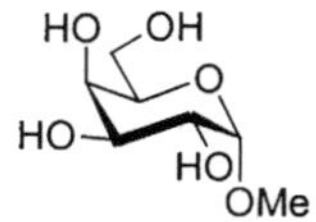 $\mathbf{S_N}$ (2%) **35**	not given	>50	32
2	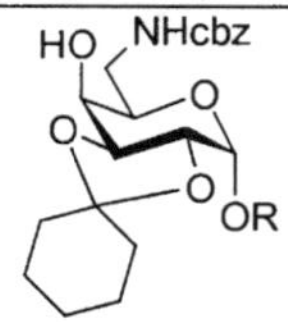 R= DOS + 3AG	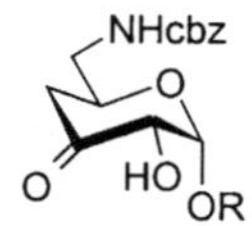 **MS**[4,H-3]+**Hy** (50%)	1. DAST, CH_2Cl_2 rt, 5 h 2. H_2O	50	37
3	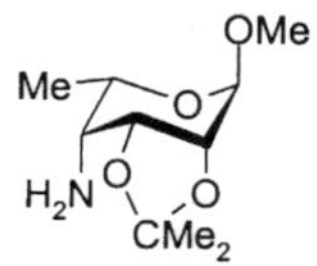	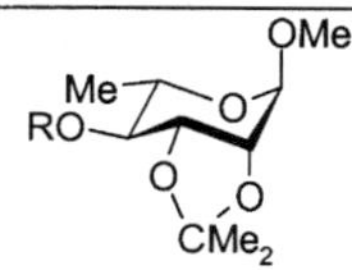 $\mathbf{S_N2}$ R = H (32%) R = Ac (50%)	$NaNO_2$, $HOAc/H_2O$ 0 °C	82	34

Table 4 (continued)

	Educt	Product(s)	Conditions	Σ (%)	Ref.
4	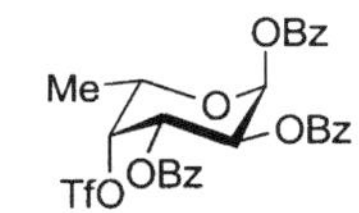	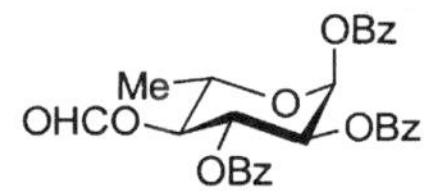 S_N2 (43%) 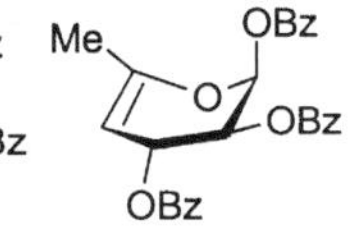 E[4,H-5] (4%)	NaF, DMF rt, 15 h	47	35
5	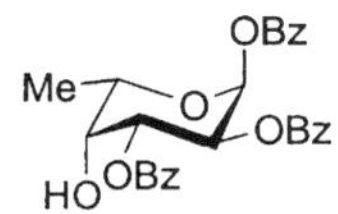	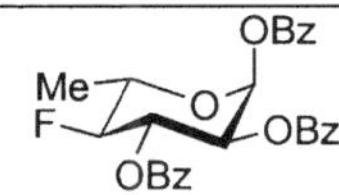 S_N2 (61%) 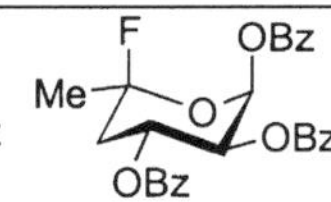 (15%) **36**	DAST, CH_2Cl_2 0 °C → rt, 1 h	75	35
6	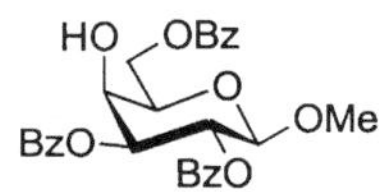	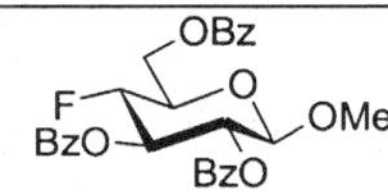 S_N2 (48%) 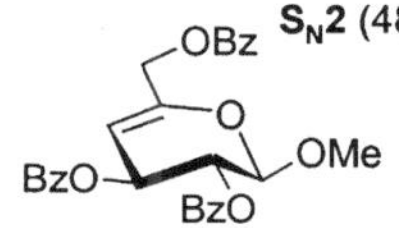 E[4,H-5] (23%) **37** 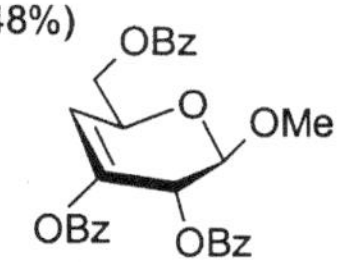 E[4,H-3] (13%) **38**	DAST dimethoxyethane 60 °C, 15 min	84	36

4.2 Reactions of Pyranose Derivatives Containing an Equatorially Oriented Leaving Group

When, under the assumption of conformational homogeneity, looking at the set of reactions possible with educts containing an equatorially oriented LG, hydride shift (**MS**[H], route **f**, Scheme 4) and elimination (**E**[H], route **h**, Scheme 4) cannot be found. However, besides the direct nucleophilic substitution with inversion of configuration (by the external nucleophile or the solvent, S_N2), all other sorts of participation, from vicinal positions, by antiperiplanarly oriented substituents – including each single ring atom of the pyranose – appear. These are in particular:

1. Participation of a (*trans*-located, likewise equatorial) neighbouring acyloxy or acylamino group, which follows the pattern outlined in Scheme 2 and with the cases of "*O*-attack" in Scheme 3, respectively.
2. Participation of the ring heteroatom according to routes (**b** + **c**), (**b** + **d**) or (**b** + **e**) (Scheme 1) becomes feasible as soon as the LG is located at C-2 and C-4, respectively. A general survey of the reaction paths with educts from the pentopyranose series, including the respective intermediate oxiranium ion (**39** or **44**), is given in Schemes 9 and 10. (Within hexopyranose derivatives, reactions of type **e** additionally cause inversion of configuration at C-5.) The individual species are:
   a. Substitution with retention of configuration at the original position of the LG [route (**b** + **c**), Schemes 9 and 10] giving products **40** (**AS**[2,rO]) and **45** (**AS**[4,rO]), respectively.
   b. Migration of the ring heteroatom and entry of the nucleophile at the origin of the migrating group with inversion of configuration at both positions [route (**b** + **d**), Schemes 9 and 10]. The net result of this rearrange-

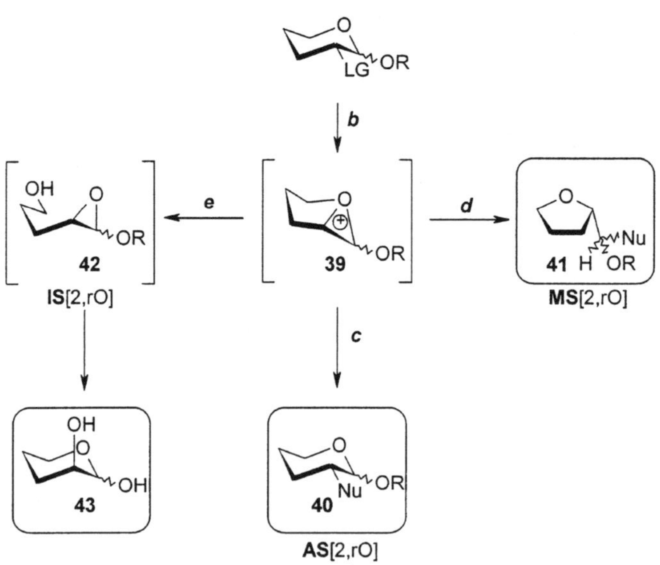

Scheme 9

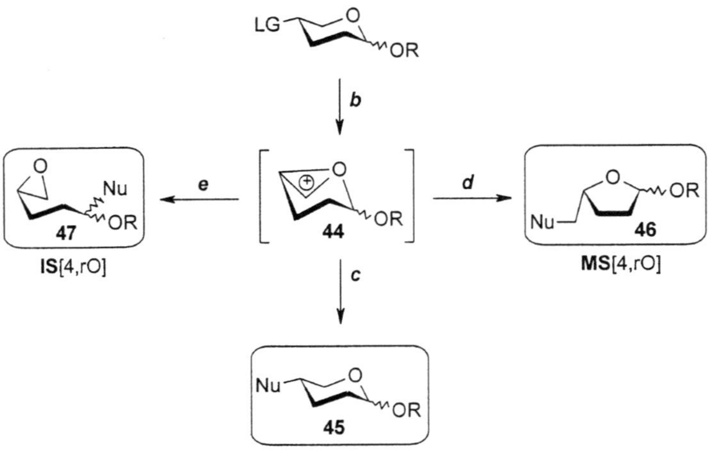

IS[4,rO]
MS[4,rO]

Scheme 10

AS[4,rO]

ment reaction is ring contraction, of the pyranose, to form a derivative of 2,5-anhydroaldose **41** (**MS**[2,rO]) and of aldofuranose **46** (**MS**[4,rO]), respectively.

c. Attack at C-1 in **44** [route (**b** + **e**), Scheme 10] opens the pyranose ring to give the 4,5-anhydro derivative of the (open-chain) aldose **47** (**IS**[4,rO]). [The corresponding reaction with **39** [route (**b** + **e**), Scheme 9] has not yet been observed; however, with water as the nucleophile, it should lead (via the highly reactive structure **42**, **IS**[2,rO]) to an aldose with inverted configuration at C-2 (**43**) and, in hexopyranoses, at C-2 and C-5, respectively.]

3. The possible involvement of ring C-atoms in side reactions of nucleophilic displacements with carbohydrate-derived educts – in the form of an alkyl migration (routes **g**, Scheme 4) or fragmentation reaction (routes **i**, Scheme 4) – is outlined in Schemes 11 and 12.

   a. As in an alkyl migration, the C-atom shifts as a nucleophile (in taking the electron pair with it), provision has to be given in the educt structure for a stabilisation of the carbocation thus created. The oxocarbenium ion, as normally involved with educts from this class of carbohydrates, leads to a formyl side chain with inversion of configuration at the terminus (cf. routes **g**, Scheme 4). Thus, in such products from the carbohydrate series, the side chain takes the same position ("above" or "below" the sugar ring) as the LG did in the educt. However, as will be demonstrated below, the product structure allows epimerisation (**Ep**). As shown in Scheme 11, with educts containing the LG at C-2, C-4 migrates (**MS**[2,C-4]) and vice versa (**MS**[4,C-2]); in each case, C-3 forms the cation and is converted into the formyl side chain. For educts containing an equatorial LG at C-3, two C-atoms, namely C-1 and C-5, could migrate (**MS**[3, C-1] or **MS**[3,C-5], routes **g**, Scheme 12) with C-2 or C-4 forming the side chain.

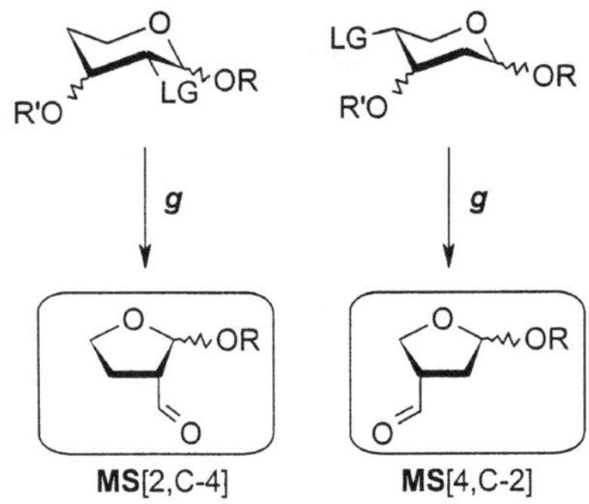

Scheme 11

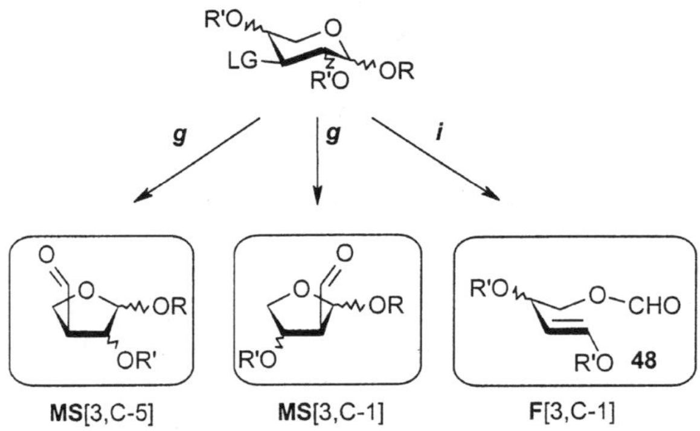

Scheme 12

b. Fragmentation affords a well-stabilised carbocation to be eliminated from the vicinal position. In carbohydrate-derived educts, the most stabile carbenium ion arises, by heterolytic cleavage of the C-1/C-2 bond, at the anomeric centre (C-1). Consequently, use of this fact is made in solvolysis reactions of certain educts containing the LG at C-3 (**F**[3,C-1], route **i**, Scheme 12). As the direct product of fragmentation, enol derivative **48**, constitutes a precursor to a 2-deoxyaldose, the net effect of this type of fragmentation consists of shortening the sugar chain by one C-atom (C-1) and introduction of a deoxy grouping at C-2.

In analogy to the presentation chosen in Sect. 4.1, an arbitrary selection of characteristic results is collected in Tables 5, 6 and 7 arranged according to the position of the LG. Whenever possible, for each product depicted, an indication concerning the mechanism of its formation is given by using the symbols explained in Sect. 4.1.

4.2.1
Reactions of Pyranose Derivatives Containing an Equatorially Oriented Leaving Group at C-2

In deamination reactions with methyl D-glucosaminides (Table 5, entries 2 [38a], 3 [38b] and 4 [39a]), attack of the ring oxygen at C-2 (**MS**[2,rO]), to form

Table 5

	Educt	Product(s)	Conditions	Σ (%)	Ref.
1	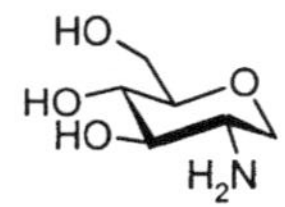	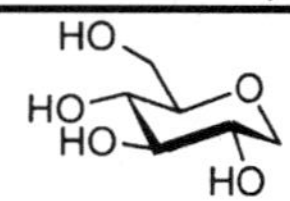 **AS**[2,rO] (26%)	$NaNO_2$, $HOAc/H_2O$ rt, 2 h	26	40
2	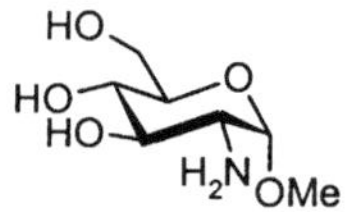	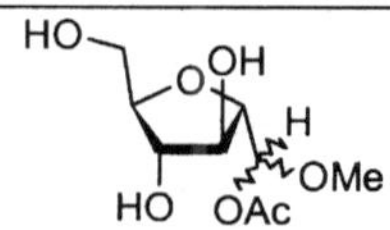 **MS**[2,rO] (76%) 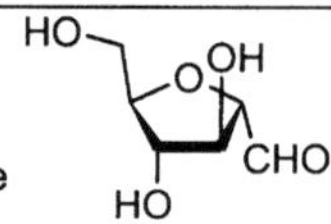 **MS**[2,rO]+**Hy** (17%)	$NaNO_2$, HOAc rt, 1 h	93	38a
3	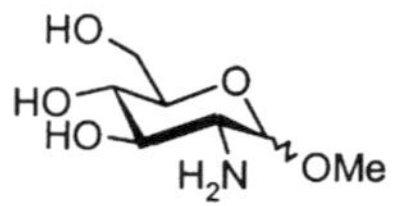 α-anomer β-anomer	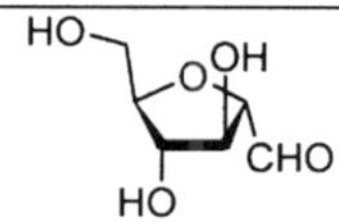 **MS**[2,rO]+**Hy** (73%) (84%) 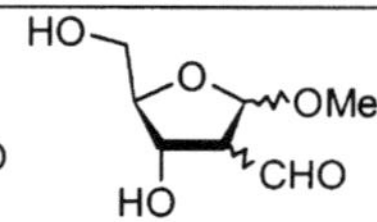 **MS**[2,C-4] (27%)+**Ep** (16%)	$NaNO_2$ [pH 3.5], rt, 12 h	100	38b
4	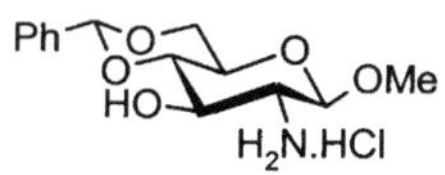	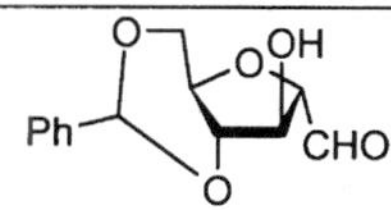 **MS**[2,rO]+**Hy** (51%)	$NaNO_2$, NaOAc, $HOAc/H_2O$ rt, 2 h	51	39a

Table 5 (continued)

	Educt	Product(s)	Conditions	Σ (%)	Ref.
5		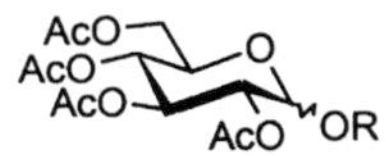 **AS**[2,rO] R = Ac (38%) R = H (12%) 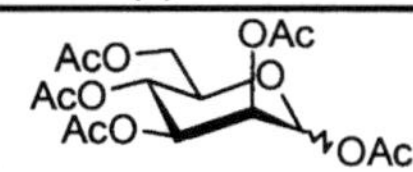 **ISC**[2,NG-1] or **S$_N$2** (4%) 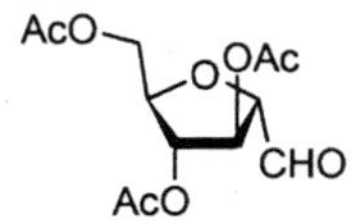 **MS**[2,rO]+**Hy** (12%) **49** 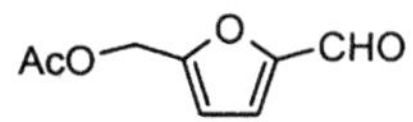 **+E** (5%) **50**	$NaNO_2$, HOAc rt, 1 h	71	38a
6	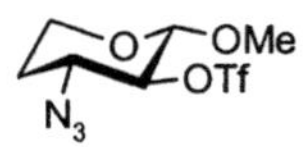	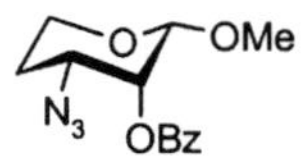 **S$_N$2** (n.g.) 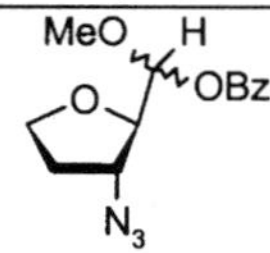 **MS**[2,rO] (62%)	NaOBz, HMPA 90 °C, 5 h	62	41
7	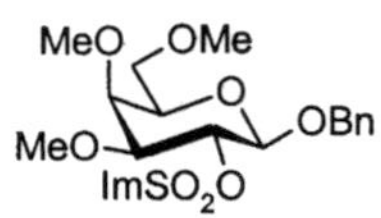	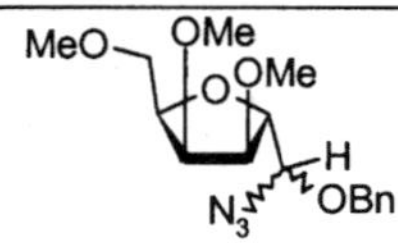 **MS**[2,rO] (59%)	$TBAN_3$, toluene 80 °C, 6 h	59	42
8	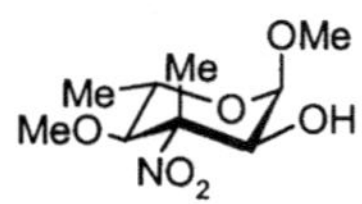	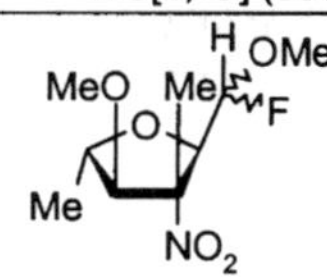 **MS**[2,rO] (70%)	DAST, CH_2Cl_2 0 °C → reflux, 2 h	70	23

Table 5 (continued)

	Educt	Product(s)	Conditions	Σ (%)	Ref.
9	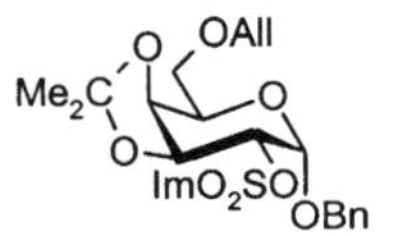	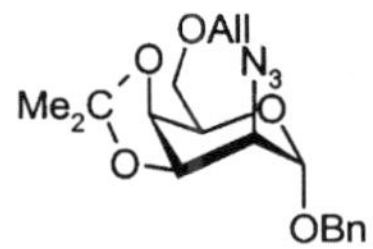 S_N2 (17%) 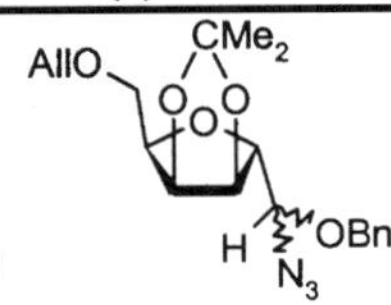 MS[2,rO] (40%)	TBAN$_3$, toluene 80 °C, 6 h	57	42
10	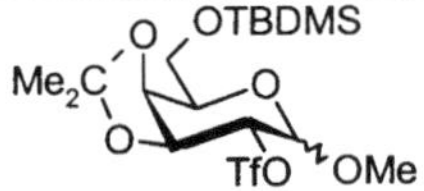 α-anomer	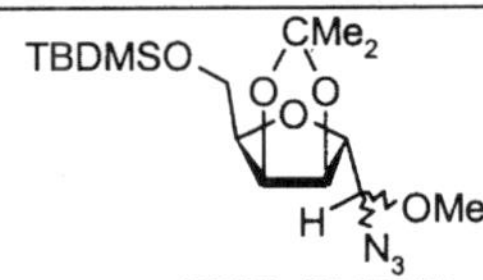 MS[2,rO] (81%)	NaN$_3$, DMF 40 °C, 15 h	81	43
11	β-anomer	(91%)		91	
12	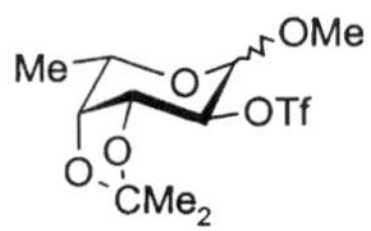 α-anomer	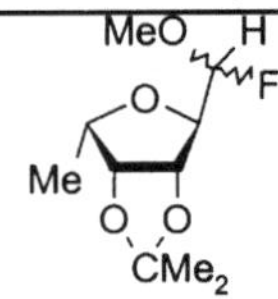 MS[2,rO] (55%)	3HF.TEA, TEA, MeCN rt, 16 h	55	44
13	β-anomer	(55%)		55	
14	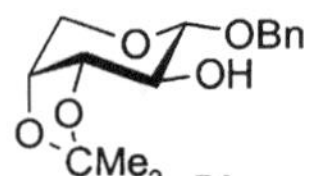 **54**	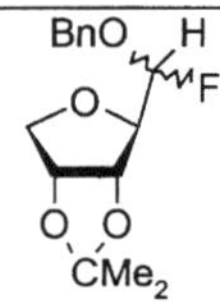 MS[2,rO] (60%) **56**	DAST, CH$_2$Cl$_2$ rt, 5 h	60	45

Table 5 (continued)

	Educt	Product(s)	Conditions	Σ (%)	Ref.
15	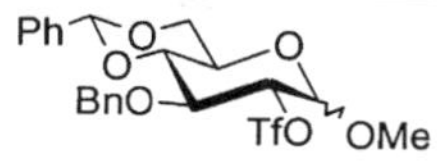 α-anomer	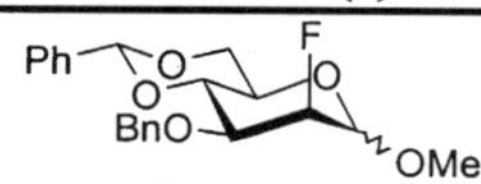 **S$_N$2** (30%)	TBAF, MeCN reflux, 2.5 h	30	46
16	β-anomer	(77%)	TBAF, MeCN reflux, 30 min	77	46
17	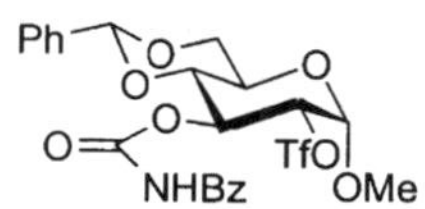	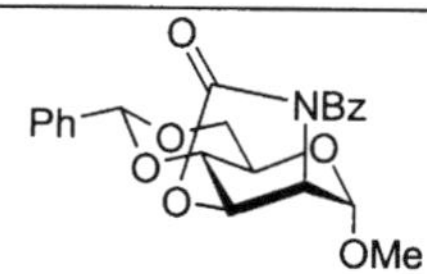 **ISC**[2,NG-3] (85%)	NaH, THF 0 °C	85	47
18	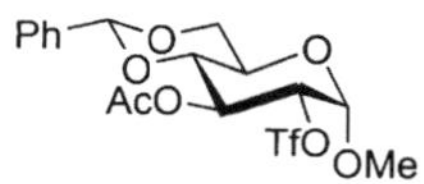	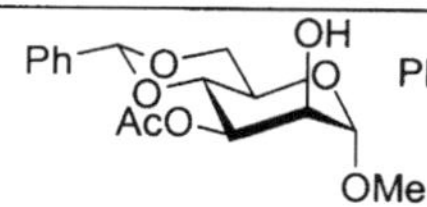 (54%) 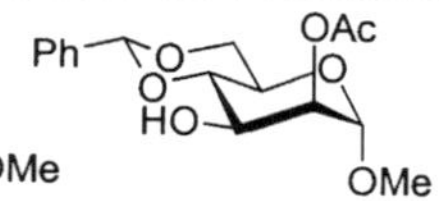 (19%)	TBANO$_2$, MeCN reflux, 6 h	73	48
19	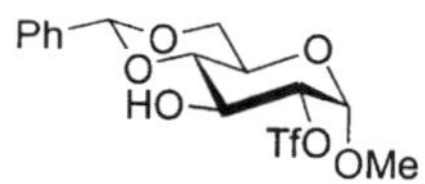	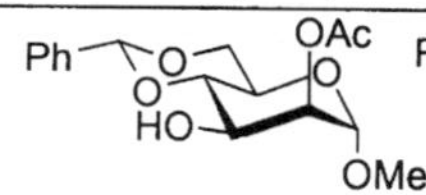 **S$_N$2** (22%)  **IS**[2,O-3] (44%) **52**	NaOAc, DMF rt, 6 d	66	49
20	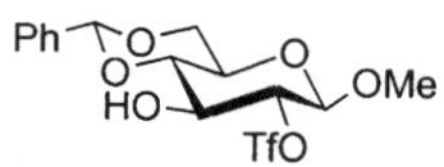	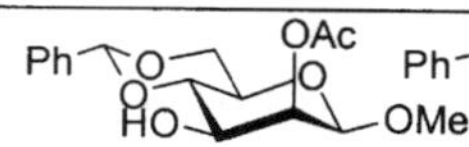 **S$_N$2** (78%) 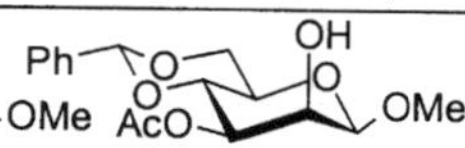 (13%)	NaOAc, DMF rt, 2 d	91	49

Table 5 (continued)

	Educt	Product(s)	Conditions	Σ (%)	Ref.
21	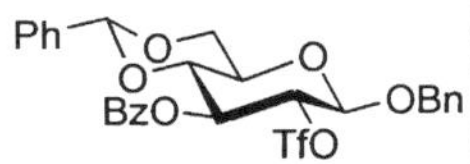	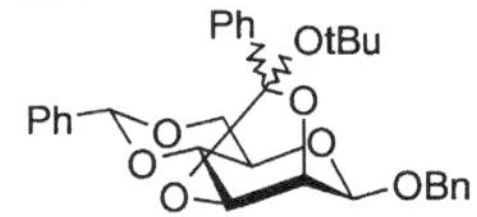 **ISC**[2,NG-3] (25%) **51**   (75%)	*t*-BuOH, *sym*-collidine, toluene reflux, 2 h	100	50
22	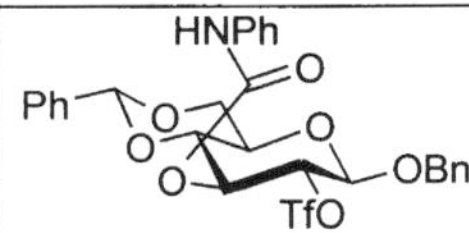	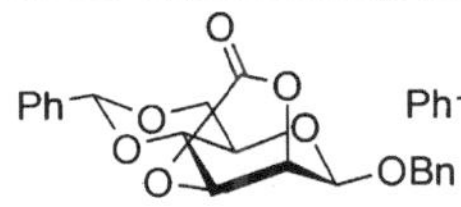 (major) **ISC**[2,NG-3] 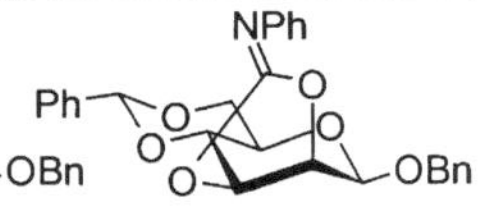 (minor) **ISC**[2,NG-3]	pyridine/DMF 75 °C	90	51
23	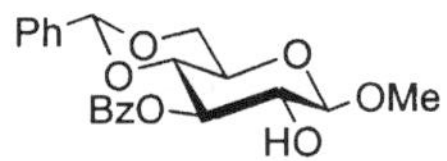	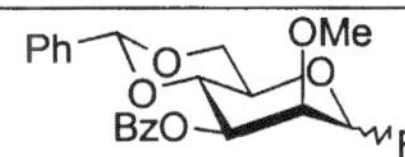 **MS**[2,NG-1] (64%) **53**	DAST, CH_2Cl_2 rt, 5 h	64	45
24	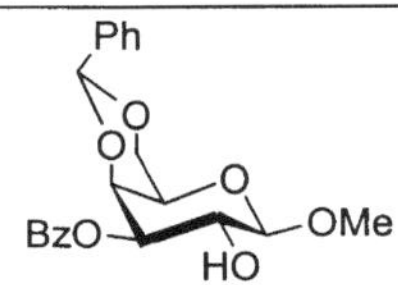 **57**	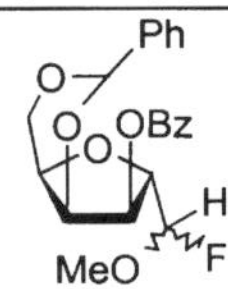 **MS**[2,rO] (60%) **58**	DAST, CH_2Cl_2 rt, 5 h	60	45

Table 5 (continued)

	Educt	Product(s)	Conditions	Σ (%)	Ref.
25	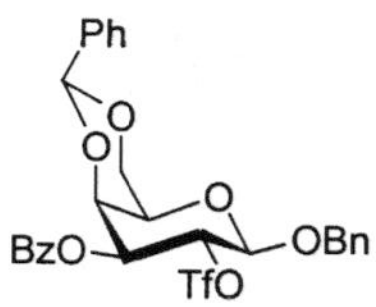 **59**	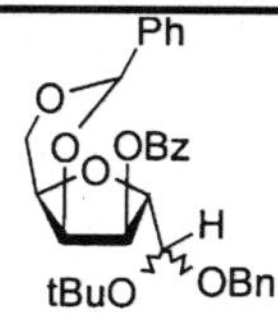 **MS**[2,rO] (12%) **61** 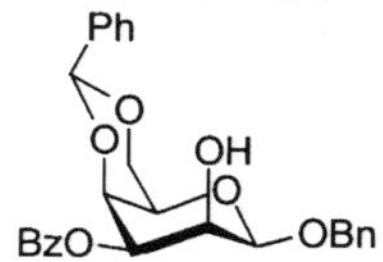 (10%) 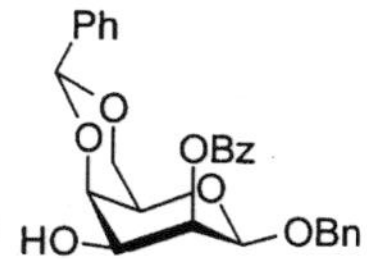 (48%)	*t*-BuOH, *sym*-collidine, toluene reflux, 6 h	70	24 50

derivatives of 2,5-anhydro-D-mannose, is the dominating reaction path. In the presence of water and independent of the anomeric configuration (entry 3), participation of C-4 (**MS**[2,C-4]) to (originally) form 2-*C*-formyl-D-arabinofuranosides, which epimerise to the *ribo*-isomers, is also observed. Of special interest is the result shown in entry 4 [39a] as this **MS**[2,rO]) ring contraction was, in no single other case, observed with educts (compare entries 15–23) containing a 4,6-benzylidene acetal protection of the *trans*-decalin type. (In this work [39a] older results that claimed formation of 4,6-*O*-benzylidene-D-mannose from the same reaction [39b] have been corrected; however, the structural depiction of the reaction product shows D-*gluco* configuration.) The essential role of the aglycon in stabilizing intermediate carbenium ions can be deduced from the reaction of 2-amino-1,5-anhydro-2-deoxy-D-glucitol (entry 1 [40]), where substitution with retention of configuration (**AS**[2,rO]) dominates. The same applies for the case of an *O*-acylated derivative (entry 5 [38a]). There, additionally, direct $\mathbf{S_N2}$, by acetate, or substitution under participation of the (*trans*-located) acetoxy group at C-1 [**ISC**[2,NG-1] according to route (**b′** + **d′**), Scheme 2] took place. Furthermore, the product of ring contraction (**49**, **MS**[2,rO]) was found to be subject to elimination to give **50**. (The deficit in the total yield was assumed to be a consequence of *O,N*-acetyl migration.)

In entries 6–24, characteristic results obtained from substitution reactions of 2-sulfonates and from the DAST reaction of educts containing an unprotected OH-group at C-2 are collected. They show dependence on the type of protecting groups present. Whereas monocyclic educts, independent of the configuration at other chiral centres (entries 6 [41], 7 [42] and 8 [23]), as well as derivatives of galactose and arabinose, which contain a 3,4-*O*-isopropylidene protecting group (entries 9 [42], 10 and 11 [43], 12 and 13 [44] as well as 14 [45]), are predominantly subject to ring contraction (**MS**[2,rO]), none of the derivatives of methyl 4,6-*O*-benzylidene-α- or -β-D-glucopyranoside (entries 15 and 16 [46], 17 [47], 18 [48], 19 and 20 [49], 21 [50], 22 [51] or 23 [45]) give that particular reaction (but see entry 4). Dependent on the type of protecting group at O-3, they show, besides straight $\mathbf{S_N2}$, other forms of neighbouring group participation. (Of these, the **ISC**[2,NG-3] mode depicted in entries 21 and 22 is of importance in the stereoselective synthesis of β-D-mannopyranosides. Formation of a cyclic carbonate and iminocarbonate in the latter case follows the "*O*-attack-route" shown in Scheme 3.) When the same reaction is carried out with a pair of anomers (entries 15/16 [46] and 19/20 [49], respectively), the β-anomer reacts faster and gives a higher yield. In entries 18 and 20, the respective product formed in minor proportion is the result of either acetyl migration following the $\mathbf{S_N2}$ or of participation of the acetoxy group in the substitution step according to the **ISC** mode [route (**b′** + **e′**) + **Hy**, Scheme 2]. The involvement of the latter mechanism is proved (entry 21) by the isolation of orthoester **51** (**ISC**[2,NG-3]) in the solvolysis reaction, with *tert*-butyl alcohol as the nucleophile. When a lower alcohol was used, the product with the orthoester structure of type **51** reached a proportion of more than 90%, as roughly estimated by TLC [50].

In this series of reactions, at C-2 of glucopyranosides, the following details are of interest:

1. Formation of 2,3-anhydro compound **52** (**IS**[2,O-3]) in the course of a triflate displacement (entry 19 [49]). Under the conditions cited, the anhydro ring formation was not observed with the corresponding β-anomer (entry 20), but occurred with both anomers almost exclusively when *tert*-butoxide in *tert*-butyl alcohol (or TBAF in toluene) was applied. As a corollary, excellent yields of S_N2 products were obtained, again with both anomers, in reactions with good nucleophiles (e.g. azide in DMF).
2. Formation of a product of aglycon migration (**53**, **MS**[2,NG-1]) in a DAST reaction (entry 23 [45]), whereas an analogous educt, but using the triflate/TBAF-tandem reaction, allowed smooth introduction of fluorine by S_N2 displacement (entry 16 [46]).

According to the general "rules" presented in the introductory sections, both results cannot be envisaged to have taken place as a result of the given diequatorial steric relationship between the LG (at C-2) and the participating group (OH at C-3 or OMe at C-1) in the educt. Thus conformational flexibility of this peculiar part of the otherwise – especially in the region of *trans*-annellation – rigid molecule has to be anticipated [49]. However, this principle is not as general as published results [28] would suggest. At least in one single re-inspected case, the DAST treatment of benzyl 3,4-*O*-isopropylidene-α-D-arabinofuranoside (**54**, the enantiomer of the original educt, entry 14 [45]), the claimed formation of a product of aglycon migration (**55**, **MS**[2,NG-1], 66%), could not be reproduced as the reaction led to a product of ring contraction (**56**, **MS**[2,rO], entry 14).

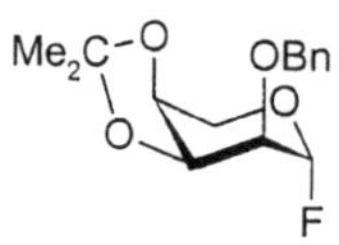

MS[2,NG-1] (66%) **55**

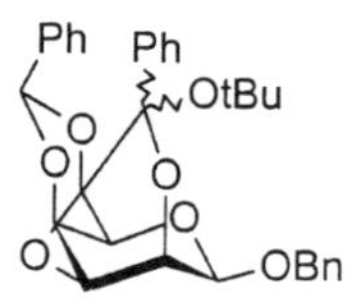

AS[2,NG-3] **60**

In contrast to the results obtained with the 4,6-*O*-benzylidene-protected educt of the *gluco* series (entry 23), methyl 3-*O*-benzoyl-4,6-*O*-benzylidene-β-D-galactopyranoside (**57**, entry 24 [45]), on treatment with DAST, underwent ring contraction to form **58** (**MS**[2,rO]) only. On the other hand, from the solvolysis reaction of the benzyl galactoside **59** (entry 25 [50]), in analogy to the situation found in the *gluco* series (entry 21), formation of a side product also containing an orthoester structure (**60**, **ISC**[2,NG-3], 12%) has been claimed. However, close inspection of the NMR data revealed [24] that the structural assignment has to be changed to **61** (**MS**[2,rO], entry 24).

From these findings, priority rules concerning the option "ring contraction vs. aglycon migration" in attempted displacements at C-2 of hexopyranosides containing an equatorially oriented LG (but no NG of the "complex type" in equatorial position at C-3) can be deduced:

1. Independent of the anomeric configuration, ring contraction (**MS**[2,rO]) is the first choice in all monocyclic educts as well as in cases where a *cis*-fused 3,4-*O*-isopropylidene or *cis*-fused 4,6-*O*-benzylidene ring is annellated. [The first situation is given in educts with *arabino* and *galacto* configuration, the second in the *galacto* (and *gulo*) series.]
2. As soon as a *trans*-fused 4,6-*O*-benzylidene ring is present (possible in the *gluco* and *allo* series), ring contraction is impeded (but see the result shown in entry 4, which might be explained by the involvement of a "hot" carbenium ion). However, in the case of a *trans*-diequatorial arrangement of the NG, at C-1, and the LG, at C-2, aglycon migration (**MS**[2,NG-1]) may take place. The latter occurs obviously with more ease than in cases of *O*-glycosides when more nucleophilic NGs (arylthio, azido, acetoxy, etc.) are located at C-1 [28].

To emphasise the importance of this ring contraction during the course of substitution reactions at C-2 discussed above, an example with preparative attraction [52] is outlined in Scheme 13 (the transformation is also operative with the α-anomer). Here advantage is taken of the (regular) ring contraction of – monocyclic – glucopyranosides under solvolytic conditions, after they had been used as chiral auxiliaries in the asymmetric cyclopropanylation of allyl alcohols.

4.2.2
Reactions of Pyranose Derivatives Containing an Equatorially Oriented Leaving Group at C-3

Concerning the results collected in Table 6, the following comments on reactions starting from educts containing an equatorially oriented LG at C-3 can be made: with good nucleophiles, S_N2 is convincingly operative (e.g., entry 4 [49]). Incidentally, participation of the solvent may appear as a side reaction if a potentially participating NG does not interfere (cf. entries 3 [53] and 8 [54a]).

In the absence of a good nucleophile, the involvement of an NG capable of participation becomes dominant. Starting from a vicinal *trans* disposition of LG and NG in the educt, besides anchimerically assisted substitution with retention of configuration (**AS**[3,NG-2], entry 8) and group migration (**MS**[3,NG-2], entry 8), participation of NGs of the "complex" (**ISC**[3,NG-2], entries 7 [54b], 11 and 12 [47], (**ISC**[3,NG-4] + **Hy**), entry 14 [55]) as well as the "simple type" (**IS**[3,O-2], entry 5 [49]) have been observed. In some instances, e.g., for the results shown in entry 6 [48], discrimination between an (S_N**2** + **AcM**) and an (**ISC** + **Hy**) mechanism cannot be made. Of special interest is an unusual "triflate rearrangement", reported by Binkley [55] during his efforts to

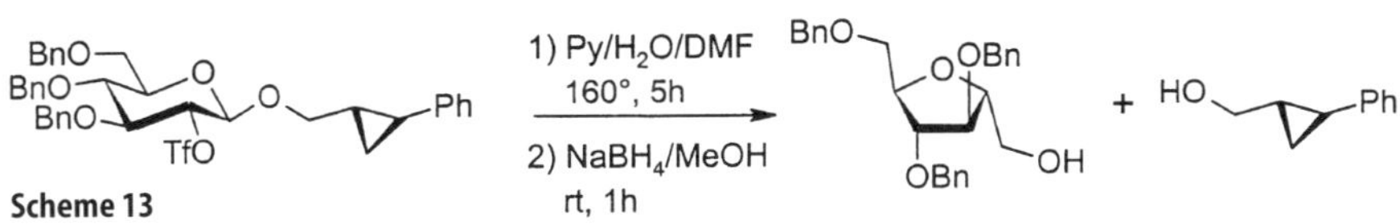

Scheme 13

fully elucidate the reactivity of carbohydrate triflates. Its embedment into the types of NG-assisted reactions is illustrated in Scheme 14.

According to Binkley's observations, treatment of methyl 4-*O*-benzoyl-2,6-dideoxy-*β*-D-*arabino*-hexopyranoside in CH_2Cl_2, with triflic anhydride in the presence of DTBMP, rapidly leads to formation of triflate **62**, which – simply upon standing at room temperature for 4–8 h – rearranges into the highly reactive "orthoester triflate" **64**. The structures of compounds **62** and **64** were deduced from their respective spectroscopic data and confirmed by a series of chemical transformations. Thus, as shown in Scheme 14, immediate treatment of the triflation mixture (containing **62**) with tetra-*n*-butylammonium bromide (TBABr) gave the S_N2 product **65** (93%), whereas the same procedure, applied to a triflation mixture which had stood for 4 h at room temperature (containing **63/64**), under retention of configuration gave the **AS** product **66** (74%). Analogous treatment with methanol led, by the **ISC** mode, to a mixture of orthobenzoates **67** (95%). Most interestingly, treatment with tributyltin hydride caused reduction to yield the 3,4-benzylidene acetal **68** (82%). (A similar acetal-forming reaction is known to occur upon $LiAlH_4$-treatment of 2-*O*-acylated aldopyranosyl bromides.) Reaction of the rearranged triflate **63/64** with water (cf. entry 14) produced, by hydrolysis of the **ISC** product with or without **AcM** (see Scheme 2), the 3-benzoate **69** with inverted configuration at C-3 (77%). This "triflate rearrangement" was also observed with the corresponding *α*-anomer [55].

For reactions producing or involving three-membered ring systems (**IS**[3, O-2] in entry 5 and **MS/AS**[3,NG-2] in entry 8, respectively), conformational flexibility is required (cf. entry 19, Table 5, and the arguments raised there).

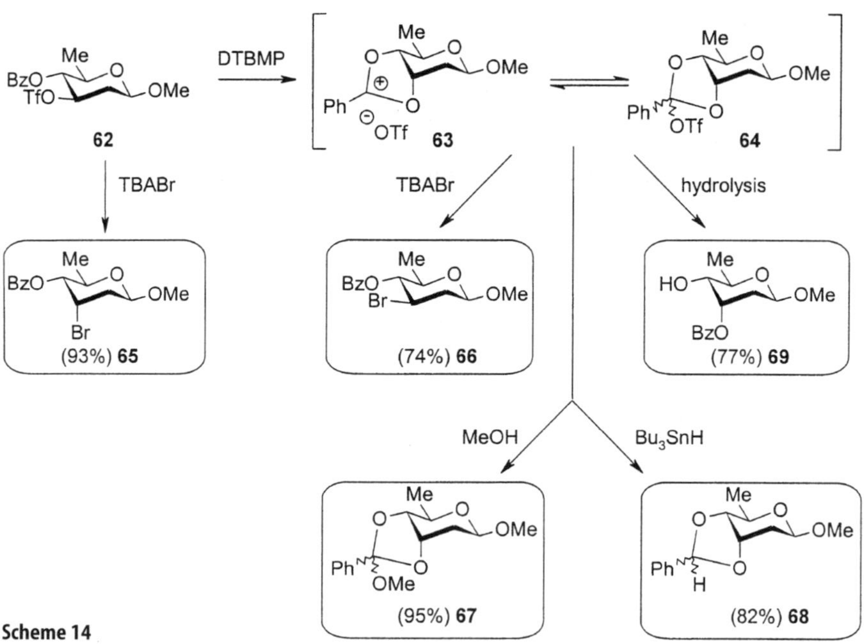

Scheme 14

Table 6

	Educt	Product(s)	Conditions	Σ (%)	Ref.
1	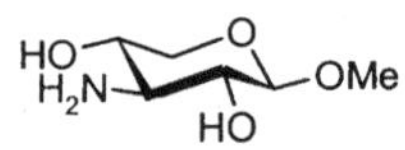	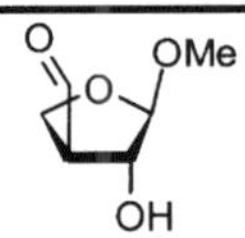 **MS**[3,C-5] (major) 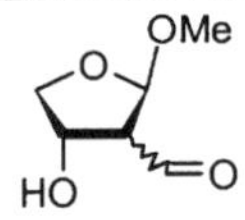 **MS**[3,C-1]+**Ep** (minor)	$NaNO_2$, HCl, H_2O 0 °C → rt, 3.5 h	84	56
2	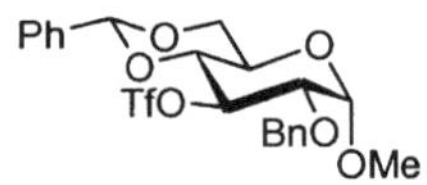	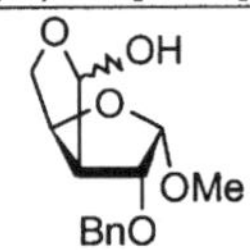 **MS**[3,C-5]+**HAF** (92%)	DTBMP/H_2O/HF*i*P reflux, 1.5 h	92	53
3	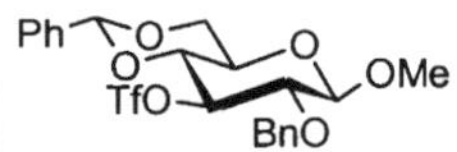	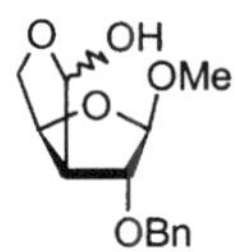 **MS**[3,C-5]+**HAF** (62%) 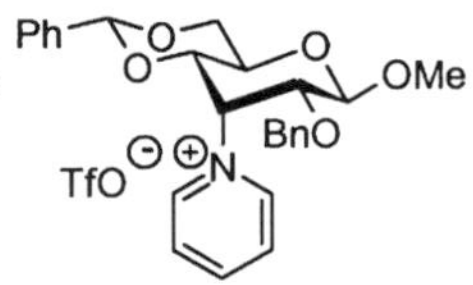 $\mathbf{S_N2}$ (5%)  **F**[3,C-1]+**Hy** (11%)	pyridine/H_2O/toluene reflux, 1 h	78	53
4	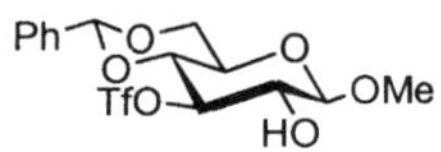	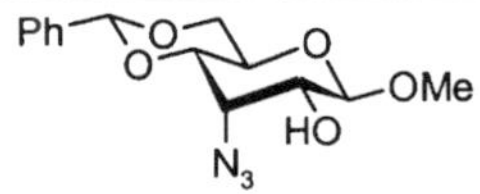 $\mathbf{S_N2}$ (85%)	NaN_3, DMF rt, 1 h	85	49

Table 6 (continued)

	Educt	Product(s)	Conditions	Σ (%)	Ref.
5	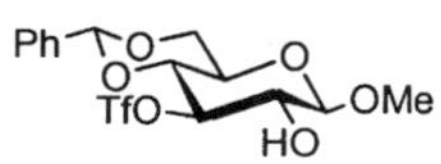	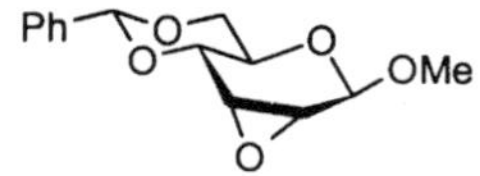 **IS**[3,O-2] (98%)	*t*-BuONa, *t*-BuOH 23 °C, 2 h	98	49
6	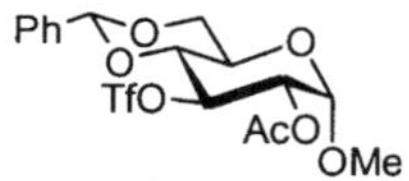	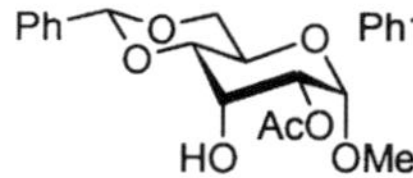 (13%) 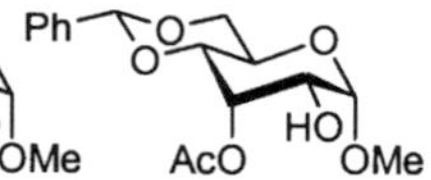 (54%)	$NaNO_2$, DMF 60 °C, 18 h	67	48
7	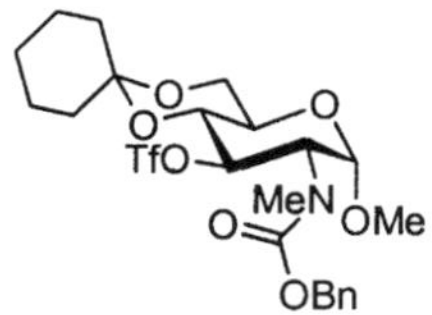	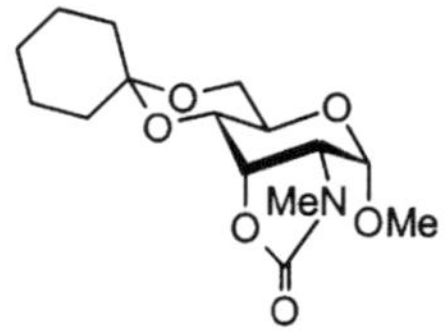 **ISC**[3,NG-2] (95%)	pyridine 65 °C, 1.5 h	95	54b
8	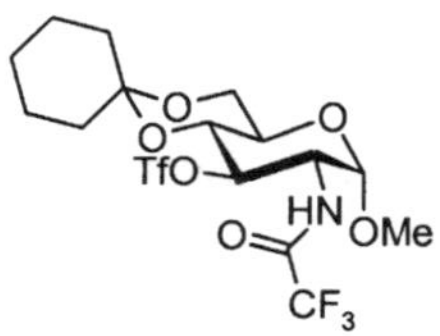	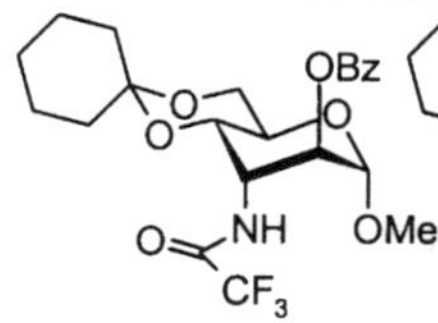 **MS**[3,NG-2] (62%) 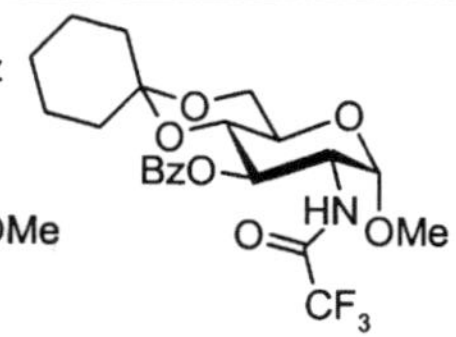 **AS**[3,NG-2] (17%) 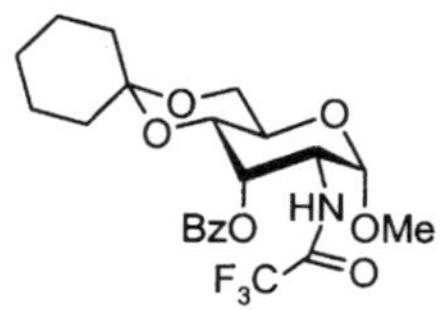 **S$_N$2** (4%)	NaOBz, DMF 60 °C, 4 h	83	54a

Table 6 (continued)

	Educt	Product(s)	Conditions	Σ (%)	Ref.
9	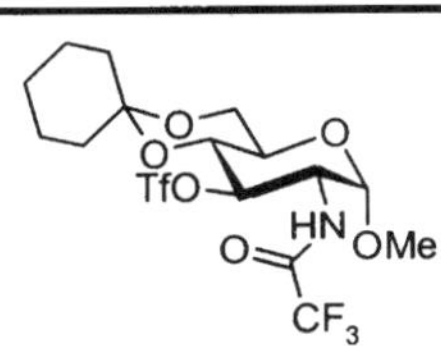	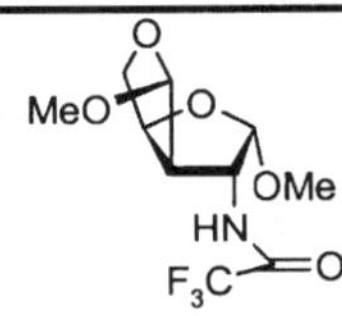 **MS**[3,C-5]+**AF** (72%)	MeOH (sealed) 100 °C, 6 min	72	54a
10	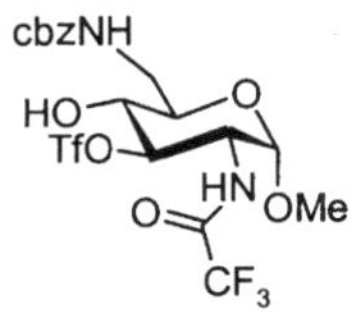	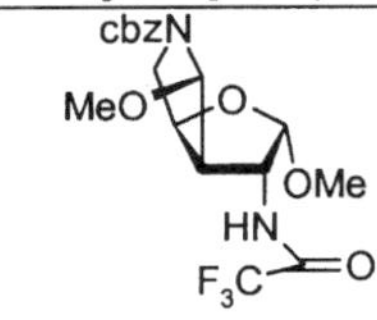 **MS**[3,C-5]+**AF** (34%)	MeOH (sealed) 100 °C, 25 min	34	54a
11	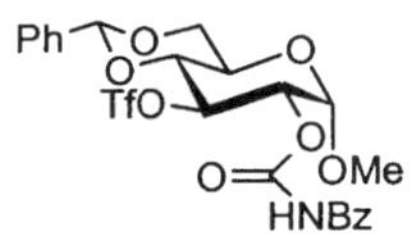	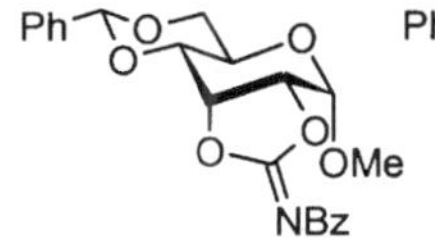 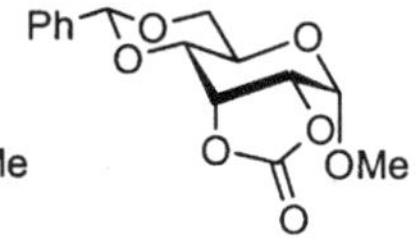 **ISC**[3,NG-2] (98%)	NaH, THF rt, 12 h	98	47
12	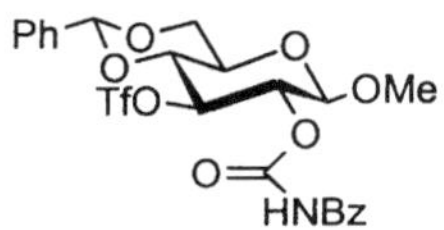	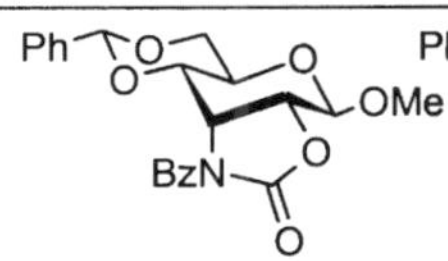 **ISC**[3,NG-2] (37%) 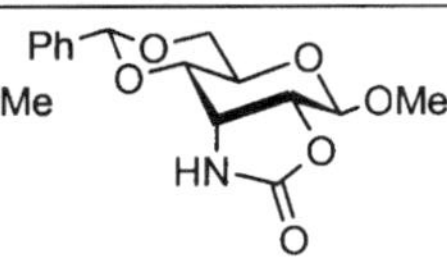 (20%)	NaH, THF 0 °C, 3 h	57	47

Table 6 (continued)

	Educt	Product(s)	Conditions	Σ (%)	Ref.
13	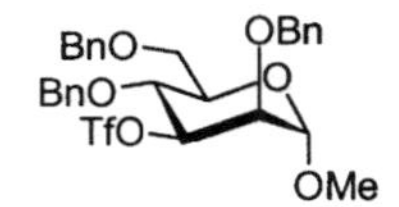	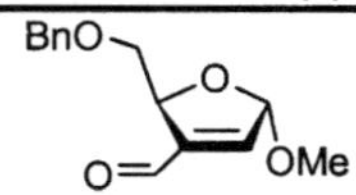 **MS**[3,C-5]+**Hy**+**E** (27%)	pyridine/H_2O/HF*i*P reflux	27	53
14		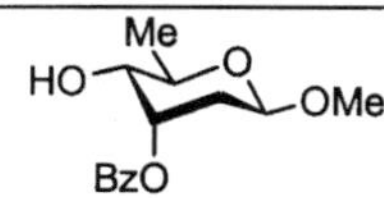 **ISC**[3,NG-4]+**Hy** (77%) **69**	DTBMP/H_2O/CH_2Cl_2 rt, 12 h	77	55
15		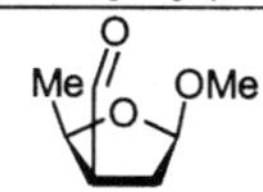 **MS**[3,C-5] (54%)	DTBMP/CH_2Cl_2 –20 °C → rt, 30 min	54	31
16	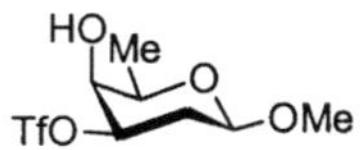	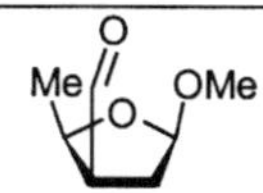 **MS**[3,C-5] (65%)	DTBMP/benzene 80 °C, 30 min	65	31
17		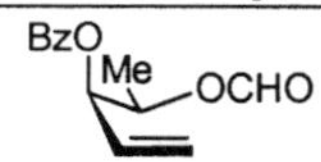 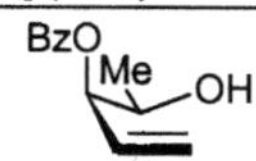 **F**[3,C-1] (45%) **75** +**Hy** (30%) **76** 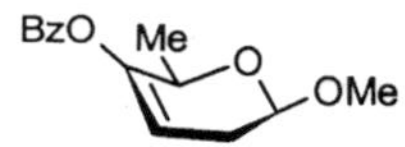 **E**[3,H-4] (9%)	TBAOBz, toluene/H_2O reflux, 6 h	84	33a

Table 6 (continued)

	Educt	Product(s)	Conditions	Σ (%)	Ref.
18	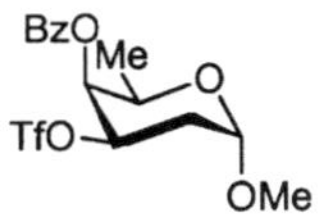	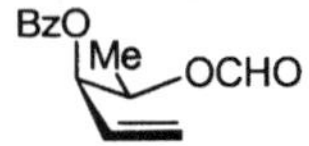 **F**[3,C-1] (4%) 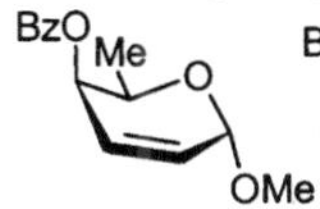 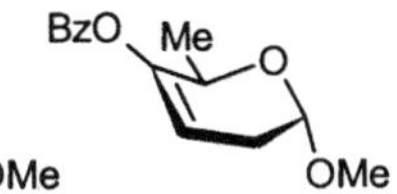 **E**[3,H-2] (15%) **E**[3,H-4] (53%)	pyridine/H_2O 100 °C, 25 min	73	33
19	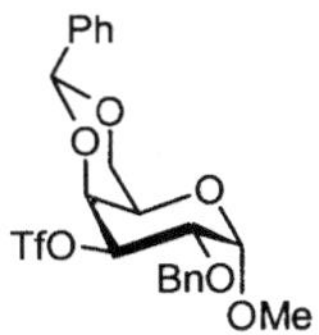	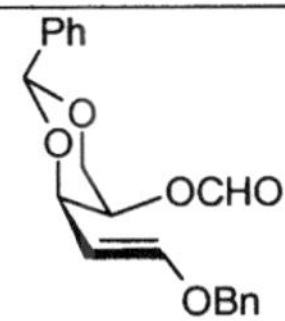 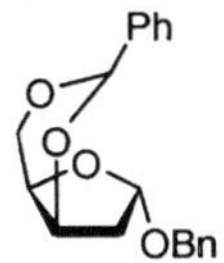 **F**[3,C-1] (38%) **77** (26%) **78**	pyridine/H_2O/HF*i*P reflux, 1 h	64	53
20	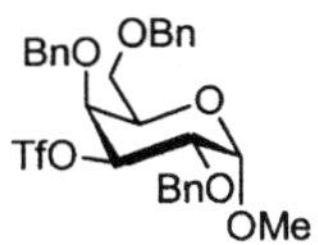	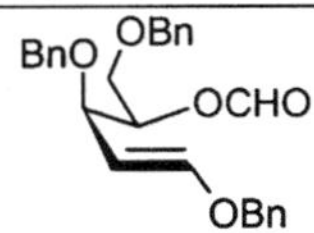 **F**[3,C-1] (58%)	pyridine/H_2O/HF*i*P reflux	58	53

Table 6 (continued)

	Educt	Product(s)	Conditions	Σ (%)	Ref.
21	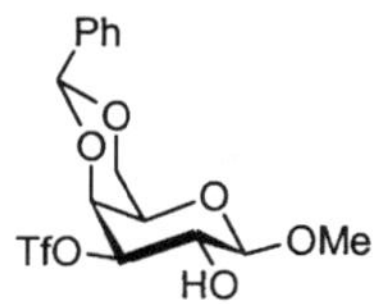	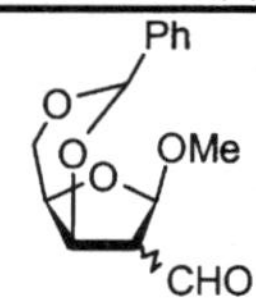 **MS**[3,C-1] (57%) +**Ep**	NaN_3, DMF rt, 18 h	57	49
22	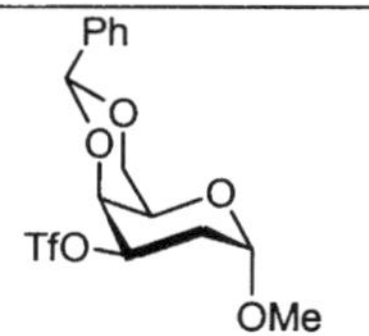	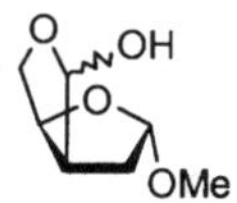 **MS**[3,C-5]+**HAF** (70%)	pyridine/H_2O/HF*i*P rt	70	53
23	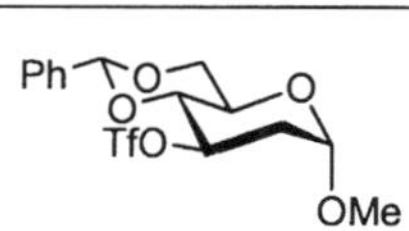	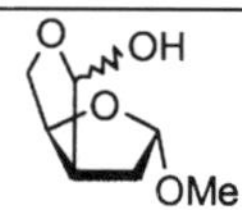 **MS**[3,C-5]+**HAF** (91%)	pyridine/H_2O/HF*i*P rt	91	53

Of further interest is that a trifluoroacetamido group, under conditions of a displacement reaction, gives "*N*-attack" only (entry 8), whereas, under conditions of solvolysis, ring contraction (see entries 9 and 10) and no participation is observed. The forced participation of a *N*-benzoylcarbamate group, depicted in entries 11 and 12, shows dependence in the mode of attack (*O*- or *N*-cyclisation) on the anomeric configuration.

Deamination reactions with educts containing an equatorially oriented LG at C-3 follow the "alkyl shift routes" **g** (Scheme 4), especially **MS**[3,C-5] and/or **MS**[3,C-1] (Scheme 12). In each case, the ring-contracting step proceeds with inversion of configuration at the position of the original activation only, so that the formyl side chain in the products takes the same steric position ("above" or "below" the plane of the sugar ring) as the LG did before in the educt. This is illustrated in Scheme 15, where the directly formed products **71** and **73**, arising from educt **70** in the $^4C_1$ conformation, have their side chain "above" the plane. A speciality of importance in this context [4] results from the tendency of these ring-contracted products containing a formyl side chain, on the one hand, to epimerise at the branching point (symbol **Ep**, see structure **74**) and, where possible, to form furanoid hemiacetals (symbol **HAF**, see structure **72**) with the OH-group (or any nucleophilic substituent) from the original side chain (the former C-6) on the other. The **MS**[3,C-1] mode is intrinsically of minor importance as it was reported from the reaction of methyl 3-amino-3-deoxy-*β*-D-xylopyranoside only (entry 1 [56]). The same results as for deaminations have also been obtained in solvolysis reactions with otherwise unprotected 3-sulfonates of methyl hexopyranosides [57].

As elaborated within the last 15 years by the pioneering work of Binkley, Castillon as well as Knapp, the choice for the reaction path in displacements, with poor nucleophiles, of educts containing an equatorial LG of high nucleofugicity (generally triflate) is primarily dependent on the configuration at C-4. Thus educts containing an equatorial substituent (*xylo*, *gluco* and *manno* configuration, entries 1 [56], 2, 3, 13 and 23 [53], 9 and 10 [54a] and 15 [31]) at C-4 are easily subject to ring contraction of the type **MS**[3,C-5] (see Scheme 15), whereas those with an axially oriented substituent (galactose derivatives, entries 17 [33], 19 and 20 [53]) give fragmentation. As far as can be jugded from results

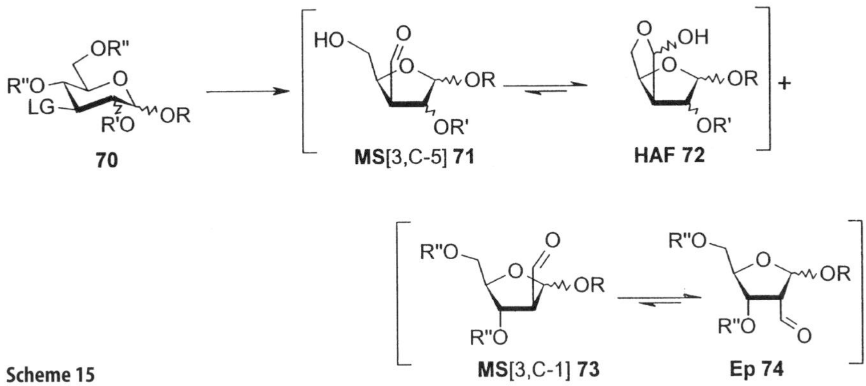

Scheme 15

obtained under different conditions, the latter generalisation is infringed in those cases where the OH-group at C-4 or C-2 is unprotected. Here ring contractions, i.e., **MS**[3,C-5] (entry 16 [31], cf. entries 15 and 17) and **MS**[3,C-1] (entry 21 [49], cf. entry 19), have been found instead of fragmentation. Furthermore, if anything at all can be deduced from the limited number of examples known so far, the options soon merge into one as soon as a 2-benzyloxy substituent together with a 4,6-*O*-benzylidene protection of the *trans*-decalin type (**F**[3,C-1] in entry 3 [53]) or a 2-deoxy function and an α configuration together with a 4,6-*O*-benzylidene protection of the *cis*-decalin type (**MS**[3,C-5] in entry 22 [53]; cf. entry 23) is present. In comparison, α-anomers show a lesser tendency to give fragmentation than their corresponding β-anomers (cf. entries 17 and 18), for which effect "proper orbital alignment" at C-1 and C-3 was thought to be essential [33a]. Generally, the ring-contraction reactions do not show remarkable sensitivity towards the type of protecting groups present, as, in the **MS**[3,C-5] course, those from C-4 and C-6, including the 4,6-alkylidene acetal, are cleaved. Castillon et al. [53a] have provided a mechanistic explanation that is based on the stability of a carbocation in the benzylidene moiety. However, in the *gluco* series, it also occurs with 4,6-cyclohexylidene acetals and may therefore simply be the consequence of the hydrolysis of the oxocarbenium ion formed with C-4 in the course of the ring contraction. The direct product of fragmentation contains a formate grouping, as in **75** and **77** (entries 17 and 19). Its hydrolysis liberates an OH-group at the former C-5 as shown with alkenol **76**. In the case of **77** (entry 19), however, the liberated OH-group directly adds to the enol ether under formation of 2-deoxypentofuranoside **78**. A further side reaction consists of elimination. This is found with conformational flexible educts of the 2,6-dideoxyhexose series (entries 17 and 18) and with 3-formylpentofuranosides containing a *trans* relationship between H-3 and the substituent at C-2 (entry 13).

In striking contrast to the general tendency of educts of the glucose series to give ring contraction **MS**[3,C-5] rather than fragmentation **F**[3,C-1], is the long-known transformation of 3-*O*-mesyl-D-glucose (**80**) into 2-deoxy-D-ribose (**81**) [58]. As outlined in Scheme 16, alkaline treatment (pH 8.0–8.5) of **80**, which is easily accessible by acid hydrolysis of 1,2:5,6-di-*O*-isopropylidene-3-*O*-mesyl-α-D-glucofuranose (**79**), causes this fragmentation and allows isolation of **81** (in the form of its anilide) in a yield of 40–50% (unfortunately, no comments concerning the possible structures of by-products were given).

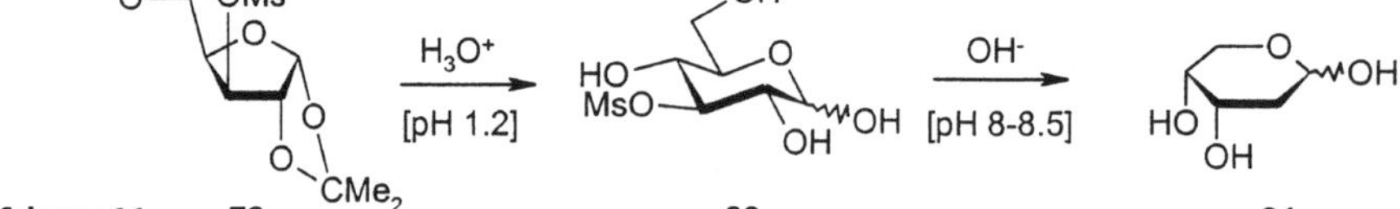

Scheme 16 **79** **80** **81**

4.2.3
Reactions of Pyranose Derivatives Containing an Equatorially Oriented Leaving Group at C-4

For educts containing an equatorial LG at C-4, the possible reaction paths involving ring members (rO and C-2, respectively) as participating groups have already been presented in Sect. 4.2 (see modes **AS**, **IS**, **MS**[4,rO] in Scheme 10 as well as **MS**[4,C-2] in Scheme 11). All of these reactions have been found operative in the deamination of methyl 4-amino-4-deoxy-*α*-D-glucopyranoside [59] (entry 1). Of special interest therein is the highly reactive compound **82** (4,5-anhydro-D-galactose, formed by **IS**[4,rO]) and the products of its hydrolysis **83** (L-altrose, by opening of the anhydro ring at C-4) and **84** (D-glucose, by opening at C-5). This **IS**[4,rO])-type of reaction was also discussed to explain the formation of D-glucose (**84**) in the solvolysis of methyl 4-*O*-*p*-nitrophenylsulfonyl-*α*-D-glucopyranoside [60] (entry 3). Recently, the stable 4,5-anhydro sugar **86** was isolated in 40% yield by chromatography from the DAST reaction of methyl 3,6-dideoxy-3-*C*-methyl-3-nitro-*α*-L-glucopyranoside [23] (**85**, entry 4). The **MS**[4,C-2] mode of ring contraction is of minor importance and the reaction product (a 3-*C*-formylpentofuranoside) is subject to **Ep + HAF** (epimerisation and hemiacetal formation, see entry 1). The most significant rearrangement to be observed with this class of educts is the **MS**[4,rO] mode of ring contraction (entry 2 [34a], entry 4 [23], entries 6–8 [61]). As this reaction (under inversion of configuration at C-4 and C-5) opens easy access to 5-substituted hexofuranosides, it has already been used for various syntheses [62]. A special case with involvement of this **MS**[4,rO] option is shown in entry 10 [31]. It has been proposed that the unusual oxetane ring in product **87** is formed by intramolecular S_N2 reaction in the rearranged triflate **89**, that in turn results, by rearrangement, from the intermediate oxiranium ion **88**. However, the formation of **87** can also be envisaged to have taken place by intramolecular attack, at C-5, of the 3-OH-group in the state of intermediate **88** with concomitant opening of the C-5/rO bond (**MIS**[4,rO/O-3]).

Of particular significance to the methodology of fluorine introduction into carbohydrate molecules by nucleophilic substitution is the work of Morishima [61]. As depicted in Table 7, the DAST route (entries 6, 7 and 8) gave **AS**[4,rO] and **MS**[4,rO] exclusively, whereas the triflate/TASF tandem allowed $\mathbf{S_N2}$ together with **AS**[4,rO] (entry 5) or **E**[4,H-3] (entry 9).

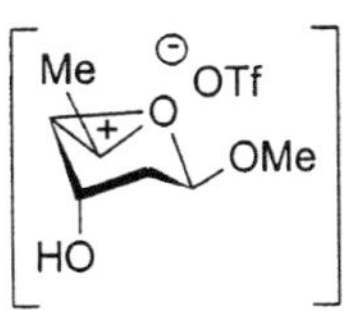

88

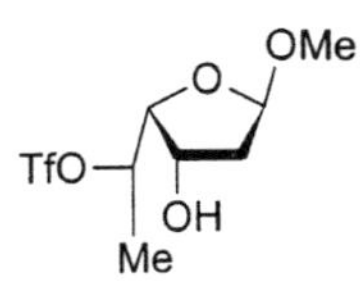

89

When good nucleophiles are applied to triflates, straight $\mathbf{S_N2}$ occurs (entries 11 and 12 [33c] and 13 [48]). In the latter case, although diequatorial orientation of LG and NG-3 was given, a single product was isolated (cf. entry 18, Table 5, and

Table 7

	Educt	Product(s)	Conditions	Σ (%)	Ref.
1	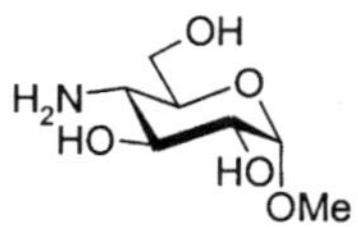	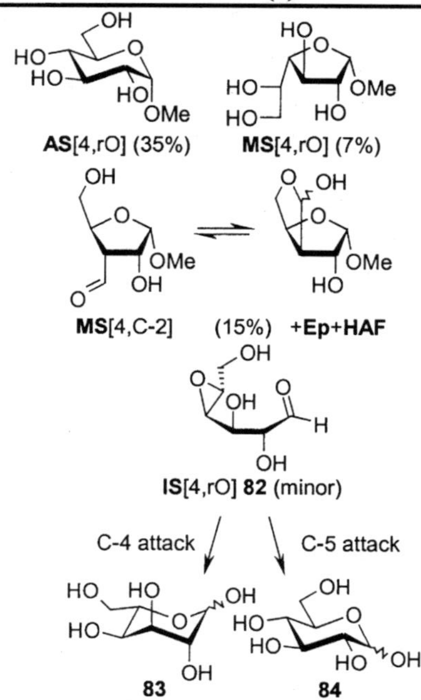	$NaNO_2$, $HOAc/H_2O$ [pH 3–4]	>57	59

Table 7 (continued)

	Educt	Product(s)	Conditions	Σ (%)	Ref.
2	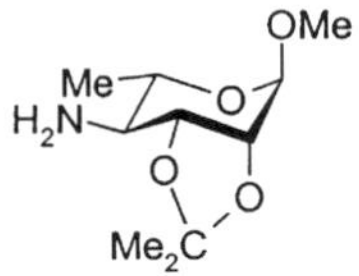	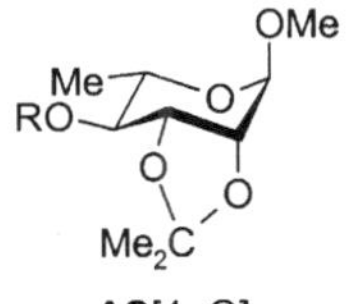 **AS**[4,rO] R = H or Ac (30%) 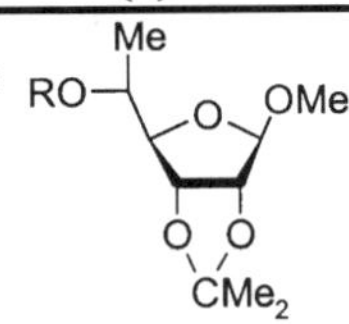 **MS**[4,rO] R = H or Ac (70%)	$NaNO_2$, $HOAc/H_2O$ 0 °C, 1.5 h	100	34a
3	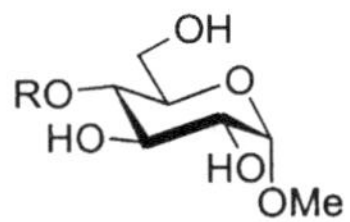 R = $p$$NO_2$$PhSO_2$	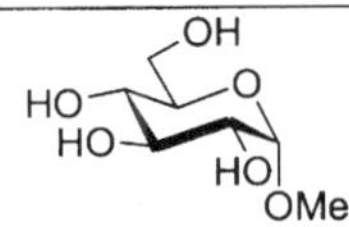 **AS**[4,rO] (50%) 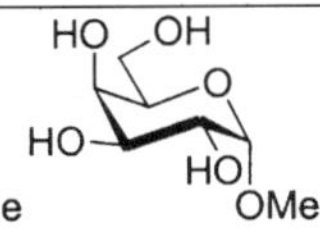 **S$_N$2** (8%) 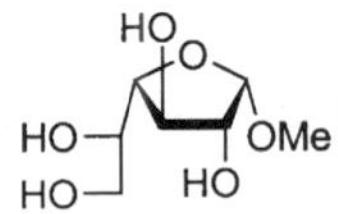 **MS**[4,rO] (8%) 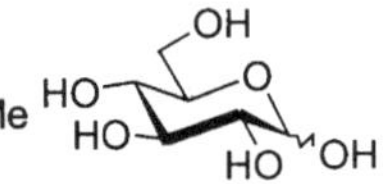 **84** (trace)	acetate buffer [pH 4–5] 100 °C, 5 h	64	60
4	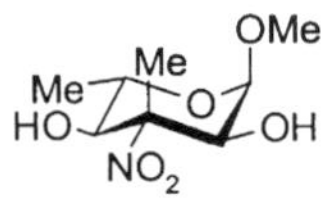 **85**	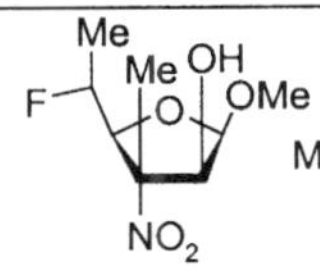 **MS**[4,rO] (46%) 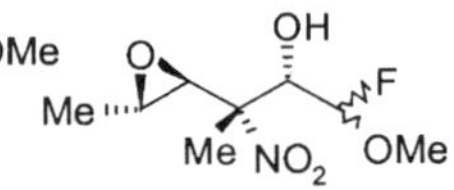 **IS**[4,rO] (40%) **86**	DAST, CH_2Cl_2 reflux, 1.5 h	86	23
5	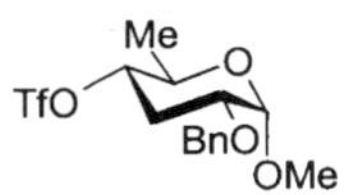	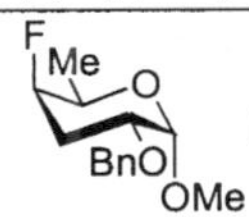 **S$_N$2** (54%) 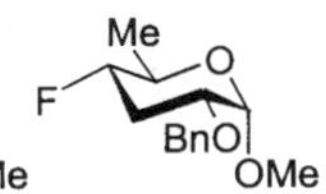 **AS**[4,rO] (13%)	TASF, CH_2Cl_2 rt, 1.5 h	67	61

Table 7 (continued)

	Educt	Product(s)	Conditions	Σ (%)	Ref.
6	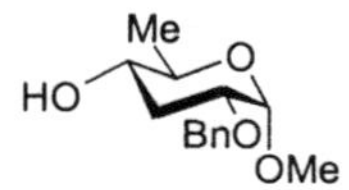	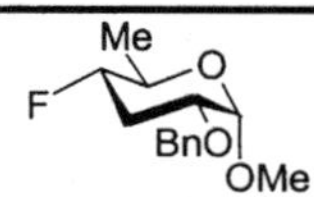 **AS**[4,rO] (66%) 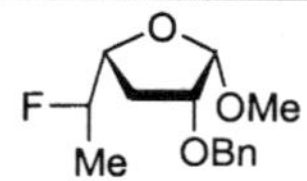 **MS**[4,rO] (17%)	DAST, CH_2Cl_2 −13 °C → rt, 2 h	83	61
7	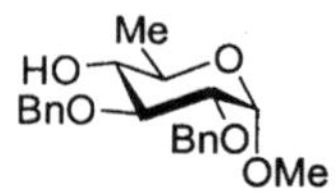	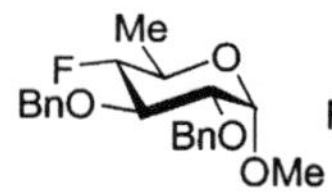 **AS**[4,rO] (2%) 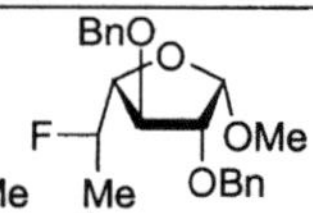 **MS**[4,rO] (38%) starting material (recovered) (39%)	DAST, CH_2Cl_2 −13 °C → rt, 2 h	79	61
8	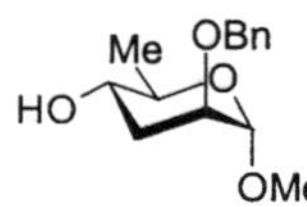	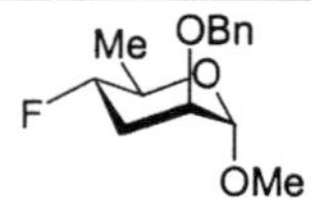 **AS**[4,rO] (28%) 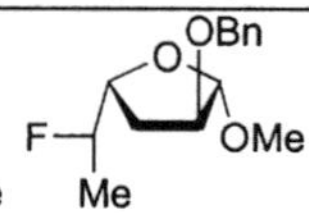 **MS**[4,rO] (21%) starting material (recovered) (34%)	DAST, CH_2Cl_2 −13 °C → rt, 2 h	83	61
9	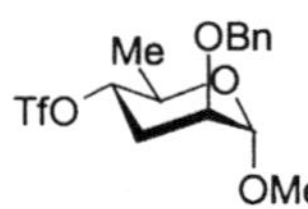	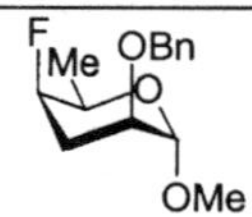 **S$_N$2** (21%) 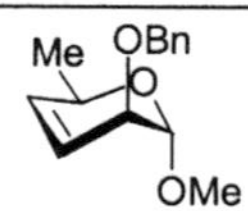 **E**[4,H-3] (24%)	TASF, CH_2Cl_2 rt, 1.5 h	45	61
10	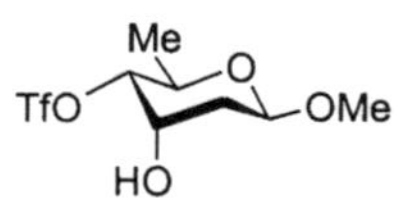	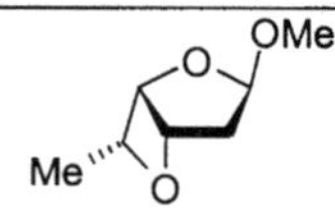 **MIS**[4,rO/O-3] (54%) **87**	DTBMP/benzene 80 °C, 30 min	54	31

Table 7 (continued)

	Educt	Product(s)	Conditions	Σ (%)	Ref.
11	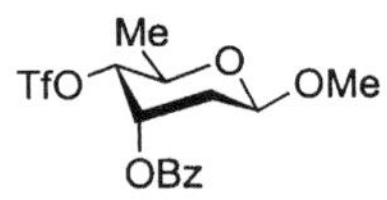	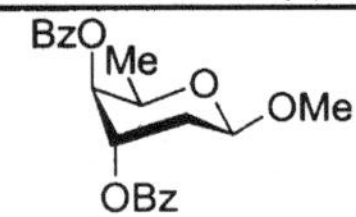 **S$_N$2** (73%)	TBAOBz, toluene reflux, 30 min	73	33c
12	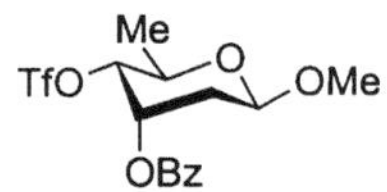	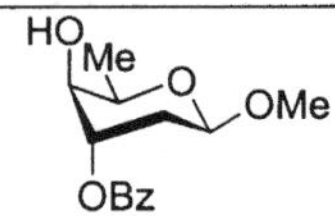 **S$_N$2** (96%)	$TBANO_2$, toluene 23 °C, 2 d	96	33c
13	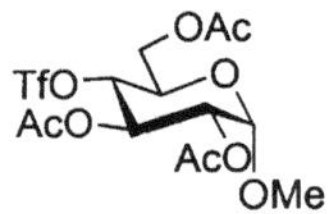	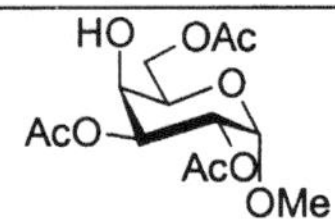 **S$_N$2** (79%)	$NaNO_2$, DMF 20 °C, 1 h	79	48
14	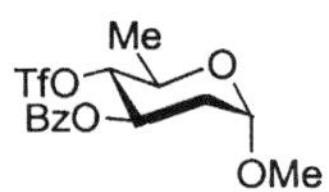	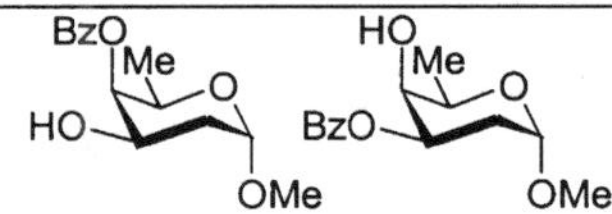 **ISC**[4,NG-3]+**Hy** (70%) (21%)	DTBMP/H_2O/CH_2Cl_2 rt, 24 h	91	55c
15	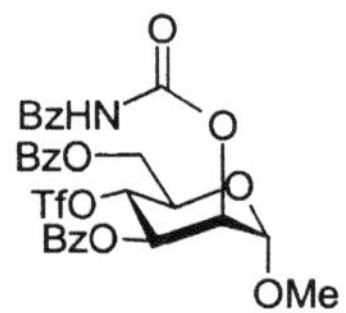	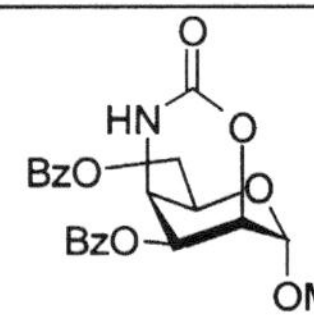 **ISC**[4,NG-2]+**Hy** (65%)	NaH, THF rt, 6 h	65	47

entry 6, Table 6). The solvolysis reaction shown in entry 14 [55c] followed the one described in the preceding section (entry 14, Table 6 and Scheme 14). The rearranged triflate involved in this transformation is of the structure **90.** A case of NG participation from an axial, non-vicinal position, namely **ISC**[4,NG-2], is shown in entry 15 [47].

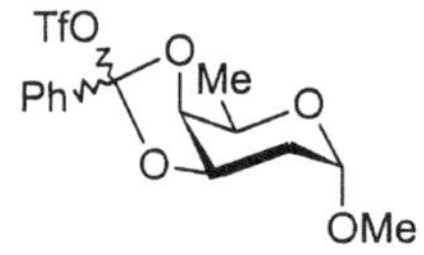

90

4.3 Reactions of Pyranose Derivatives at the Primary Position

For attempts at S_N reactions which start from educts containing their LG at a primary position, a first general reduction in the number of possible product structures arises from the lack of chirality at the reaction centre. Thus, besides **S_N1** vs. **S_N2**, different, per se competing mechanisms lead to the same product. This is exemplified in Scheme 17 with the participation of the NG from C-4, where either **S_N2** or **AS**[6,NG-4] and, in the case of DAST reaction, (**ISC**[6,NG-4] + **Hy**) and **AcM**, respectively, may be operative.

A further reduction in the number of products to be expected has its origin in the lack of importance of the participation of the ring oxygen. An intermediate oxiranium ion of structure **91**, comprising atoms C-6, C-5 and rO, would give rise (see Scheme 18) to the mechanistic types **AS**[6,rO] (by attack at C-6 to produce **93**), **MS**[6,rO] (by attack at C-5 with inversion of configuration to give ring enlargement to the septanose derivative **92**) and **IS**[6,rO] (by attack at C-1 under production of 5,6-anhydroaldose **94**). However, product formation other than by **AS**[6,rO] (indistinguishable from straight S_N) has not been observed (but see reactions of 5-thioaldopyranose derivatives, Sect. 6.1). Nevertheless, to explain certain results from the deamination reaction of 1-amino-2,6-anhydro-1-deoxy-alditols, an intermediate, similar to **91** but involving C-1, C-2 and rO, has been suggested [63].

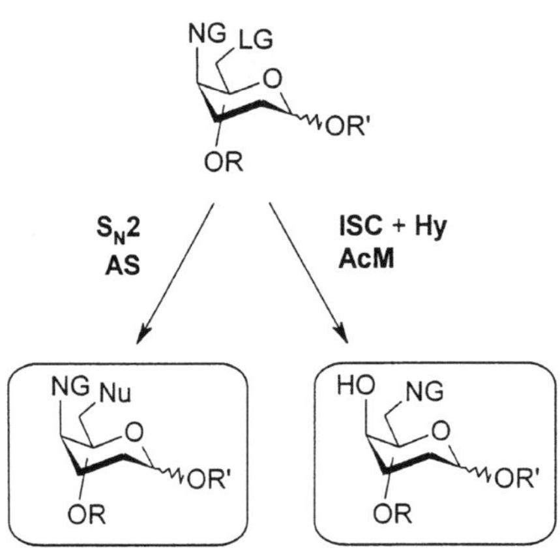

Scheme 17

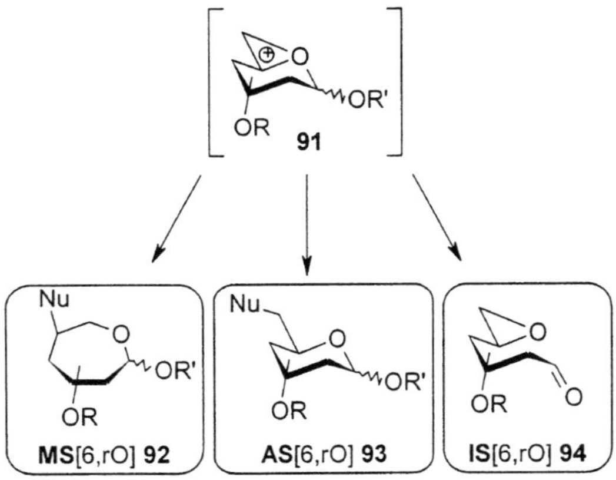

Scheme 18

In contrast to these restrictions, the special position of the reaction centre above the ring plane together with a conformational flexibility opens additional ways for participation of NGs (Table 8). As in educts from the D-series (adopting the $^4C_1$-conformation), the side-chain orientation changes from equatorial to axial, *cis*-located (likewise axial) substituents in position 1 or 3 may interfere. This is demonstrated with the 1,6-aglycon shift **MS**[6,NG-1] (out of a β-D-configuration exclusively, entries 1 [64] and 3 [65]) as well as 3,6-anhydro ring formation **IS**[6,O-3] (with a benzyl ether as participating NG, entry 2 [64]). As has been discussed in detail [64] the course of the reaction depends on the nature of the protecting group at O-2. The ester functionality (entries 1 and 3) eases the rearrangement (by formation of an intermediate acyloxonium ion of type **16**) and causes entry of the nucleophile (fluoride) from the β-face. Of some interest are the results shown in entry 4 [65], where on attempted DAST fluorination at C-6, products **95** and **96** were isolated instead of the expected 6-deoxy-6-fluoro compound (analogous to **97** in entry 5 [65]). As an explanation, the migration of the 2,4-dimethylbenzoyl protecting group from O-4 to O-6 with, in part, subsequent S_N2 fluorination was given.

According to Scheme 17, the product ensemble **95/96** may – via an intermediate acyloxonium ion (see Scheme 2) – also be the result of **MS**[6,NG-4] and (**ISC**[6,NG-4] + **Hy**), respectively. However, straight S_N, including solvent participation, together with hydride shift **MS**[6,H-5] and elimination **E**[6,H-5] (for deaminations and sulfonate displacements, respectively), is the general standard reaction repertoire (entry 6 [66a]). This latter work is of special value as a subtle study on the influence of the kind of leaving and protecting groups, fluorinating agent, solvent and other parameters on a single transformation (the introduction of fluorine at C-6 of D-galactose). In this context, another thorough investigation from the same laboratory, describing scope and limitations of the DAST fluorination of benzyl-protected D-glucopyranose derivatives, is worthy of attention [66b].

Table 8

	Educt	Product(s)	Conditions	Σ (%)	Ref.
1	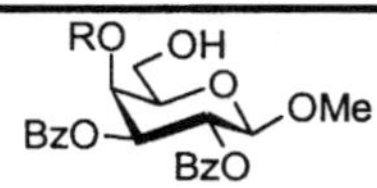 R = protected α-D-Gal*p*	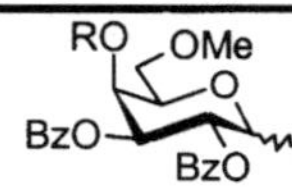 **MS**[6,NG-1] (α 10%, β 50%)	DAST, CH_2Cl_2 –45 °C → rt, 3 h	60	64
2	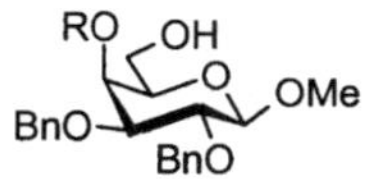 R = protected α-D-Gal*p*	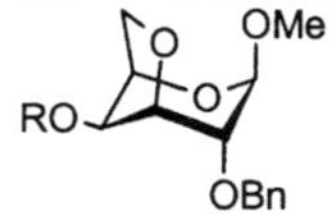 **IS**[6,O-3] (73%)	DAST, CH_2Cl_2 –45 °C → rt	73	64
3	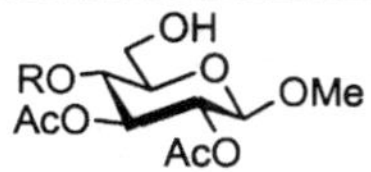 R = protected β-D-Gal*p*	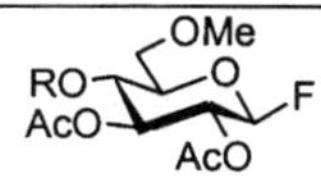 **MS**[6,NG-1] (36%) 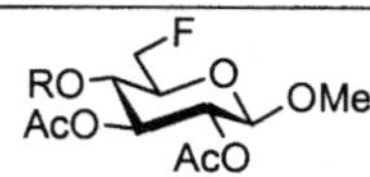 **S$_N$2** (32%)	DAST, CH_2Cl_2 0 °C → rt, 1 h	68	65
4	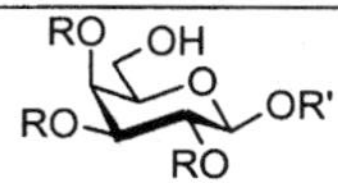 R = 2,4-Me_2Bz R' = protected 4-β-D-Glc*p*	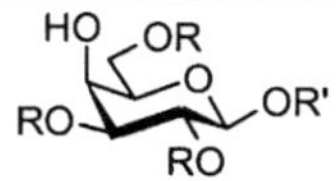 **AcM** (33%) **95** or **ISC**[6,NG-4]+**Hy** 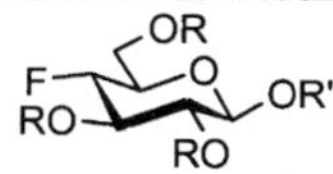 +**S$_N$2** (24%) **96**	DAST, CH_2Cl_2 0 °C → rt, 12 h	57	65
5	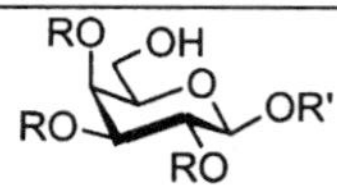 R = 2,4,6-Me_3Bz R' = protected 4-β-D-Glc*p*	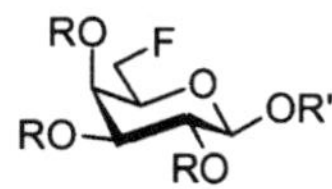 **S$_N$2** (94%) **97**	DAST, CH_2Cl_2 0 °C → rt, 5 d	94	65
6	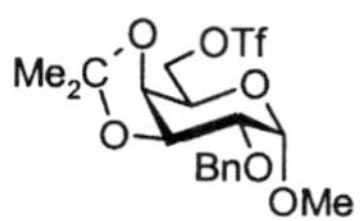	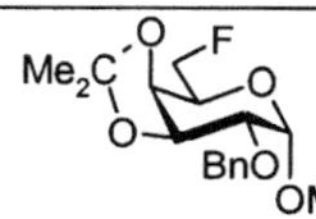 **S$_N$2** (83%) 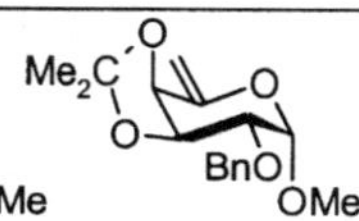 **E**[6,H-5] (15%)	TASF, CH_2Cl_2 –30 °C → rt, 12 h	98	66a

5
Reactions of Furanose and Septanose Derivatives

5.1
Reactions of Furanoid Educts

When turning our attention to the situation of nucleophilic substitution reactions in furanose derivatives, an important simplification arises from the fact that, according to our present knowledge, no single product has been isolated for which participation of the ring oxygen atom or another ring member had been anticipated. A further reduction in the quantity of side reactions results from the poor number of possible *trans* relationships (known to be essential for participation) within LGs and vicinal NGs located in the furanose ring. In contrast to this, the probability of the involvement of NGs from the side chain is drastically raised, as all ring positions (except the branching point C-4) are prone to interference.

In Table 9 a few characteristic types of reactions are collected. They comprise, besides straight $\mathbf{S_N2}$ (entries 2 [67], 8 [68], 9 [69] and 10 [70]) and various eliminations (entries 3 [67], 4 [71], 7 and 8 [68] as well as 10), **IS** (entries 1 and 6 [72] as well as 9), **MS**[NG] (entries 1 and 5 [71] as well as 11 [73]) and **AS** (entry 7).

The most important features are: in attempted reactions at C-2, straight $\mathbf{S_N2}$ is operative in cases of *cis* configuration at C-1 and C-2 only (entry 2); for educts containing a *trans*-relationship of aglycon and LG, aglycon migration **MS**[2, NG-1] is the major reaction path in deamination and DAST reaction (entries 1 and 5), whereas the triflate/fluoride tandem sequence leads to elimination only (entries 3 and 4). As expected, a *trans* relationship between the LG and the OH-group located in positions 2/3 and 3/2, respectively, gives rise to formation of 2,3-anhydro compounds (**IS**[2,O-3], entry 1, and **IS**[3,O-2], entry 6).

Nucleophilic substitution reactions at C-5 of hexofuranoses are known to be prone to NG participation, either by attack of an acyl group located at C-6 [74] or the OH-group at C-3 [75]. Elucidation of the mechanistic features, which are operative in attempted fluorine introduction at C-5 in 3-*O*-benzyl-6-deoxy-1,2-*O*-isopropylidenehexofuranoses, brought the following results [68]: the triflate/fluoride tandem shows inclination to trigger elimination, whereas the DAST route rather mediates substitution by fluoride. However, depending on the orientation of the 3-*O*-benzyl group with respect to the side chain (entries 7 and 8), dramatic differences in the stereochemical course of this S_N reaction can be observed as, in educts with a *trans* relationship, inversion of configuration ($\mathbf{S_N2}$ to form **99**, entry 8) and, in the case of *cis* orientation, retention of configuration (**AS**[5,NG-3] to yield **98**, entry 7), predominate.

In entry 9, the general reaction pattern from the DAST treatment of 1,2-*O*-isopropylidene-α-D-glucofuranose and its derivatives with the OH-3 position free (or protected as an ether) is depicted. Besides the $\mathbf{S_N2}$ product **100** [69], varying proportions of 3,6-anhydro compounds are additionally formed by **IS**[6,O-3] (when starting from a 3-benzyl ether, formation of benzyl fluoride is observed by ^{19}F NMR spectroscopy). In the case of the 3-acetate shown in entry 10, no NG participation but (to a minor extent) **E**[5,H-6] was observed [70].

Table 9

	Educt	Product(s)	Conditions	Σ (%)	Ref.
1	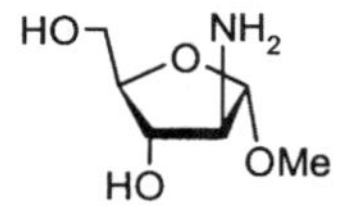	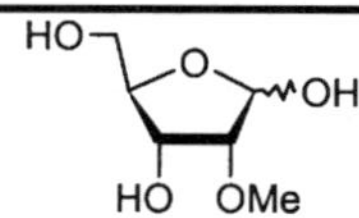 **MS**[2,NG-1] (major) 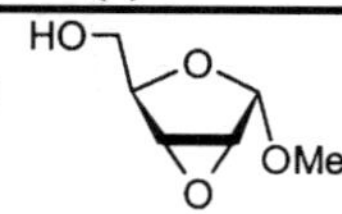 **IS**[2,O-3] (minor)	$NaNO_2$, $HOAc/H_2O$	not given	72
2	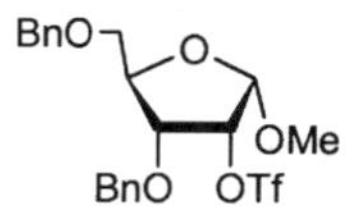	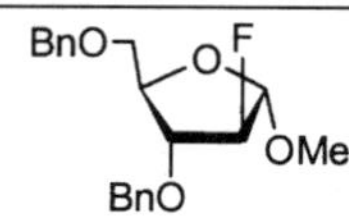 $\mathbf{S_N2}$ (62%)	TBAF, THF −10 °C → 0 °C, 5 h	62	67
3	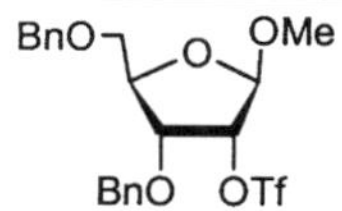	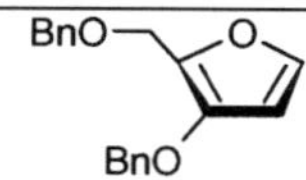 **E** (76%)	TBAF, THF −10 °C → 0 °C, 5 h	76	67
4	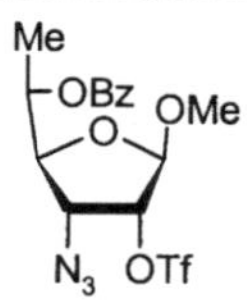	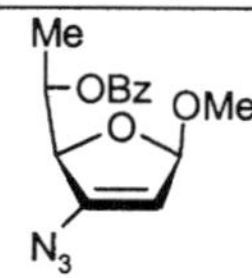 **E**[2,H-3] (51%)	TBAF, THF −10 °C, 2 h	51	71
5	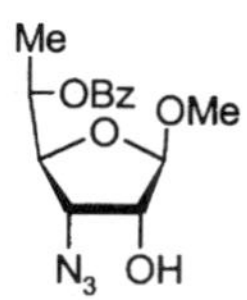	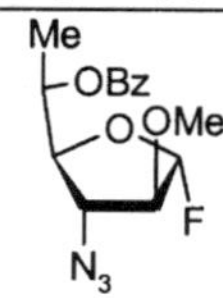 **MS**[2,NG-1] (50%)	DAST, toluene 60 °C, 2 h	50	71

Table 9 (continued)

	Educt	Product(s)	Conditions	Σ (%)	Ref.
6	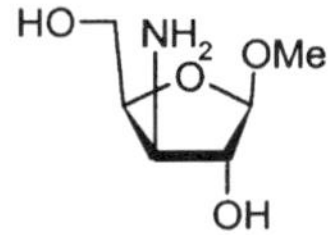	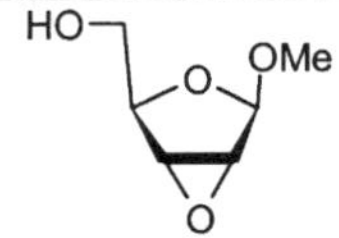 **IS**[3,O-2]	$NaNO_2$, $HOAc/H_2O$	not given	72
7	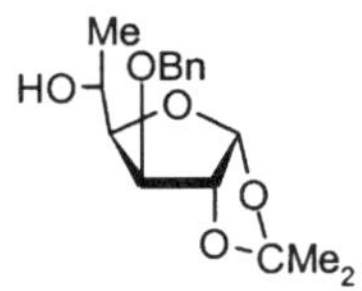	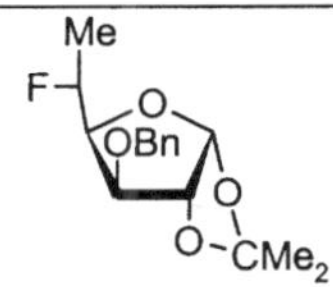 **AS**[5,NG-3] (61%) **98** 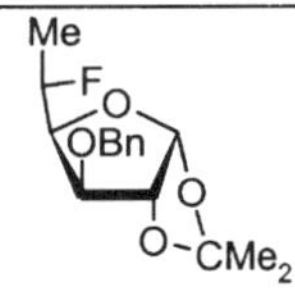 **S$_N$2** (3%) 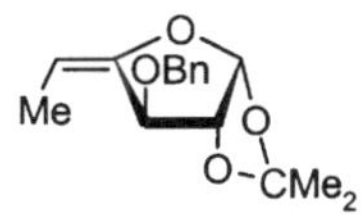 **E**[5,H-4] (5%) 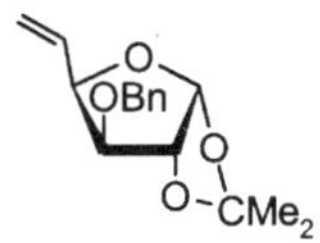 **E**[5,H-6] (8%)	DAST, CH_2Cl_2 −10 °C → rt, 90 min	77	68
8	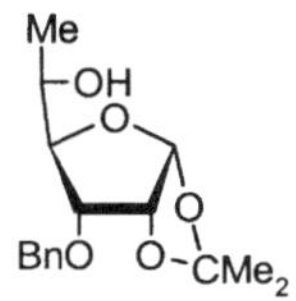	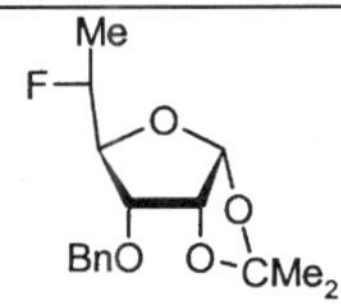 **S$_N$2** (61%) **99** 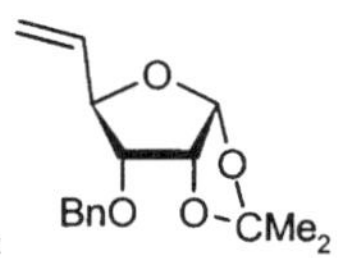 **E**[5,H-6] (2%)	DAST, CH_2Cl_2 −10 °C → rt, 90 min	63	68

Table 9 (continued)

	Educt	Product(s)	Conditions	Σ (%)	Ref.
9	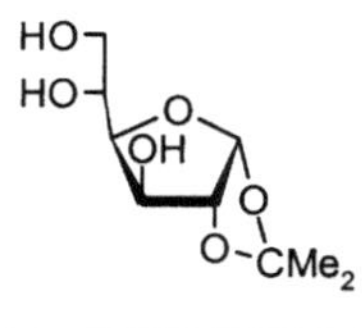	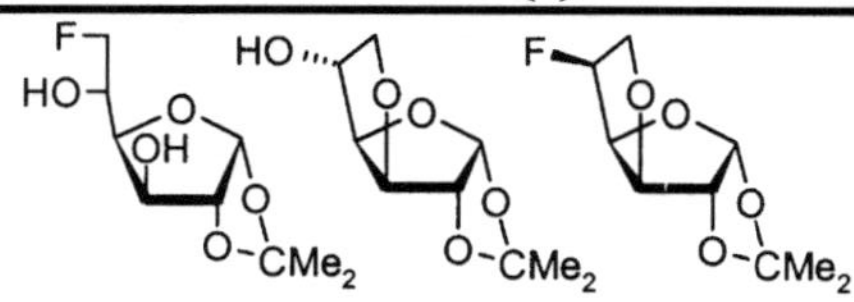 **S**$_N$ (70%) **100** **IS**[6,O-3] +**S**$_N$**2**[5]	DAST, CH_2Cl_2 –40 °C → rt, 2 h	>70	69 45
10	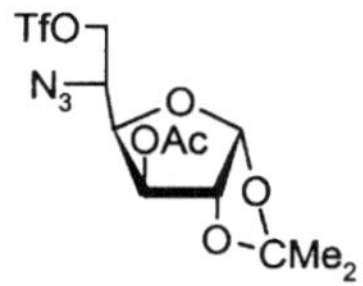	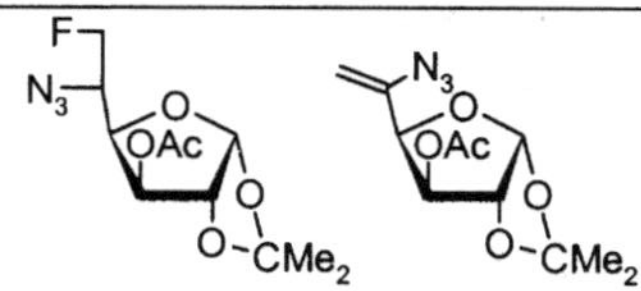 **S**$_N$ (75%) **E**[6,H-5] (minor)	TBAF, MeCN rt, 14 h	>75	70
11	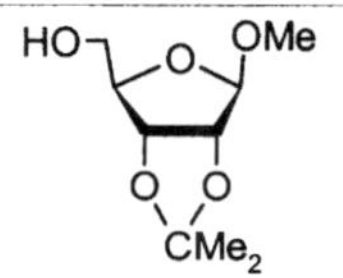 **101**	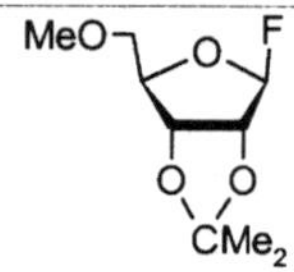 **MS**[5,NG-1] (55%) **102**	DAST, CH_2Cl_2 –15 °C → rt, 24 h	55	73

MS[5,NG-1], an aglycon migration different from **MS**[2,NG-1] outlined in entries 1 and 5, was observed on attempted fluorination by DAST of methyl 2,3-*O*-isopropylidene-*β*-D-ribofuranoside **101** [73] (entry 11). The *β* configuration found in the final ribofuranosyl fluoride **102** is rather the result of the thermodynamic equilibration than a mechanistic consequence.

5.2 Reactions of Septanoid Educts

Only a few attempts at nucleophilic substitution reactions with educts derived from (conformationally flexible) septanoses are known [76]. As can be seen from Table 10, besides straight S_N2, only elimination reactions have been reported [77]. Of interest is that in attempted substitutions of the epimeric 5-tosylates **103** and **105** (entries 1–3), the **E**[5,H-4] reaction (to give **104** containing a bridge head double bond) is preferred over the **E**[5,H-6] mode (which produces the glycal type **106**); however, the latter product predominates when bulky bases (*t*-BuOK in *t*-BuOH, entry 4) are applied.

6 Reactions of Educts Containing Sulfur or Nitrogen as Ring Heteroatom

The change in the type of ring heteroatom from oxygen to sulfur or nitrogen gives rise to characteristic shifts in the reaction pattern in those cases where participation of the ring heteroatom represents one of the possible paths (see Sects. 4.2.1 and 4.2.3 concerning educts with an equatorial LG in positions 2 and 4 in pyranosides). As compared to the situation with oxygen in the ring, the more powerful nucleophilicity of nitrogen and, especially, sulfur gives reason for a dominating role of **AS**- and **MS**-types of side reactions in attempted nucleophilic substitutions. Additional to the rS participation in educts activated at positions 2 and 4, the hitherto unprecedented involvement of position 6 in anchimerically assisted reactions comes into play.

6.1 Reactions of Educts Containing Sulfur as Ring Member

The results obtained with educts from a series of 5-thiopentopyranoses carrying an equatorial LG at position 2, 3 or 4 only show **AS**[4,rS] (entries 1, 2 and 4, Table 11), **MS**[4,rS] (entry 1) and **MS**[2,rS] (entry 3), the latter independent of the anomeric configuration [78a]. Activated positions incapable of rS participation, as given for C-3 in entry 1, do not react under the mild conditions of these anchimerically assisted transformations. Of further interest is that educt **107**, which contains a 2,3-acetal protection out of a *trans* configuration, is subject to **AS**[4,rS] reaction only after it has lost this grouping (therefore the ring strain was thought to inhibit formation of the [rS/C-5/C-4]-thiiranium ion [78a]). Contradictory are the results observed with educts bearing a *cis*-acetal protection. Whereas dominance of the ring-contraction **MS**[rS] modes in entries 1 and 3 fulfills general expectations (concerning ring strain and carbenium ion stability, respectively), this type is missing in entry 2.

Table 10

	Educt	Product(s)	Conditions	Σ (%)	Ref.
1	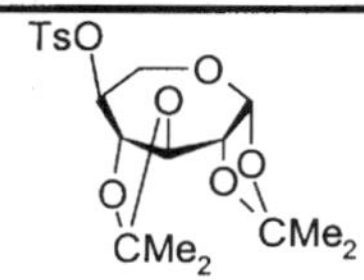 **103**	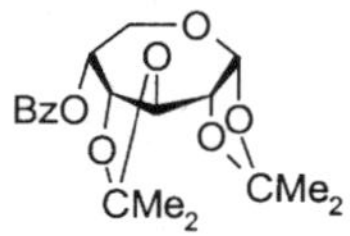 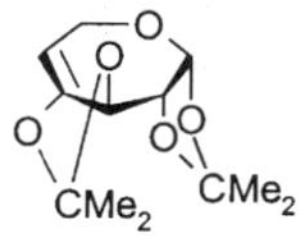 **S$_N$2** (major) **E**[5,H-4] (minor) **104**	LiOBz, DMF 140 °C, 24 h	not given	77
2	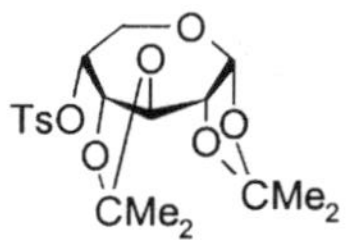 **105**	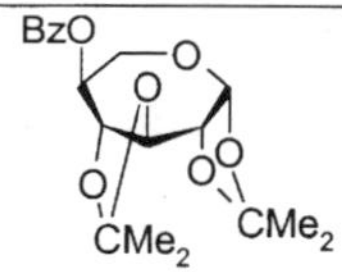 recovered starting material **S$_N$2** (10%) (45%) **104** (28%) **105**	LiOBz, DMF 125 °C, 40 h	55	77
3	**105**	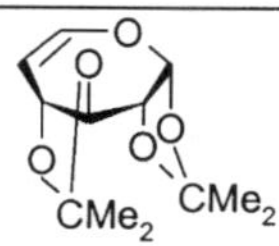 (28%) **104** **E**[5,H-6] (9%) **106**	MeONa, MeOH reflux, 50 h	37	77
4	**105**	**104** : **106** = 1: 2	*t*-BuOK, *t*-BuOH 80 °C, 48 h	not given	77

Table 11

	Educt	Product(s)	Conditions	Σ (%)	Ref.
1	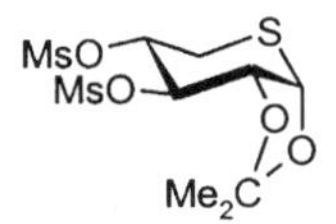	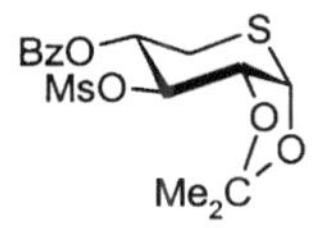 **AS**[4,rS] (7%) 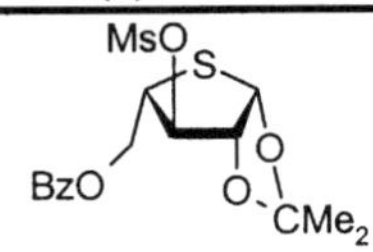 **MS**[4,rS] (41%)	NaOBz, MeOH/CH_2Cl_2 rt, 12 h	48	78a
2	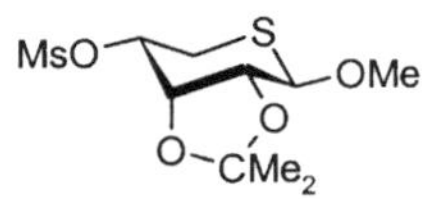	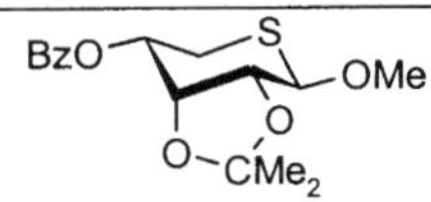 **AS**[4,rS] (43%)	NaOBz, MeOH rt, 2 d	43	78a
3	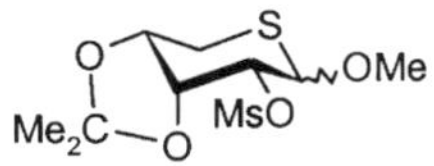	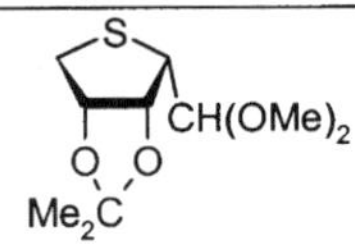 **MS**[2,rS] α-anomer (47%) β-anomer (32%)	MeOH rt, 12 h	47 32	78a
4	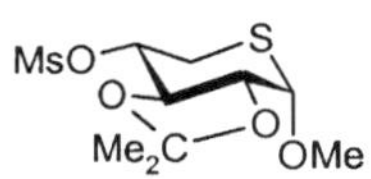 **107**	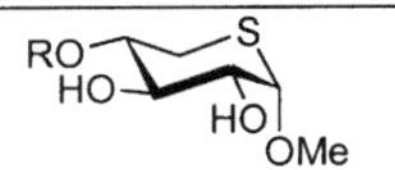 **AS**[4,rS] R = Ac (67%) R = H (7%)	NaOAc, HOAc/H_2O 100 °C, 45 min	74	78a
5	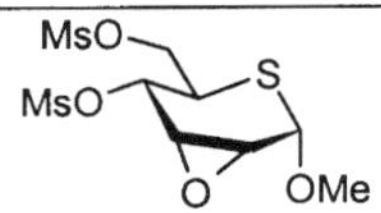 **108**	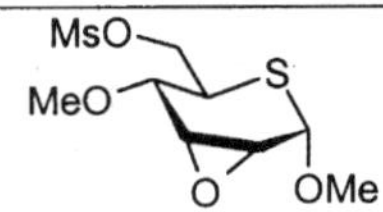 **AS**[4,rS] (82%) **109**	$BaCO_3$, MeOH/CH_2Cl_2 20 °C, 12 h	82	78b

Table 11 (continued)

	Educt	Product(s)	Conditions	Σ (%)	Ref.
6	**109**	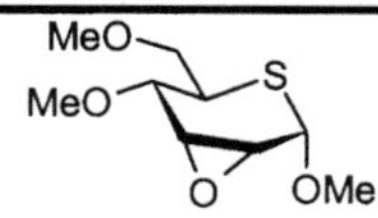 **AS**[6,rS] (74%) **110** 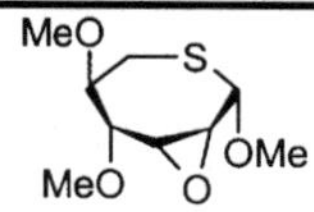 **MS**[6,rS] (13%) **111**	$BaCO_3$, MeOH reflux, 2 h	87	78b
7	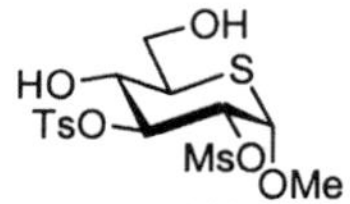 **112**	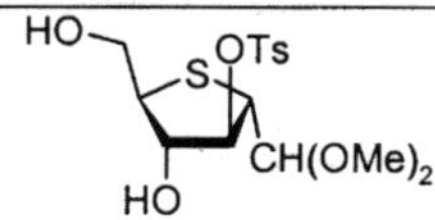 **MS**[2,rS] (96%)	HCl, MeOH 20 °C, 24 h	96	78b
8	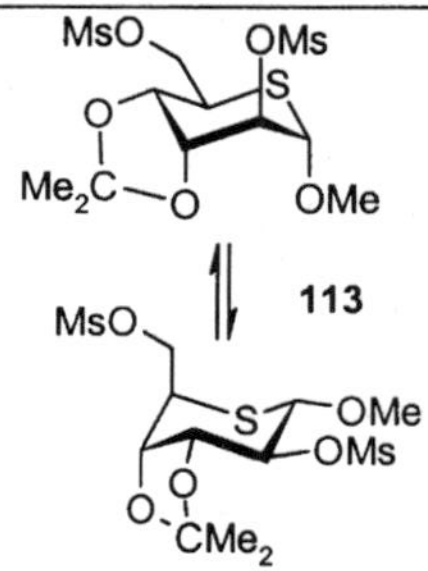 **113**	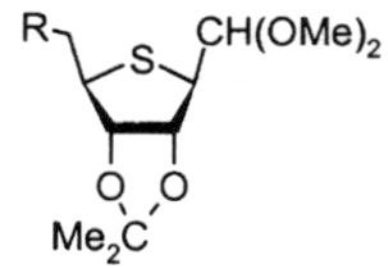 **MS**[2,rS]+**AS**[6,rS] **114** R = Cl (36%) R = OMe (12%) 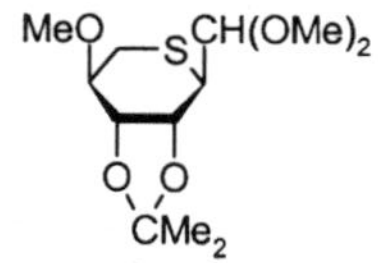 **MS**[2,rS]+**MS**[6,rS] (6%) **115**	$BaCO_3$, TEA.HCl MeOH	54	78b
9	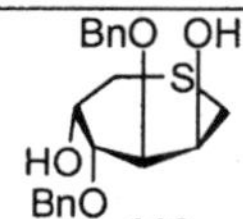 **116**	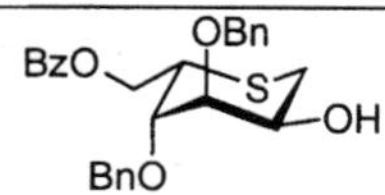 **MS**[5,rS] (80%) **117**	BzOH/TPP/DEAD THF 0 °C	80	79

Table 11 (continued)

	Educt	Product(s)	Conditions	Σ (%)	Ref.
10	**116**	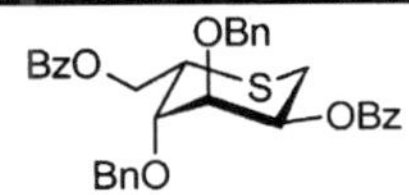 **MS**[5,rS]+**AS**[2,rS] (43%) **118** 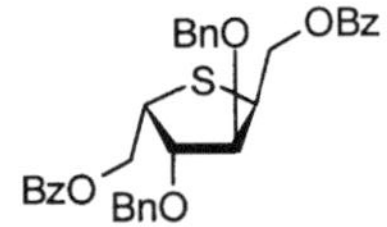 **MS**[5,rS]+**MS**[2,rS] (33%) **119**	BzOH/TPP/DEAD THF 20 °C	76	79
11	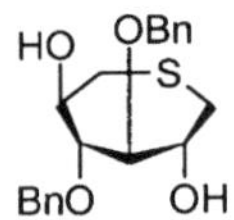 **120**	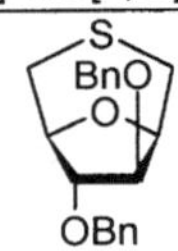 **IS**[5,O-2] (61%) **121** 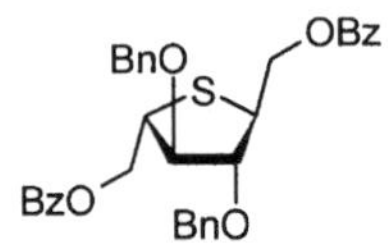 **MS**[5,rS]+**MS**[2,rS] (24%) **122**	BzOH/TPP/DEAD THF 20 °C	85	79

Essential findings on the exceptional role of rS participation came from Hughes's and Le Merrer's work with 5-thiohexopyranosides (entries 5–8 [78b]) and polyhydroxylated thiepanes (entries 9–11 [79]), respectively. First, the enhanced nucleophilicity of sulfur over oxygen was shown by the selective methanolysis (as well as benzoate displacement) of the 4-mesyloxy group in the 4,6-dimesylate **108** (to give **109**, entry 5), whereas the corresponding educt with oxygen in the ring (D-allopyranoside), under the same conditions, did not react, but gave selective substitution at C-6 when treated with sodium benzoate in boiling DMF [78b]. When treating 6-mesylate **109** under more forced conditions (entry 6), evidence for the involvement of a thiiranium ion comprising rS, C-5 and C-6 arose from the isolation of (ring-enlarged) septanoside **111** besides the **AS/S$_N$2** product **110**. An explanation for this behaviour was deduced from conformational studies undertaken with dimesylate **108**. According to these, the LG at C-6 adopts an "axial" orientation (depicted as **108a** in Scheme 19) inappropriate for rS participation (as the latter requires the arrangement shown in **108b**).

Second, methanolysis with the 2,3-disulfonate **112** (entry 7) gave reaction at C-2 only (ring-contraction **MS**[2,rS]) in excellent yield. Third, treatment of the (conformationally flexible) 2,6-dimesylate **113** initially also gave ring contraction involving C-2 (**MS**[2,rS]), but this was immediately followed by [rS/C-5/C-6]-thiiranium formation and opening of the latter, either at C-6 (**AS**[6,rS]) to give ring-contracted product **114**, or at C-5 (**MS**[6,rS]) to produce **115** (as a consequence of ring contraction and ring expansion involving rS shift from C-1 to C-2 and from C-5 to C-6).

The peculiar participating and migrating disposition of sulfur as the ring heteroatom was employed in the synthesis of various thiosugars from 2,5-dihydroxythiepanes **116** and **120** (showing C_2 symmetry) by a Mitsunobu reaction [79]. Thus, with **116** in the presence of benzoic acid (BzOH), triphenyl phosphane (TPP) and diethyl azodicarboxylate (DEAD) at 0 °C, **MS**[5,rS] was operative to form ring-contracted product **117**. At 20 °C, this ring contraction was followed either by **AS**[2,rS] to produce the 2,6-dibenzoate **118** or by **MS**[2,rS] to give 1,6-dibenzoate **119**. The same reaction sequence, a twofold ring contraction leading to 1,6-dibenzoate **122**, was also observed to occur in a minor proportion with the (diastereomeric) educt **120**. The main reaction consisted of an intramolecular substitution with inversion of configuration of the first Mitsunobu-activated OH-group (at C-5) by the unaffected other (**IS**[5,O-2]) to give 2,5-anhydrothiepane **121**. Interestingly, from the series of "normal" sugar derivatives, no single case has come to our attention where a Mitsunobu reaction had led to side reactions of the **MS**-type. [Zamojski et al. [80] studied the corresponding reaction with otherwise unprotected methyl α- as well as β-D-glucopyranoside. The products formed in the β-series showed involvement of **S$_N$2** (at C-6 and C-3, 91%,

Scheme 19 **108a** **108b**

thus opening a new entry in the D-*allo* series), which in part was followed either by **AcM** (from O-3 to O-4) or **AS**[4,rO] and (**AS**[4,rO] + **AS**[2,rO]), respectively; additionally, the product of (**IS**[6,O-3] + $\mathbf{S_N2}$[4]) was formed in trace amounts. From the α-series, these workers isolated 11 products (82% total yield), which all showed $\mathbf{S_N2}$[6] followed (either consecutively or alternatively) by **IS**[3,O-2], **IS**[4,O-3], $\mathbf{S_N2}$[3], $\mathbf{S_N2}$[4], **AS**[4,rO]), **AS**[2,rO]) and **AcM**, respectively.]

In relation to the depiction of products formed by ring contractions of septanoid educts in Tables 11 and 12 it is annotated, that the presentation in the "neutral" Haworth's projection under the issues given is too crowded to be easily read. The alternative Mills depiction also does not allow clear presentations when structures are involved which contain bridged systems. Therefore, the pyranoid structures are shown in $^1C_4$ conformation with the side chain at C-5 in equatorial orientation. This may be justified by the fact that this latter conformation had been found to predominate in sulfoxides obtained from structures of type **117** [79b].

6.2
Reactions of Educts Containing Nitrogen as Ring Member

Similar reactions to those performed with hydroxylated thiepanes [79] (depicted in Table 11, entries 9–11), have also been carried out with polyhydroxylated azepanes of C_2 symmetry [81] (entries 1–4, Table 12). The results obtained either under conditions of a Mitsunobu reaction (entries 2 and 4) or of mesylation (entries 1 and 3) are essentially the same as described above. Under the latter conditions, the intermediate [rN/C-5/C-6]-aziridinium ion is opened, at C-6, by chloride liberated during the mesylation step. The reaction shown in entry 4 again consists of diol monoactivation followed by intramolecular substitution by the *trans*-located, free OH-group.

In entries 5–11 results from attempted nucleophilic displacements at C-2 and C-6 with educts from a series of *N*-protected 2,5-dideoxy-2,5-iminohexitols are depicted. The methods used were Mitsunobu reaction (entries 6, 7 and 9 [81b, c]), sulfonate/nucleophile tandem (entries 5 and 8 [81b,c] as well as 11 [82]) and DAST reaction (entry 10 [83]). As can be seen, in most cases reactions at C-2 involve rN participation (**AS** or **MS**-type). Concerning the mechanism for substitution at C-6, the **AS**-type became evident only by isolation of the ring-expanded **MS** product **125** (entry 8), which is formed via the same intermediate aziridinium ion. To explain the rN participation (**MS**-mode) in the reaction shown in entry 11, conformational change had to be anticipated [82]. Of further interest to note is that rN participation in *N*-benzyl-protected educt **123** in the course of a Mitsunobu reaction (entry 6) could be avoided by changing the *N*-protection to benzyloxycarbonyl (cbz in **124**, entry 7), whereas in the DAST reaction of benzyl- as well as cbz-protected educt **126** containing an equatorial OH-group at C-4, only the **AS**-type of reaction was observed (entry 12 [84]). In this contribution, participation of the cbz protecting group was thought to be responsible for the observed substitution with retention of configuration. By using the triflate/TASF tandem, only elimination, but in both directions, was observed [84].

Table 12

	Educt	Product(s)	Conditions	Σ (%)	Ref.
1	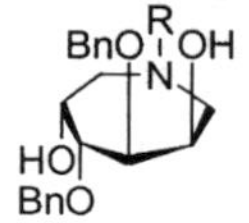 R = CH_2CH_2Ph	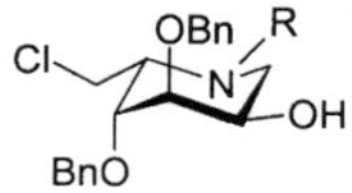 **MS**[5,rN] (60%)	MsCl, TEA, CH_2Cl_2 0 °C	60	81a 81c
2	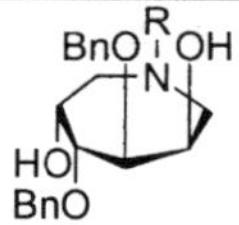 R = Bn	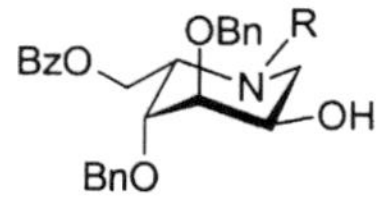 **MS**[5,rN] (65%) 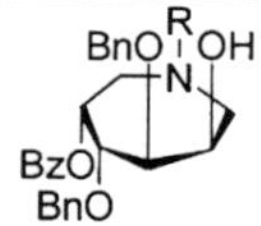 **AS**[5,rN]	BzOH/TPP/DEAD THF 0 °C	65	81a 81c
3	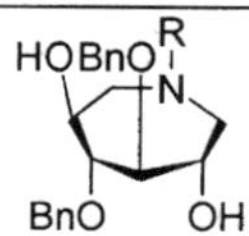 R = CH_2CH_2Ph	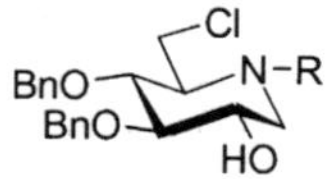 **MS**[5,rN] (56%)	MsCl, TEA, CH_2Cl_2 0 °C	56	81a 81c
4	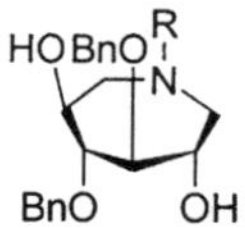 R = Bn	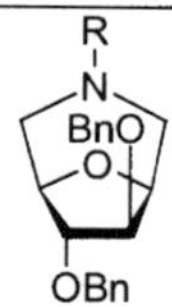 **IS**[5,O-2] (80 or 94%)	BzOH/TPP/DEAD THF, 0 °C	80	81a 81c
			TPP/DEAD THF 0 °C	94	
5	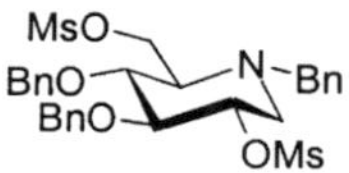	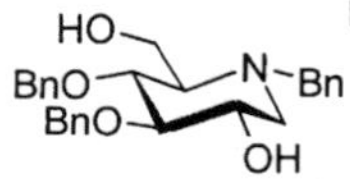 $\mathbf{S_N/AS}$[6]+ **AS**[2,rN] (31%) 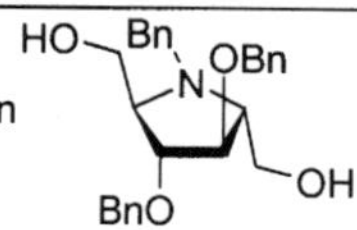 $\mathbf{S_N/AS}$[6]+ **MS**[2,rN] (51%)	1. CsOAc, DMF 40 °C–50 °C 2. K_2CO_3, MeOH	82	81b 81c

Table 12 (continued)

	Educt	Product(s)	Conditions	Σ (%)	Ref.
6	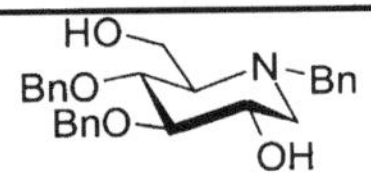 **123**	(29%) (64%)	1. BzOH/TPP/DEAD THF, 0 °C 2. K_2CO_3, MeOH	93	81b 81c
7	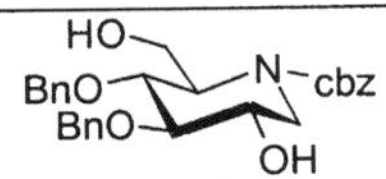 **124**	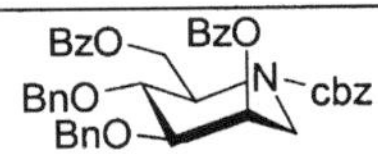 **S$_N$2**[6]+**S$_N$2**[2] (91%)	BzOH/TPP/DEAD THF 0 °C	91	81b 81c
8	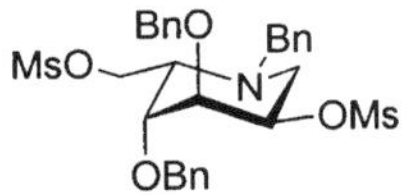	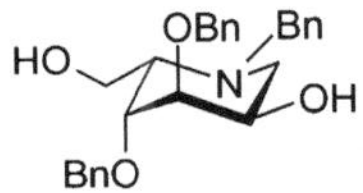 **AS**[6,rN]+ **AS**[2,rN] (18%) 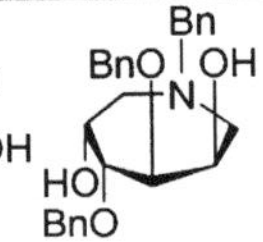 **AS**[2,rN]+ **MS**[6,rN] (8%) **125** 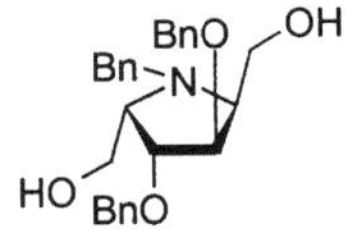 **AS** [6,rN]+**MS**[2,rN] (45%)	1. CsOAc, DMF 40 °C–50 °C 2. K_2CO_3, MeOH	71	81b 81c
9	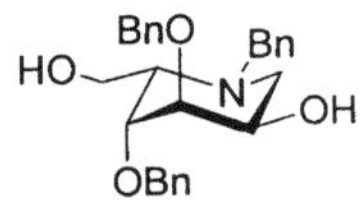	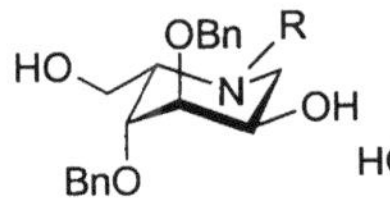 **S$_N$/AS**[6]+ **AS**[2,rN] (70%) 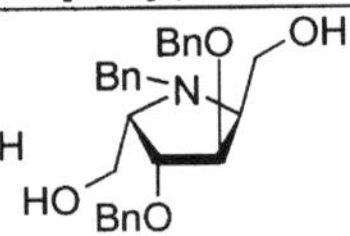 **S$_N$/AS**[6]+ **MS**[2,rN] (18%)	1. BzOH/TPP/DEAD THF, 0 °C 2. K_2CO_3, MeOH	88	81b 81c
10	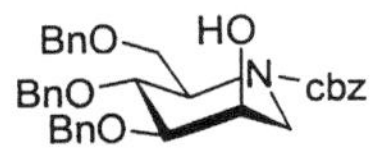	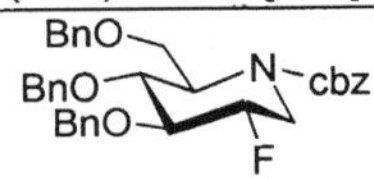 **S$_N$2** (53%)	DAST, CH_2Cl_2/pyridine 0 °C → rt, 48 h	53	83

Table 12 (continued)

	Educt	Product(s)	Conditions	Σ (%)	Ref.
11	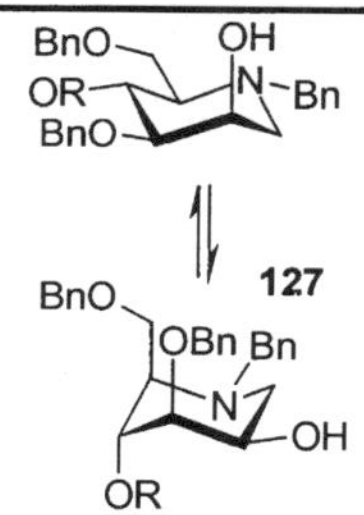 127 R = 4-MeOBn	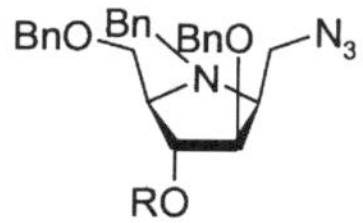 **MS**[2,rN] (72%)	1. MsCl, TEA, CH_2Cl_2 20 °C, 4 h 2. LiN_3, DMF 60 °C, 1 h	72	82
12	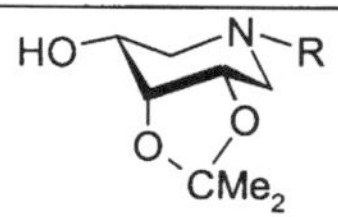 R= Bn or cbz **126**	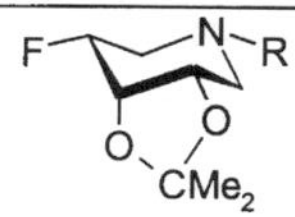 **AS**[4,rN] (66% R = cbz)	DAST, CH_2Cl_2	66	84
13	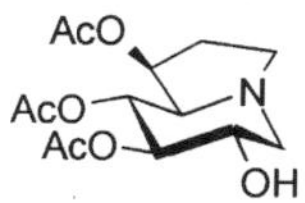 **128**	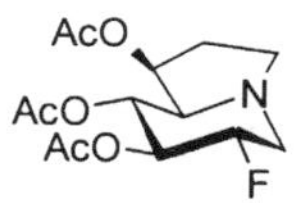 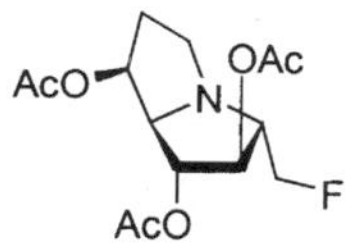 **AS**[2,rN] (49%) **129** **MS**[2,rN] (19%) **130**	DAST, CH_2Cl_2 –78 °C → reflux, 2 h	68	85a
14	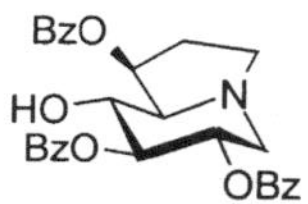 **131**	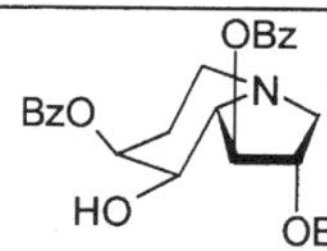 **MISC**[5,rN/NG-6]+**Hy** (52%) **134**	DAST, CH_2Cl_2 0 °C, 1 h	52	85b

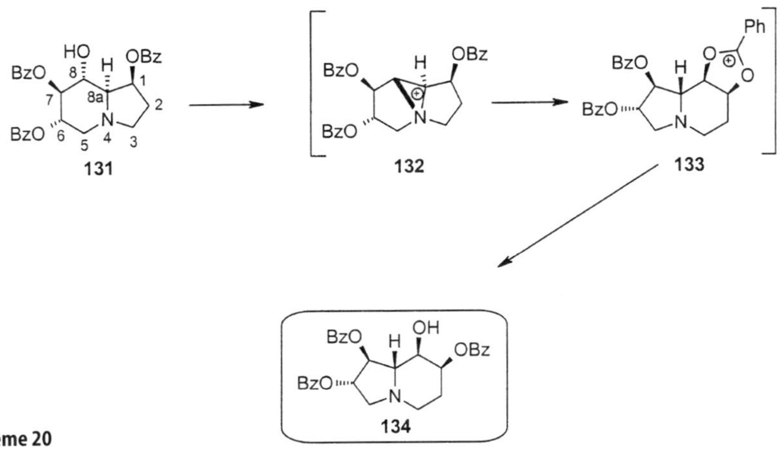

Scheme 20

Of special interest are the results obtained from DAST reactions of castanospermine derivatives **128** and **131**, respectively (entries 13 and 14 [85]). In the first case [85a], the usual mechanistic principles **AS**[2,rN] (to form **129**) and **MS**[2,rN] (to yield ring-contracted product **130**) were operative [*Note*: Here the carbohydrate-like numbering is applied instead of the one that generally exists for indolizine derivatives, see structure **131**]. In entry 14, the result [85b] from attempted introduction of fluorine at C-4 of the piperidine ring (C-8 in the indolizine numbering) is shown. As illustrated in detail in Scheme 20, by using the more distinct Mill's depiction, formation of aziridinium ion **132** (with inversion of configuration at C-4) initially occurs. This is opened at C-5 under inversion of configuration according to the **ISC** path, by participation of the benzoyl group from C-6 (C-1 in the indolizine nomenclature), giving the tricyclic acyloxonium ion **133**; the hydrolysis (**Hy**) of the latter leads to the final product **134**. The net result of the transformation (**131** into **134**) is piperidine–pyrrolidine interconversion with inversion of configuration at C-4 and C-5 (C-8 and 8a in indolizine numbering) [85b].

7
Reactions of Educts Stemming from Sugar Derivatives Other Than Aldo(Pyrano/Furano)ses

Various educts which do not contain the basic structure of an aldose (as, e.g., 1,5-anhydroalditols) have, to some extent, already been treated together with their parent compounds. Herein, the reactivity pattern of two further groups of educts, the 2-triflates of aldonolactones in solvolysis reactions, and the carbonyl-containing aldose derivatives on treatment with DAST, will be discussed separately.

7.1
Reactions of 2-Triflates of Aldono-1,4- and -1,5-lactones

Over the last ten years, the ring-contraction **MS**[2,rO]-type of side reaction, originally observed with educts containing an equatorially oriented LG at C-2 in pyranosides (see Sect. 4.2 and especially 4.2.1), in its application in the field of aldono-1,5- and, in particular, -1,4-lactones, has been brought to synthetic importance by the studies of Fleet and his group. As the main part of this work was recently presented in a comprehensive way [86a], a few entries only are given in Table 13 to illustrate the pattern of the possible transformations (and to admit our mechanistic interpretations, since – in the original work – participation of the lactone-ring oxygen was not brought into discussion).

In entries 1–3 characteristic reactions of 3,4-*O*-isopropylidene-D-altrono-1,5-lactone (**135**) and the 2-triflate of the corresponding cyclohexylidene derivative **137**, respectively, are shown [86]. When Mitsunobu conditions to form a cyclic ether from a diol, were applied to **135**, the 2,6-anhydro compound **136** was formed as expected (entry 1; by activation at C-6 and **IS** reaction of unprotected OH-2, of course without any configurational change). When starting from triflate **137** (entries 2 and 3), according to our mechanistic view, participation of the lactone-ring oxygen comes into play. This results in formation of the intermediate oxiranium ion **141**. The latter is opened either, by methanol, at C-1 (to give **MS**[2,rO]-product **138** with inverted configuration at C-2) or, by OH-6, at the same position (to produce **MIS**[2,rO/O-6] product **139**, also of D-*allo* configuration) and alternatively at C-2 (under overall retention of configuration to generate the **AIS**[2,rO/O-6]-product 2,6-anhydro-D-altrono-1,5-lactone **140**). [For the formation of **140** originally [86b] base-catalysed epimerisation at C-2 of triflate **137** was anticipated.]

A fascinating scenario was opened when Fleet started to explore the reactions of 2-triflates of hexono-1,4-lactones (entries 4–8 [86]). Here, first of all, the **IS**[2,O-6] path, as observed with educt **142** of the D-*gluco* configuration, is presented. The direct product of intramolecular substitution with inversion of configuration at C-2, 2,6-anhydro-D-mannono-1,4-lactone **143**, was isolated from reaction in aprotic media only (entry 4) as its lactone ring, in the presence of methanol (entry 5), was subject to transesterification (**TE**, to yield **144**). (An inverted sequence of reaction, namely transesterification to form **145** followed by intramolecular substitution, was originally proposed.)

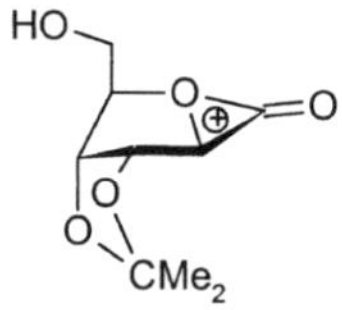

141

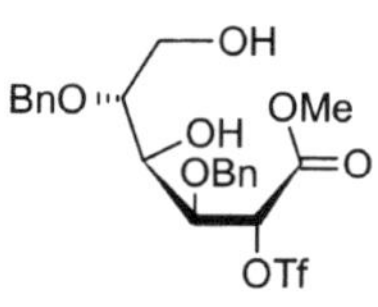

145

When OH-6 is protected (as in educt **146**, entry 6), a new, exciting type of **MS**[2,rO] reaction, unprecedented in the furanose series, appears: Contraction of the five-membered ring, with inversion of configuration at C-2, to form the oxetane derivative **147**. Furthermore, in acidic medium (entry 7), besides participation of O-6 (**IS**[2,O-6] + **TE**) to give **144**, attack of O-5 (out of the benzyl ether protection) also occurs. [The dependence of the product ratio in this reaction was studied in detail by using methanol and/or tetrahydrofuran as solvent(s) and HCl as well as camphorsulfonic acid (CSA) as proton source.] Formation of product **148** can be envisaged to happen by rO participation via oxonium ion **149**, which, by reaction with methanol, is subject to cleavage of the benzyl group from O-5 as well as transesterification (**TE**).

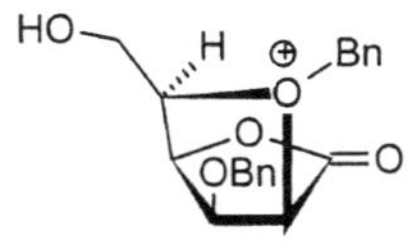

149

In the case of the D-allono-1,4-lactone derivative **150**, only the (**IS**[2,O-5] + **TE**) route (giving **152**, entry 8) and not the corresponding sequence involving O-6 was found to be operative. Interestingly, for the ring-contracted product **151**, without giving any reasons, D-*allo* configuration (**151a**) was reported. [According to our mechanistic understanding, it is the product of a **MS**[2,rO] reaction, which includes inversion of configuration at C-2, and should therefore be of the D-*altro* structure **151b**. Is epimerisation involved?]

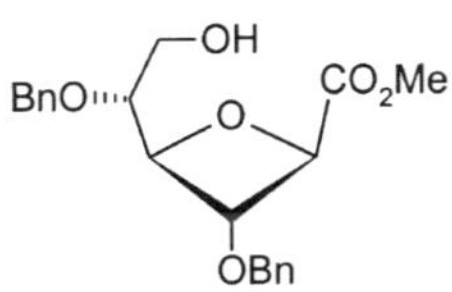

151a

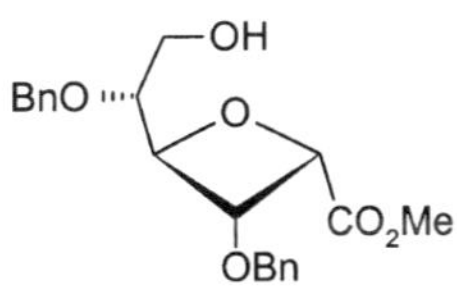

151b

When educt **150** is kept in DMF or methanol (the latter containing 1% HCl), it gives an S_N2 reaction, either with the solvent or chloride, with formation of a 2-*O*-formyl- and 2-chloro-2-deoxy-D-altrose derivative, respectively.

These types of reactions of 2-triflates of aldono-1,4- and -1,5-lactones, as elaborated by Fleet and co-workers, have been used extensively in the synthesis of different classes of natural products such as oxetanocin nucleosides [87] and *C*-glycofuranosides [88].

7.2
Reactions of Carbonyl Groups in Dialdoses and Aldosuloses with DAST

The reaction of carbonyl groups with excess DAST under formation of geminal difluorides (Scheme 21) resembles the transformation of alcohols into fluoroalkanes by the DAST route (see Scheme 6). Inasmuch as similar side reactions have also been observed in both cases, this special section is included here. As will be seen, many questions are still left unanswered.

In Table 14, the results as obtained by Castillon (entries 1–9 [89]) and by Walker (entry 10 [73]) are depicted. Geminal difluorination was observed in entries

Table 13

	Educt	Product(s)	Conditions	Σ (%)	Ref.
1	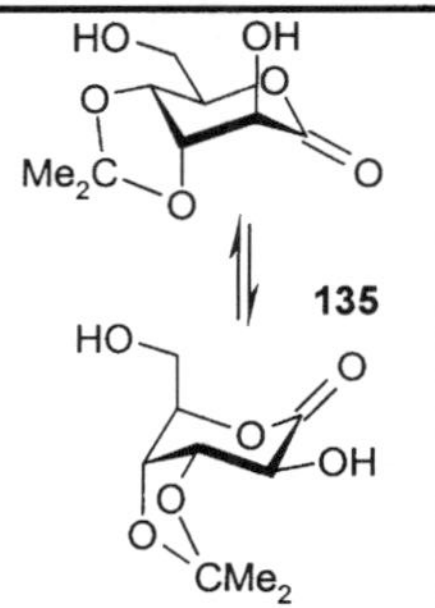 **135**	 **IS**[6,O-2] (60%) **136**	TPP/DEAD THF rt, 1 h	60	86a
2	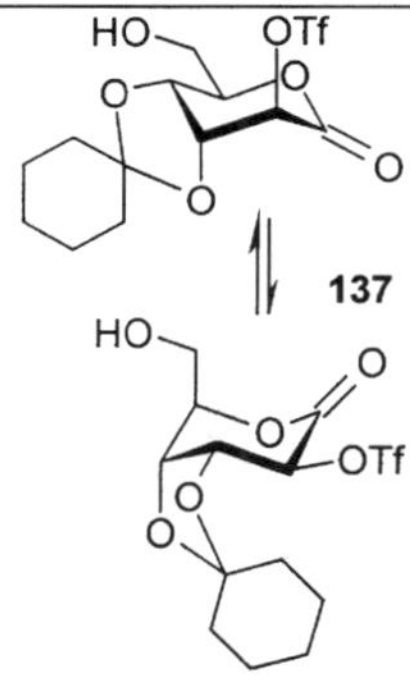 **137**	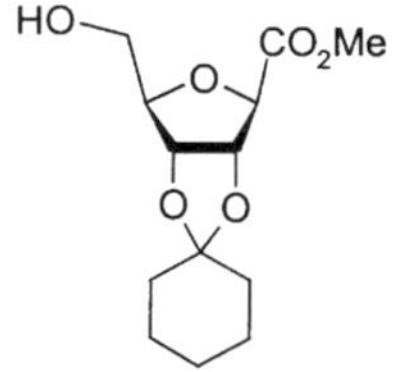 **MS**[2,rO] (94%) **138**	K_2CO_3, MeOH rt, 10 min	94	86a
3	**137**	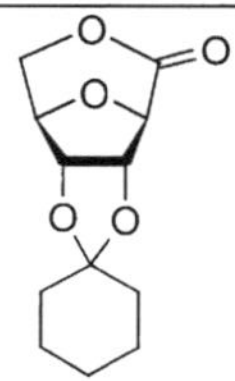 **MIS**[2,rO/O-6] (46%) **139** 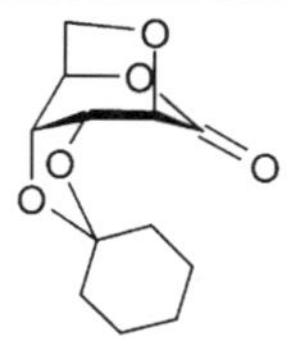 **AIS**[2,rO/O-6] (26%) **140**	NaOAc, DMF rt, 3 h	72	86a

Table 13 (continued)

	Educt	Product(s)	Conditions	Σ (%)	Ref.
4	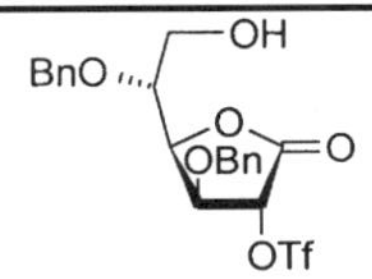 **142**	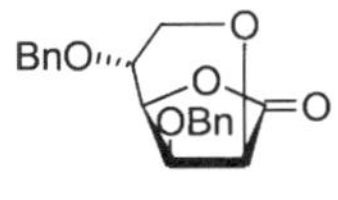 **IS**[2,O-6] (64-88%) **143**	a) pyridine/acetone rt, 8 h (88%) b) NaOAc, DMF, rt 12 h (64%) c) DMF, rt, 24 h (85%)	64–88	86
5	**142**	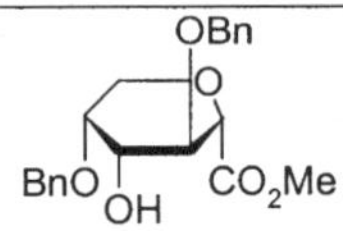 **IS**[2,O-6]+**TE** (65 or 90%) **144**	a) pyridine/MeOH, rt, 24 h (65%) b) CSA, THF/MeOH rt, 6 h (90%)	65 or 90	86
6	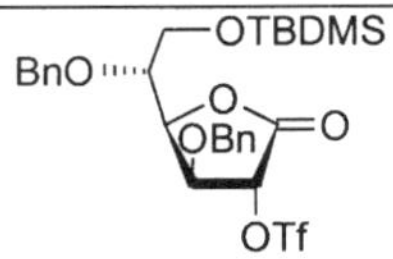 **146**	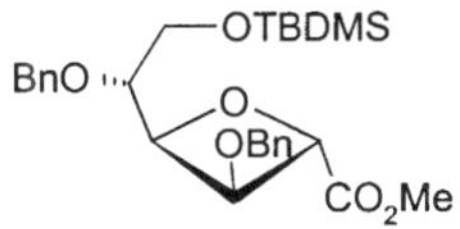 **MS**[2,rO] (75%) **147**	K_2CO_3, MeOH rt, 10 min	75	86
7	**142**	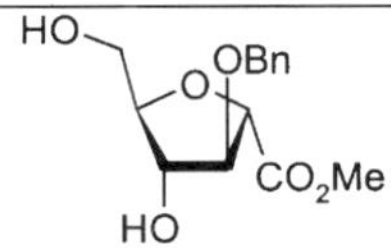 **IS**[2,O-5]+**TE** (52%) **148** (30%) **144**	HCl, MeOH rt, 12 h	82	86
8	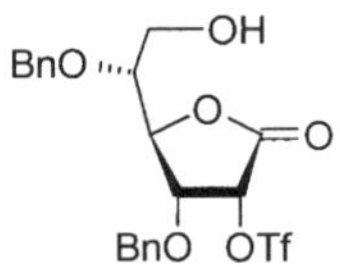 **150**	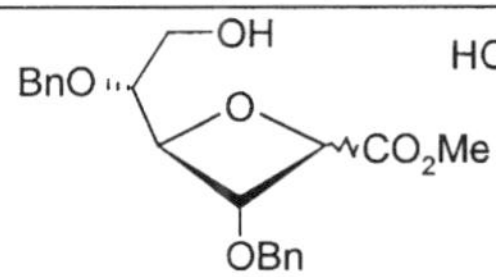 **MS**[2,rO] (41%) **151** 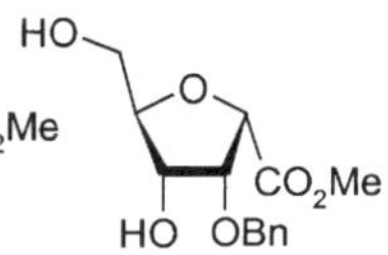 **IS**[2,O-5]+**TE** (45%) **152**	K_2CO_3, MeOH rt, 15 h	86	86

Table 14

	Educt	Product(s)	Conditions	Σ (%)	Ref.
1	 **153**	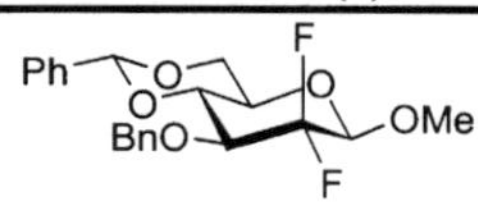 (80%)	DAST, CH_2Cl_2 rt, 24 h	80	89a
2	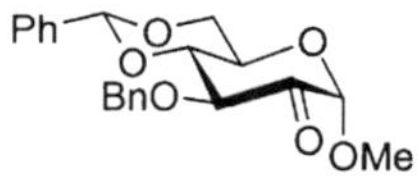 **154**	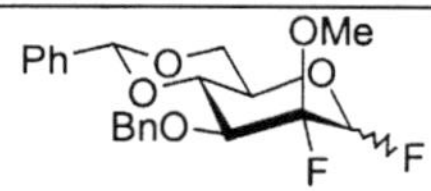 (43%) **155**	DAST, benzene reflux, 10 h	43	89a
3	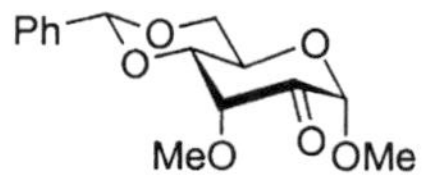 **156**	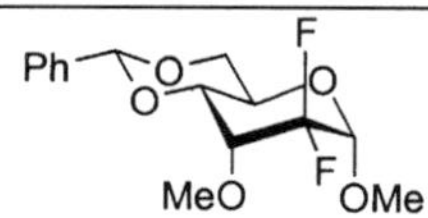 (39%)	DAST, CH_2Cl_2 reflux, 7 h	39	89a
4	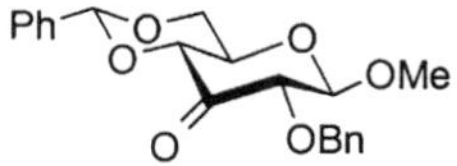 **157**	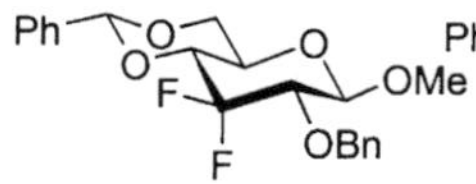 (40%) 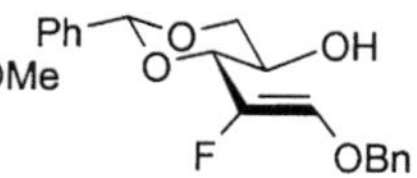 (17%) **158**	DAST, benzene reflux, 2 h	57	89a
5	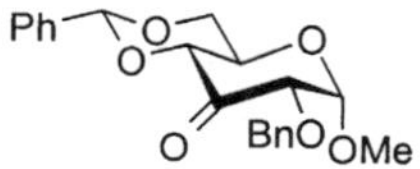 **159**	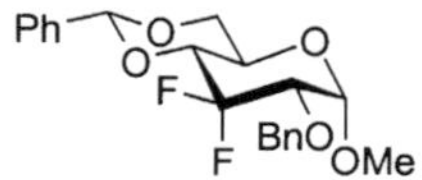 (48%) (10%) **158**	DAST, benzene rt, 10 h	58	89a

Table 14 (continued)

	Educt	Product(s)	Conditions	Σ (%)	Ref.
6	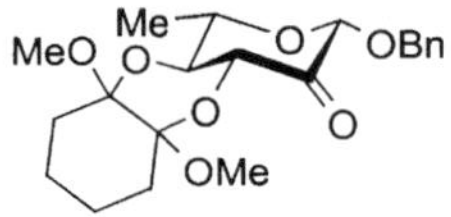 **160**	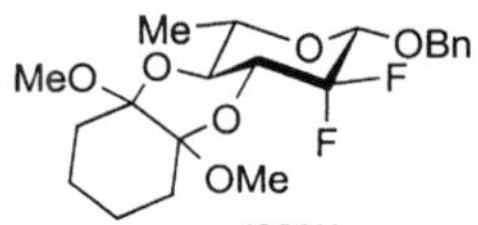 (69%)	DAST, CH_2Cl_2 rt, 6 h	69	89b
7	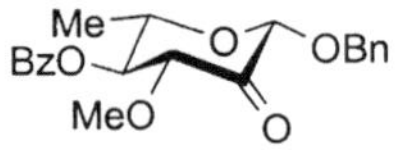 **161**	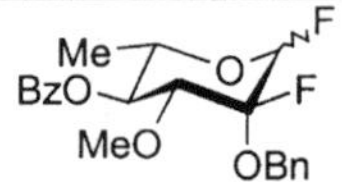 (50%) **162a**	DAST, CH_2Cl_2 rt, 24 h	50	89b
8	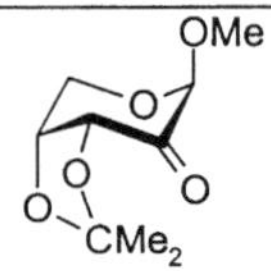 **163**	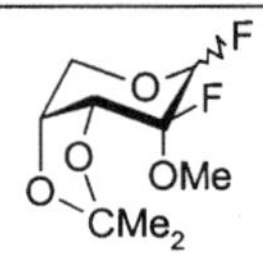 (72%) **164a**	DAST, CH_2Cl_2 rt, 8 h	72	89a
9	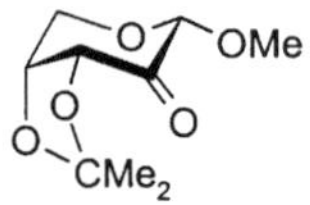 **165**	(n.g.) **164a**	DAST, CH_2Cl_2	not given	89a
10	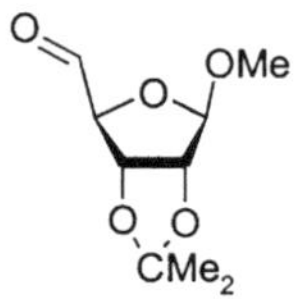 **166**	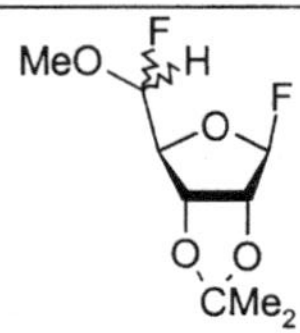 (56%) **167**	DAST, CH_2Cl_2 rt, 15 min	56	73

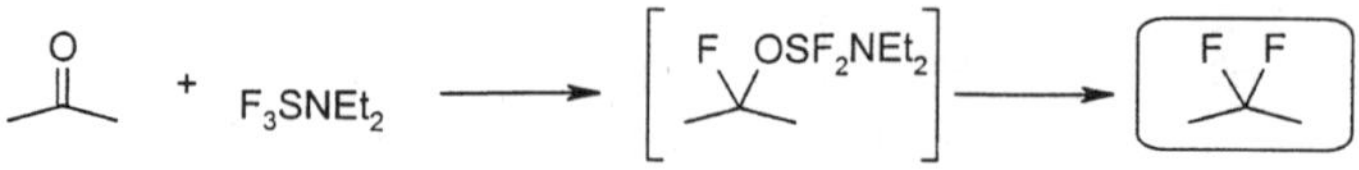

Scheme 21

1, 3, 4, 5 and 6 only. When looking for structural features common to the respective educts, only "a rigid system containing a *trans*-acetal protection" can be found (which is in good agreement with the reduced inclination of such educts to give side reactions, see Sect. 4). Furthermore, conflicting characteristics predominate:

1. The alkoxy shift observed in the reaction of **154** (to form product **155**) has no parallel in the reaction of educt **156**.
2. The fragmentation found to occur with 3-uloses **157** and **159** (to form product **158**) comes as some surprise, as, in the hexopyranose series, educts of *gluco* configuration were subject to ring contraction **MS**[3,C-5] rather than fragmentation **F**[3,C-1] (see Table 6, entries 2 and 3, where with the β-anomer, under soloolytic conditions, fragmentation was observed only).
3. The 1,2-aglycon shifts reported for reactions of 2-uloses **161** and **163/165** (to form products **162** and **164**, respectively) are in striking contrast to the results obtained with educts containing an unprotected alcohol instead of a carbonyl group (Sect. 4.2.1, Table 5, entries 12–14). An inspection of their ^{19}F NMR data revealed [45] that they might also be products of ring contraction; their possible structures are depicted in **162b** and **164b**. [It should be mentioned that thus far no pure, defined product has been crystallised or separated out of the mixture of diastereomers formed in these reactions (e.g. **162a** or **b**).]

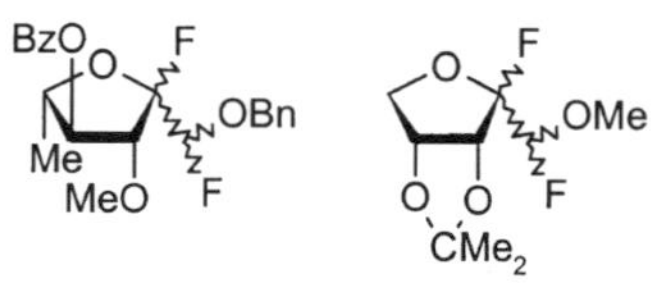

162b **164b**

The 1,5-aglycon shift observed with pentodialdofuranoside **166**, to yield "glycos-1,5-diyl difluoride" **167** (entry 10 [73]), is in complete agreement with the findings made with the corresponding 5-*O*-unprotected furanoside (Table 9, entry 11).

8 Miscellaneous Reactions Related to the Topic

Within this section (including Table 15) a rough survey of transformations which show some relation to those described in previous sections is given.

First of all, the pioneering work of Lemieux and Fraser-Reid [90] has to be mentioned, who, as early as 1963, in their investigation of the brominolysis of

2-deoxy-2-iodoaldosides, had described many of the side reactions discussed in this compilation (e.g., aglycon migration **MS**[2,NG-1], ring contraction **MS**[2,rO], anhydro ring formation **IS**[2,O-3] and **IS**[3,O-6] as well as the combination of the latter two steps). Within recent years, the chemistry of 2-deoxy-2-halo sugars again was subject to studies in the group of Spilling [91]. These workers elucidated the dependence (on the nature of the protecting groups present as well as reaction conditions) of the ratio 1,2-anhydro ring formation **IS**[2, NG-1] vs. ring contraction **MS**[2,rO] (followed by elimination). The **MS**[3,C-5] ring-contracting mode (followed by **E1cB**) was found by Magnusson [92] to occur when 4-*O*-unprotected 2,3-anhydrohexo- and -pentopyranosides were treated with LiBr in the presence of tetramethylurea (TMU). The mechanism was deduced as a sequence of steps comprising *trans*-diaxial epoxide opening by bromide, conformational inversion, attack of C-5 at C-3 (according to Scheme 12, route **g**) followed by elimination; an example is shown in entry 1. Bols and Thomsen [93] (entry 3) have recently disclosed a very interesting new type of rearrangement with carbohydrate-derived educts. Although base treatment of 1,6-anhydro-2,3,4-tri-*O*-tosyl-β-D-glucopyranose (**168**) is known to produce epoxide **169** in high yield [94] (entry 2), reaction of **168** with methoxide in chloroform produced mainly ring-contracted product **170** (entry 3). As mechanism, a tandem elimination/Favorskii rearrangement was anticipated. Divergent results were reported [95] from DAST reaction of 1,6-anhydrohex-2-enopyranose **171** under different conditions. Whereas in CH_2Cl_2 smooth introduction of fluorine, at C-4, under inversion of configuration occurred at room temperature, ring contraction together with an allylic rearrangement, leading to 2,5-anhydrohex-3-enoseptanosyl fluorides **172**, took place merely by changing the reaction temperature to –80 °C (entry 4). In DMF, at –50 °C, a hetero-Cope rearrangement, starting from the isolable 4-*O*-formyl derivative of educt **171** (which was generated by reaction with the solvent), led to the corresponding hex-3-enopyranose **173** (entry 5).

Furthermore, some transformations involving special LGs and unusual participation of certain NGs should be mentioned. There is, firstly, the (**MS**[3,C-5] + **HAF**) sequence observed with 2*N*,3*O*-benzene-1,2-disulfonyl-protected glucosaminide **174** (entry 6 [96]). Secondly, an unusual sequence of an O-4/S-6-benzoyl and O-3/O-4-TBDPS migration followed by the **IS**[2,O-3] path. This was observed when educt **175** was treated with NaOMe and *manno*-epoxides **176** were isolated (entry 7 [97]) instead of the "planned" **IS**[2,S-6] product **177**.

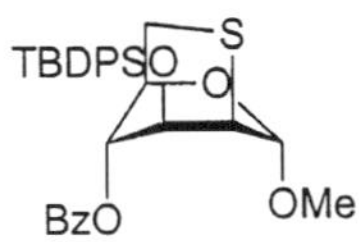

177

In contrast to this, an intramolecular substitution according to **IS**[3,S-6] was reported from the NaHS treatment of the 3,6-dimesylate **178** of a 2,5-anhydrohexose (entry 8 [98]). Interestingly, when trying to invert the configuration at C-4 (via the triflate **179**), an unexpected participation of an oxygen atom from

Table 15

	Educt	Product(s)	Conditions	Σ (%)	Ref.
1	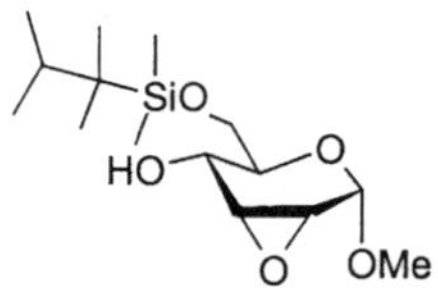	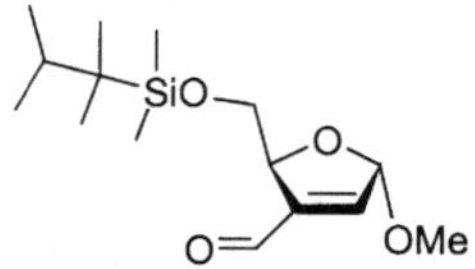 **MS**[3,C-5]+**E1cB** (57%)	LiBr, TMU/toluene reflux, 10 min	57	92a
2	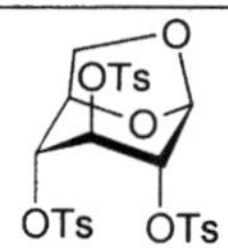 **168**	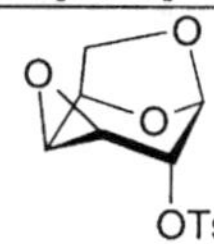 **IS**[4,O-3] (85%) **169**	TBAOH $MeO(CH_2)_2OH$/benzene rt, 10 min	85	94
3	**168**	(minor) **169** 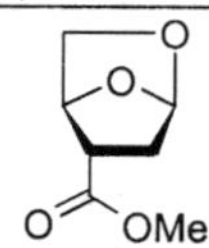 (44%) **170**	NaOMe, MeOH/$CHCl_3$ rt, 30 min	>44	93
4	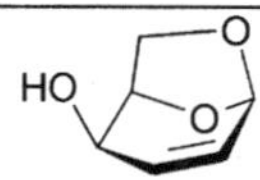 **171**	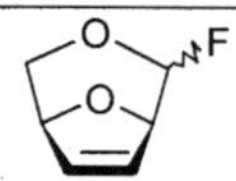 (48%) **172**	DAST, CH_2Cl_2 –80 °C, 10 min	48	95
5	**171**	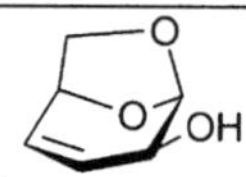 (50%) **173**	DAST, DMF –50 °C, 20 min	50	95

Table 15 (continued)

	Educt	Product(s)	Conditions	Σ (%)	Ref.
6	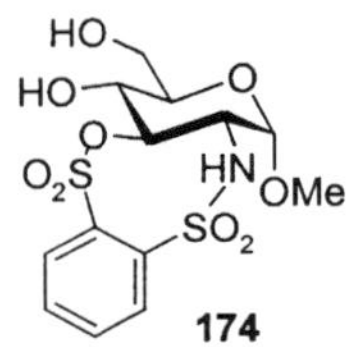 **174**	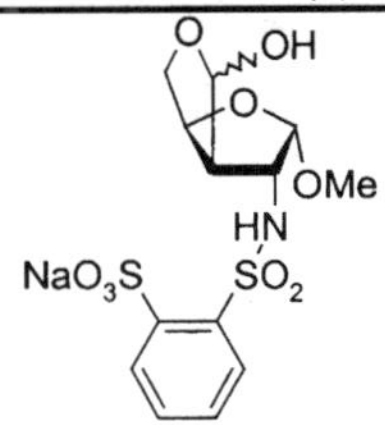 **MS**[3,C-5]+**HAF** (82%)	NaOMe, MeOH 50 °C, 36 h	82	96
7	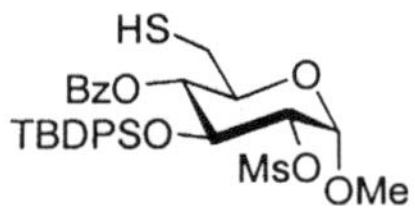 **175**	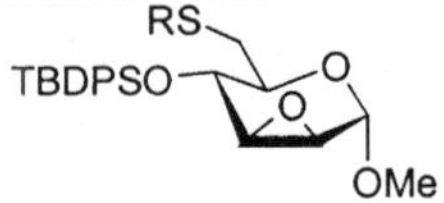 (96%) **176** R = Bz (20%) R = H (76%)	NaOMe, MeOH rt, 21 h; reflux, 30 min	96	97
8	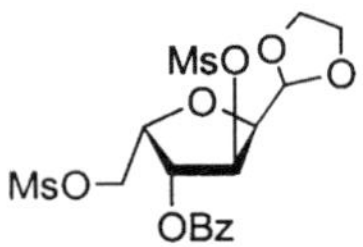 **178**	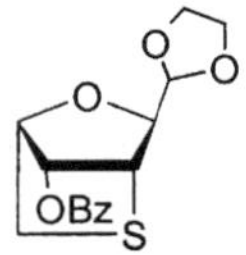 $\mathbf{S_N}$[6]+**IS**[3,S-6]	NaHS, DMF 80 °C, 40 h	not given	98
9	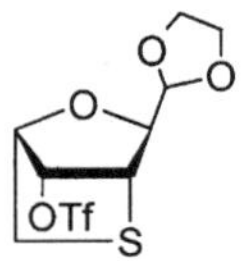 **179**	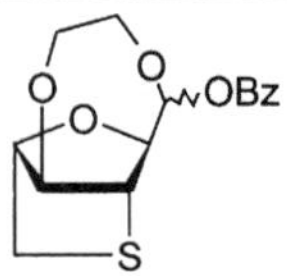 **IS**[4,NG-1] (42%) **180**	LiOBz, DMF reflux, 24 h	42	98

Table 15 (continued)

	Educt	Product(s)	Conditions	Σ (%)	Ref.
10	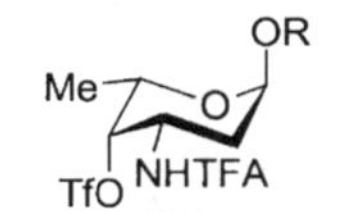 **181**	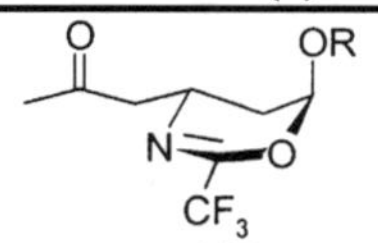 **182**	TBAI	not given	99
11	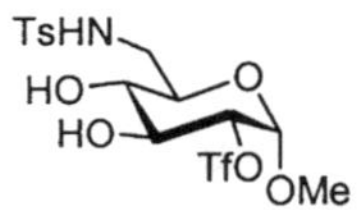	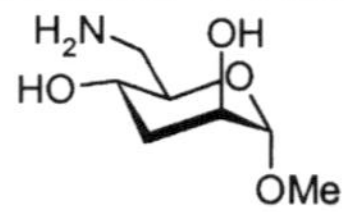 **IS**[2,O-3]+**S$_N$2**[3] (9%) 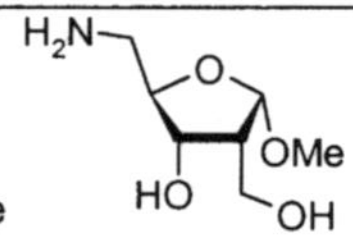 **MS**[2,C-4] (45%)	Na/NH_3, THF –50 °C, 1 h	54	100
12	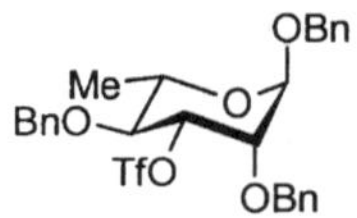	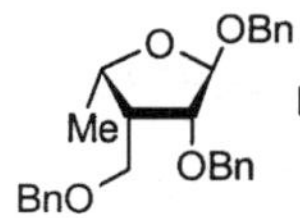 **MS**[3,C-5] (40%) 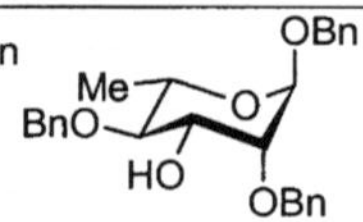 (n.g.)	$LiBEt_3H$, dioxane	>40	101
13	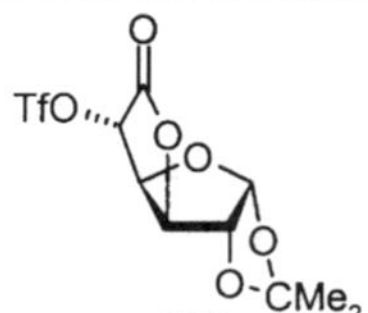 **183**	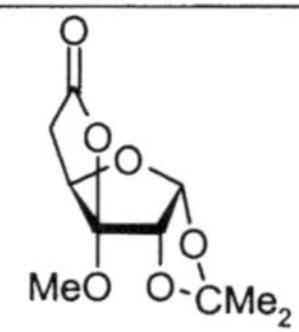 (15%) **184**	pyridine/MeOH rt	not given	102

the acetal protecting group at C-1, to form the tricyclic structures **180**, was observed (entry 9 [98]).

Another surprising result, caused by an unusual kind of NG participation, was found [99] in the course of studies towards the chemical modification of an anthracycline antibiotic. As shown in entry 10, triflate **181**, on treatment with TBABr, underwent a hitherto unique rearrangement to form compound **182**. The mechanistic explanation for this was stated to consist of **E**[4,H-5] followed by opening of the pyranoid enol-acetal system thus formed with simultaneous ring closure, between C-1 and the carbonyl oxygen of the *N*-protecting TFA group. The ring-contracting type of side reactions under involvement of ring carbon atoms also became evident when hydride was used as nucleophile (**MS**[2,C-4] and **MS**[3,C-5] in entries 11 [100] and 12 [101], respectively). Finally, mention is made of another interesting kind of rearrangement we have observed [102] with 1,2-*O*-isopropylidene-5-*O*-triflyl-α-D-glucofuranurono-6,3-lactone **183** (and other derivatives activated at C-5, including those of L-*ido* configuration). On storing **183** in pyridine that contained water or methanol (entry 13), formation of 5-deoxy-3-ulose derivative **184** was observed, albeit in low yield. The mechanism obviously consists of a two-step hydride shift (from C-4 to C-5 followed by that from C-3 to C-4). Both the 4-*C*-methylglucuronic acid derivative **185** and the carbocyclic educt **186** did not give this rearrangement.

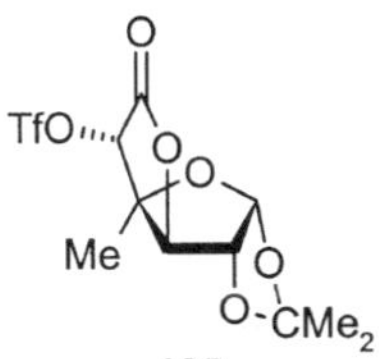

185

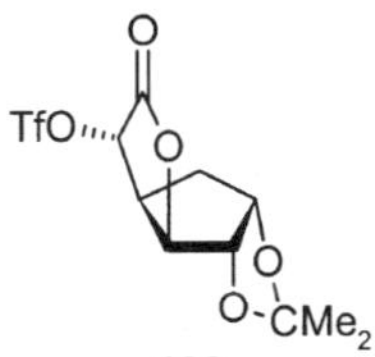

186

9
Concluding Remarks

Although we have performed careful and extensive searches of the available literature, it is possible that some essential contributions have been overlooked or were not cited in their full meaning. For this we apologise. Furthermore, we once again refer to a statement made in the introductory section, that "the significance of each single reaction cited finds its measure in the given total yield of isolated products". For the further development of the present understanding of the types of reactions involved, it will be essential (besides having to clarify certain contradictory results) to design experiments which will answer the more far-reaching questions still left unanswered.

Acknowledgements. Financial support by the Fonds zur Förderung der wissenschaftlichen Forschung, Vienna (P-11021 OECH) is gratefully acknowledged.

10
References

1. For more general information on all types of reactions involved see any textbook of organic chemistry, e.g., March J (1992) Advanced organic chemistry, 4th edn. Wiley, New York (chaps 5, 10, 17 and 18)
2. Cherest M, Felkin H, Sicher J, Sipos F, Tichy M (1965) J Chem Soc 2513
3. Hammer CF, Heller SR, Craig JH (1972) Tetrahedron 28:239; see also: Cossy J, Dumas C, Michel P, Gomez Pardo D (1995) Tetrahedron Lett 36:549
4. Williams JM (1975) Adv Carbohydr Chem Biochem 31:9
5. Goodman L (1967) Adv Carbohydr Chem 22:109
6. Ball DH, Parrish FW (1969) Adv Carbohydr Chem 24:139
7. Binkley RW, Ambrose MG (1984) J Carbohydr Chem 3:1
8. Defaye J (1970) Adv Carbohydr Chem Biochem 25:181
9. Hanessian S, Pernet AG (1976) Adv Carbohydr Chem Biochem 33:111
10. Yoshimura J (1984) Adv Carbohydr Chem Biochem 42:69
11. de Lederkremer RM, Varela O (1994) Adv Carbohydr Chem Biochem 50:125
12. Redlich H (1994) Angew Chem 106:1407
13. Tsuchiya T (1990) Adv Carbohydr Chem Biochem 48:91
14. Dax K, Albert M, Ortner J, Paul BJ (2000) Carbohydr Res 327:47
15. (a) Toshima K, Tatsuta K (1993) Chem Rev 93:1503 and reviews cited therein; (b) Paulsen H (1996) In: Khan SH, O'Neill RA (eds) Modern methods in carbohydrate synthesis. Harwood, Amsterdam, p 1; (c) Schmidt RR (1996) In: Khan SH, O'Neill RA (eds) Modern methods in carbohydrate synthesis. Harwood, Amsterdam, p 20; (d) Norberg T (1996) In: Khan SH, O'Neill RA (eds) Modern methods in carbohydrate synthesis. Harwood, Amsterdam, p 82; (e) Mehta S, Pinto BM (1996) In: Khan SH, O'Neill RA (eds) Modern methods in carbohydrate synthesis. Harwood, Amsterdam, p 107; (f) Sinay S, Mattet J-M (1996) In: Khan SH, O'Neill RA (eds) Modern methods in carbohydrate synthesis. Harwood, Amsterdam, p 130; (g) Madsen R, Fraser-Reid B (1996) In: Khan SH, O'Neill RA (eds) Modern methods in carbohydrate synthesis. Harwood, Amsterdam, p 155; (h) Barresi F, Hindsgaul O (1996) In: Khan SH, O'Neill RA (eds) Modern methods in carbohydrate synthesis. Harwood, Amsterdam, p 251; (i) Boons G-J (1996) Tetrahedron 52:1095; (j) Stöckl WP, Weidmann H (1989) J Carbohydr Chem 8:169
16. Compare the discussion on "Migratory Aptitude" in ref 1, p 1058
17. Ref 4, p 48
18. (a) Hase S, Matsushima Y (1972) J Biochem 72:1117; (b) Hase S, Tsuji Y, Matsushima Y (1972) J Biochem 72:1549; (c) Ref 4, p 44
19. Binkley RW, Ambrose MG (1987) J Carbohydr Chem 6:203
20. Karpiesiuk W, Banaszek A, Zamojski A (1989) Carbohydr Res 186:156
21. Llewellyn JW, Williams JM (1973) J Chem Soc Perkin Trans 1 1997
22. Kovac P, Yeh HJC, Jung GL, Glaudemans CPJ (1986) J Carbohydr Chem 5:497
23. Borrachero-Moya P, Cabrera-Escribano F, Gomez-Guillen M, Madrid-Diaz F (1997) Tetrahedron Lett 38:1231
24. Albert M, Dax K (1999) unpublished results
25. Micheel F, Neier W, Riedel T (1967) Chem Ber 100:2401
26. Wiggins LF (1946) Nature 157:300
27. (a) Castillon S, Dessinges A, Faghih R, Lukacs G, Olesker A, Thang TT (1985) J Org Chem 50:4913; (b) Faghih R, Cabrera Escribano F, Castillon S, Garcia J, Lukacs G, Olesker A, Thang TT (1986) J Org Chem 51:4558
28. Nicolaou KC, Ladduwahetty T, Randall JL, Chucholowski A (1986) J Am Chem Soc 108:2466
29. (a) Griffith DA, Danishefsky SJ (1990) J Am Chem Soc 112:5811; (b) Zheng C, Seeberger PH, Danishefsky SJ (1998) Angew Chem 110:831
30. Sharma M, Bernacki RJ, Hillman MJ, Korytnyk W (1993) Carbohydr Res 240:85

31. Binkley RW (1992) J Org Chem 57:2353
32. Ng Ying Kin NMK, Williams JM (1971) J Chem Soc Chem Commun 1123
33. (a) Binkley RW (1991) J Carbohydr Chem 10:417 (the NMR data given do not provide conclusive evidence for the assigned structure of the products containing the pyridinio moiety); (b) Binkley RW (1990) J Carbohydr Chem 9:771; (c) Binkley RW (1991) J Org Chem 56:3892
34. (a) Al-Radhi AK, Brimacombe JS, Tucker LCN (1972) J Chem Soc Perkin Trans I 315; (b) Al-Radhi AK, Brimacombe JS, Tucker LCN (1970) J Chem Soc Chem Commun 1250
35. Lindhorst TK, Thiem J (1991) Carbohydr Res 209:119
36. Petrakova E, Yeh HJC, Kovac P, Glaudemans CPJ (1992) J Carbohydr Chem 11:407
37. Takahashi Y, Tsuneda S, Tsuchiya T, Koyama Y, Umezawa S (1992) Carbohydr Res 232:89
38. (a) Gouda I, Larm O, Mosihuzzaman M (1983) Acta Chem Scand B 37:72; (b) Erbing C, Lindberg B, Svensson S (1973) Acta Chem Scand 27:3699
39. (a) Akiya S, Osawa T (1959) Chem Pharm Bull 7:277; (b) Irvine JC, Hynd A (1914) J Chem Soc 105:698
40. Ng Ying Kin NMK, Williams JM, Horsington A (1971) J Chem Soc (C) 1578
41. Grouiller A, Bazin H, Gagnieu C (1982) Tetrahedron Lett 23:2559
42. El Sayed Ahmed FM, David S, Vatele J-M (1986) Carbohydr Res 155:19
43. Fleet GWJ, Seymour LC (1987) Tetrahedron Lett 28:3015
44. Baer HH, Hernandez Mateo F, Siemsen L (1989) Carbohydr Res 187:67
45. Dax K, Albert M, Ortner J, Paul BJ (1999) unpublished result (cited in ref 14)
46. Haradahira T, Maeda M, Omae H, Yano Y, Kojima M (1984) Chem Pharm Bull 32:4758
47. Knapp S, Kukkola PJ, Sharma S, Murali Dhar TG, Naughton ABJ (1990) J Org Chem 55: 5700
48. Albert R, Dax K, Link RW, Stütz AE (1983) Carbohydr Res 118:c5
49. Knapp S, Naughton ABJ, Jaramillo C, Pipik B (1992) J Org Chem 57:7328
50. Ivanova IA, Nikolaev AV (1998) J Chem Soc Perkin Trans 1 3093
51. Kunz H, Günther W (1988) Angew Chem 100:1118
52. (a) Charette AB, Cote B (1993) J Org Chem 58:933; (b) Charette AB, Cote B, Marcoux J-F (1991) J Am Chem Soc 113:8166
53. (a) Kassou M, Castillon S (1995) J Org Chem 60:4353; (b) Kassou M, Castillon S (1997) J Org Chem 62:3696
54. (a) Tsuchiya T, Ajito K, Umezawa S, Ikeda A (1984) Carbohydr Res 126:45; (b) Tsuchiya T, Shitara T (1982) Carbohydr Res 109:59
55. (a) Binkley RW, Sivik MR (1986) J Carbohydr Chem 5:647; (b) Binkley RW, Sivik MR (1986) J Org Chem 51:2621; (c) Binkley RW, Abdulaziz MA (1987) J Org Chem 52:4713
56. (a) Reist EJ, Calkins DF, Goodman L (1968) J Am Chem Soc 90:3852; (b) Reist EJ (1967) Chem Ind 1957
57. (a) Inoue S, Ogawa H (1960) Chem Pharm Bull 8:79; (b) Austin PW, Buchanan JG, Saunders RM (1965) J Chem Soc Chem Commun 146; (c) Austin PW, Buchanan JG, Saunders RM (1967) J Chem Soc (C) 372
58. Recondo E, Rinderknecht H (1960) Helv Chim Acta 1653 and references cited therein
59. Ng Ying Kin NMK, Williams JM, Horsington A (1969) J Chem Soc Chem Commun 971
60. Austin PW, Buchanan JG, Large DG (1967) J Chem Soc Chem Commun 418
61. (a) Mori Y, Morishima N (1992) Chem Pharm Bull 40:826; (b) Mori Y, Morishima N (1991) Chem Pharm Bull 39:1088
62. (a) Hanessian S (1966) J Chem Soc Chem Commun 796; (b) Stevens CL, Glinski RP, Taylor KG, Sirokman F (1970) J Org Chem 35:592; (c) Stevens CL, Glinski RP, Gutowski GE, Dickerson JP (1967) Tetrahedron Lett 649
63. Ref 4, p 50
64. Kihlberg J, Frejd T, Jansson K, Magnusson G (1988) Carbohydr Res 176:287
65. Cai S, Hakomori S, Toyokuni T (1992) J Org Chem 57:3431
66. (a) Mulard LA, Kovac P, Glaudemans CPJ (1994) Carbohydr Res 259:117; (b) Kovac P, Yeh HJC, Jung GL (1987) J Carbohydr Chem 6:423
67. Su T-L, Klein RS, Fox JJ (1981) J Org Chem 46:1790

68. Mori Y, Morishima N (1994) Bull Chem Soc Jpn 67:236
69. Card PJ, Reddy GS (1983) J Org Chem 48:4734
70. Dax K, Grassberger V, Stütz AE (1990) J Carbohydr Chem 9:903
71. Hasegawa A, Goto M, Kiso M (1985) J Carbohydr Chem 4:627
72. Ref 4, p 57
73. Lloyd AE, Coe PL, Walker RT, Howarth OW (1993) J Fluorine Chem 60:239
74. Chalk RC, Ball DH, Lintner MA, Long L (1970) J Chem Soc Chem Commun 245
75. Nayak UG, Whistler RL (1968) J Org Chem 33:3582
76. Pakulski Z (1996) Polish J Chem 70:667
77. Driver GE, Stevens JD (1990) Aust J Chem 43:2063
78. (a) Hughes NA, Wood CJ (1986) J Chem Soc Perkin Trans 1 695; (b) Al-Masoudi NAL, Hughes NA (1987) J Chem Soc Perkin Trans 1 2061
79. (a) Fuzier M, Le Merrer Y, Depezay J-C (1995) Tetrahedron Lett 36:6443; (b) Le Merrer Y, Fuzier M, Dosbaa I, Foglietti M-J, Depezay J-C (1997) Tetrahedron 53:16731
80. Kim M, Grzeszczyk B, Zamojski A (1999) Carbohydr Res 320:244
81. (a) Poitout L, Le Merrer Y, Depezay J-C (1996) Tetrahedron Lett 37:1613; (b) Poitout L, Le Merrer Y, Depezay J-C (1996) Tetrahedron Lett 37:1609; (c) Le Merrer Y, Poitout L, Depezay J-C, Dosbaa I, Geoffroy S, Foglietti M-J (1997) Bioorg Med Chem 5:519
82. Schumacher-Wandersleb MHMG, Petersen S, Peter MG (1994) Liebigs Ann Chem 555
83. Andersen SA, Ebner M, Ekhart CW, Gradnig G, Legler G, Lundt I, Stütz AE, Withers SG, Wrodnigg T (1997) Carbohydr Res 301:155
84. Di J, Rajanikanth B, Szarek WA (1992) J Chem Soc Perkin Trans 1 2151
85. (a) Furneaux RH, Gainsford GJ, Mason JM, Tyler PC (1994) Tetrahedron 50:2131; (b) Furneaux RH, Gainsford GJ, Mason JM, Tyler PC (1995) Tetrahedron 51:12611
86. (a) Estevez JC, Fairbanks AJ, Fleet GWJ (1998) Tetrahedron 54:13591 and their own work cited therein; (b) Estevez JC, Fairbanks AJ, Hsia KY, Ward P, Fleet GWJ (1994) Tetrahedron Lett 35:3361
87. (a) Wilson FX, Fleet GWJ, Vogt K, Wang Y, Witty DR, Choi S, Storer R, Myers PL, Wallis CJ (1990)) Tetrahedron Lett 31:6931; (b) Wilson FX, Fleet GWJ, Witty DR, Vogt K, Wang Y, Storer R, Myers PL, Wallis CJ (1990) Tetrahedron Asymmetry 1:525; (c) Wang Y, Fleet GWJ, Storer R, Myers PL, Wallis CJ, Doherty O, Watkin DJ, Vogt K, Witty DR, Wilson FX, Peach JM (1990) Tetrahedron Asymmetry 1:527; (d) Saksena AK, Ganguly AK, Girijavallabhan VM, Pike RE, Chen Y-T, Puar MS (1992) Tetrahedron Lett 33:7721
88. (a) Bichard CJF, Brandstetter TW, Estevez JC, Fleet GWJ, Hughes DJ, Wheatley JR (1996) J Chem Soc Perkin Trans 1 2151; (b) Estevez JC, Saunders J, Besra GS, Brennan PJ, Nash RJ, Fleet GWJ (1996) Tetrahedron Asymmetry 7:383; (c) Mantell SJ, Ford PS, Watkin DJ, Fleet GWJ, Brown D (1993) Tetrahedron 49:3343; (d) Mantell SJ, Ford PS, Watkin DJ, Fleet GWJ, Brown D (1992) Tetrahedron Lett 33:4503; (e) Myerscough PM, Fairbanks AJ, Jones AH, Bruce I, Choi SS, Fleet GWJ, Al-Daher SS, Cenci di Bello I, Winchester B (1992) Tetrahedron 48:10177; (f) Smith MD, Long DD, Marquess DG, Claridge TDW, Fleet GWJ (1998) J Chem Soc Chem Commun 2039
89. (a) El-Laghdach A, Echarri R, Matheu MI, Barrena MI, Castillon S (1991) J Org Chem 56:4556; (b) Barrena MI, Matheu MI, Castillon S (1998) J Org Chem 63:2184; (c) Echarri R, Matheu MI, Castillon S (1994) Tetrahedron 50:9125
90. (a) Lemieux RU, Fraser-Reid B (1963) Can J Chem 42:539; (b) Lemieux RU, Fraser-Reid B (1963) Can J Chem 42:547
91. (a) Kozlowski JS, Marzabadi CH, Rath NP, Spilling CD (1997) Carbohydr Res 300:301; (b) Marzabadi CH, Spilling CD, Tyler LM (1994) Tetrahedron 50:6783; (c) Marzabadi CH, Spilling CD (1993) J Org Chem 58:3761
92. (a) Ponten F, Magnusson G (1997) J Org Chem 62:7972; (b) Rehnberg N, Magnusson G (1990) J Org Chem 55:5467 and their own work cited therein
93. Bols M, Thomsen IB (1998) J Chem Soc Chem Commun 1869
94. Szeja W (1988) Carbohydr Res 183:135
95. Oberdorfer F, Haeckel R, Lauer G (1998) Synthesis 201
96. Tatsuta K, Miyashita S, Akimoto K, Kinoshita M (1982) Bull Chem Soc Jpn 55:3254

97. Izquierdo I, Rodriquez M, Plaza MT, Pleguezuelos J (1998) J Carbohydr Chem 17:61
98. Popsavin V, Beric O, Popsavin M, Csanadi J, Vujic D, Hrabal R (1999) Tetrahedron Lett 40:3629
99. Suarato A, Penco S, Vigevani A, Arcamone F (1981) Carbohydr Res 98:c1
100. Tsuchiya T, Nakamura F, Umezawa S (1979) Tetrahedron Lett 2805
101. Pozsgay V (1980) Tetrahedron Lett 21:211
102. Albert R, Dax K, Stütz AE (1985) In: Defaye J (ed) Abstracts of the third European symposium on carbohydrates. Grenoble, France, p 131

Direct Conversion of 5,6-Unsaturated Hexopyranosyl Compounds to Functionalized Cyclohexanones

Robert J. Ferrier

Industrial Research Limited, PO Box 31–310, Lower Hutt, New Zealand
E-mail: r.ferrier@irl.cri.nz

Hexopyranoid glycosides and glycosyl esters that have an exocyclic double bond at C-5 (with or without an O-linked substituent at C-6) can be converted directly, efficiently and under mild conditions to cyclohexanone derivatives of the inosose or deoxyinosose categories. Usually mercury(II) salts are used to promote the process, but Pd(II) compounds can also be employed. Carbon-6 of the starting materials bonds to C-1 of aldehydic metal-containing intermediates, the initial anomeric centres become secondary alcohols, and the aglycone of the starting materials are lost during the reaction. The C-5 centre becomes a carbonyl group. Otherwise, titanium, aluminium or Grignard compounds can be used to promote the conversions but, very significantly, in these instances the products retain the C-1 substituents of the starting materials. Also, when triisobutylaluminium is used, the products isolated are the alcohols derived by reduction of the carbonyl groups of the initially formed cyclohexanones. The reactions have been used in the synthesis of a wide range of inositols, their derivatives, and other compounds containing functionalized cyclohexane rings.

Keywords: Hexopyranoid-5-enes, Carbasugars, Inositols, Inososes, Deoxyinososes, Inosamines, Conduritols

Topics in Current Chemistry, Vol. 215

1 Introduction

It was during the search for a new and direct route from readily available monosaccharide derivatives to functionalized cyclopentanes which would serve as prostaglandin synthesis precursors that the reaction discussed in this chapter, and shown in outline in Scheme 1, was first encountered [1]. From the 6-de-

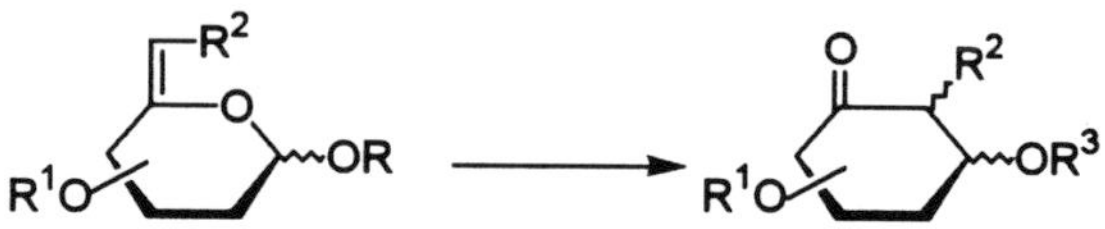

Scheme 1

oxyhex-5-enoside derivative **1** (Scheme 2), the methoxymercuration adduct **2** was readily obtained, but attempts to induce the metallated C-6 to displace the tosyloxy leaving group from C-2 and give the 2-oxa-bicyclo[2.2.1]heptane derivative **3** were unsuccessful. Recognition that hydroxymercuration of the same alkene would give the hemiacetal **4**, which would ring open and lose methanol as illustrated to give the acyclic **5** in which C-6 would now be doubly activated as a nucleophile by the bonded metal atom and the carbonyl group at C-5, led to compound **1** being heated in aqueous acetone with mercury(II) chloride. The cyclopentane derivative **6** was not formed; instead, the cyclohexanone **7** cystallized in high yield when the solution cooled down to room temperature [1]. Rather, therefore, than C-6 of the osulose **5** bonding to the poorly electrophilic C-2 [2], it had taken part in an aldol-like reaction with the released C-1 aldehydic group. The process is therefore very similar to the ring closure **8** → **9** which, under the influence of inositol cyclase, is the key step in the biosynthetic conversion of D-glucose 6-phosphate into *myo*-inositol 1-phosphate (Scheme 3a) [3]. Likewise, there is a strong resemblance to the ring-closure reaction by which the unstable hemiacetal **10** is converted to the hydroxycyclohexanone **12** via enolate **11** during the bioconversion of 3-deoxy-D-*arabino*-heptulosonic acid 7-phosphate to shikimic acid and the benzenoid rings of the aromatic amino acids (Scheme 3b) [4].

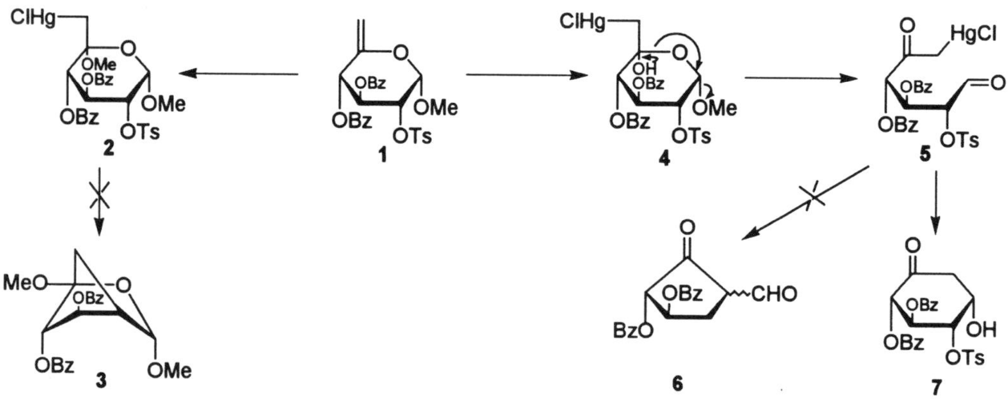

Scheme 2

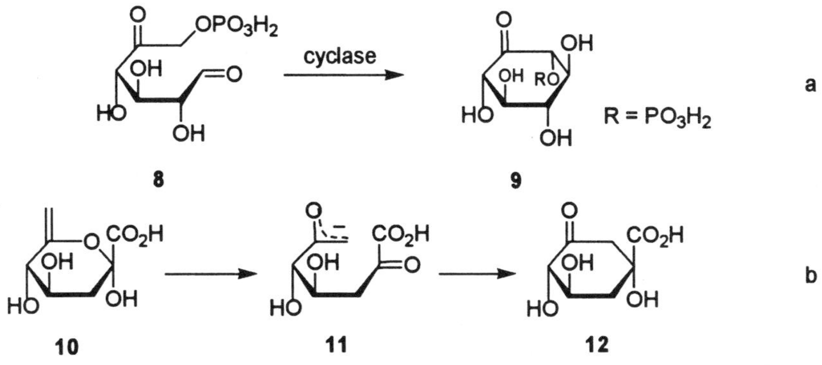

Scheme 3

When it was found that **10** rearranges spontaneously in solution to **12** the requirement for the dehydroquinate synthase in the cyclization step was questioned [5], but then it was discovered that a small proportion of the epimer at the tertiary centre of **12** was also produced in the non-catalyzed reaction while the enzyme-catalyzed process was stereospecific [6]. It is clear therefore that the intermediate **11** is retained in the active site of the enzyme until cyclization is complete.

In the 20 years since its discovery the reaction under consideration here has often been used in its original form; however, it has been modified and extended in several important ways. Consequently, 5,6-unsaturated hexopyranosyl compounds are well-recognized starting materials for the synthesis of not just carbocyclic analogues of pyranoid sugars the – carba-sugars – but inositols and related compounds such as inososes (pentahydroxycyclohexanones), inosamines (pentahydroxycyclohexylamines) and conduritols (tetrahydroxycyclohexenes). Beyond that, several natural products and pharmaceutically significant compounds with oxygenated cyclohexane components in their structures have been made by the reaction, the products being enantiomerically pure in all cases. This survey covers the basic features of the original reaction, developments of it (notably those utilizing 6-substituted 5-enose compounds and those that allow retention of the aglycone of the starting material) and applications in synthesis. A detailed review of the earlier parts of the story has been published [7].

2 Use of 6-Deoxyhex-5-enose Derivatives with Mercury(II) Salts

2.1 General

That mercury-containing intermediates such as **5** are involved in the reaction has been established by their isolation [8] and their conversion to cyclohexanones by treatment with sulfide [8a] or thiourea [8b]. Otherwise, acid-catalyzed hydrolysis of products of methoxymercuration such as **2** gives intermediates that react further to afford the hydroxycyclohexanones [9]. From the starting alkenes no such carbocyclic products are formed by acid hydrolysis, rather 6-deoxyhexos-5-uloses are obtained, establishing that the mercury salt is required for the carbocyclization to occur. Because the starting alkenes and conceivably the mercury-containing intermediates are subject to acid hydrolysis to give undesired hexos-5-ulose products [10], attention was given to minimizing the acidity of the reaction media. Consequently, mercury(II) acetate was found to improve the yield in the reaction of tetra-*O*-benzoyl-6-deoxy-*β*-D-*xylo*-hex-5-enose affording 93% of the tri-*O*-benzoyl analogue of compound **7**, whereas 55% was reported when the chloride was used [8a].

A very useful development came with the recognition that the reaction can proceed efficiently with catalytic proportions of mercury(II) salts even at room temperature; the activity of the salts decreasing in the order trifluoroacetate, chloride and acetate, oxide [11], the trifluoroacetate impressively being efficient at 1 mol% [12].

In the context of the severity of the reaction conditions it has to be recognized that the ketonic products commonly have good leaving groups at C-3 (carbohydrate numbering) and may be isolated as derived enones, e.g. **13** (see Table 1, entry 4), which was formed in 80% yield from an *α*-D-*ribo*-hex-5-enoside [13]. This isomer is particularly subject to *β*-elimination because of the axial nature of both H-4 and the benzoyloxy group at C-3 (carbohydrate numbering), and conditions much milder than those employed in the initial work ($HgSO_4$, H_2SO_4, 80 °C) have to be used if elimination is to be avoided. Otherwise, the alternative type of enone (e.g. **14**) can be readily made from the normal hydroxyketone products, e.g. **7**, following methanesulfonylation [7, 14] or acetylation [7, 15] of the hydroxy groups.

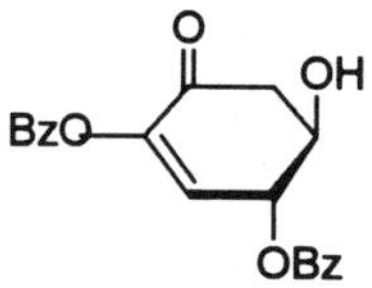

13

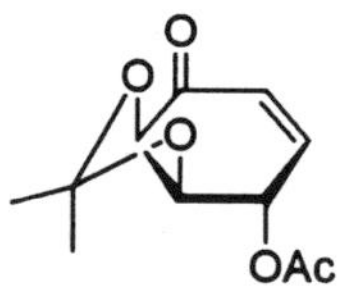

14

2.2 Stereochemistry and Mechanism

A significant feature of the reaction outlined in Scheme 1 (R, R^1 = alkyl, acyl; R^2 = H, OR; R^3 = H) is the appreciable stereoselectivity shown at the new chiral centre, there normally being strong bias in favour of diastereomers with the new hydroxyl group *trans*-related to the substituent at C-3, compound **7** for example being obtainable in 89% yield [8a]. The mercury(II)-based reaction, however, despite claims to the contrary [16], is not a stereospecific process.

The main evidence (apart from the characterization of the mercuriated intermediates) from which proposed mechanisms for the ring closure have been developed is the impact of the orientation of the substituent at C-3 of the starting materials on the stereochemical outcomes. Table 1 indicates that the

Table 1

Entry	Starting 5-ene	Main product	Main product (%)	Ref.
1	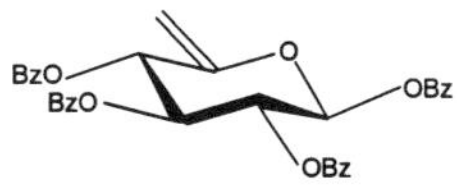	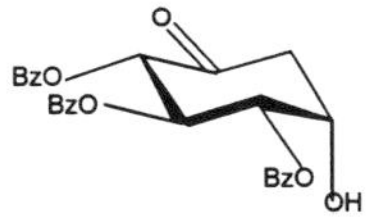	93	8a
2	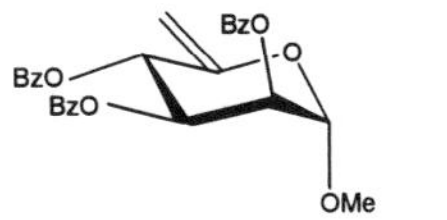	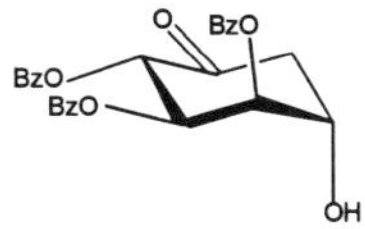	83	13
3	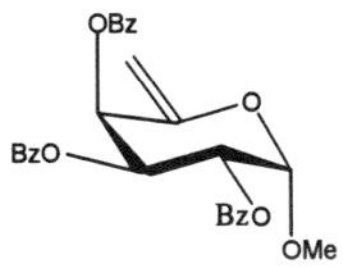	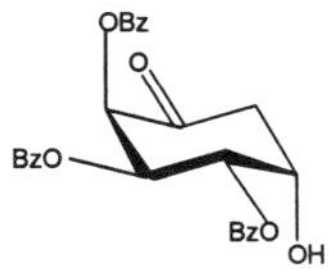	99	13
4	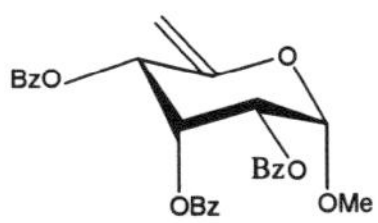	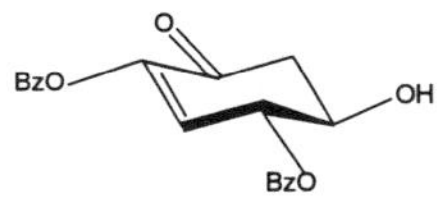	80	13
5	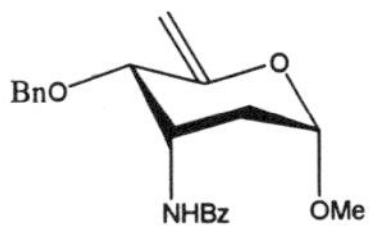	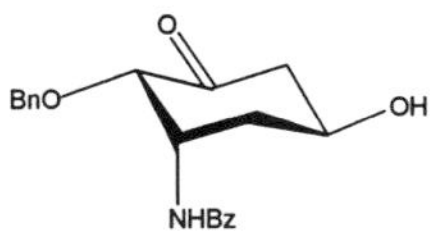	100	16
6	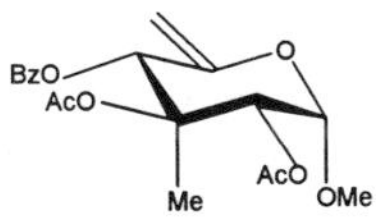	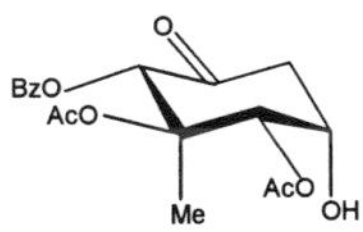	75	19
7	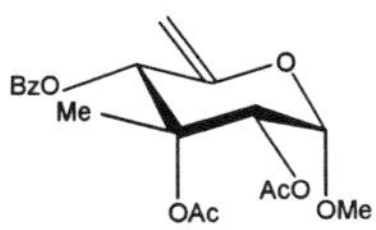	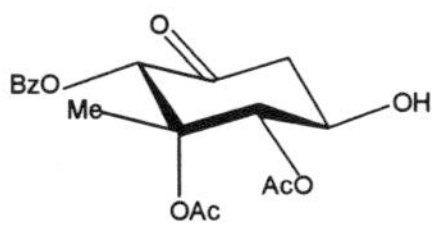	83	19
8	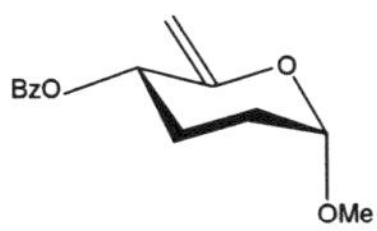		57	18

reaction can be very selective in favour of the α-products formed not just from D-glucose-derived starting materials (entry 1) but also from analogous D-mannose- and D-galactose-based alkenes (entries 2 and 3). Inversion of configuration at C-2 or C-4, therefore, does not affect the stereochemical outcomes. When, however, the starting alkenes have the inverted stereochemistry at C-3 the consequences are dramatically different, and β-alcohols are the main products (Table 1, entries 4 and 5), and can be formed with high stereoselectivity.

Early attempts to rationalize the stereochemistry of the reaction correlated the configurations of the products with the conformations of the starting al-

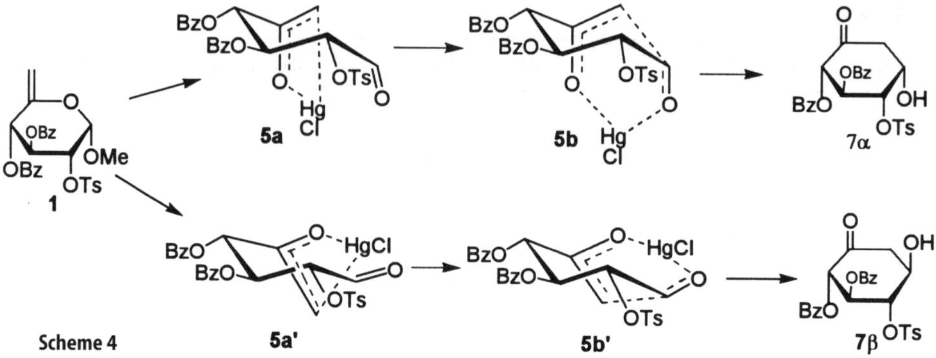

Scheme 4

kenes [13], but when the importance of the configuration of the substituent at C-3 was recognized two hypotheses invoking intramolecular coordination of the mercury atom were developed. One was based on the concept of coordination of delocalized mercury enolates with O-1 of the aldehydic group in either of two rotamer states (Scheme 4) [7, 8b, 17]. Thus, intermediate **5** (Scheme 2) leads to the enolates **5a** and **5a′** and then to the transition states **5b** and **5b′** of which the chair-like former is favoured relative to the latter. In consequence, the isomer **7a** is the major product of ring closure. Clearly, a large axial substituent at C-3 would destabilize **5b** and *β*-products would be relatively favoured. In the other proposed coordination-dependent path the direction of attack at C-1 by C-6 is controlled by binding of the mercury atoms to the electronegative atoms at C-3 and products with the C-1 and C-3 substituents *trans*-related are formed [18].

It is difficult to choose between these two possibilities; however, ring closures of the epimeric 5-enes **15** and **16** (Scheme 5), which have both a *C*-methyl and an acetoxy substituent at C-3 (Table 1, entries 6 and 7) [19], suggest strongly that, at least in these cases, O-3 coordination is the determining factor. Since single products are obtained in good yields in both examples, each having the new hydroxyl group *trans* to the acetoxy groups at C-3, a mechanism involving transition states like those illustrated in Scheme 5 has to be favoured. In the case of the 2,3-dideoxy alkene in entry 8 of Table 1, marked loss of stereoselectivity is

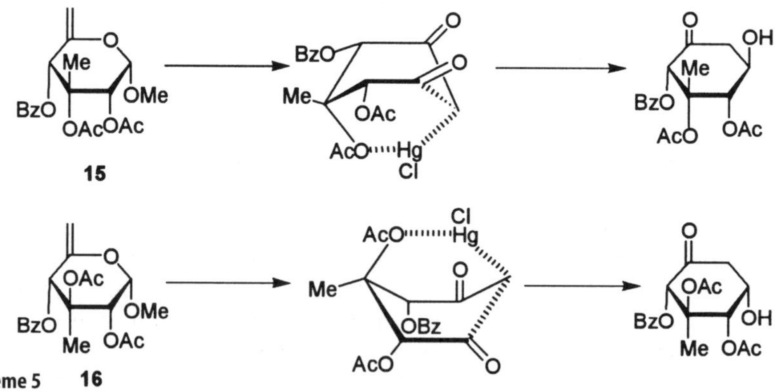

Scheme 5

shown [18], and from methyl 4-*O*-benzyl-3,6-dideoxy-2-*O*-methyl-D-*ribo*-hexopyranoside, a further 3-deoxy substrate, again both the α- and β-hydroxycyclohexanones are formed in significant proportions [20].

Such features as reaction conditions, reagent proportions and the *O*-substituents on the 5-enes can affect the α,β ratios of the products [8b], but they usually do so to a lesser extent than does the configuration at C-3. There, however, appears to be one aspect of the *O*-substituents that has a dramatic influence on product formation: the presence of silyl ethers that can result in inverted product ratios (see Sect. 4).

One study committed to investigating the stereochemical details of the reaction by the use of the specifically deuterated alkene **17** showed that the stereochemical integrity of the C-6 methylene group is lost during the reaction (Scheme 6), and this has been taken, along with the absence of exchange of the isotope with the solvent and with results of molecular orbital calculations, as evidence that C-6 free radicals are the intermediates involved in the ring-closure step of the process [8c]. If this were the case, mercury(II), however, is unlikely to be recoverable from the ring-closure step as is required by its ability to function in catalytic proportions. To account otherwise for the above findings would seem to require that C-5–C-6 bond rotation can occur at some stages in the process – perhaps within intermediate/transition states analogous to **5a** and **5b**.

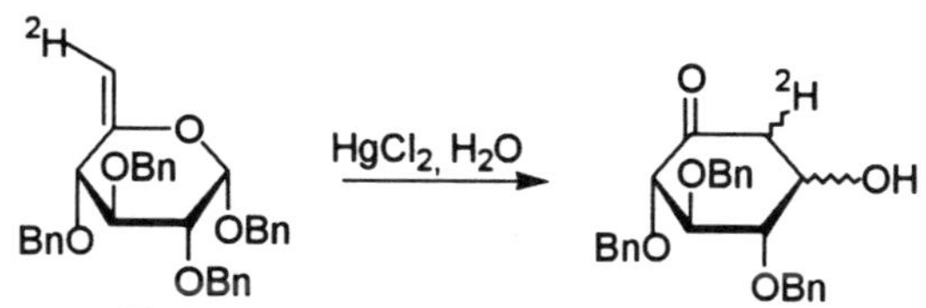

Scheme 6 **17**

2.3 Applications

Of the several reactions available for making cyclohexane derivatives from carbohydrates the present is amongst the most useful. It has been applied in many instances – mainly to the preparation of sugars with a methylene group in place of the ring oxygen atom (carbasugars) and of inositols and inositol derivatives. On occasion, however, it has been used in syntheses related to natural compounds outside these areas [7].

Direct application of the reaction gives 2-deoxyinosose derivatives (e.g. **7**) which can be applied in many ways to afford natural products or analogues required for bioorganic purposes – particularly to act as specific enzyme inhibitors and bioactive inositol phosphates. From compounds like **7**, enones such as **13** or **14** are readily obtainable, and by reduction of the carbonyl groups and hydroxylation of the double bonds, derivatives of the parent inositols can be made. An interesting example uses lactose as starting material, and via a derivative having the 6-deoxy-5-ene function in the glucose moiety, a galactosylated cyclohexenone is made and hence derivatives of the naturally occurring isomeric β-D-galactosyl-*myo*-inositols **18** and **19** [21]. Related work can allow access to partially substituted inositols such as 2,3,6-tri-*O*-benzyl-D-*myo*-inositol from which important inositol phosphates – especially the secondary messenger of calcium mobilization D-*myo*-inositol 1,4,5-trisphosphate (**20**) – can be made [22].

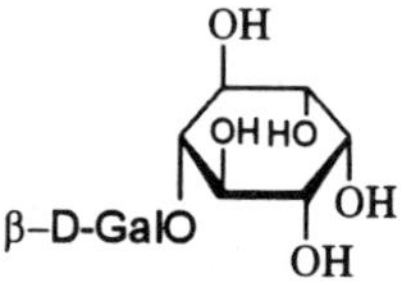

18

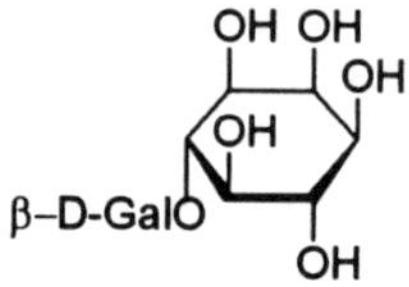

19

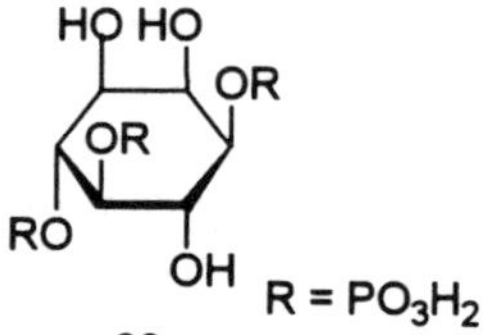

20

A good example of the application of the reaction to the preparation of carbasugars that have a methylene group in place of the ring oxygen atom [23] is in the synthesis of the carbasugar nucleotide uridine 5′-(carba-α-D-galactopyranosyl) diphosphate (**21**) which is a competitive inhibitor of a galactosyl transferase. The carbocyclic component was made from methyl α-D-galactopyranoside via the normal deoxyinosose, the carbonyl group of which was methylenated to give an *exo*-alkene and hence a carba-α-D-galactose compound by hydroboration [24]. In an alternative approach to carbasugar-like compounds, the exocyclic carbon atom of the β-glucosidase inhibitor cyclophellitol (**22**) was introduced at C-2 of a methyl glucoside derivative prior to making the 5-ene and effecting mercury-induced cyclization [25]. By a similar approach the *C*-methyl-*myo*-inositol laminitol **23,** which is found in algae and has antifungal properties, was made from a 4-*C*-methylglucoside via the 6-deoxy-5-ene and then the corresponding cyclohexenone derivative [22].

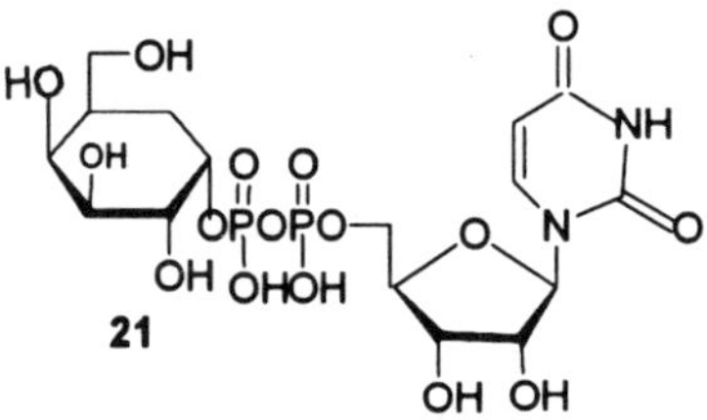

21

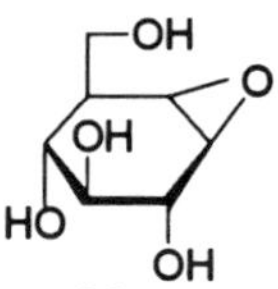

22

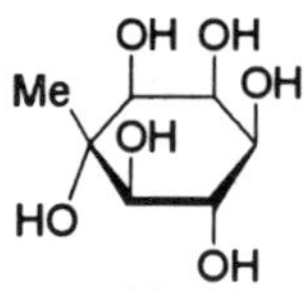

23

The ready availability of enones such as **14** from hexose derivatives has led to the production of specific conduritols, e.g. (+)-conduritol C (**24**) and its enantiomer – ultimately from D-galactose [15].

Perhaps, however, aminocyclitol derivatives are the most important group of compounds to have become accessible by application of the mercury-induced reaction, since they figure prominently in natural products – notably the aminoglycoside antibiotics. Central to many of these are compounds **25** (streptidine) and **26** (2-deoxystreptamine) that occur in glycosylated forms in streptomycin and the neomycins, respectively, and compounds of this type have frequently been made by the mercury-based cyclization procedure. The amino groups can be introduced by starting with aminodeoxyhex-5-enoses or by aminating cyclohexanones derived from hexoses, or by applying both approaches [7]. Other aminocyclitol compounds to have been made are **27**, which was designed as a potential *N*-acetylhexosaminidase inhibitor [26], **28**, an analogue of a key component of the potent insecticidal chitinase inhibitor allosamidin [27], and **29** which was designed as a glycosidase transition state analogue and intended as a hapten for generating catalytic antibodies capable of promoting specific glycosylations [28].

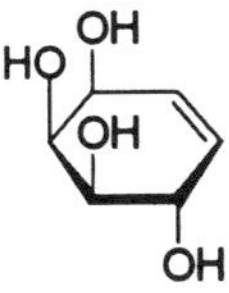

24

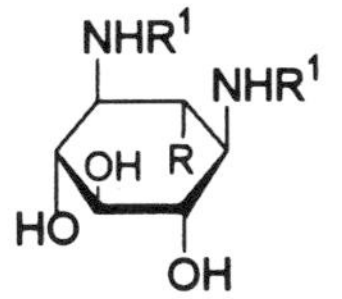

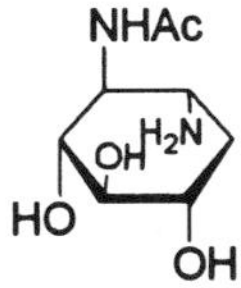

25 R = OH, R^1 = NH(C=NH)NH_2 **27**

26 R = R^1 = H

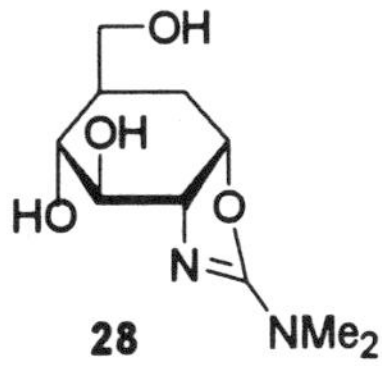

28

Some natural products outside the carbohydrate field – particularly those with highly hydroxylated cyclohexane ring components – have been subject to synthetic studies which are dependent on the mercury-based rearrangement reaction: the alkaloid (+)-lycoricidine (**30**) [12] and the cycloheptane-based (+)-calystegine B_2 (**31**) which stimulates growth of nitrogen-fixing *Rhizobia* [29] are examples, and compound **32** which offers novel access to the anthracyclinone components of anthracyclin anti-cancer compounds, has been produced by cycloaddition of a naphthalene-based *o*-xylylene to a 2,3-unsaturated hex-4-uloside followed by carbocyclization of the product by use of the mercuration procedure [30]. Studies on HMG-CoA reductase inhibitors like compactin have afforded the tetra-carbon-substituted **33** made from a hex-5-enopyranoside with deoxy-branch chains at C-2, C-3 and C-4 [31].

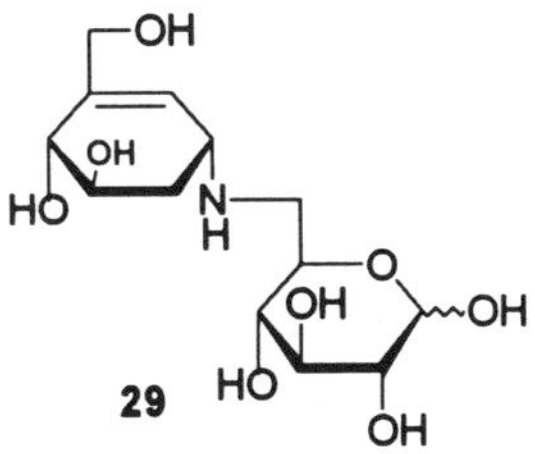

29

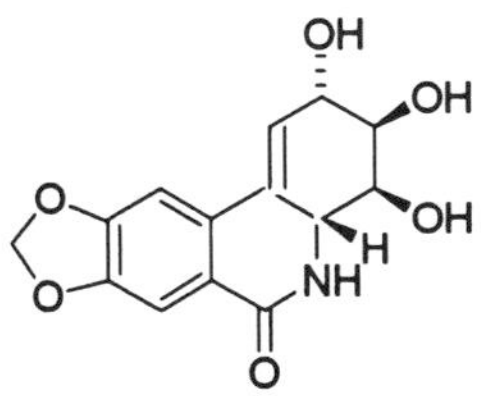

30

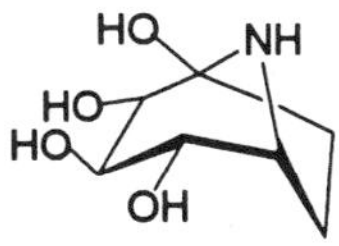

31

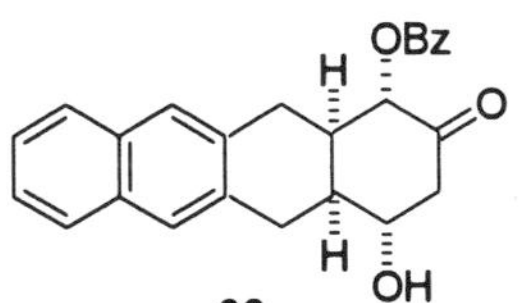

32

33

3
Use of 6-Deoxyhex-5-enose Derivatives with Palladium Salts

In 1988 it was first reported that palladium salts can be used to bring about the carbocyclization reaction [32] and, while this has been employed effectively with 5 mol% of palladium chloride to make a cyclohexenone and hence cyclophellitol (**22**) [14], reports have indicated that it gives poorer yields of carbocyclic products than do mercury(II) salts, and that stereoselectivity is lower [33] or even reversed [34]. However, used catalytically in aqueous dioxane at 60 °C, palladium chloride gives good to excellent yields of α-products from the 5-enes derived from methyl 2,3,4-tri-*O*-benzyl or -benzoyl-α-D-glucoside, -galactoside or -mannoside. The sulfuric acid used in earlier work is not required, and therefore the conditions can be extremely mild [35].

Since 1,5-dienes can be cyclized by palladium salts via σ,π-complexes [36], it appears probable that similar intermediates are involved in the carbohydrate cyclizations.

4
Use of 6-Substituted Hex-5-enose Derivatives

A very significant development, particularly with respect to the synthesis of the inositols and their biologically important derivatives, came with the finding that the metal-ion-promoted cyclization is very effective with 5-enes that retain an oxygen substituent at C-6. Such compounds are best prepared from the corresponding glycosides that have a formyl-substituted C-5, and the derived enol acetates are the best-known members. Treatment of compound **35**, derived from aldehyde **34**, with mercury(II) trifluoroacetate in aqueous acetone affords the inosose **36** in 57% yield (together with the epimer at the hydroxy centre in the ratio 6:1), and reduction of the major product with sodium triacetoxyborohydride gives the *myo*-inositol derivative **37** stereospecifically (Scheme 7) [37]. The clear value of such compounds for the preparation of specific inositol phosphate derivatives was concurrently illustrated by the synthesis of a D-*myo*-inositol 1,3,4,5-tetrakisphosphate derivative [38].

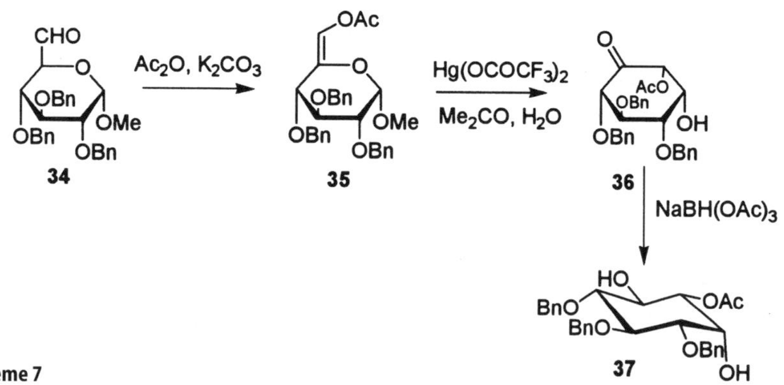

Scheme 7

Although the mercury(II) salt method can be used satisfactorily for the cyclization [39], comparative experiments have shown that the palladium-salt-based procedure can be more effective in giving cyclohexanes, although all four products are formed [40]. Thus the latter method converts compound **35** into **36** (49%) together with the other *cis*-related product (24%) and the *trans*-isomers (17 and 10%) in 77% overall yield, while mercury trifluoroacetate gives only the *cis*-products, but in poor yield – at least under the conditions used. The stereochemistry of the alkene in compounds such as **35** does not affect the product ratios. In the case of the L-*arabino* C-4 epimer of alkene **35** (derived from methyl α-D-galactopyranoside), the favoured products are the two inososes with the acetoxy groups at the new chiral centre *cis* to the benzyloxy group at C-4 of the starting compound, and the hydroxyl groups *cis* or *trans* to the acetates. Again use of palladium chloride results in all four products, while mercury(II) trifluoroacetate affords only two; in this case, the yield using the latter reagent is not anomalously low at 75% [40]. From the D-mannose-derived alkene (epimer of **35** at C-2), the only product formed with either reagent is the analogous epimer of compound **36** [40].

Ketone **36** and its diastereomers are reduced to the corresponding alcohols – in high yield and sometimes with dramatically different and useful selectivities [41]. For example, while tetramethylammonium triacetoxyborohydride gives the same *myo*-inositol **37** from **36** as does the analogous sodium salt with >99% selectivity, sodium borohydride effects the reduction with similar stereoefficiency, but in the opposite sense, to give the *allo*-inositol epimer, the products being formed in high yields and retaining their chirality. The method therefore is of considerable value for making many specifically substituted inositol derivatives; a naturally occurring membrane phosphoinositide, for example, was made by reduction of the carbonyl group of a compound closely related to **36** followed by protecting group manipulation and phosphorylation [39a].

Little is known of the reasons behind the stereochemistry observed in the Hg(II)- and $PdCl_2$-induced carbocyclizations, but some relevant observations were made during the first study of the cyclization of 6-acetoxy-5-enes [37]. At least from compounds with the D-*xylo* configuration (e.g. **35**), and in keeping with their 6-deoxy-analogues (Sect. 2.2), the main products have the α configuration (axial OH) at the new alcohol centre. On the other hand, the acetoxy group has the α (equatorial) orientation. Furthermore, mercury(II) salts apparently favour the formation of *cis*-products at the two new chiral centres [40]. When, however, the C-2–C-4 hydroxyl groups carry triethylsilyl substituents (e.g. compound **38**) the situation changes and the main product again has the hydroxy group and the acetoxy group *cis*-related to each other *but now both β*. A striking feature of the reaction is that the product **39** prefers the conformation with the silyloxy groups all axial, so again the new hydroxy and acetoxy groups are axial and equatorial, respectively (Scheme 8) [37]. This observation therefore parallels that noted with a 6-deoxy-2,4-di-*O*-silylated analogue of compound **38** which anomalously gives 75% yield of the β-alcohol when subjected to the $PdCl_2$-catalyzed reaction [35]. It appears therefore that the use of silyl protecting groups in place of alkyl or acyl substituents

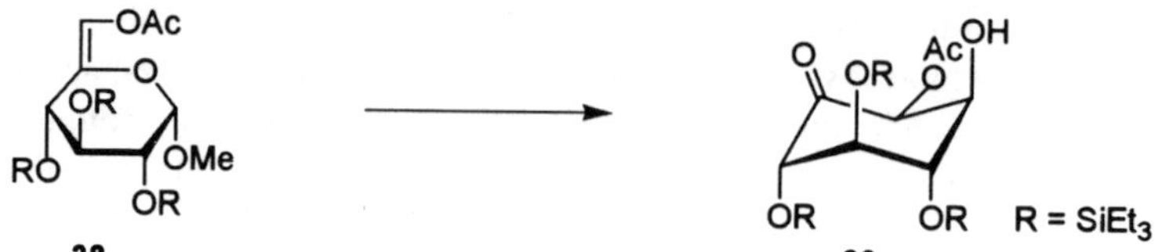

38 39

Scheme 8

may afford a means of altering the stereochemistry of the carbocyclization processes.[1]

An extension of the reaction of 6-acetoxy-5-enes leads to the use of alkenes with different substituents at C-6, a notable example being the extended chain sugar derivative **40** made from the D-galactose-based C-4 epimer of aldehyde **34** by an aryllithium addition reaction. Treatment with mercury(II) chloride in aqueous acetonitrile gives the expected alcohols in 75% yield, and these on acetylation then afford the enone **41** (70%) of relevance to the synthesis of pancreastatin (**42**) which shows promising anti-cancer properties [42]. A very similar reaction, involving a 6-*C*-phenyl-5-ene, but one in which the C-1 substituent is not lost, is covered in the following section.

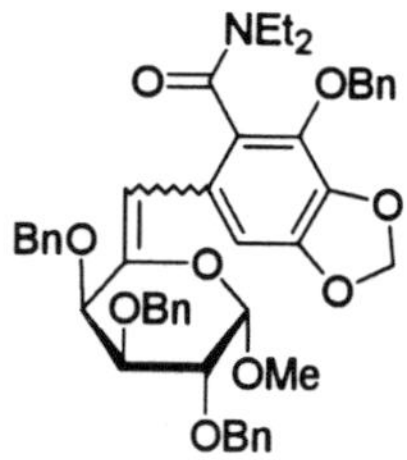

40

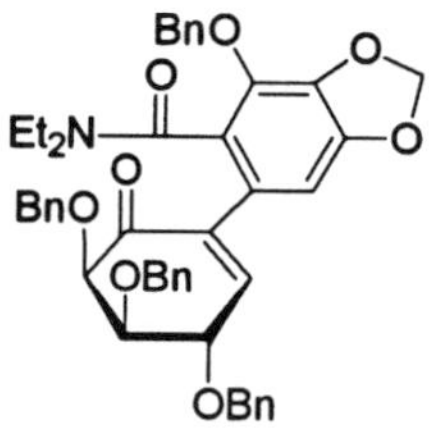

41

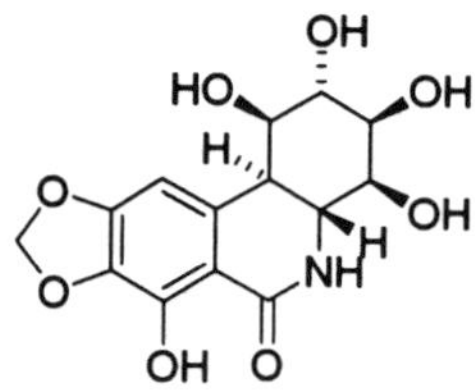

42

5
Use of 6-Deoxy- and 6-Substituted Hex-5-enose Derivatives with Titanium, Aluminium and Grignard Reagents

Recognition of the vinyl acetal character of hex-5-enopyranosides and of the convertibility of such compounds by reductive rearrangement to carbocyclic products by treatment with triisobutylaluminium led to the modified route to cyclohexanes from 6-deoxyhex-5-enoside derivatives illustrated in Scheme 9 [43]. In this way compound **43** is converted in 79% yield to the deoxyinositol **44**, there being two important differences relative to the mercury(II)-promoted ring closure: the anomeric centre substituent is retained, and specific reduction is effected at the carbonyl centre. From the β-anomer of compound

[1] For related examples of conformational inversion of pyranoid rings by the introduction of bulky silyl ether groups see Shuto S, Terauchi M, Yahiro Y, Abe H, Ichikawa S, Matsuda A (2000) Tetrahedron Lett 41:4151.

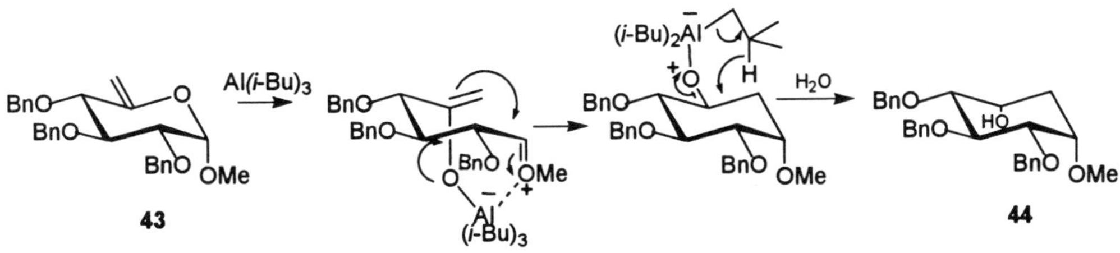

Scheme 9

43 the C-1 epimer (carbohydrate numbering) of **44** is obtained in 70% yield together with 10% of the all-equatorial compound and 6% of **44**, these minor products showing that the 6-*exo-trig* ring-closure step and the carbonyl reduction are not altogether specific. The main products derived from the analogous enes made from methyl β-D-galactopyranoside and methyl α-D-mannopyranoside have retained configuration at C-1 and *cis*-related hydroxy, methoxy groups [43]. Applied in the maltose and isomaltose series the method gives pseudo-disaccharide derivatives **45** and **46** in 54 and 65%, respectively [44].[2]

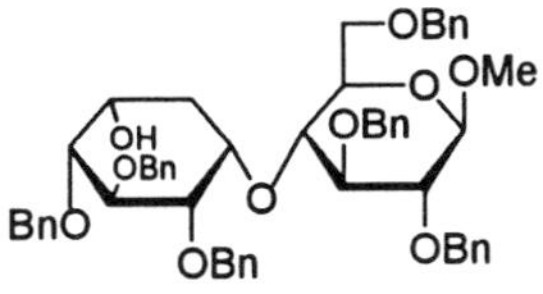

45

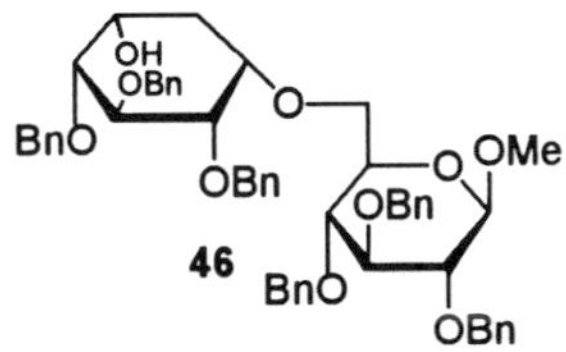

46

A further ring-rearrangement reaction, which is intermediate in effect between the mercury(II)- and triisobutylaluminium-promoted methods in that it gives cyclohexanones with retained aglycons, is dependent on the use of the mild Lewis acid isopropyltitanium trichloride. From 5-enoside **43** it gives the ketone derivative **47** (Scheme 10) with retained stereochemistry at the methoxylated carbon atom in near quantitative yield. Similar results are obtained with the analogous 5-enes derived from methyl α-D-galactopyranoside and -mannopyranoside, but mixed products arise from the corresponding β-glycosides [45].

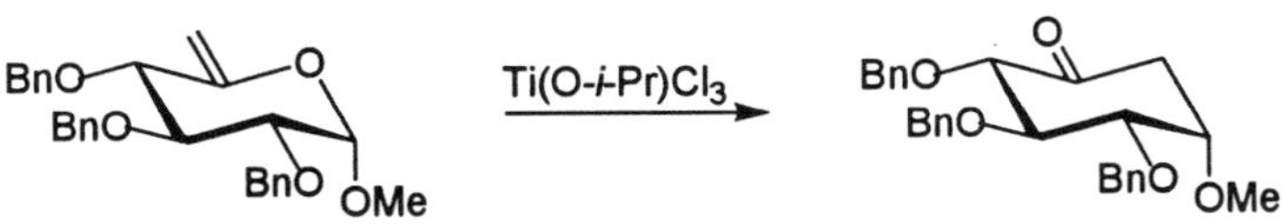

Scheme 10 **43** **47**

[2] For important extensions of this reaction see Sollogaub M, Mallet J-M, Sinaÿ P (2000) Angew Chem Int Ed Engl 39:362 and Sollogaub M, Pearce AJ, Hérault A, Sinaÿ P (2000) Tetrahedron: Asymmetry 11:283.

In agreement with the results obtained with the titanium alkoxide, titanium tetrachloride converts the 6-substituted alkene orthoester **49** into the expected cyclohexanone **48** in 88% yield, and Grignard reagents promote the carbocyclization reaction and then react with the carbonyl functions of the products. In this way compound **49**, on treatment with phenylmagnesium bromide, affords the diphenyl product **50** that has both of the phenyl groups equatorial (Scheme 11) [46].

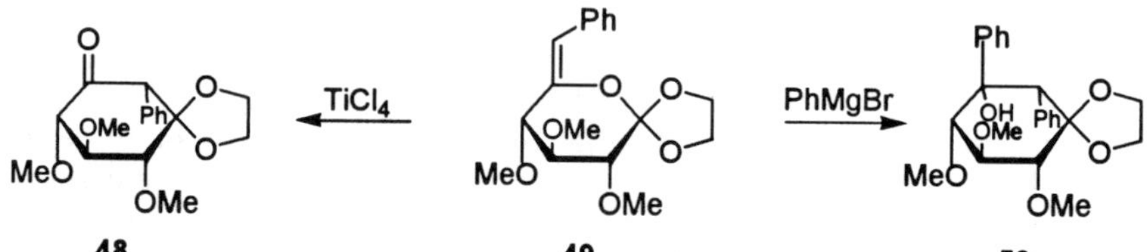

Scheme 11

6 References

1. Ferrier RJ (1979) J Chem Soc Perkin Trans 1:1455
2. Ferrier RJ, Prasit P (1980) Carbohydr Res 82:263
3. (a) Chen C H-J, Eisenberg F (1975) J Biol Chem 250:2963; (b) Billington D C (1993) The inositol phosphates – chemical synthesis and biological significance. VCH, New York, p 6
4. (a) Ganem B (1978) Tetrahedron 34:3353; (b) Knowles JR (1989) Aldrichim Acta 22:59
5. Bartlett PA, Satake K (1988) J Am Chem Soc 110:1628
6. Bartlett PA, McLaren KL, Marx MA (1994) J Org Chem 59:2082
7. Ferrier RJ, Middleton S (1993) Chem Rev 93:2779
8. (a) Blattner R, Ferrier RJ, Haines SR (1985) J Chem Soc Perkin Trans 1 2413; (b) Dubreuil D, Cleophax J, De Almeida MV, Verre-Sebrié C, Liaigre J, Vass G, Gero SD (1997) Tetrahedron 53:16747; (c) Yamauchi N, Terachi T, Eguchi T, Kakinuma K (1994) Tetrahedron 50:4125
9. Mádi-Puskás M, Pelyvás I, Bognár R (1985) J Carbohydr Chem 4:323
10. Chretien F, Chapleur Y (1984) J Chem Soc Chem Commun 1268
11. Chida N, Ohtsuka M, Ogura K, Ogawa S (1991) Bull Chem Soc Jpn 64:2118
12. Chida N, Ohtsuka M, Ogawa S (1991) Tetrahedron Lett 32:4525
13. Machado AS, Olesker A, Lukacs G (1985) Carbohydr Res 135:231
14. Takahashi H, Iimori T, Ikegami S (1998) Tetrahedron Lett 39:6939
15. Mereyala HB, Gaddam BR (1994) J Chem Soc Perkin Trans 1:2187
16. Machado AS, Olesker A, Castillon S, Lukacs G (1985) J Chem Soc Chem Commun 330
17. Machado AS, Dubreuil D, Cleophax J, Gero SD, Thomas NF (1992) Carbohydr Res 233:C5
18. Laszlo P, Pelyvas IF, Sztaricskai F, Szilagyi L, Somogyi A (1988) Carbohydr Res 175:227
19. Sato K-i, Sakuma S, Nakamura Y, Yoshimura J, Hashimoto H (1991) Chem Lett 17
20. Blattner R personal communication
21. Mereyala HBV, Guntha S (1993) J Chem Soc Perkin Trans 1 841
22. Sato K-i, Bokura M, Taniguchi M (1994) Bull Chem Soc Jpn 67:1633
23. Suami T, Ogawa S (1990) Adv Carbohydr Chem Biochem 48:21
24. Yuasa H, Palcic MM, Hindsgaul O (1995) Can J Chem 73:2190
25. Sato K-i, Bokura M, Moriyama H, Igarashi T (1994) Chem Lett 37
26. Wang L-X, Sakairi N, Kuzuhara H (1995) Carbohydr Res 275:33
27. Corbett DF, Dean DK, Robinson SR (1993) Tetrahedron Lett 34:1525

28. Tagmose TM, Bols M (1997) Chem Eur J 3:453
29. Boyer F-D, Lallemand J-Y (1994) Tetrahedron 50:10443
30. Chew S, Ferrier RJ (1984) J Chem Soc Chem Commun 911
31. Ermolenko MS, Olesker A, Lukacs G (1994) Tetrahedron Lett 35:711
32. Adam S (1988) Tetrahedron Lett 29:6589
33. Barton DHR, Augy-Dorey S, Camara J, Dalko P, Delaumény JM, Géro SD, Quiclet-Sire B, Stütz P (1990) Tetrahedron 46:215
34. Miyamoto M, Baker ML, Lewis MD (1992) Tetrahedron Lett 33:3725
35. Iimori T, Takahashi H, Ikegami S (1996) Tetrahedron Lett 37:649
36. Moberg C, Sutin L, Heumann A (1991) Acta Chem Scand 45:77
37. Bender SL, Budhu RJ (1991) J Am Chem Soc 113:9883
38. Estevez VA, Prestwich GD (1991) J Am Chem Soc 113:9885
39. (a) Peng J, Prestwich GD (1998) Tetrahedron Lett 39:3965; (b) Chen J, Feng L, Prestwich GD (1998) J Org Chem 63:6511; (c) Jia ZJ, Olsson L, Fraser-Reid B (1998) J Chem Soc Perkin Trans 1 631
40. Takahashi H, Kittaka H, Ikegami S (1998) Tetrahedron Lett 39:9703
41. Takahashi H, Kittaka H, Ikegami S (1998) Tetrahedron Lett 39:9707
42. Park TK, Danishefsky SJ (1995) Tetrahedron Lett 36:195
43. Das SK, Mallet J-M, Sinaÿ P (1997) Angew Chem Int Ed Engl 36:493
44. Pearce AJ, Sollogoub M, Mallet J-M, Sinaÿ P (1999) Eur J Org Chem 2103
45. Sollogoub M, Mallet J-M, Sinaÿ P (1998) Tetrahedron Lett 39:3471
46. Collins DJ, Hibberd AI, Skelton BW, White AH (1998) Aust J Chem 51:681

Claisen Rearrangements in Carbohydrate Chemistry

Barbara Werschkun, Joachim Thiem

Institut für Organische Chemie, Universität Hamburg, Martin-Luther-King-Platz 6, 20146 Hamburg, Germany
E-mail: thiem@chemie.uni-hamburg.de

The application of the Claisen rearrangement in carbohydrate chemistry has proven a valuable method for the realization of complex synthetic schemes. Apart from some interesting findings of a more principal nature, which will also be highlighted, many of the presented examples have culminated in the synthesis of novel carbohydrate mimetics or important representatives of other classes of natural products. For each case, the respective synthetic context will be given and most of the target structures are shown to give an idea of the variety and complexity of molecular frameworks that are accessible with this reaction. The material is organized into sections with respect to which portion of the Claisen system is of carbohydrate origin: the allylic alcohol moiety, the enol ether functionality, or both. Particular emphasis is placed on the discussion of practical issues such as synthetic accessibility of the respective starting materials, frequently encountered experimental problems, generality of the approach, and potential extension of the synthetic utilization.

Keywords: Carbohydrates, Rearrangements, Enol ethers, Ketene acetals, Chiral building blocks

Topics in Current Chemistry, Vol. 215

1 Introduction

Within the large and diverse field of synthetic carbohydrate chemistry, the stereocontrolled formation of carbon–carbon bonds represents an important and at times demanding task. This problem has been widely addressed in connection with the total synthesis of natural products and the utilization of sugars as versatile chiral starting materials [1]. In the closer context of carbohydrate research, extensive efforts have been directed towards the synthesis of *C*-glycosides [2], carbasugars [3], and branched-chain derivatives [4], both as naturally occurring compounds and as potential mimetics for biologically active glycostructures. Substantial difficulties are frequently imposed on such endeavors by the structural complexity of saccharides with their high density of reactive oxygen functionalities.

The Claisen rearrangement [5], the intramolecular reaction of allyl enol ethers **1** to γ,δ-unsaturated carbonyl compounds **3**, has become a valuable tool for organic synthesis [6]. The sigmatropic process allows a significant alteration of the molecular framework within a single step. The concerted mechanism involves a highly organized transition state **2** that often directs the stereochemical course in the reaction of substituted derivatives and enables the simultaneous formation of two asymmetric centers (Scheme 1).

Participation of an oxygen functionality renders the reaction particularly interesting for an application to carbohydrate substrates. Major disadvantages of the original Claisen rearrangement are the very high reaction temperatures and the limited synthetic access to enol ethers as starting materials. A number of modifications have been introduced to overcome these problems. Those that will also be encountered in many of the examples presented later in this chapter are briefly outlined in the following. A procedure developed by Eschenmoser et al. [7] starts from an allylic alcohol **4** and generates first a mixed amide acetal **5a** to provide after alcohol elimination the ketene *N,O*-acetal **6a** which can undergo sigmatropic rearrangement to the unsaturated amide **7a** (Scheme 2). The method introduced by Johnson et al. [8] utilizes orthoesters **5b** for the forma-

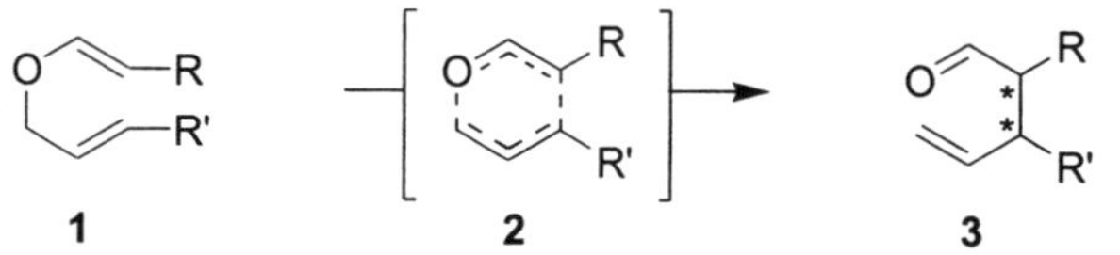

Scheme 1. The Claisen rearrangement [5, 6]

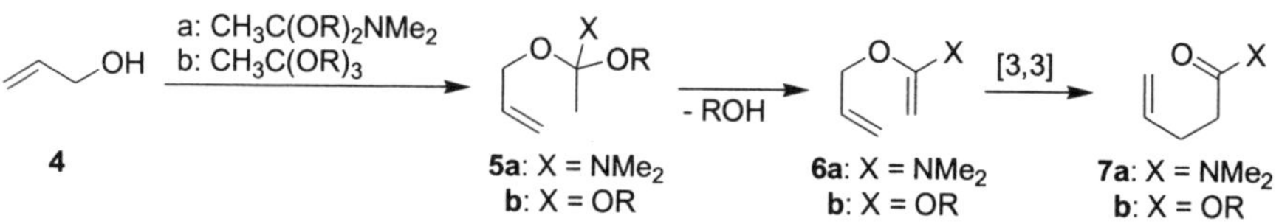

Scheme 2. The Eschenmoser–Claisen and Johnson–Claisen rearrangements [7, 8]

tion of ketene acetals **6b** and rearrangement of the latter to unsaturated esters **7b** in a very similar fashion.

Both processes provide an experimentally simple means for the introduction of a C_2 or C_3 fragment. However, they still require reaction temperatures of 140–200 °C and suffer from the restricted availability of orthoesters and amide acetals, respectively. A more general solution is offered by the methodology developed by Ireland et al. [9]. Deprotonation of an allyl ester **8** and immediate *O*-silylation of the enolate **9** furnishes a silyl ketene acetal **10**, which is less prone to undergo unwanted side reactions, as compared to the enolate anion itself, but which still rearranges easily to the silyl ester **11** at temperatures often below 60 °C (Scheme 3).

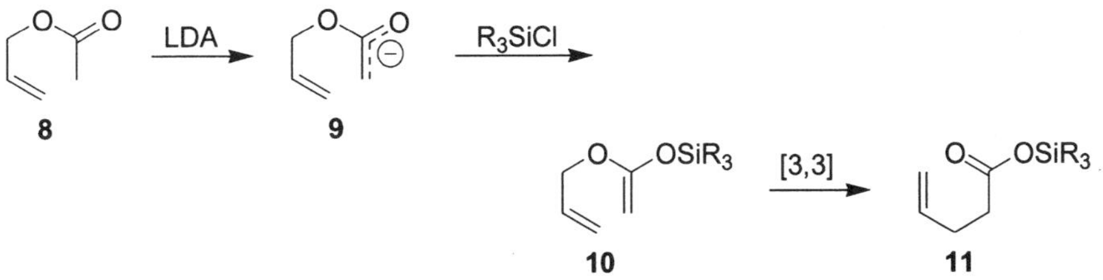

Scheme 3. The Ireland–Claisen rearrangement [9]

A major advantage of the Ireland reaction lies in the employment of allyl esters as starting materials which are readily prepared in virtually unlimited structural variety. Moreover, during the enolization step, the choice of reaction conditions permits control over double-bond geometry in the case of substituted substrates (Scheme 4). While under kinetic control formation of the (*Z*)-enolate (*Z*)-**14** is generally favored, its isomer (*E*)-**14** can be obtained by the addition of a chelating agent such as HMPA [10].

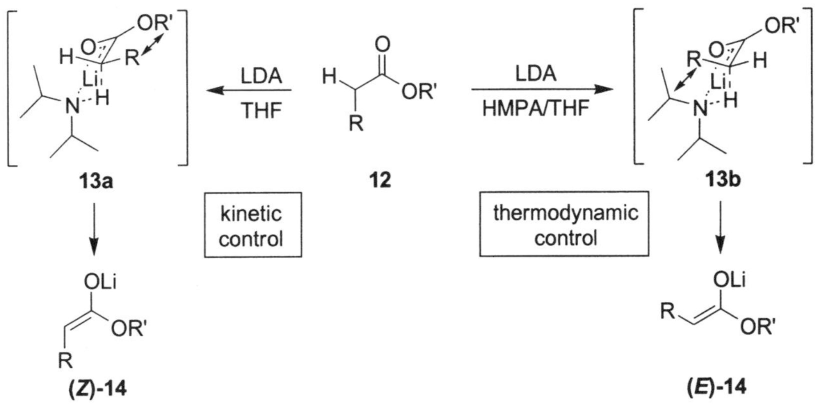

Scheme 4. Control of enolate geometry [10]

2 Carbohydrate-Derived Allylic Alcohols

2.1 Glycals

Glycals [11] are by far the most common unsaturated carbohydrate derivatives and have proven valuable starting materials in a large variety of synthetic endeavors. In a program directed towards the stereoselective synthesis of *α*-*C*-glycopyranosides, Fraser-Reid et al. explored the sigmatropic rearrangement of reactive intermediates derived from 4,6-*O*-benzylidene-D-allal **15** [12]. The study comprised investigation of the original enol ether Claisen rearrangement, as well as several synthetic modifications (Scheme 5). A major problem originated from the general high reactivity of glycals towards nucleophiles leading to allylic substitution. In an attempt to apply the Johnson orthoester procedure, the starting material thus reacted with the catalyst propionic acid in an unwanted fashion to give the unsaturated esters **16** and **17**, while no Claisen rearrangement could be observed.

The alternatively explored thermal rearrangement of vinyl ether **18** avoided such side reactions; however, in this case, a high sensitivity towards the reaction temperature became apparent. After no reaction could be achieved at 145 °C, an

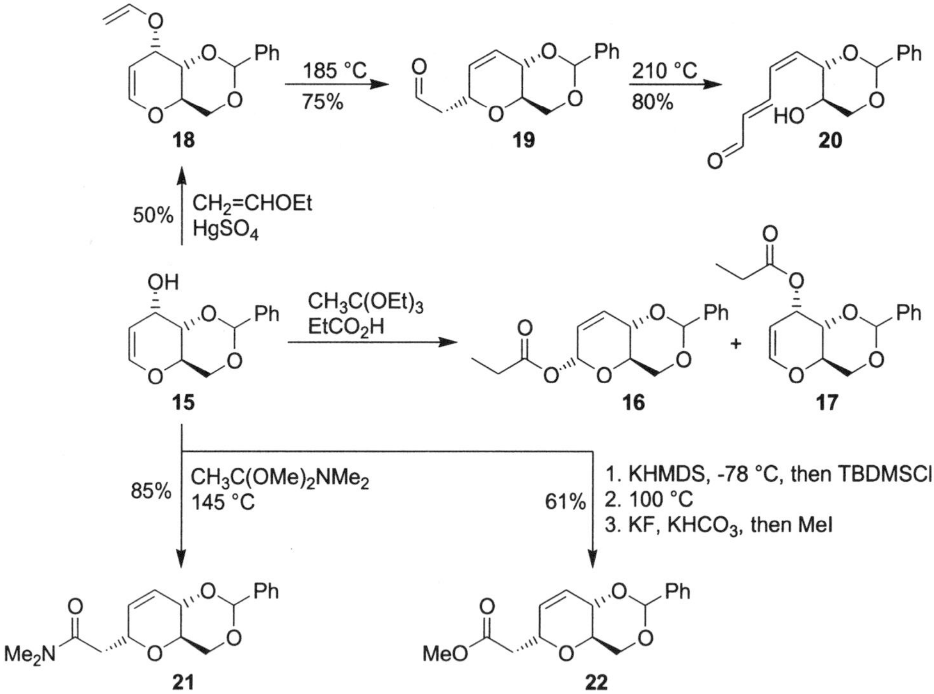

Scheme 5. The first glycal Claisen rearrangements by Fraser-Reid [12] and Ireland [13]

increase to 210 °C induced ring opening of the primarily formed rearrangement product **19** to give the dienic aldehyde **20.** At 185 °C, **19** was finally obtained as the sole product in good yield, but the overall process suffered substantially from the low conversion of only 50% during the formation of enol ether **18** from the allylic alcohol starting material. Eventually, the most satisfying results were reported for the Eschenmoser amide acetal procedure, which afforded the rearranged dimethyl amide **21** in a high yield of 85%.

The same D-allal starting material was also subjected to the silyl ketene acetal rearrangement protocol by Ireland et al. [13]. After several work-up steps, the rearrangement product, methyl ester **22**, could be obtained in 60% yield, but the rearrangement was reported to be unusually slow, probably due to steric restrictions of the bicyclic framework. The reaction required prolonged heating at 100 °C to proceed to completion, thus lacking the particular advantages of short rearrangement times and low temperatures normally associated with the Ireland procedure. This method was applied more successfully to less-constrained starting structures (Scheme 6). Both pyranoid and furanoid glycal propionates **23**–**26**, after conversion to the silyl ketene acetals, rearranged smoothly to the *C*-glycosidic acids **27–30**, with particularly fast reaction rates being observed for the furanoid series ($t_{1/2}$ = 3 min at 35 °C). Significant stereoselectivities were obtained during the formation of the new asymmetric center within the *C*-glycosidic side chain. The product ratio reached values of around 4:1 and could be reversed by changing the solvent system from pure THF to a THF/HMPA mixture.

As the absolute configurations were not determined, no conclusion could be drawn about the involved transition state until a later study, where, in an elegant series of experiments, the glycal-derived rearrangement products **33**, **36**, and **39** were transformed to chiral target molecules of known stereochemistry [14]. Correlation with the (–)- and (+)-nonactic acids **34** and **37** and the Prelog-

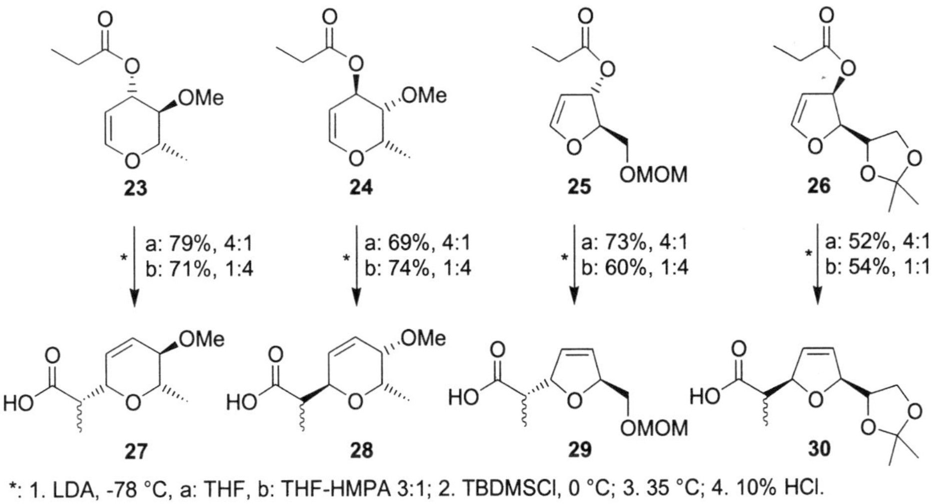

Scheme 6. Ireland–Claisen rearrangement of pyranoid and furanoid glycals [13]

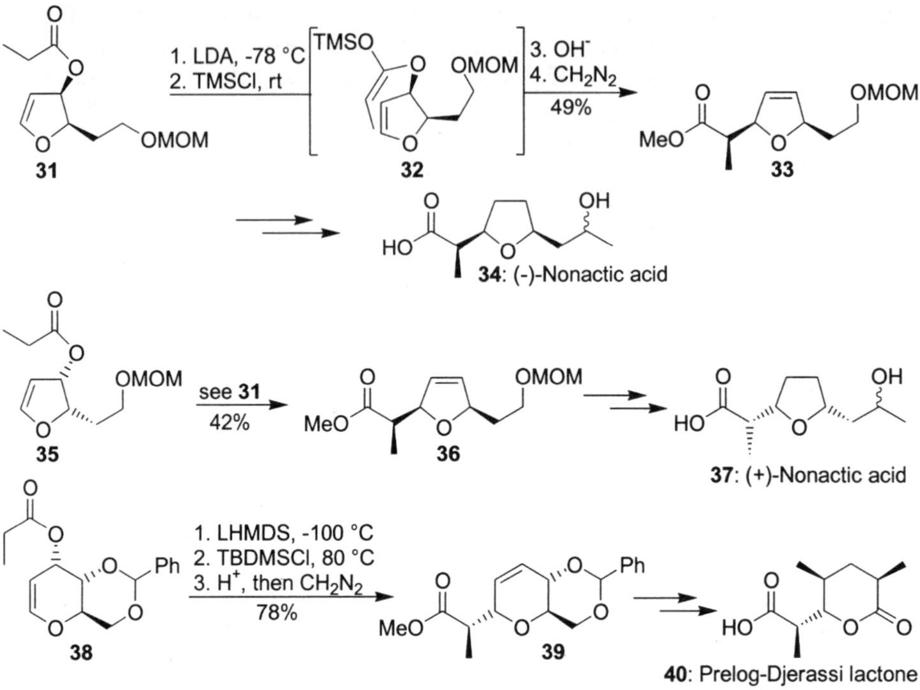

Scheme 7. Assignment of the absolute configuration by conversion to chiral target molecules revealed a boatlike transition state for glycal Claisen rearrangements [14]

Djerassi lactone **40**, respectively, revealed the absolute configurations formed during Claisen rearrangement. In this way, it could be demonstrated that the transition state in all cases adopts a boatlike conformation as is shown for the (*E*)-ketene acetal **32** in Scheme 7.

Having thus established the main stereochemical features, the Claisen rearrangement of glycal esters could be further utilized to provide key intermediates for a number of natural product syntheses: Carbohydrates served as starting materials in the total synthesis of the ionophore antibiotics lasalocid A **43** by Ireland et al. [15] and indanomycin **46** by Ley et al. [16] (Scheme 8), as well as the 3-acyl tetramic acid antibiotics tirandamycic acid **49** [17] and (+)-streptolic acid **52** [18] (Ireland et al., Scheme 9). Several further examples have been reported in the literature [19–21].

Again, in all studies, rearrangement through a boatlike transition state was observed. This seems remarkable considering the general energetic stabilization of the chair conformation for six-membered cyclic assemblies. An underlying rationale, hinting at the glycal ring oxygen atom as the major directing factor, was later proposed. Its guiding principles, however, are better understood against the background of related studies reported by Curran et al. [22]. The bis(ketene acetal) **54** (Scheme 10), derived from di-*O*-acetyl-L-arabinal **53**, upon heating to 60 °C underwent Claisen rearrangement to provide, after work-up, the unsaturated acid **56** as the sole reaction product. However, the primarily formed

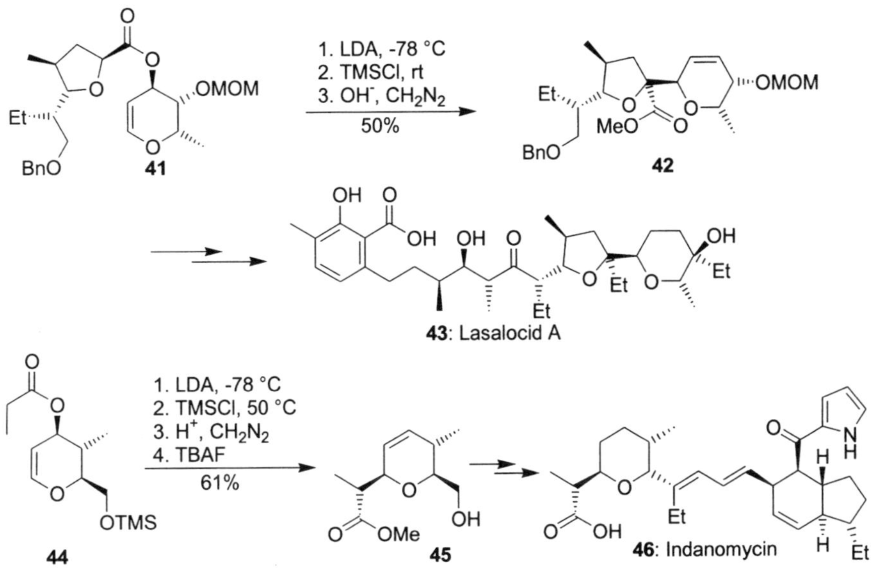

Scheme 8. Glycal Claisen rearrangement products as intermediates for the total synthesis of ionophore antibiotics [15, 16]

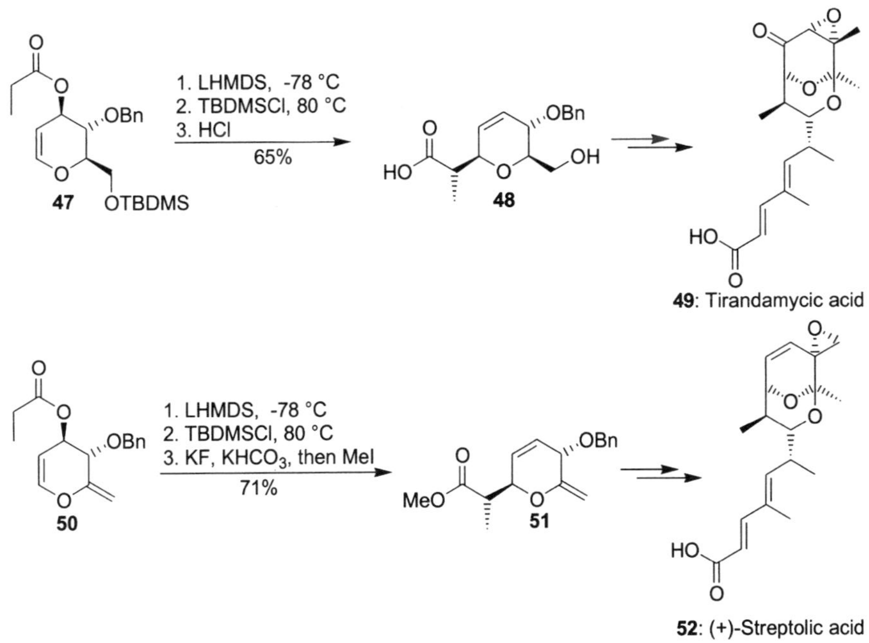

Scheme 9. Glycal Claisen rearrangement products as intermediates for the total synthesis of 3-acyltetramic acid antibiotics [17, 18]

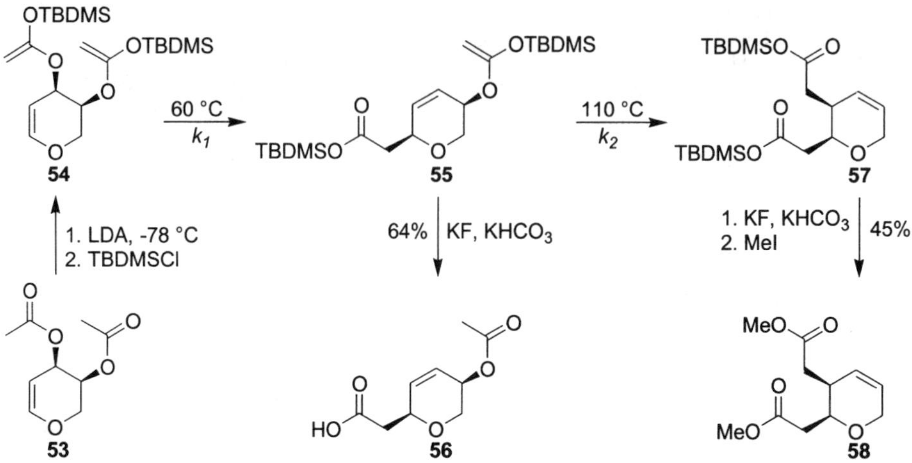

Scheme 10. Chemoselective mono- vs. di-Claisen rearrangement of a glycal bis(ketene acetal) [22]

silyl derivative **55** still contained an allyl ketene acetal moiety and, by heating the crude reaction mixture to 110 °C, a second sigmatropic rearrangement could be initiated to give bis(silyl ester) **57**, isolated as its dimethyl derivative **58**.

Kinetic measurements of the reaction rates k_1 and k_2 for the individual rearrangement steps revealed that the first Claisen rearrangement proceeded 20 times faster than the second one. This unexpected chemoselectivity proved useful for the preparation of key intermediate **59** in the synthesis of pseudomonic acid C **60** (Scheme 11).

Similar selective mono-Claisen rearrangements could be achieved with the ketene acetals derived from glycal di- and triacetates **61**–**64** (Scheme 12) with the ratios k_1/k_2 reaching values of 35–575.

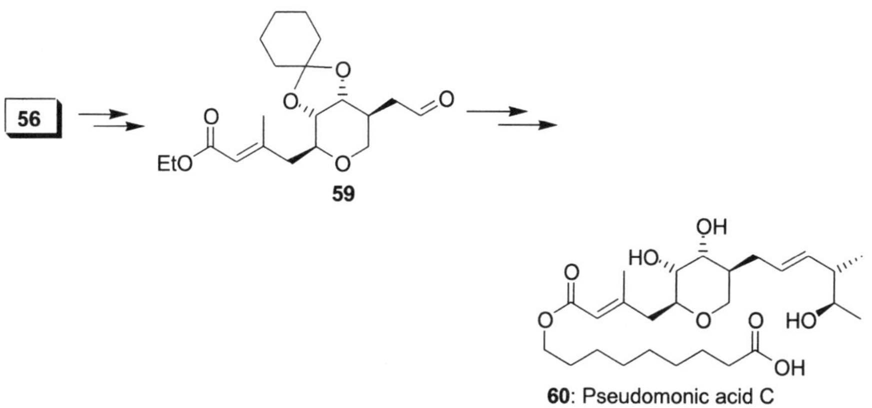

Scheme 11. Utilization of the mono-Claisen product for the total synthesis of pseudomonic acid C [22]

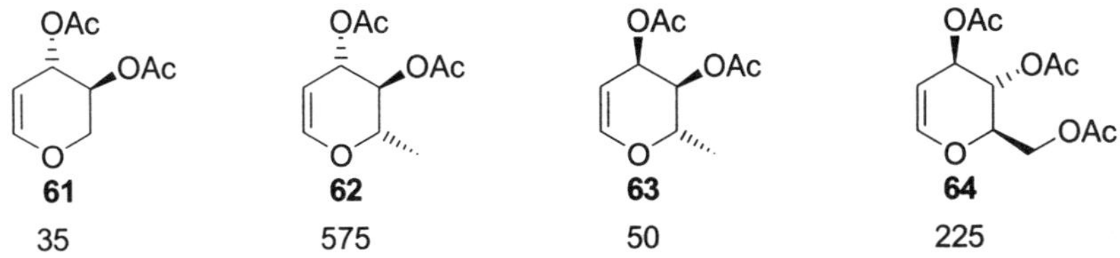

Scheme 12. Ratio of reaction rates for mono- and di-Claisen rearrangements of glycal bis(ketene acetals) [22]

The main structural difference between the substrates in both rearrangement steps is provided by the ring oxygen atom, which only in the first case substitutes the allylic double bond involved in the sigmatropic rearrangement and might display an accelerating influence. With an exchange of the oxygen atom for a methylene group the sharp distinction between both rearrangements should disappear. This was indeed observed for the reaction of cyclohexenyl derivative **65**, which showed reaction rates k_1 and k_2 of virtually the same value, inevitably leading to a mixture of mono- and di-Claisen products **66** and **67** (Scheme 13). Further evidence for a decisive effect arising from oxygen substitution was provided by direct comparison of the reaction rates of glycal **68a** and its carbocyclic counterpart **68b**, which showed a difference in the order of one magnitude.

The studies of such closely related structures also ruled out any possible steric effects and the driving influence for reaction rate enhancement has to be seen in the oxygen atom in the γ-allylic position (C6 of the Claisen system, Scheme 14). In previous reports [23], Carpenter and Burrows had developed a model to predict the influence of substituents on various pericyclic reactions based on Hückel orbital energy calculations. According to this approach, a π-

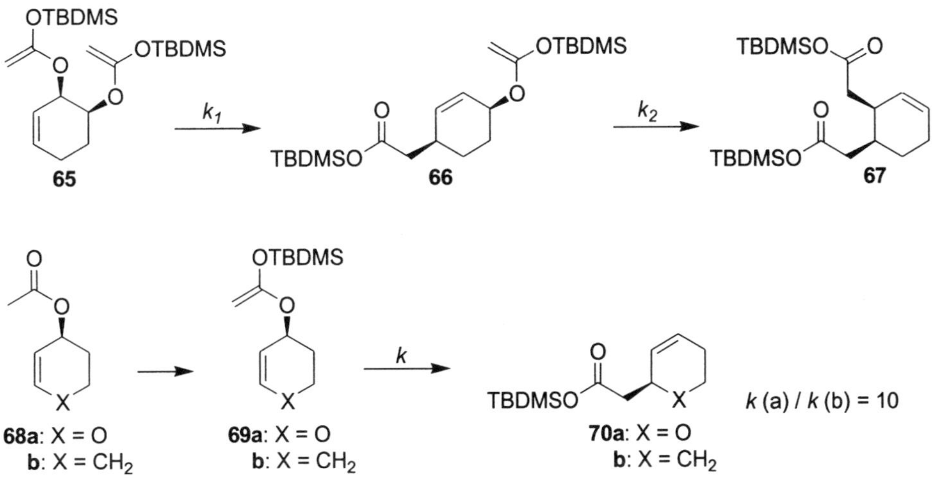

Scheme 13. Unselective mono- and di-Claisen rearrangement of a carbocyclic bis(ketene acetal) and different reaction rates for a glycal ketene acetal and its carbocyclic counterpart [22]

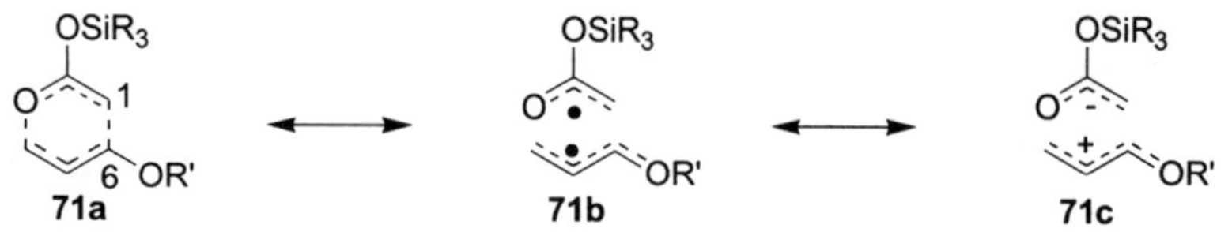

Scheme 14. Resonance structures for the polar transition state during Claisen rearrangement [24, 25]

donor substituent in the 6-position should lead to a loss of resonance energy in proceeding from the reactant to the transition state and thus effect a deceleration of the rearrangement. The transition state itself had been shown to be of a rather polar nature with bond breaking considerably more advanced than bond formation [24], which can be described by a diradical resonance structure **71b** (Scheme 14). With the silyloxy group in position 2, the polarity will be increased even further [25] and an ionic resonance structure **71c** also has to be considered. Given these circumstances, it can be seen how an additional electron-donating substituent at C6 might lower the transition state energy by stabilizing the "allyl cation" substructure C4 to C6 and thus accelerate the reaction by facilitating cleavage of the oxygen–carbon bond.

The prevailing principle has been termed "vinylogous anomeric effect" as a molecular orbital rationalization may be provided by a $\pi \rightarrow \sigma^*$ interaction in analogy to the anomeric effect [26]. Moreover, the vinylogous anomeric effect, which was independently introduced by Denmark et al. to explain the unusual diaxial conformation of cyclic α-chloroketoximes [27], also seems to account for conformational preferences of certain glycals in the ground state [22c].

This stereoelectronic effect was also suggested as the driving force behind the predominating occurrence of boatlike transition states observed in glycal ketene acetal rearrangements (vide supra). A comprehensive study by Ireland et al. [28] compared the stereochemical behavior of pyranoid, furanoid, and acyclic glycal starting materials with the respective all-carbon analogs. The propionates **72** were selectively converted to the (*E*)- or the (*Z*)-ketene acetals **73**, which rearranged to the methyl-branched acids **78** in varying diastereomeric ratios (Scheme 15). Cyclohexenyl derivative (*E*)-**73a** gave acid (*S*)-**78a** as the major product, thereby following the usual course through a chairlike transition state **74a**. From isomer (*Z*)-**73a**, the same (*S*)-configured product was formed, which in this case required a boatlike transition state **77a**. This was readily explained by the large difference in steric interactions for the two transition state structures in this particular example. The glycal-derived ketene acetals **73b**, however, in both cases preferred rearrangement through a boatlike transition state **75b** and **77b**, respectively, which could not be sufficiently accounted for by steric reasons.

A similar picture was obtained from the cyclopentenyl/furanoid glycal series of starting materials. The carbocyclic compounds showed the expected preference for a chairlike transition state except for situations with extreme steric hindrance, whereas all of the investigated glycals favored a boatlike transition state irrespective of steric influences. Two stereoelectronic explanations were

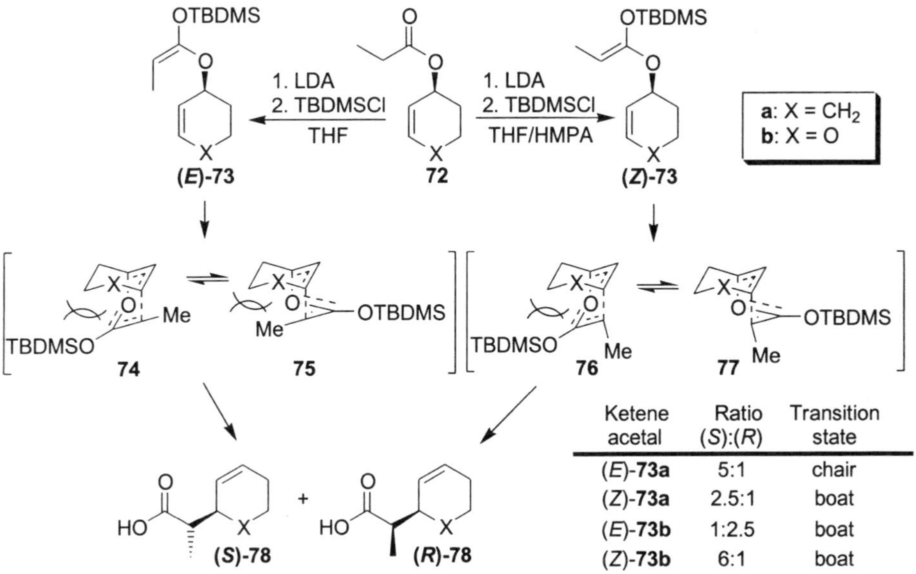

Ketene acetal	Ratio (S):(R)	Transition state
(E)-**73a**	5:1	chair
(Z)-**73a**	2.5:1	boat
(E)-**73b**	1:2.5	boat
(Z)-**73b**	6:1	boat

Scheme 15. Different transition state geometries for the Claisen rearrangement of cyclohexenyl and glycal ketene acetals [28]

suggested: The glycal oxygen atom stabilizes the dipolar transition state in either conformation; however, the boat form, which is assumed to have a "looser" geometry with a higher degree of C–O bond cleavage [29], will be significantly more affected by the stabilizing donor substituent. Additionally, the Coulombic attraction between the two ends of the dipole induced by the vinylogous anomeric effect is maximized in the boatlike transition state as compared to the chair form. Energy calculations based on kinetic rate determinations indicated an increased relative stabilization of the boatlike transition state in the glycal series in the order of 1 – 2 kcal mol^{-1}.

Rearrangement of a glycal-derived *N,O*-ketene acetal was reported by Colombo et al. [30] during the synthesis of *C*-glycosyl α-amino acids (Scheme 16). Following the Steglich methodology [31] for the synthesis of γ-dehydro-α-amino acids, *N*-benzoylalanine ester **79** was dehydrated to an intermediate oxazole **80** that rearranged at room temperature to the oxazolone **81** as a 3:1 diastereomeric mixture. Further manipulations resulted in the synthesis of the two *C*-glycosyl amino acids **82** and **83**, the latter being obtained as the γ-lactone.

An enzyme-catalyzed Claisen rearrangement of a glycal derivative was presented in an interesting contribution by Berchtold et al. [32]. The sigmatropic rearrangement of chorismate **84** to prephenate **85** in the shikimate biosynthetic pathway (Scheme 17) is one of the most intriguing transformations found in nature and its mechanistic details are little understood [33]. In a program to investigate the structural requirements for chorismate mutase catalysis, 6-oxa-5,6-dihydrochorismate **86** was shown to be an excellent substrate for chorisma-

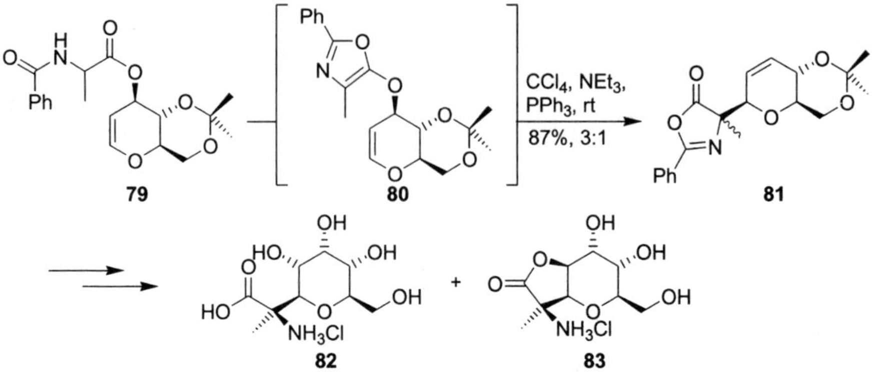

Scheme 16. Synthesis of *C*-glycosyl α-amino acids by glycal Claisen rearrangement [30]

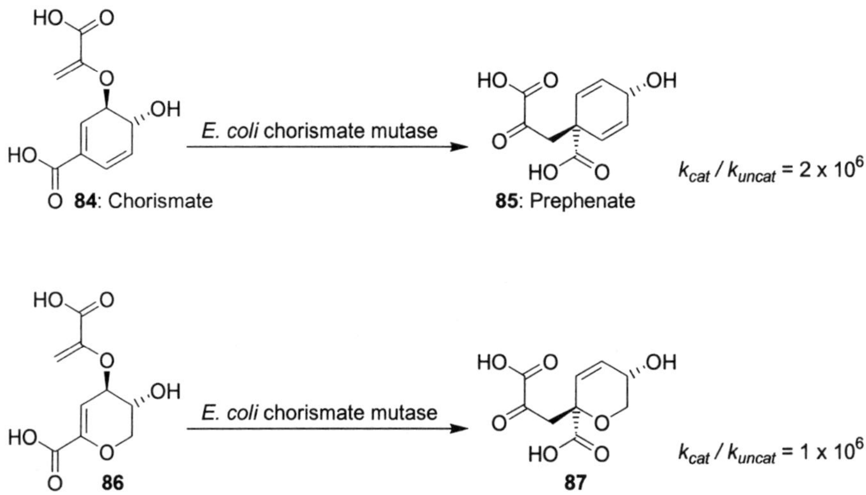

Scheme 17. Enzyme-catalyzed Claisen rearrangement of chorismate and a glycal structural analog [32]

te mutase-prephenate dehydrogenase from *Escherichia coli*. The rearrangement of glycal **86** to **87**, which without catalysis proceeded with a half-life of 1200 h at 30 °C, was accelerated by the enzyme by a factor $k_{cat}/k_{uncat} = 1 \times 10^6$, compared to $k_{cat}/k_{uncat} = 2 \times 10^6$ for the natural substrate.

2.2 Other Endocyclic Unsaturated Derivatives

2,3-Unsaturated glycosides, which are easily obtained by the reaction of glycals with nucleophiles [34], served as the starting materials for the first reported Claisen rearrangement in carbohydrate chemistry [35] by Ferrier and Vethaviyasar (Scheme 18). Heating vinyl ethers **88a** and **88b** at 185 °C resulted in the

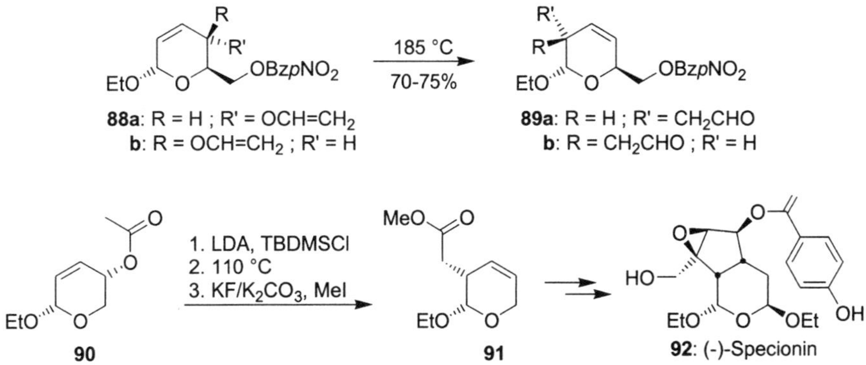

Scheme 18. Claisen rearrangement of 2,3-enopyranosides to C2-branched 3,4-enopyranosides [35, 36]

formation of the branched aldehydes **89a** and **89b** in 75 and 70% yield, respectively. The *threo*-compound **88b** rearranged significantly more rapidly than its *erythro*-analog **88a**, which was explained by the axial orientation of the migrating residue. Curran et al. applied the Ireland procedure to the structurally related acetate **90** to obtain unsaturated branched-chain ester **91**. This compound as well as similar others served as an intermediate in the synthesis of (–)-specionin **92** and related members of the iridoid family of natural products [36].

Rearrangements involving the anomeric position were reported by Heyns and Hohlweg [37], Descotes et al. [38], and de Raadt and Ferrier [39] for the propenyl and vinyl glycosides **93a** and **93b**, respectively (Scheme 19). Chandrasekhar et al. [40] investigated a series of analogous phenyl glycosides **95** and **97** and found remarkable differences in the reaction rates of α- and β-anomers. In this only example of an aromatic Claisen rearrangement in carbohydrate chemistry to date, sigmatropic rearrangement was followed by tautomerization of the initially formed cyclohexadienones to give phenols **96** and **98** as the isolated products.

Claisen rearrangement of 3,4-unsaturated glycosides to C4-branched derivatives provided useful intermediates for the total synthesis of thromboxanes (Scheme 20). First reported independently by Corey et al. [41] and Hernandez [42], both the Eschenmoser amide acetal and the Johnson orthoester procedure afforded good results for the conversion of allylic alcohol **99** (R^1 = Me, R^2 = H) to the respective amide or ester **100**. Corey et al. further transformed the former compound to lactone **101**, previously described as a precursor for the total synthesis of thromboxane B_2 (**102**). A number of further derivatives **100** (R^1 = allyl, R^2 = H or TBDMS, R^3 = NMe_2 or OEt) and its C2-epimer were prepared in a similar manner during later thromboxane synthetic studies [43, 44].

The same conversion of **99** to **100** was also utilized by Fleet et al. [45] for the enantiospecific synthesis of (*S*)-quinuclidinol (**103**, Scheme 21). White et al. [46] employed the thermal rearrangement of the related vinyl ether **104** to aldehyde **105** for the total synthesis of monic acid C (**106**).

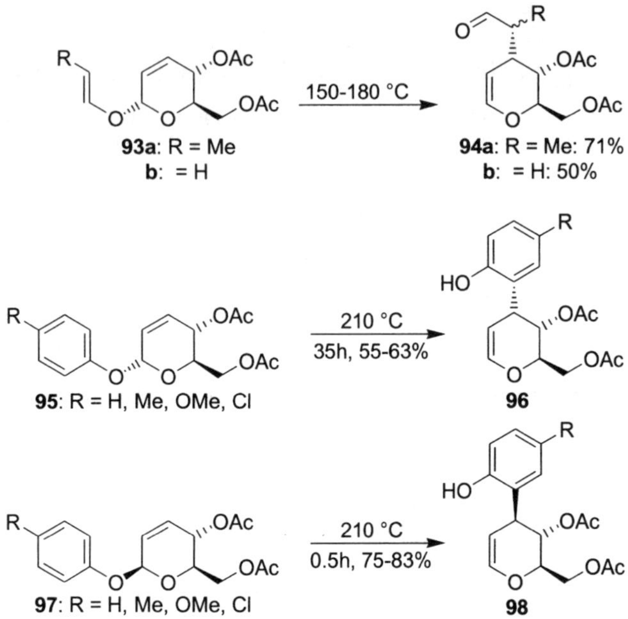

Scheme 19. Claisen rearrangement of 2,3-enopyranosides to C3-branched glycals [37–40]

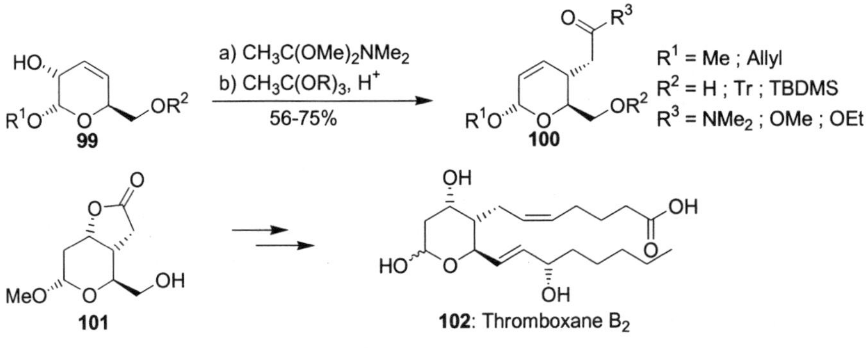

Scheme 20. Claisen rearrangement of 3,4-enopyranosides to C4-branched 2,3-enopyranosides and application for the total synthesis of thromboxanes [41–44]

Aiming at the total synthesis of the antibiotic and antitumor agent nogalomycin, Vatele [47] studied the Claisen rearrangement of 4,5-unsaturated glycosides **107** and **109** as an efficient method for the stereoselective formation of a quarternary chiral center (Scheme 22). The *α,β*-unsaturated ester **110** was formed by elimination of the sulfinyl group subsequent to rearrangement of ketene acetal **109**.

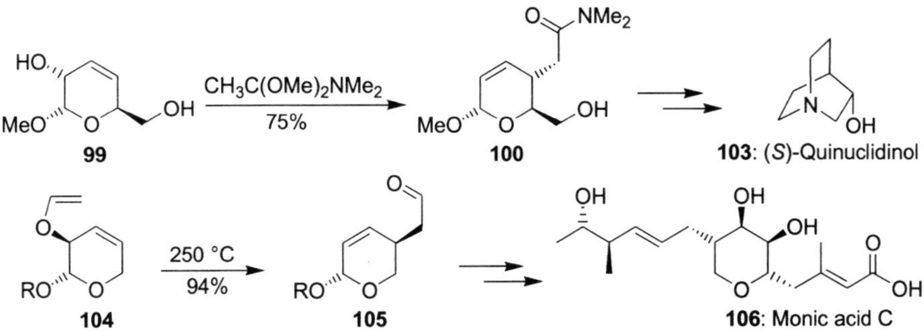

Scheme 21. Claisen rearrangement of 3,4-enopyranosides to C4-branched 2,3-enopyranosides and application for natural product syntheses [45, 46]

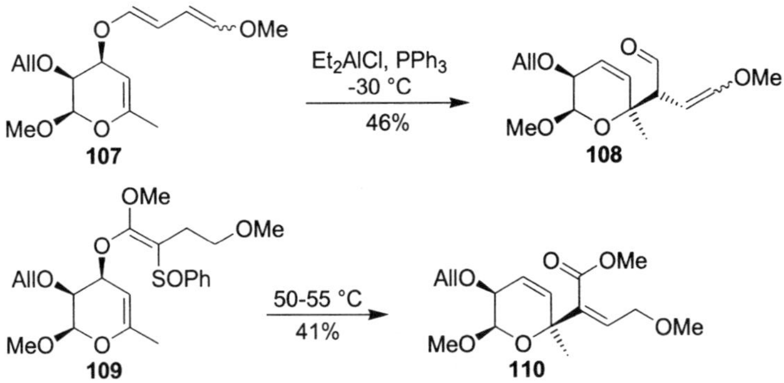

Scheme 22. Claisen rearrangement of 4,5-enopyranosides to C5-branched 3,4-enopyranosides [47]

2.3
Exocyclic Unsaturated Derivatives

Stereoselective syntheses of geminal dialkyl sugars were described in a series of reports [48] by Fraser-Reid et al. (Scheme 23). Claisen rearrangement of exocyclic allylic alcohols **111**–**113** in all cases proceeded with bond formation taking place from the β-face, irrespective of ring substitution. For the C2- and C4-alkylidene derivatives **111** and **113**, respectively, this had to involve an axial orientation of the newly formed bond, an unusual finding, which is in marked contrast to the equatorial folding generally observed for analogous carbocyclic structures. An explanation might be provided by electronic interaction of the ring oxygen lone pair with the electron-deficient orbital at the spiro carbon, as shown schematically for the C4-substituted derivative **117**. This methodology was successfully incorporated into several synthetic projects [49, 50].

The exclusive β-folding selectivity was lost, however, when proceeding to substituted allylic alcohols [51]. Probably due to some steric hindrance caused by the adjacent methyl group, Eschenmoser–Claisen rearrangement of **118** in-

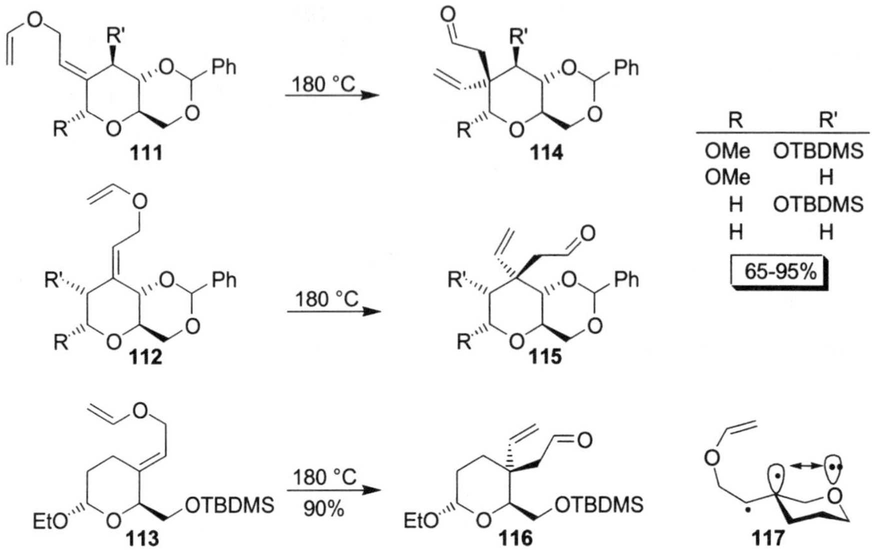

Scheme 23. Exocyclic Claisen rearrangement of C2-, C3-, and C4-alkylidenepyranosides to geminal dialkyl sugars: Exclusive β-face stereoselectivity [48]

volved bond formation from the α-face resulting in the formation of amide **119** (Scheme 24). The different configuration of side-chain epimer **120** apparently avoided such steric interference and reacted in the previously encountered way. Substitution in the vinylic portion of the Claisen system, as exemplified in the Johnson orthopropionate rearrangement starting from primary allylic alcohol **122**, did not affect the β-folding pattern. Additionally, an exclusive diastereoselectivity could be observed for the asymmetric center in the side chain, leading to ester **123** as the only isolated product.

Similar rearrangements were reported by Tadano et al. employing furanoid starting materials. Although these generally exhibited a greater stereochemical ambiguity, with the highly shielded diacetonide **124** (Scheme 25) selective β-folding was achieved during Johnson orthoester rearrangement [52]. The branched ester **125** obtained served as an intermediate in the synthesis of (+)-asteltoxin (**126**) [53]. Reaction of the same starting material with orthopropionate yielded the chiral ester **127** in a 5:1 diastereomeric ratio with its side-chain epimer [54] and resulted in the synthesis of the antibiotic (–)-acetomycin (**128**) [55]. Reaction of **124** with the cyclic acetals **129** and **131**, alcohol elimination, and rearrangement of the intermediate enol ethers resulted in the chiral cyclohexanone **130** and cyclohexenone **132**, respectively. Subsequent hydride reduction and 1,4-conjugate additions of these highly functionalized products succeeded with exclusive facial selectivity [56].

Even with the reaction site placed in lower proximity to the furanoside ring, as in the side-chain allylic alcohols **133** and **135** (Scheme 26), a stereodirecting influence could be achieved in some cases as, for example, with the C3-*gem*-dial-

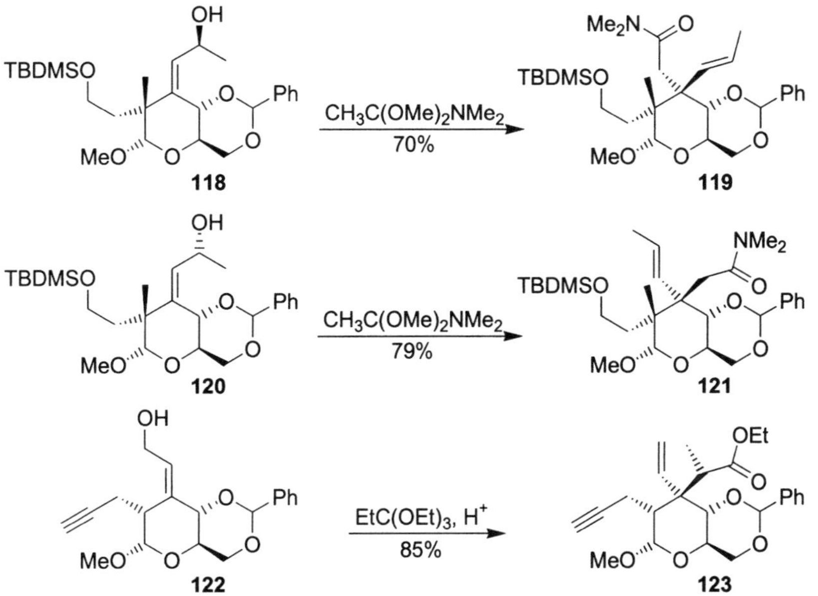

Scheme 24. Exocyclic Claisen rearrangement of substituted C3-alkylidenepyranosides: Varying stereoselectivities [51]

kyl derivative **135** [57]. Although high selectivities were also observed during the formation of branched esters **134** from (*Z*)-allylic alcohol **133**, both the C3-epimer and the (*E*)-configured geometrical isomer resulted in 1:1 diastereomeric mixtures [58].

Water-promoted Claisen rearrangement in the aglycone moiety of unprotected glucosides was reported by Augé et al. [59]. Reaction of allyl vinyl ethers **137a** and **140a** in alkaline aqueous solution proceeded at 60 °C to give, after in situ reduction, the unsaturated alcohols **138** and **141**, respectively (Scheme 27). In comparison, thermal rearrangement of the corresponding acetates in toluene required, even at a higher temperature, a much longer reaction time for the transformation of **137b** to **139**. α-Anomer **140b** did not withstand these harsher conditions and eliminated immediately following rearrangement to conjugated aldehyde **143** and tetra-*O*-acetylglucose **142**. The rate-enhancement of reactions in aqueous solution through entropy-driven association of the hydrophobic parts of the glyco-organic substrates had previously also been demonstrated for [4+2] cycloadditions [60]. The main purpose of the saccharide residue was to increase the water solubility of the respective substrates. At the same time, the chirality of the remote pyranoside ring induced a moderate stereoselectivity during the reactions.

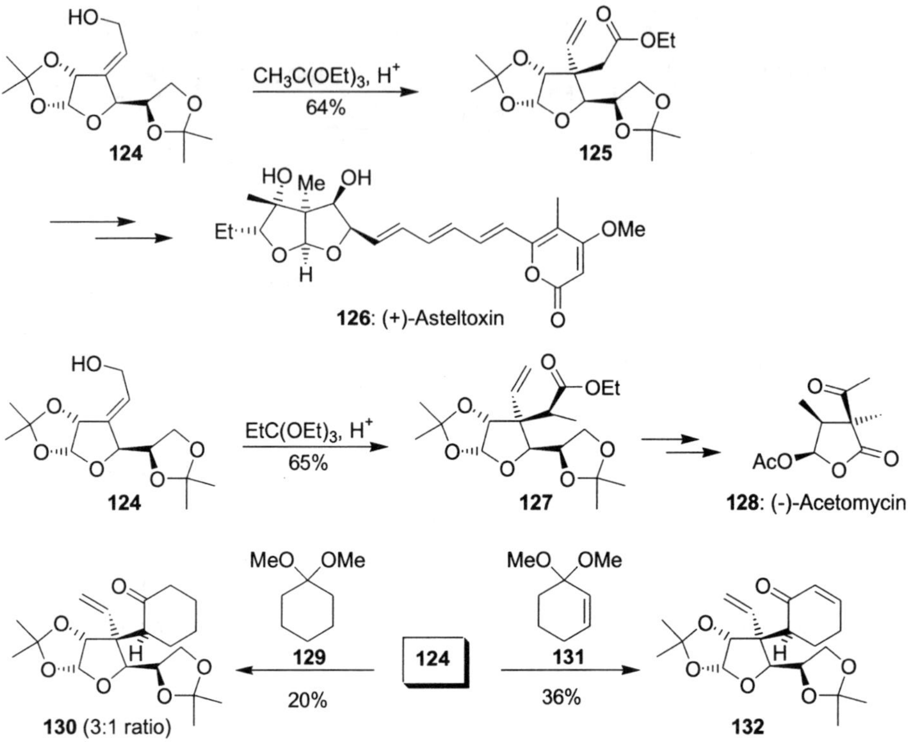

Scheme 25. Stereoselective exocyclic Claisen rearrangements of a C3-alkylidene furanose and applications in natural product syntheses [52–56]

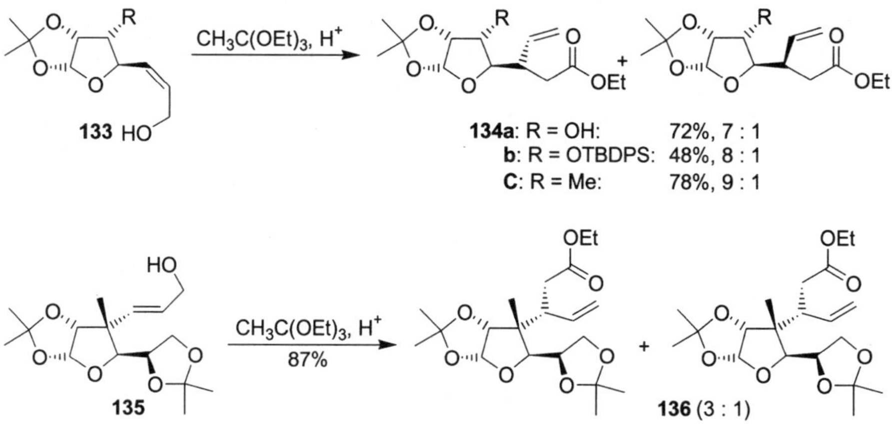

Scheme 26. Stereochemical induction in the remote-site exocyclic Claisen rearrangement of furanoses [57, 58]

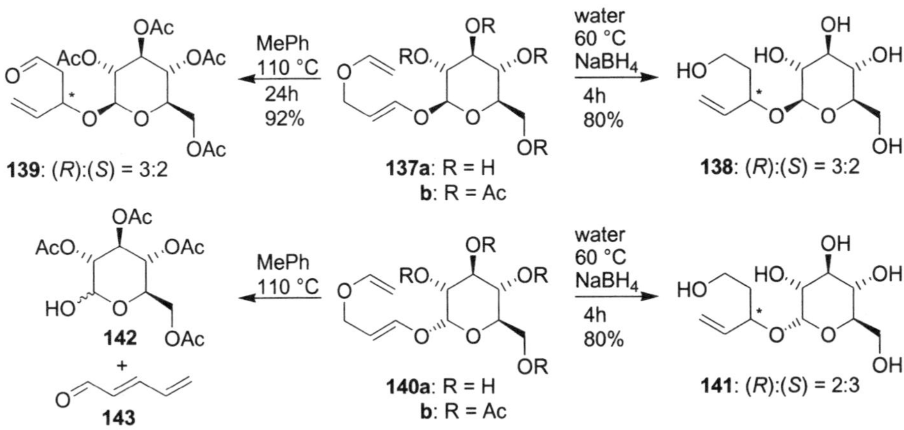

Scheme 27. Water-promoted Claisen rearrangement in the aglycone moiety: Rate enhancement and stereochemical induction [59]

3 Carbohydrate-Derived Enol Ethers and Ketene Acetals

3.1 Enol Ethers

Application of the Claisen rearrangement to systems in which the vinylic portion is of carbohydrate origin is considerably restricted by the difficulty in preparing the respective starting enol ethers. Only one study has been reported so far. This was by Furuichi et al. [61] as an extension of their previous synthesis of vicinal deoxy-carbonyl sugars by the elimination of β-selenoxy alcohols [62]. The β-hydroxyphenylselenides **145** and **149** were accessible by diaxial ring opening of the anhydro sugars **144** and **148** and were transformed into the corresponding allyl ethers **146** and **150** (Scheme 28).

Oxidation to the intermediate selenoxides **147** and **151** and β-elimination provided the allyl enol ethers **152** and **154**, which under mild conditions (elimination in refluxing benzene) could be isolated and characterized. When the elimination was performed in refluxing xylene, it was immediately followed by sigmatropic rearrangement to the *C*-allyl keto sugars **153** and **155**. As is shown in Scheme 29, the main products in each case were those with the side chain in axial orientation. The thermodynamically more stable equatorial isomers, however, were readily formed upon treatment with silica gel, so that both configurations can be obtained selectively with this methodology.

The reaction of crotyl ether **156** resulted in the predominant formation of (*R*)-configured *C*-butenyl derivative **158**, thus demonstrating that sigmatropic rearrangement in this system preferably proceeded through a chairlike transition state **157** (Scheme 30). The deoxy branched-chain derivatives obtained contain an array of functional groups suitable for selective further manipulations and thus appear to be attractive chiral building blocks for

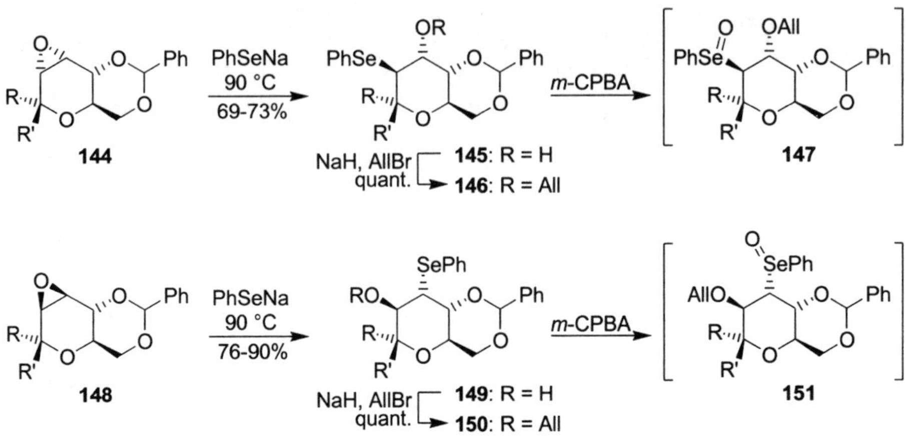

a: R = OMe ; R' = H. b: R = H ; R' = OMe.

Scheme 28. Intermediate sugar selenoxides for the synthesis of enol ethers [61, 62]

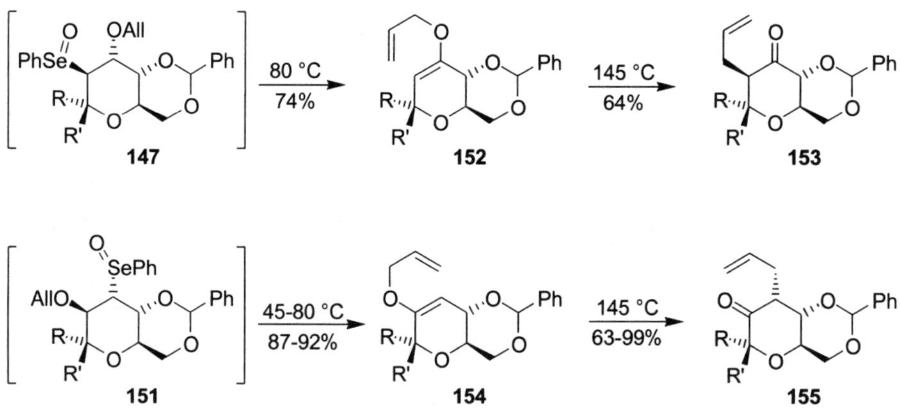

a: R = OMe ; R' = H. b: R = H ; R' = OMe.

Scheme 29. Synthesis and Claisen rearrangement of sugar enol ethers [61, 62]

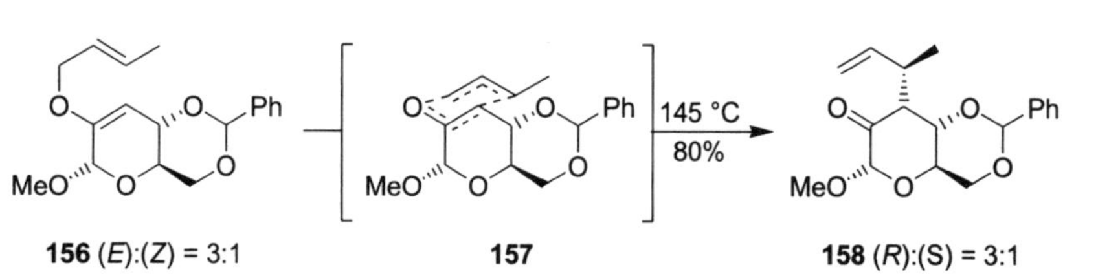

Scheme 30. Transition state geometry and stereoselectivity in the Claisen rearrangement of a crotyl sugar enol ether [61]

natural product synthesis even though no such applications have yet been reported.

3.2 Ketene Acetals

Uronic acids and their derivatives seem to be particularly suitable starting materials for the Ireland ketene acetal Claisen rearrangement. Allyl uronates can be easily prepared in virtually unlimited variety. However, their conversion to ketene acetals as required for sigmatropic rearrangement is associated with serious fundamental complications; these are summarized in Scheme 31. A typical allyl uronate **159** contains a suitable leaving group in the β-position to the ester carbonyl. Thus, following α-deprotonation, the enolate anion **160**, which is supposed to provide ketene acetal **161** by *O*-silylation, will show a marked tendency to undergo β-elimination to the conjugated ester **162**. Under conditions that suppress this process or with deoxygenated substrates that cannot react along a β-elimination pathway, α-elimination of the allylic alkoxide resulting in the formation of ketene **163** and a cascade of ensuing side reactions may become a relevant problem (vide infra).

In a program aimed at the total synthesis of polyether antibiotic monensin, Ireland et al. devised a procedure for the in situ silylation of an ester enolate with a β-leaving group, first executed with the model substrate **165** (Scheme 32). When added to a premixed solution of LDA and TMSCl in 10% HMPA/THF at – 100 °C, crotyl ester **165** produced after rearrangement at room temperature, desilylation, and treatment with diazomethane, the diastereomeric Claisen products **166a** and **166b** in almost 80% combined yield [63]. The success of this three-component competition experiment depended on the relative rates of *N*-silylation vs. enolization and β-elimination vs. *O*-silylation, respectively, with all of these processes occurring on a subminute time scale at – 100 °C.

Although sometimes difficult to reproduce (see below), this experimental approach was extended to glycal uronate **167** (Scheme 33) to obtain bis(tetrahy-

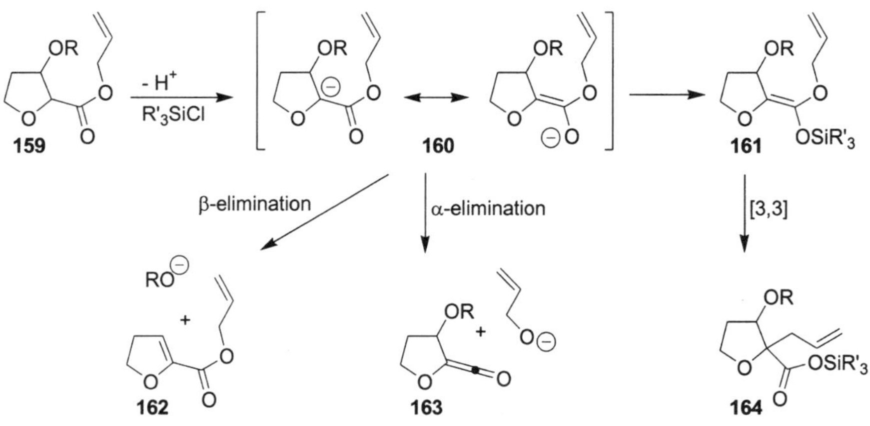

Scheme 31. Competing reaction pathways for β-alkoxy ester enolates

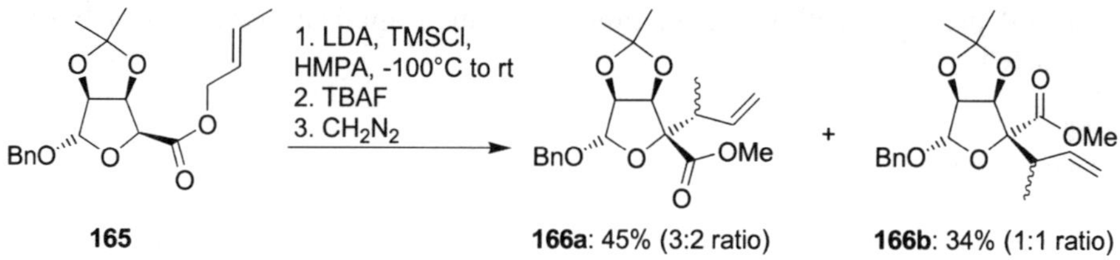

Scheme 32. Simultaneous enolization-silylation and subsequent Claisen rearrangement of a crotyl penturonate [63]

drofuran) building block **168** as a diastereomeric mixture [64]. Omission of HMPA from the reaction mixture in an attempt to alter the diastereomeric ratio caused the rate of *O*-silylation to drop far below the rate of β-elimination and no Claisen products were detected in that instance. Elaboration of the primary hydroxyl to the monensin tetrahydropyran moiety and reductive elimination within the terminal carbohydrate unit furnished the tricyclic glycal building block **169**, suitable for a second Claisen rearrangement after coupling to spiroketal acid **170** [65]. This substrate was expected to require considerably less demanding reaction conditions since, without a leaving group in the neigh-

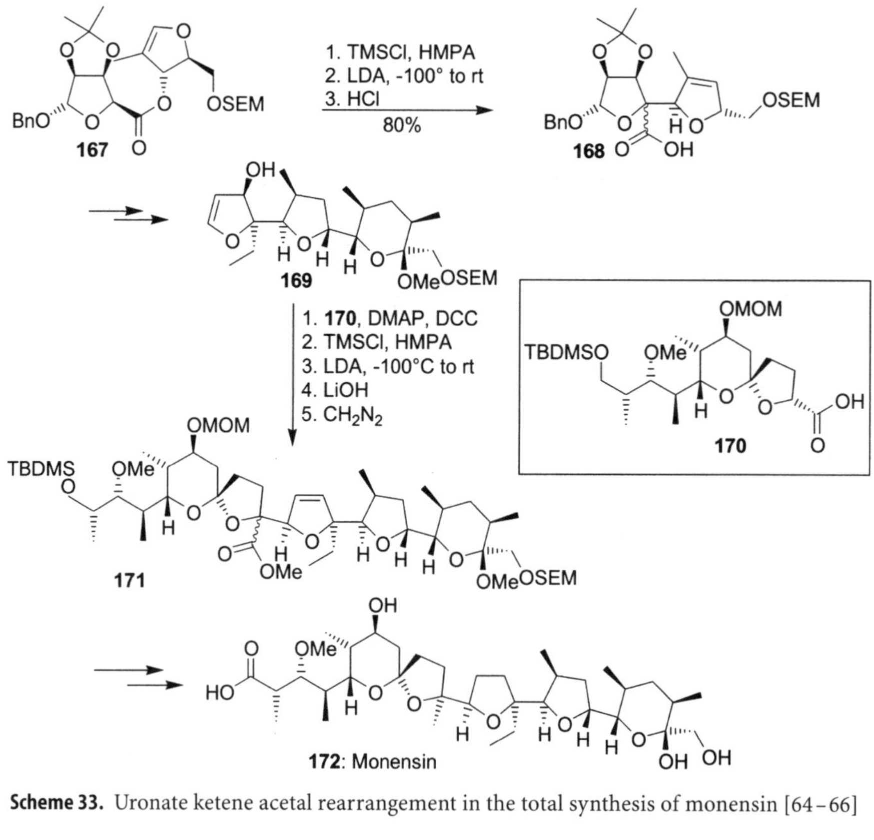

Scheme 33. Uronate ketene acetal rearrangement in the total synthesis of monensin [64–66]

boring position, β-elimination can be excluded as a possible side reaction. However, attempts to perform the enolization-silylation step at higher temperatures (–78 °C) and with less reactive silyl chlorides (TBDMSCl, TESCl) led to the isolation of the *O*-silylated glycal as the only defined reaction product. On the basis of model studies, α-elimination and ketene formation were assumed to be the likely cause. Subsequent fragmentations of the highly reactive ketene would also account for the complete loss of the acid portion of the starting material. At a lower temperature and with the use of the more reactive TMSCl as silylating agent, ketene acetal formation and Claisen rearrangement succeeded to give a diastereomeric mixture of the complex methyl esters **171** in 58% yield. This advanced intermediate finally permitted the total synthesis of monensin (**172**) to be completed [66].

Rizzacasa et al. [67] further modified the experimental protocol for the reaction of allyl ester **173** (Scheme 34), a closely related analog of the original model substrate in the Ireland study. In this case, the inherent problem of *N*-silylation of the dialkylamide base prior to enolization was very successfully diminished by changing the order in which the reagents were mixed. Namely, the base had to be added separately to a premixed solution of the ester substrate and the silylating agent. Following this procedure, a yield of 74% was reported for the branched rearrangement products **174a** and **174b**, with a substantial diastereomeric excess of the D-*lyxo* epimer. Assignment of the absolute configuration at the quaternary carbon atom was possible by X-ray structural analysis of alcohol **175b** [68].

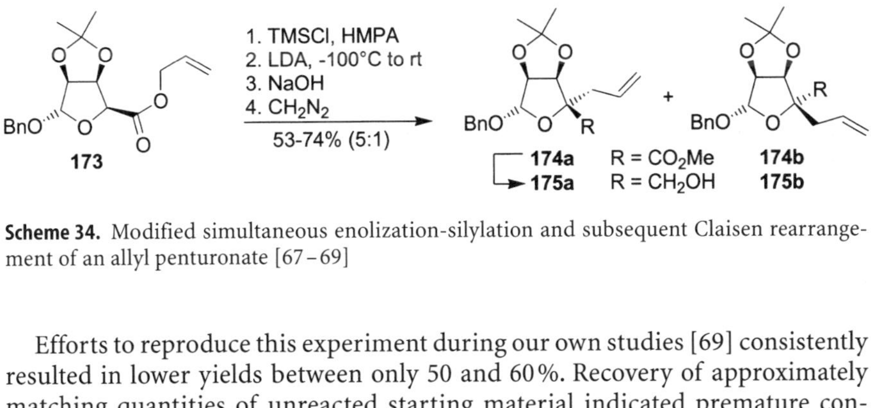

Scheme 34. Modified simultaneous enolization-silylation and subsequent Claisen rearrangement of an allyl penturonate [67–69]

Efforts to reproduce this experiment during our own studies [69] consistently resulted in lower yields between only 50 and 60%. Recovery of approximately matching quantities of unreacted starting material indicated premature consumption of the base by *N*-silylation as the main side reaction, whereas no loss of the ester by β-elimination was observed. Similar experiences were reported by Rizzacasa et al. when changing to a different substrate structure [70]: Benzyloxy-substituted uronate **176** gave the rearrangement products **177a** and **177b** in only 47% total yield, also with complete recovery of the corresponding amount of unchanged starting material (Scheme 35). Even less successful was the conversion of dibenzyloxy derivative **178a** in our own study. Attempts to increase C–H acidity by more electron-withdrawing protecting groups, as in dibenzoate **178b**, resulted in fragmentation.

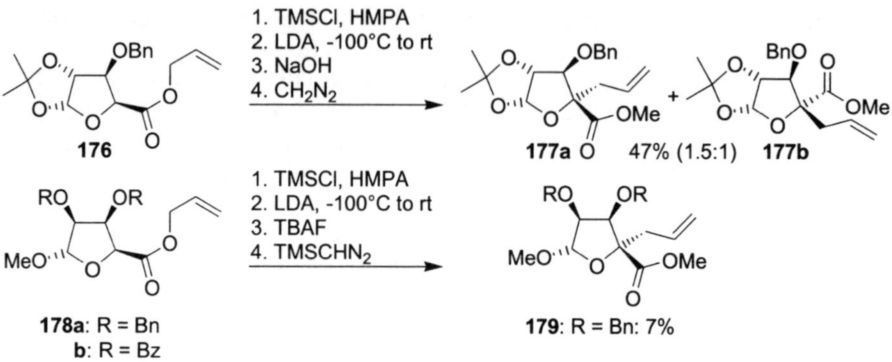

Scheme 35. Low yields in the ketene acetal rearrangement of β-benzyloxy uronates [69, 70]

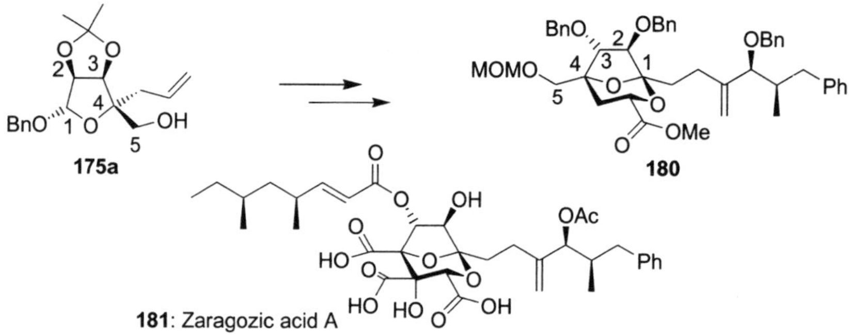

Scheme 36. Utilization of a uronate rearrangement product for the synthesis of zaragozic acid A [70]

The main synthetic purpose of the Rizzacasa group, the preparation of intermediate **180** for the synthesis of zaragozic acid A (**181**), thus had to be reached by employing the more accessible 2,3-*cis*-derivative **175a** and inverting the stereochemistry of one hydroxyl (Scheme 36).

More recently [71], Rizzacasa et al. achieved a very good yield of 81% in the reaction of "1-deoxy uronate" **182**, which also showed a quite remarkable stereoselectivity (Scheme 37). The main reaction product **183** was further converted to the microbial metabolite sphydrofuran (**184**). In a similar reaction sequence, we had previously and independently prepared the spirobicyclic dialdo compound **185** [72].

In our recently reported systematic investigation [69], the dependency of total yields on the C-H acidity of the respective esters could also be demonstrated in the reaction of diastereomeric uronates **186** and **188** (Scheme 38). The enantiomeric ketene acetals obtained after deprotonation were expected to show the same chemical behavior during rearrangement, which resulted in the formation of enantiomeric mixtures of C4-epimers **187a** and **187b** and **189a** and **189b**, respectively, in virtually the same product ratio. The varying total yields could thus be attributed to different deprotonation rates for substrates with 3,4-*cis* or 3,4-*trans* ring substitution.

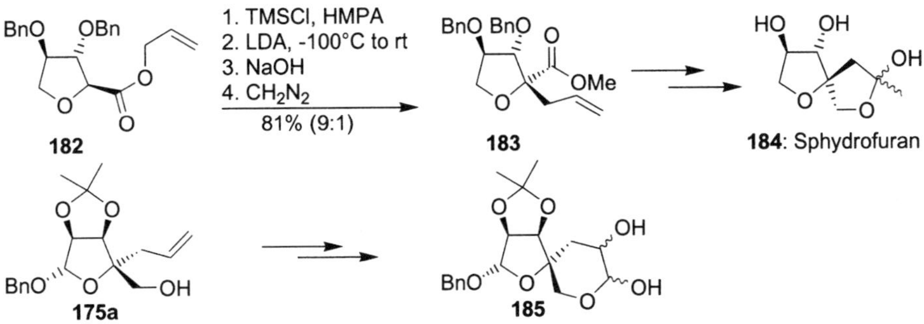

Scheme 37. Conversion of uronate rearrangement products to spirobicyclic derivatives [71,72]

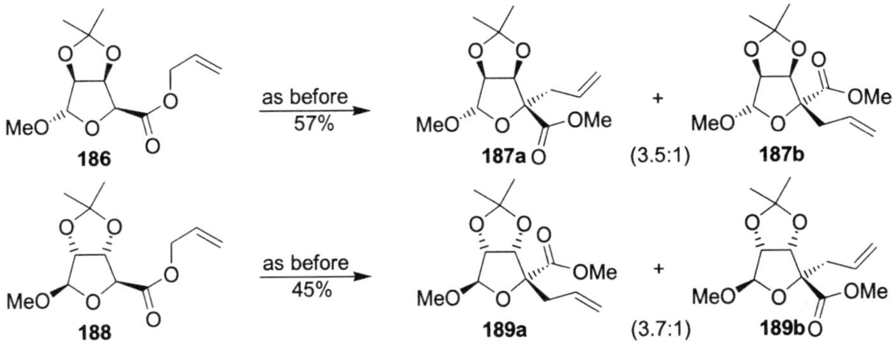

Scheme 38. Claisen rearrangement of diastereomeric penturonates to enantiomeric product mixtures [69]

In the same study, information about the transition state was obtained by means of *cis*-pentenyl uronate **190**, which gave rise to a mixture of four rearrangement products, two of which are shown in Scheme 39. The side-chain epimers **192a** and **192b** were formed in a ratio of 5:1. X-ray structural analysis of the major product **192a** allowed assignment of the absolute configurations, leading to the conclusion that, during rearrangement, a chairlike transition state **191a** is favored.

In order to facilitate the experimental procedure and possibly improve the yields, we also employed β-deoxygenated starting materials (Scheme 40). However, an attempt to perform enolization of ester **193** prior to the addition of a silylating agent and at a temperature of −78 °C only led to the isolation of a dimeric product **194**, apparently resulting from an ester condensation of the highly reactive enolate. Therefore, the simultaneous enolization-silylation protocol had to be applied even in the deoxygenated series. In accordance with the previous results, the same mixture of rearrangement products **194** was obtained in varying yields from both diastereomeric uronates **193** and **195**. Unfortunately, deoxygenation resulted in a complete loss of stereoselectivity. The lack of a directing influence in the immediate proximity to the reaction center could not be compensated for either by a *cis* configuration of the remaining ring

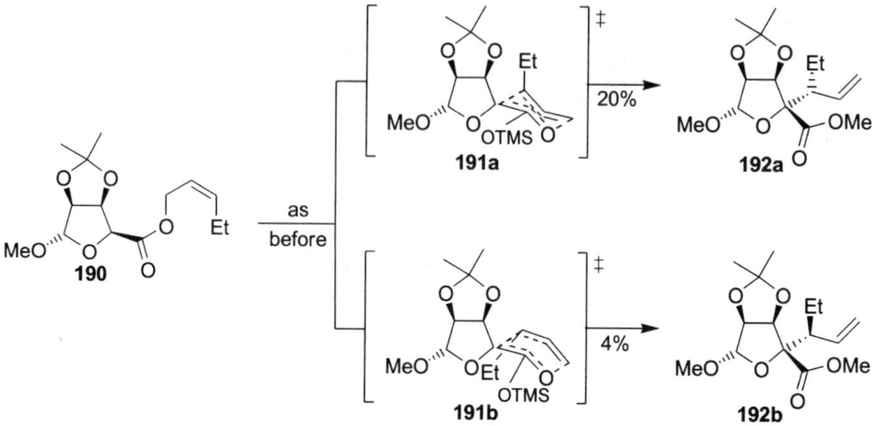

Scheme 39. Transition state geometry in the Claisen rearrangement of a pentenyl uronate [69]

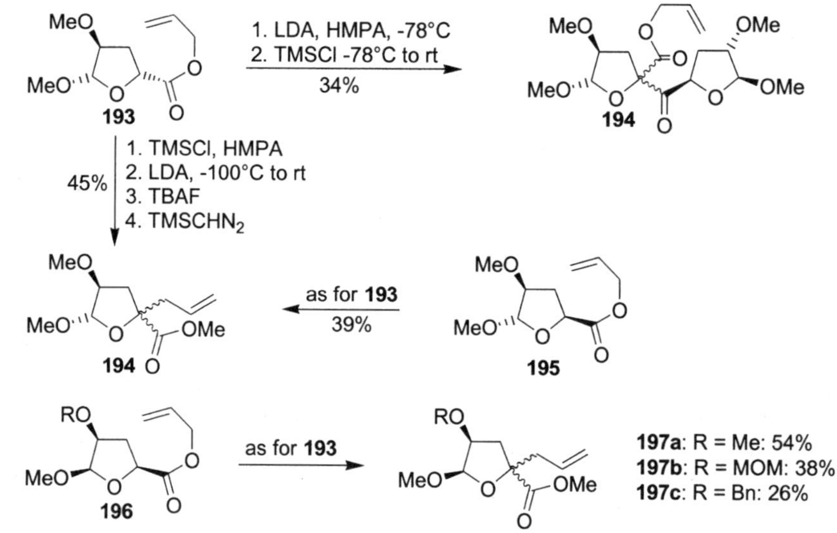

Scheme 40. Claisen rearrangement of deoxy penturonates [69]

substituents, or by increasing the steric demand of the O3-protecting group as in derivatives **196a–c**.

Prior to our investigation, no attempt had been reported to apply the Ireland–Claisen rearrangement to pyranoid uronate derivatives. It soon became apparent that these are associated with even greater difficulties (Scheme 41). The very poor results obtained with *galacto*- and *gluco*-esters **198a** and **198b**, respectively, could not be significantly improved by simple C4-deoxygenation, as was seen in the low-yielding transformation of **200a** to **201**. The rather rigid chair conformation of the pyranoside ring was assumed to provide a major obstacle for C5-deprotonation and sp^3-sp^2 rehybridization. Thus, additional

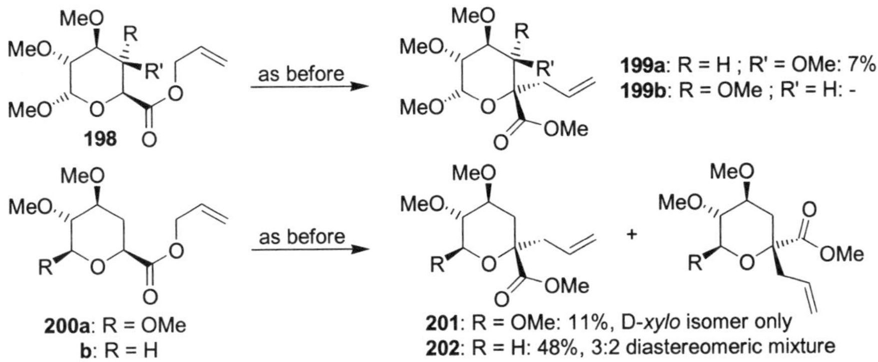

Scheme 41. Claisen rearrangement of hexuronates [69]

deoxygenation in the anomeric position was expected to facilitate ketene acetal formation by increasing the conformational flexibility. Indeed, reaction of 1,4-dideoxy derivative **200b** gave the epimeric rearrangement products **202** in a substantially improved yield of 48%; however, with virtually no stereochemical discrimination.

The outlined fundamental problems, and structural limitations notwithstanding, ketene acetal rearrangement of allyl uronates could provide the basis for synthetic transformations of considerable complexity. We recently succeeded in the preparation of methylene-linked saccharide dimers via the readily accessible uronates **203** and **205** (Scheme 42, unpublished results). Particularly gratifying was the reaction of methyl glycoside **205b** that gave the corresponding rearrangement products **207a** and **207b** in a quite remarkable yield of 67%. The very poor stereoselectivity seemed a little surprising considering

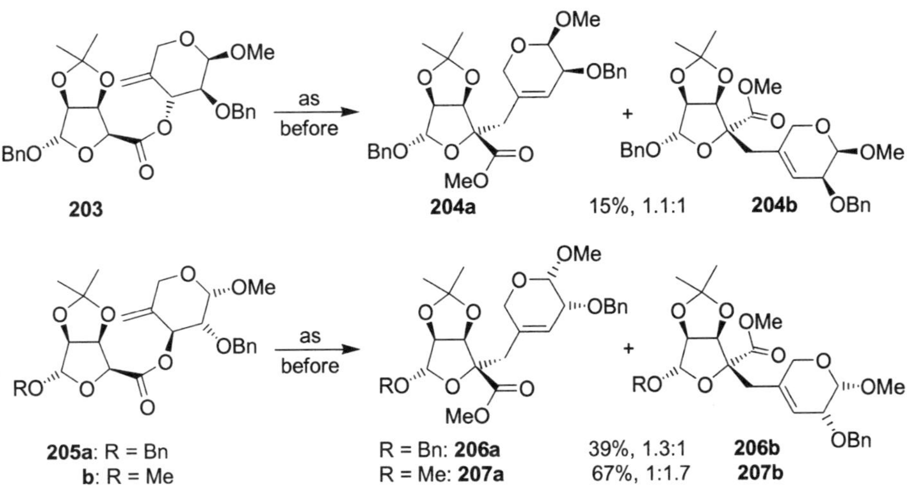

Scheme 42. Claisen rearrangement of pyranosyl uronates to methylene-linked saccharide dimers

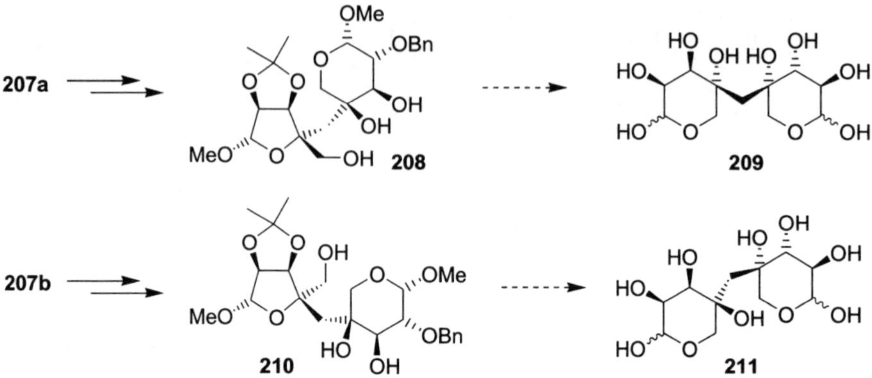

Scheme 43. Conversion of the Claisen rearrangement products to disaccharide mimetics

the much higher discrimination achieved with the simple allyl esters (vide supra).

Both epimers **207a** and **207b** were further processed to obtain the fully oxygenated species **208** and **210**, which, after deprotection, should give rise to the methylene-linked disaccharide mimetics **209** and **211** (Scheme 43).

4 Skeletal Rearrangements

In the mid-1980s, Paquette et al. reported the synthesis of cyclooctanoid natural products based on the ring enlargement of 2-methylene-6-vinyltetrahydropyrans by sigmatropic rearrangement [73]. Analogous oxygen-substituted starting materials should be easily accessible from simple carbohydrate precursors and give rise to highly functionalized chiral cyclooctenones. In a preliminary communication [74], we reported the synthesis and reaction of an exocyclic enol ether **214a** (Scheme 44) starting from acetobromoglucose **212** via the known

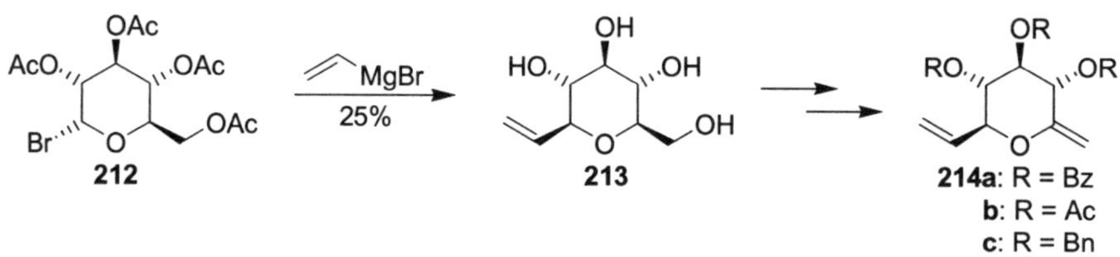

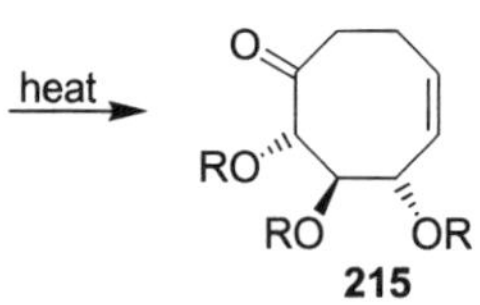

	a	b	b	b	c	c
T (°C)	145	110	145	200	20[a]	200
t (h)	12	15	15	1	20	1
yield (%)	60	30	70	85	-	75

[a]: Using $(i\text{-Bu})_3Al$ as catalyst

Scheme 44. Preparation of exocyclic sugar enol ethers and skeletal rearrangement to cyclooctenones [74]

vinyl *C*-glycoside **213** [75]. At an advanced stage in the synthetic sequence, different protecting groups could be introduced. In an initial attempt, tribenzoate **214a** was heated at 145 °C and provided, after 12 h, carbocycle **215a** as the only isolated product in a yield of 60 %. The expected strong dependency of the thermal rearrangement on the activation energy was demonstrated in the reaction of triacetate **214b** at different temperatures. Gratifyingly, the rather likely yet undesirable isomerization of the exocyclic enol ether to the thermodynamically favored endocyclic derivative was not observed under any of these conditions. No conversion could be achieved in an attempt to perform the reaction of benzyl ether **214c** at room temperature catalyzed by triisobutylaluminum, which had been successfully employed by Paquette et al. for similar substrates [76].

Only very recently [77], Sinaÿ et al. presented a series of Ferrier-type rearrangements [78] of 5,6-unsaturated sugars to cyclohexenones. In this study, the same catalyst, applied to the reaction of furanyl *C*-glycoside **216**, yielded exclusively, after reduction of the intermediate ketone, cyclohexanol **217** (Scheme 45). A conceivable Claisen rearrangement to a bicyclic product **218** was apparently not observed.

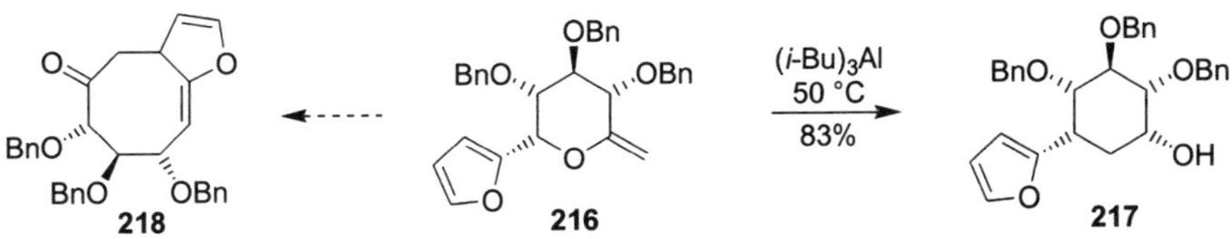

Scheme 45. Aluminum-mediated carbocyclizations: Ferrier-type reaction vs. Claisen rearrangement [77]

In the chiral cyclooctenones, the conformational flexibility that is often encountered for eight-membered carbocycles appeared to be considerably reduced by the three sp^2 carbon atoms. Very sharp and highly resolved proton NMR signals as well as unambiguously observed transannular NO effects permitted tentative deduction of a boat-chair geometry as the favored conformation in solution. X-ray crystallography of triacetate **215b** later confirmed the same conformation for the solid state [72]. For such rigid cyclic structures, a pronounced stereoselectivity during subsequent chemical transformations could be anticipated [79]. Two exemplary reactions employing simple, non-sterically demanding reagents indeed resulted in the exclusive formation of a single stereoisomer in each case (Scheme 46). Reduction of the carbonyl group, however, led to a conformationally more flexible structure **220** which no longer allowed assignment of the configuration by NMR spectroscopic methods and is therefore drawn in an ambiguous fashion.

Nagarajan and Sudha [80] employed the skeletal Claisen rearrangement of endocyclic enol ethers for the synthesis of pseudo-sugars (Scheme 47). Starting from glycal **221**, a vinyl group could be installed in two steps to provide the starting material enol ether **222**. Heating at 240 °C then smoothly afforded cyclohexene derivative **223**, which allowed easy access to *gluco-* and *manno*-configured pseudo-sugars **224a–c** by subsequent dihydroxylation.

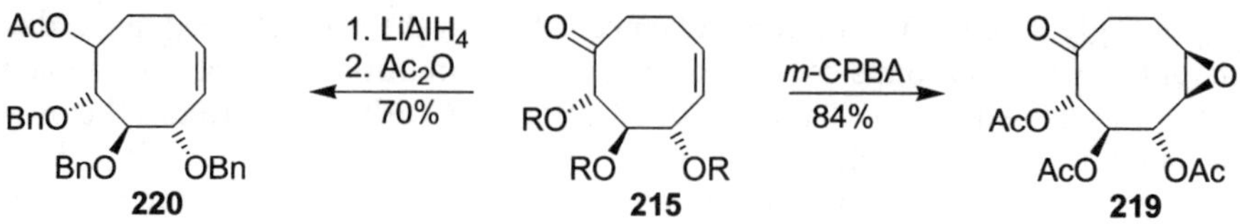

Scheme 46. Stereoselective transformations of the conformationally restricted chiral cyclooctenones [72]

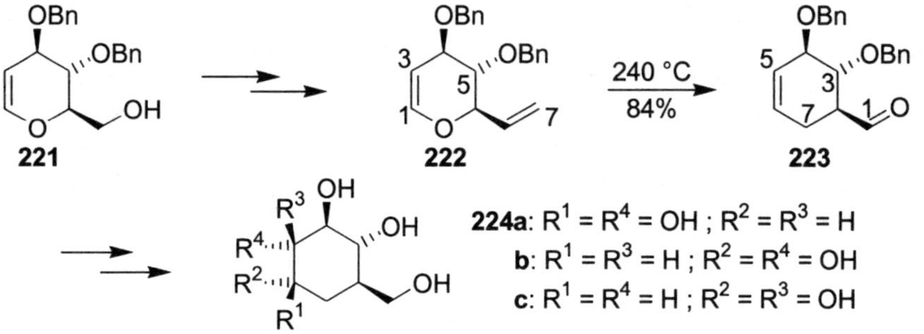

Scheme 47. Preparation of a vinyl-branched glycal, skeletal rearrangement, and conversion to carbasugars [80]

5
Perspectives

The preceding collection shows the wide range of unsaturated carbohydrate structures that can be subjected to a Claisen rearrangement to result in an equally diverse array of complex products which can serve as valuable intermediates for a variety of synthetic targets such as *C*-glycosides, alkyl-branched sugars, pseudo-sugars, saccharide mimetics and non-carbohydrate natural products. The synthetic work has been accompanied by interesting findings of a more general nature. In particular, the often high and sometimes exclusive stereoselectivities further increase the appeal of carbohydrate-based Claisen rearrangements for the solution of complex preparative tasks. The most frequent use has been made of glycals and other endocyclic unsaturated derivatives as the most easily accessible starting materials. Continuous application throughout the years within synthetic programs of many research groups from different areas puts these transformations in the rank of a standard tool of organic synthesis and allows the expectation that they will continue to serve important purposes in future endeavors. Studies employing other types of unsaturated carbohydrate structures have been refined in a limited number of laboratories. In particular, the ketene acetal rearrangement starting from uronic acid esters can be associated with substantial experimental difficulties that might discourage its more extensive use. However, since this method also provides an entry into the most complex types of product structures, it can be hoped that it will nonetheless stimulate further expansion and elimination of existing problems in the course of future applications. The most recent development has been the skeletal rearrangement of vinyl-branched sugar enol ethers that is associated with a particularly uncomplicated and high-yielding experimental procedure. Its value for the synthesis of sugar mimetics has already been demonstrated. The synthesis of highly functionalized chiral carbocycles in a more general context also seems an attractive future application.

6 References

1. (a) Hanessian S (1983) Total synthesis of natural products. The chiron approach. Pergamon, Oxford; (b) Inch TD (1984) Tetrahedron 40:3161; (c) Kunz H, Rück K (1993) Angew Chem Int Ed Engl 32:336; (d) Bols M (1996) Carbohydrate building blocks. Wiley, New York; (e) Hultin PG, Earle MA, Sudharshan M (1997) Tetrahedron 53:14823; (f) Lundt I (1997) Top Curr Chem 187:117
2. (a) Postema MHD (1992) Tetrahedron 48:8545; (b) Postema MHD (1995) *C*-Glycoside synthesis. CRC, London; (c) Levy D, Tang C (1995) The chemistry of *C*-glycosides. Pergamon, Oxford; (d) Beau JM, Gallagher T (1997) Top Curr Chem 187:1; (e) Nicotra F (1997) ibid 55; (f) Du Y, Linhardt RJ (1998) Tetrahedron 54:9913
3. Suami T, Ogawa S (1990) Adv Carbohydr Chem Biochem 48:21
4. Yoshimura J (1984) Adv Carbohydr Chem Biochem 42:69
5. Claisen L (1912) Ber Dtsch Chem Ges 45:3157
6. For recent reviews, see (a) Ziegler FE (1988) Chem Rev 88:1423; (b) Blechert S (1989) Synthesis 71; (c) Wipf P (1991) In: Trost BM, Fleming I (eds) Comprehensive organic synthesis, vol 5. Pergamon, Oxford, p 827; (d) Ganem B (1996) Angew Chem Int Ed Engl 35:936; (e) Gajewski JJ (1997) Acc Chem Res 30:219; (f) Ito H, Taguchi T (1999) Chem Soc Rev 28:43
7. Wick AE, Felix D, Steen K, Eschenmoser A (1964) Helv Chim Acta 47:2425
8. Johnson WS, Werthemann L, Bartlett WR, Brocksom TJ, Li T, Faulkner DJ, Petersen MR (1970) J Am Chem Soc 92:741
9. (a) Ireland RE, Mueller RH (1972) J Am Chem Soc 94:5897; (b) Ireland RE, Mueller RH, Willard AK (1976) ibid 98:2868; (c) for a comprehensive review, see: Pereira S, Srebnik M (1993) Aldrichim Acta 26:17
10. Ireland RE, Wipf P, Armstrong JD (1991) J Org Chem 56:650
11. Ferrier RJ (1969) Adv Carbohydr Chem Biochem 24:199
12. (a) Fraser-Reid B, Dawe RD, Tulshian DB (1979) Can J Chem 57:1746; (b) Tulshian DB, Fraser-Reid B (1984) J Org Chem 49:518
13. (a) Ireland RE, Wilcox CS, Thaisrivongs S, Vanier NR (1979) Can J Chem 57:1743; (b) Ireland RE, Thaisrivongs S, Vanier N, Wilcox CS (1980) J Org Chem 45:48
14. (a) Ireland RE, Vevert JP (1980) J Org Chem 45:4259; (b) Ireland RE, Vevert JP (1981) Can J Chem 59:572; (c) Ireland RE, Daub JP (1981) J Org Chem 46:479
15. (a) Ireland RE, Thaisrivongs S, Wilcox CS (1980) J Am Chem Soc 102:1155; (b) Ireland RE, McGarvey GJ, Anderson RC, Badoud R, Fitzsimmons BJ, Thaisrivongs S (1980) ibid 102:6178; (c) Ireland RE, Anderson RC, Badoud R, Fitzsimmons BJ, McGarvey GJ, Thaisrivongs S, Wilcox CS (1983) ibid 105:1988
16. (a) Edwards MP, Ley SV, Lister SG, Palmer BD (1983) J Chem Soc Chem Commun 630; (b) Edwards MP, Ley SV, Lister SG, Palmer BD, Williams DJ (1984) J Org Chem 49: 3503
17. Ireland RE, Wuts PGM, Ernst B (1981) J Am Chem Soc 103:3205
18. Ireland RE, Smith MG (1988) J Am Chem Soc 110:854
19. Kozikowski A, Lee J (1990) J Org Chem 55:863
20. Varelis P, Graham AJ, Johnson BL, Skelton BW, White AH (1994) Aust J Chem 47:1735
21. Schaus SE, Brånalt J, Jacobsen EN (1998) J Org Chem 63:4876

22. (a) Curran DP, Suh YG (1984) J Am Chem Soc 106:5002; (b) Curran DP, Suh YG (1984) Tetrahedron Lett 25:4179; (c) Curran DP, Suh YG (1987) Carbohydr Res 171:161
23. (a) Carpenter BK (1978) Tetrahedron 34:1877; (b) Burrows CJ, Carpenter BK (1981) J Am Chem Soc 103:6984
24. (a) Gajewski JJ, Conrad ND (1979) J Am Chem Soc 101:6693; (b) Gajewski JJ, Jurayi J, Kimbrough DR, Gande ME, Ganem B, Carpenter BK (1987) ibid 109:1170
25. Gajewski JJ, Emrani J (1984) J Am Chem Soc 106:5733
26. (a) Deslongchamps P (1983) Stereoelectronic effects in organic chemistry. Pergamon, Oxford; (b) Kirby AJ (1983) The anomeric and related stereoelectronic effects at oxygen. Springer, Berlin Heidelberg New York; (c) Tvaroska I, Bleha T (1989) Adv Carbohydr Chem Biochem 47:45; (d) Thatcher GRJ (ed) (1993) The anomeric effect and associated stereoelectronic effects. ACS Symp Series 539, ACS, Washington, DC
27. (a) Denmark SE, Dappen MS (1984) J Org Chem 49:798; (b) Denmark SE, Dappen MS, Sear NL, Jacobs RT (1990) J Am Chem Soc 112:3466
28. Ireland RE, Wipf P, Xiang JN (1991) J Org Chem 56:3572
29. Gajewski JJ, Jimenez JL (1986) J Am Chem Soc 108:468
30. Colombo L, Casiraghi G, Pittalis A, Rassu G (1991) J Org Chem 56:3897
31. (a) Kübel B, Höfle G, Steglich W (1975) Angew Chem Int Ed Engl 14:58; (b) Engel N, Kübel B, Steglich W (1977) ibid 16:394
32. Delany JJ, Padykula RE, Berchtold GA (1992) J Am Chem Soc 114:1394
33. (a) Romero RM, Roberts MF, Philipson JD (1995) Phytochemistry 40:1015; (b) Haslam E (1993) Shikimic acid metabolism and metabolites. Wiley, New York
34. Ferrier RJ, Prasad NJ (1969) J Chem Soc (C) 570
35. Ferrier RJ, Vethaviyasar N (1973) J Chem Soc Perkin Trans 1 1791
36. (a) Curran DP, Jacobs PB (1985) Tetrahedron Lett 26:2031; (b) Curran DP, Jacobs PB, Elliott RL, Kim BH (1987) J Am Chem Soc 109:5280; (c) Kim BH, Jacobs PB, Elliott RL, Curran DP (1988) Tetrahedron 44:3079
37. Heyns K, Hohlweg R (1978) Chem Ber 111:1632
38. Cottier L, Remy G, Descotes G (1979) Synthesis 711
39. De Raadt A, Ferrier RJ (1991) Carbohydr Res 216:93
40. Balasubramanian KK, Ramesh NG, Pramanik A, Chandrasekhar J (1994) J Chem Soc Perkin Trans 2 1399
41. Corey EJ, Shibasaki M, Knolle J (1977) Tetrahedron Lett 1625
42. Hernandez O (1978) Tetrahedron Lett 219
43. Pelyvas I, Lindhorst T, Thiem J (1990) Liebigs Ann Chem 761
44. Fernandez Cirelli A, Moradei O, Thiem J (1997) Liebigs Ann/Recl 1983
45. Fleet GWJ, James K, Lunn RJ (1986) Tetrahedron Lett 27:3053
46. (a) Kuroda C, Theramongkol P, Engebrecht JR, White JD (1986) J Org Chem 51:956; (b) White JD, Theramongkol P, Kuroda C, Engebrecht JR (1988) ibid 53:5909
47. Vatele JM (1986) Tetrahedron 42:4443
48. (a) Fraser-Reid B, Tsang R, Tulshian DB, Sun KM (1981) J Org Chem 46:3764; (b) Tulshian DB, Tsang R, Fraser-Reid B (1984) ibid 49:2347; (c) Tulshian DB, Fraser-Reid B (1984) Tetrahedron 40:2083; (d) Fraser-Reid B, Tulshian DB, Tsang R, Lowe D, Box VGS (1984) Tetrahedron Lett 25:4579
49. Tsang R, Fraser-Reid B (1985) J Org Chem 50:4659
50. Tsang R, Fraser-Reid B (1986) J Am Chem Soc 108:2116
51. Pak H, Dickson JK, Fraser-Reid B (1989) J Org Chem 54:5357
52. (a) Tadano K, Idogaki Y, Yamada H, Suami T (1985) Chem Lett 1925; (b) Tadano K, Idogaki Y, Yamada H, Suami T (1987) J Org Chem 52:1201
53. (a) Tadano K, Yamada H, Idogaki Y, Ogawa S, Suami T (1988) Tetrahedron Lett 29:655; (b) Tadano K, Yamada H, Idogaki Y, Ogawa S, Suami T (1990) Tetrahedron 46:2353
54. Tadano K, Ishihara J, Yamada H, Ogawa S (1989) J Org Chem 54:1223
55. Tadano K, Ishihara J, Ogawa S (1990) Tetrahedron Lett 31:2609
56. Murata T, Yanagisawa Y, Aoyama M, Tsushima H, Totani K, Ohba S, Tadano K (1998) Tetrahedron Asymmetry 9:4203

57. Tadano K, Isshiki Y, Kumagai T, Ogawa S (1993) J Carbohydr Chem 12:1
58. (a) Tadano K, Shimada K, Miyake A, Ishihara J, Ogawa S (1989) Bull Chem Soc Jpn 62: 3978; (b) Tadano K, Shimada K, Ishihara J, Ogawa S (1991) J Carbohydr Chem 10:1
59. (a) Lubineau A, Augé J, Bellanger N, Caillebourdin S (1990) Tetrahedron Lett 31:4147; (b) Lubineau A, Augé J, Bellanger N, Caillebourdin S (1992) J Chem Soc Perkin Trans 1 1631
60. (a) Lubineau A, Queneau Y (1985) Tetrahedron Lett 26:2653; (b) Lubineau A, Queneau Y (1987) J Org Chem 52:1001; (c) Lubineau A, Queneau Y (1989) Tetrahedron 45:6697; (d) Lubineau A, Augé J, Lubin N (1990) J Chem Soc Perkin Trans 1 3011
61. Furuichi K, Hashimoto H, Miwa T (1991) Carbohydr Res 220:63
62. Furuichi K, Yogai S, Miwa T (1980) J Chem Soc Chem Commun 66
63. Ireland RE, Norbeck DW (1985) J Am Chem Soc 107:3279
64. (a) Ireland RE, Norbeck DW, Mandel GS, Mandel NS (1985) J Am Chem Soc 3285; (b) Ireland RE, Armstrong JD, Lebreton J, Meissner RS, Rizzacasa MA (1993) ibid 115:7152
65. Ireland RE, Häbich D, Norbeck DW (1985) J Am Chem Soc 107:3271
66. Ireland RE, Meissner RS, Rizzacasa MA (1993) J Am Chem Soc 115:7166
67. McVinish LM, Rizzacasa MA (1994) Tetrahedron Lett 35:923
68. Gable RW, McVinish LM, Rizzacasa MA (1994) Aust J Chem 47:1537
69. Werschkun B, Thiem J (1999) Synthesis 121
70. Mann RK, Parsons JG, Rizzacasa MA (1998) J Chem Soc Perkin Trans 1 1283
71. Di Florio R, Rizzacasa MA (1998) J Org Chem 63:8595
72. Werschkun B (1997) PhD thesis, Universität Hamburg
73. (a) Kinney WA, Coghlan MJ, Paquette LA (1984) J Am Chem Soc 106:6868; (b) Kinney WA, Coghlan MJ, Paquette LA (1985) ibid 107:7352; (c) Paquette LA, Sweeney T (1990) J Org Chem 55:1703; (d) Paquette LA, Sweeney T (1990) Tetrahedron 46:4487; (e) Kang HJ, Paquette LA (1990) J Am Chem Soc 112:3252; (f) Paquette LA, Kang HJ (1991) ibid 113: 2610
74. Werschkun B, Thiem J (1997) Angew Chem Int Ed Engl 36:2793
75. Shulman ML, Shiyan SD, Khorlin AY (1974) Carbohydr Res 33:229
76. (a) Paquette LA, Friederich D, Rogers RD (1991) J Org Chem 56:3841; (b) Philippo CMG, Vo NH, Paquette LA (1991) J Am Chem Soc 113:2762; (c) Paquette LA, Philippo CMG, Vo NH (1992) Can J Chem 70:1356
77. Sollogoub M, Pearce AJ, Hérault A, Sinaÿ P (2000) Tetrahedron Asymmetry 11:283
78. Ferrier RJ (1979) J Chem Soc Perkin Trans 1 1455
79. (a) Still WC, Galynker I (1981) Tetrahedron 37:3981; (b) Hardinger SA, Fuchs PL (1987) J Org Chem 52:2739; (c) Schreiber SL, Smith DB, Schulte G (1989) ibid 54:5994
80. Sudha AVRL, Nagarajan M (1998) Chem Commun 925

Miscellaneous

Anna de Raadt, Martin H. Fechter

Spezialforschungsbereich F001 Biokatalyse, Institut für Organische Chemie der Technischen Universität Graz, Stremayrgasse 16, 8010 Graz, Austria
E-mail: deraadt@orgc.tu-graz.ac.at

Selected isomerisation, epimerisation and rearrangement reactions will be presented with a view to preparative applications.

Keywords: Isomerisation, Epimerisation, Nucleophilic rearrangements, Radicals, Ring contractions

1 Introduction

The aim of this account is to outline synthetically interesting rearrangements selected from the literature which have not been covered in previous chapters. The following areas will be discussed: Unusual cases of isomerisations and epimerisations as well as rearrangements involving nucleophiles, radical species and reactions leading to ring-contracted products.

2 Isomerisation and Epimerisation

The Meerwein–Ponndorf–Verley/Oppenauer reaction [1] induced by air-oxidised samarium(II) iodide has been found to be a means to directly isomerise aldohexopyranoses into ketohexopyranoses. Iadonisi and co-workers [2] report-

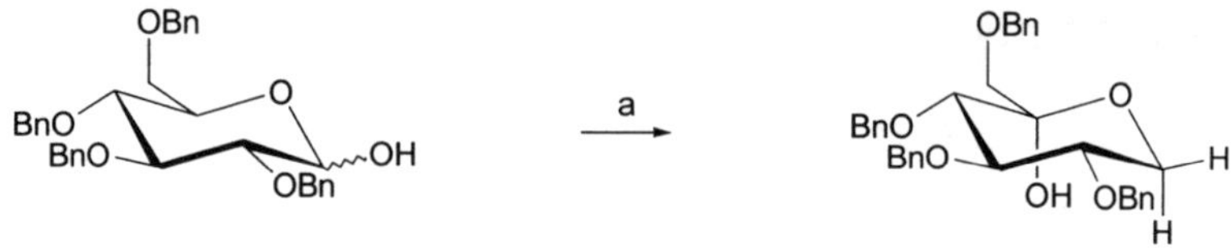

1 2

Scheme 1. (a) SmI_2/O_2, THF, reflux

ed that compounds such as **1** could be converted into derivative **2** in 81% yield by treating with air-oxidised samarium(II) iodide in refluxing tetrahydrofuran (Scheme 1). The yield of **2** could be improved by changing the reaction solvent from tetrahydrofuran to tetrahydropyran. In addition, it was also found that reaction times could be thereby reduced [3]. This methodology has been used to isomerise the D-*galacto* and D-*manno* equivalents of **1** with similar success.

During an investigation into the reactions and interactions which occur when the sugar moieties of *C*-glycosides are submitted to hydrolysis, Kumazawa and co-workers made an interesting observation involving a D-*gluco* to D-*fructo* isomerisation [4]. Treatment of β-D-*C*-glucopyranosyl compound **3** with a catalytic amount of *p*-toluenesulfonic acid in boiling water gave spiroketal derivative **4** in 35% yield, the structure of which was confirmed by X-ray crystallographic structural analysis of **5** (Scheme 2). This unusual reaction is still under investigation by these workers.

During their studies concerning C-^{1}H – C-^{2}H exchange reactions with Raney nickel catalyst in deuterium oxide, Perlin and Balza made a number of observations concerning the regio- and stereoselective aspects of this reaction [5]. In

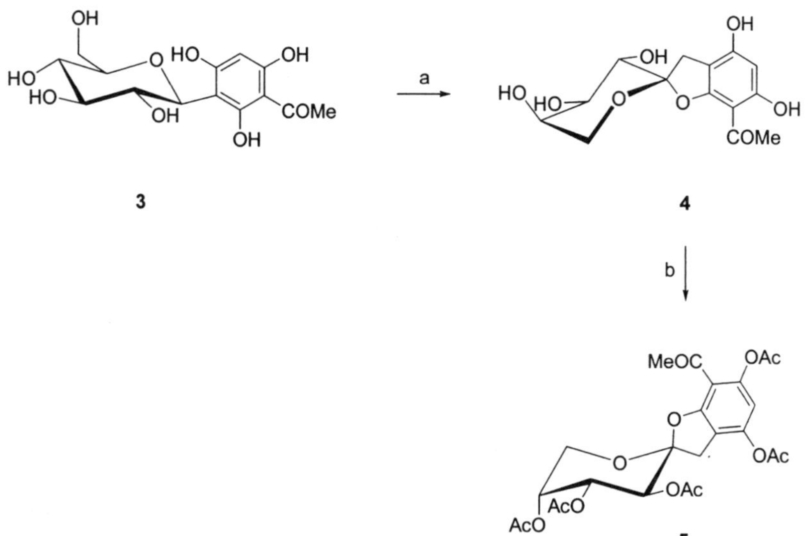

Scheme 2. (a) *p*-TsOH, H_2O, reflux; (b) Ac_2O, pyridine, DMAP

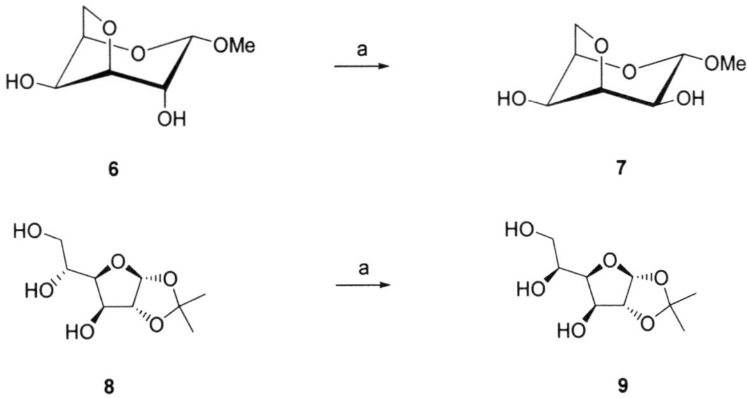

Scheme 3. (a) Raney nickel, H_2O, reflux

particular, the epimerisation of a range of carbohydrates with Raney nickel in boiling water was examined. Endocyclic epimerisation was observed with compounds such as D-galactopyranoside **6**, as here D-*talo* derivative **7** was obtained in 20% yield. In addition, exocyclic epimerisation was also observed. For example, D-glucofuranose **8** furnished L-idofuranose **9** in 25% yield after Raney nickel treatment (Scheme 3).

Miethchen and co-workers have published a number of papers concerning the use of chloral and dicyclohexylcarbodiimide (DCC) as a means to invert configuration of chiral centres in carbohydrates. These reagents were found to react with bis-*vicinal* triols with a *cis*, *trans* sequence of hydroxyl groups and this resulted in the formation of cyclic acetals in which the central atom of the triol had been inverted. For example, the rare monosaccharide D-tagatose has been easily prepared from D-fructose (Scheme 4) [6]. Concerning the mechanism of this reaction, chloral and DCC react with fructose derivative **10** to give intermediates **11** and **12**. The latter mentioned intermediate then reacts intramolecularily in an S_N2-type reaction as shown. This furnished product **13** in 59% yield.

This methodology has been applied to a number of differently configured monosaccharides including D-*lyxo*, L-*arabino*, D-*manno*, D-*galacto* and L-*rhamno* sugars. These were converted into the corresponding C-3 epimers, namely D-*arabino*, L-*lyxo*, D-*altro*, D-*gulo* and 6-deoxy-L-*altro* derivatives, respectively [7]. Furthermore, this approach has been applied to the selective epimerisation of one of the monosaccharide moieties in disaccharides [8], as shown in Scheme 5, as well as to cyclitol derivatives [9], glycosyl azides and *N*-acetylglycosylamines of rare monosaccharides [10].

It has been recently shown [11] that a Swern oxidation followed by an intramolecular Tischenko reaction gives access to L-idose or L-altrose derivatives from the corresponding protected hexos-5-uloses. As can be seen from Scheme 6, D-glucitol **14** was oxidised to aldulose **15** via a Swern oxidation. Subsequently, this compound was treated with a trivalent samarium reagent (*tert*-BuOSmI_2), thereby inducing a Tischenko reaction, which led to L-*ido*-configured *tert*-butyl

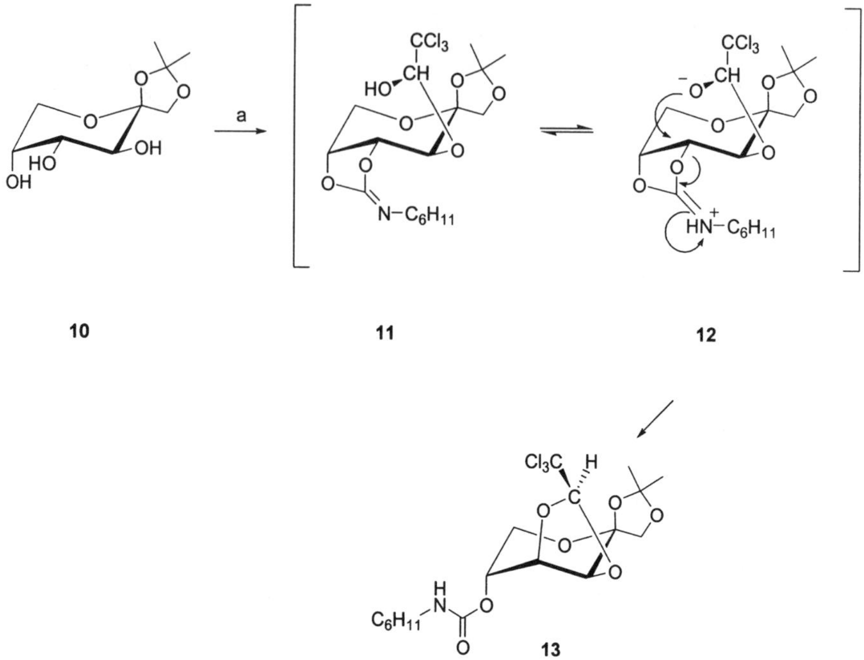

Scheme 4. (a) Chloral/DCC, $ClCH_2CH_2Cl$, reflux

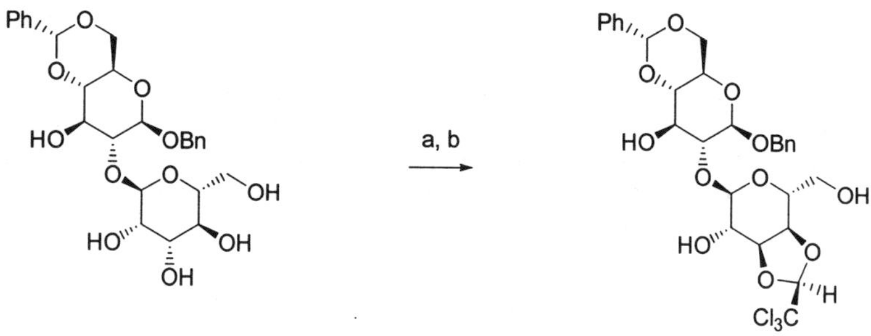

Scheme 5. (a) Chloral/DCC, $ClCH_2CH_2Cl$, reflux; (b) Et_3N/MeOH, reflux

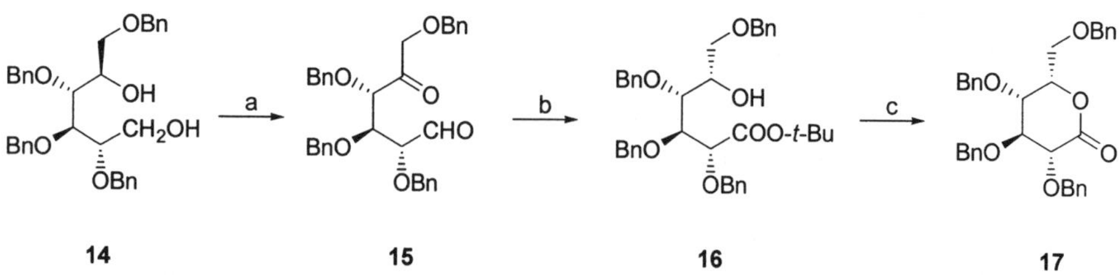

Scheme 6. (a) Swern oxidation; (b) *t*-BuOSmI_2, THF; (c) CF_3CO_2H/CH_2Cl_2 (1:1), 0 °C

ester **16**. This compound was then lactonised to **17** in 65% yield overall yield. Iadonisi and co-workers also applied this reaction sequence to the D-*galacto* analog of **14** and obtained a mixture of D-*galacto* and L-*altro* lactones (1:5) in 78% overall yield. A mechanistic explanation for the observed diastereoselectivities of the Tischenko step was also presented.

3 Rearrangements Involving Nucleophiles

On account of their biological acitivity, pyrazolidinones and hydroxypyrrolidines have sparked much interest in the field of synthetic chemistry [12]. On the one hand, the former represent a new class of antibacterial agents and intermediates for the preparation of β-lactams. On the other, the latter and bicyclic derivatives thereof are interesting in their own right as glycosidase inhibitors. In this context, Chmielewski and co-workers have published a series of papers dealing with the conjugate addition-rearrangement of hydrazines and hydroxylamines to α,β-unsaturated δ-lactones. These reactions proceed *anti* to the terminal substituent of the enelactone to afford 5-substituted pyrazolidin-3-ones or 3-substituted isoxazolidin-5-ones, respectively.

As depicted in Scheme 7, lactone **18** was reacted with *N*-benzylhydrazine to afford pyrazolidin-3-one **19** which was acetylated to give **20** in 88% overall yield as the sole product [13]. Employing lactone **21** and *N*-benzylhydroxylamine instead of *N*-benzylhydrazine furnished 3-substituted isoxazolidin-5-one **22** in good yield [14]. This strategy has been applied to a variety of δ-lactone derivatives with similar success [15]. Chmielewski and co-workers have taken this addition-rearrangement reaction one step further to allow the synthesis of bicyclic pyrazolidinones, as shown in Scheme 8. Treating D-*erythro* derivative **23** with hydrazine gave intermediate **24**, which then rearranged to **25** as before. However, owing to the leaving group at former C-6, this compound underwent intramolecular ring closure to give bicyclic structure **26** in 60% yield.

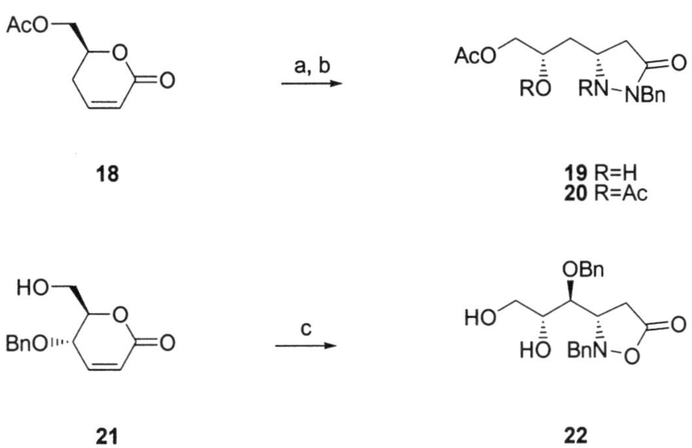

Scheme 7. (a) $BnNHNH_2$, EtOH; (b) Ac_2O, pyridine; (c) BnNHOH, EtOH

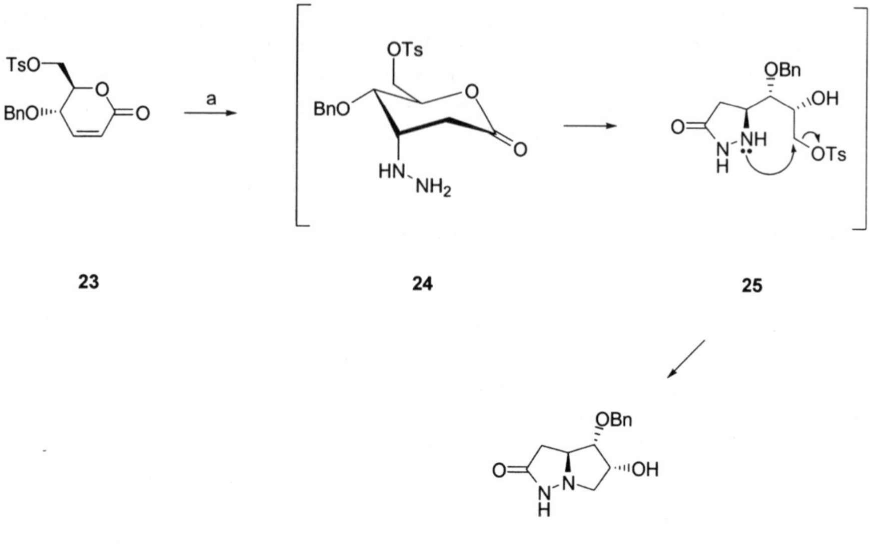

23 24 25

26

Scheme 8. (a) H_2NNH_2, EtOH

Concerning the synthesis of potential chiral building blocks, Bols and Thomsen recently reported [16] an unusual tandem elimination-Favorskii rearrangement involving tritosylate **27**. Upon treatment of this compound with sodium methoxide in chloroform, strained compound **31** was obtained in 44% isolated yield. This result was explained by the intermediates depicted in Scheme 9. Elimination of tosylate was thought to give structure **28** which then formed ketone **29** after attack by methoxide. The rearrangement of **29**, via **30**, resulted in the formation of product **31**. The use of other tritosylates was found to have a detrimental effect on the yields of Favorskii products.

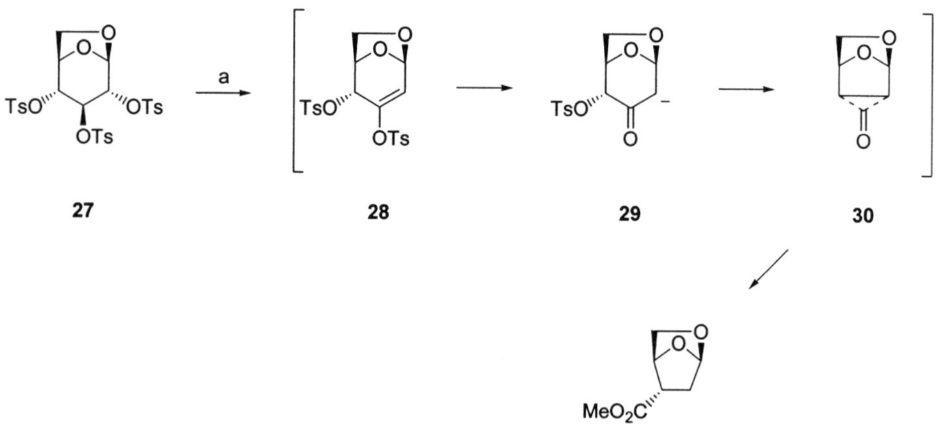

27 28 29 30

31

Scheme 9. (a) NaOMe, $CHCl_3$

Aziridino sugars are interesting synthetic targets because of their biological activity and utility as synthetic intermediates. In this context, Voelter and co-workers [17] exploited an aza-Payne rearrangement [18] to develop a mild one pot synthesis of these compounds from 2,3-anhydro sugars (Scheme 10). Dissolving α-aminooxirane **32** in a mixture of trimethylsilyl azide and diethyl ether, and treating this with diethyl ether-boron trifluoride to aid intramolecular oxirane ring opening from the *trans*-disposed nitrogen moiety, led to α-hydroxyepimine **33** in 53% yield. This methodology was successfully applied to other aminosugars in order to obtain the corresponding aziridino products.

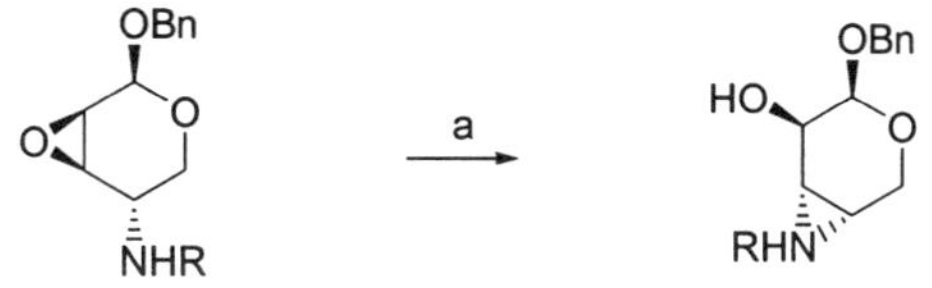

32 **33**

Scheme 10. $R = p\text{-}C_6H_4\text{-Boc}$ (a) $Me_3SiN_3, BF_3 \cdot OEt_2, Et_2O$

4
Rearrangements Involving Radical Species

The application of intramolecular radical cyclisations in synthetic organic chemistry is a valuable approach to obtain important synthetic targets such as natural product precursors and biologically active substances. The use of carbohydrate starting materials in this field is no exception to this because of their well-defined stereochemistry, highly functionalised nature and availability.

Alduronic acid 4,1- and 5,1-lactones, also termed pseudolactones, are useful starting materials in the synthesis of carbocycles. Suárez and co-workers recently reported [19] a so called "tandem strategy", that is a combination of consecutive reactions in one synthetic step, which allowed access to alduronic acid lactones from the corresponding uronic acids. This approach has the added advantage that it is compatible with the stability of protective groups frequently used in carbohydrate chemistry. In the first step of this methodology (Scheme 11), an alkoxy anomeric radical (**35**) is formed from alduronic pyranose **34** by the action of a hypervalent iodine reagent/iodine system [(diacetoxyiodo)benzene, iodine] which also promotes a β-fragmentation reaction of the C-1–C-2 bond to give **36**. In the second step, the intermediate C-2 radical is oxidised by an excess of reagent to give the oxonium ion **37** which then reacts intramolecularly with the nucleophilic carboxyl group to give lactone **38**. Alduronic acid lactone **38** was furnished in a 70% yield (α/β; 2.7:1) from uronic acid **34**. This approach has been applied to a range of pentose and hexose uronic acids with either the threose or erythrose configuration. These substrates also possessed a number of different protecting groups that remained intact after the transformation.

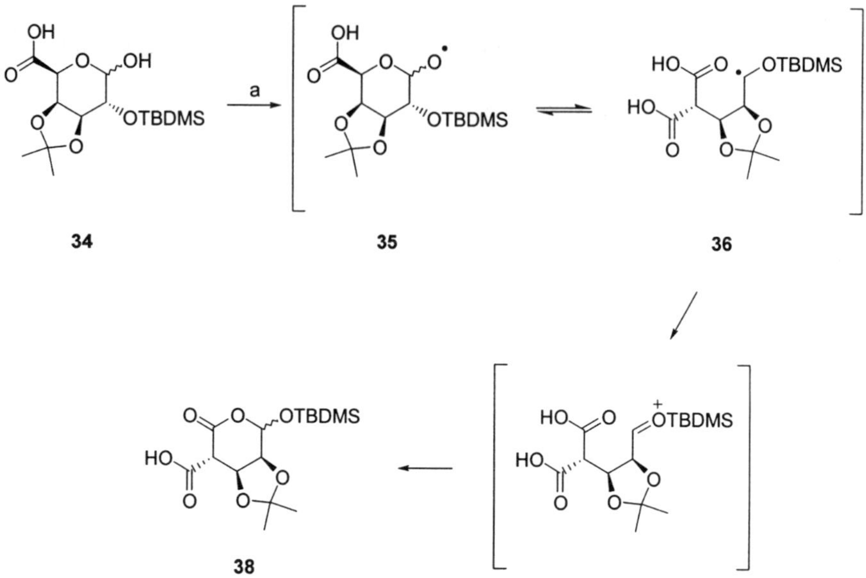

34 35 36 37 38

Scheme 11. (a) (Diacetoxyiodo)benzene, I_2, CH_2Cl_2

Photochemically induced rearrangements have also been reported to be a means to obtain lactone derivatives. Collins and co-workers investigated the photoisomerisation of a number of hexopyrano-3-uloses into their respective 1,5-lactones [20]. Taking 3-ulose **39** (Scheme 12) as an example, this was converted into a single product after UV irradiation in acetonitrile, namely lactone **40**, which was isolated in 45% yield. An insight into the mechanism of this transformation was gained by irradiating the C-1 deuterated analog of **39**. It was found that the C-3 deuterated analog of **40** was the sole product. On the basis of this result and other experiments, a reaction mechanism was proposed.

The regio- and stereoselective synthesis of branched sugars has been mainly spurred by their occurrence in a large range of natural products, in particular antibiotics. To this end, Sinaÿ and co-workers reported [21] a stereoselective approach to this class of compounds by applying a silylmethylene radical cyclisation (Scheme 13). Under basic conditions, allylic alcohol **41** was converted into derivative **42** employing (bromomethyl)chlorodimethylsilane. Radical cyclisation of this compound was effected in refluxing benzene with tributyltin hydride and azobis(isobutyronitrile) (AIBN) to furnish compound **43**. This was

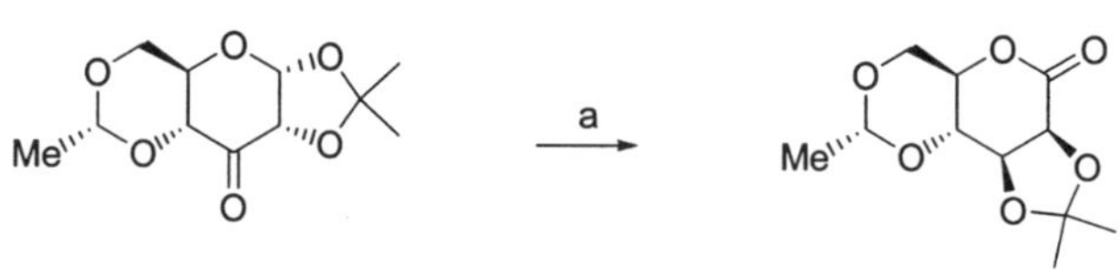

39 40

Scheme 12. (a) hν, CH_3CN

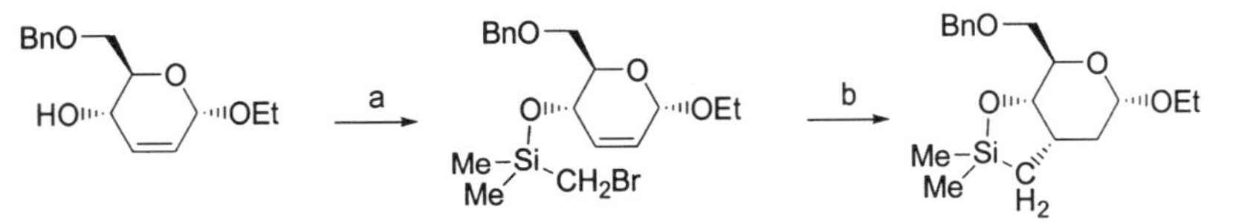

41 42 43

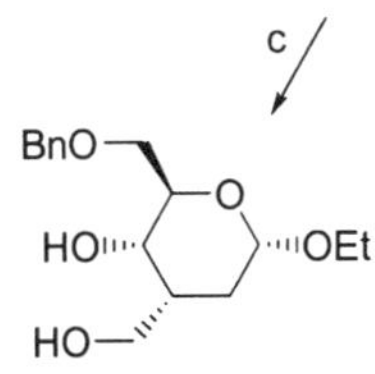

44

Scheme 13. (a) $BrCH_2(CH_3)_2SiCl$, Et_3N, CH_2Cl_2; (b) Bu_3SnH, AIBN, benzene, reflux; (c) H_2O_2, Na_2CO_3, MeOH, THF

then oxidised according to the Tamao–Kumada method [22] to give diol **44** in 73% yield. This method was also applied to a number of other allylic alcohol substrates to obtain the corresponding branched sugars in respectable yields.

Fraser-Reid and co-workers recently published an interesting radical cyclisation–intermolecular trapping sequence to obtain highly functionalised *C*-glycosyl derivatives in one step [23], as depicted in Scheme 14. After homolytic clevage of the carbon–bromine bond of glycal **45**, effected with tri-*n*-butyltin hydride and AIBN, the resulting carbon-centred radical subsequently added to C-2 in a intramolecular fashion thereby generating a stabilised α-oxy radical at C-1. This species was then trapped with unsaturated ulose **46** to furnish *C*-glycosides **47** in 70% yield. A number of different derivatives were made in this way and conclusions were drawn concerning the influence of the stereochemistry of starting material on the product obtained.

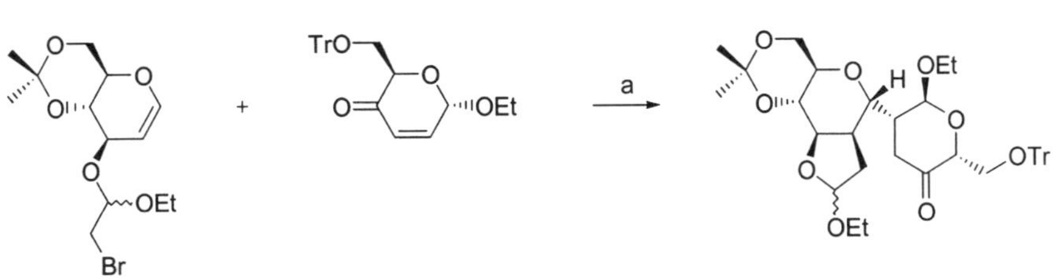

45 46 47

Scheme 14. (a) Bu_3SnH, benzene, AIBN, Δ

5 Rearrangements Leading to Ring Contractions

The ring contraction of carbohydrates constitutes a very useful route to functionalised and enantiomerically pure carbocycles. These compounds play a significant role in the construction of biologically active compounds such as enzyme regulators [24], the Corey lactone [25] and related prostaglandin intermediates [26], just to mention a few.

In particular, the growing importance of organometallic compounds in this field is evidenced by the increasing number of reports in the literature. One reagent that deserves special attention on the basis of its synthetic utility is the one-electron reducing agent samarium(II) iodide [27]. Known as a means to cyclise non-carbohydrate aldehydes or ketones tethered to olefins, Enholm and Trivellas [28] applied this strategy in the late 1980s to open-chain carbohydrate templates, as demonstrated in Scheme 15.

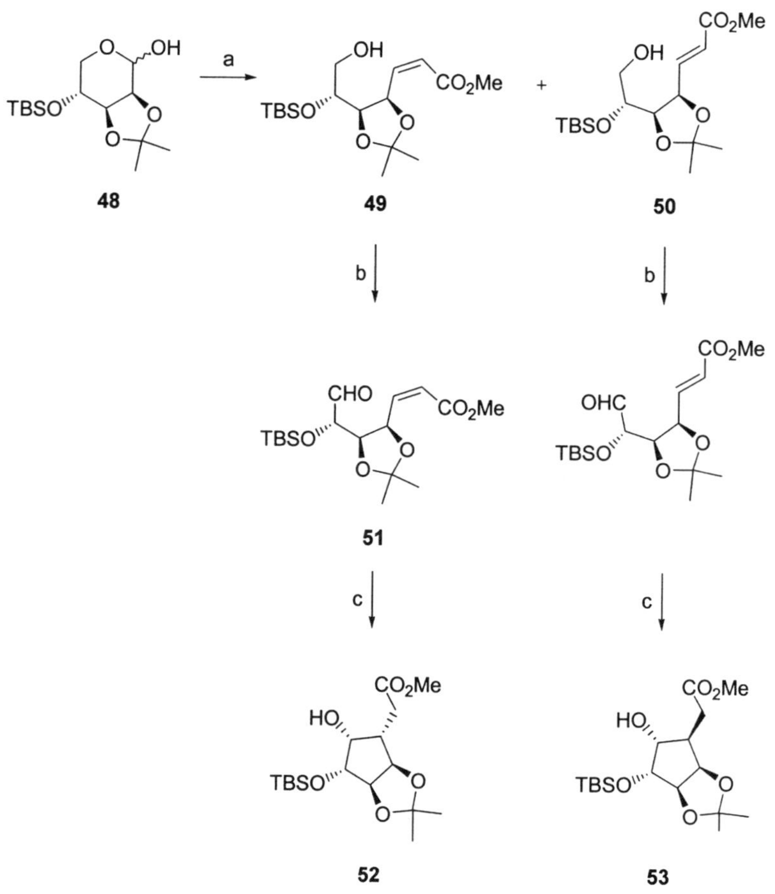

Scheme 15. (a) $Ph_3P = CHCO_2Me$, CH_2Cl_2, $PhCO_2H$; (b) PDC, CH_2Cl_2, 3Å sieves, HOAc; (c) SmI_2, THF, MeOH, –78 °C

Subjecting the D-lyxose derivative **48** to a Wittig reaction gave a mixture of chromatographically separable compounds, namely *Z*- and *E*-isomers **49** and **50** (3:1), respectively, in 80% total yield. Oxidation of the major *Z*-isomer **49** with pyridinium dichromate provided open-chain aldehyde **51**, which was then cyclised with two equivalents of SmI_2 at low temperature. This procedure furnished two products **52** and **53** (100:1), the former being provided in 73% isolated yield. The diastereoselectivity of this cyclisation step is indeed noteworthy. Interestingly, subjecting the *E*-isomer **50** to this reaction sequence provided the same products **52** and **53** (1:4), although a reversal of product ratio was observed. Apparently, the *Z*-isomer **49** favoured the formation of *cis*-product **52**, while the *E*-isomer **50** favoured the *trans*-product **53**. These authors mentioned that this observed influence of olefin geometry on the product distribution had been previously reported in similar non-carbohydrate compounds. Enholm and Trivellas later extended this approach to the synthesis of the C-ring of Anguidine [29] as well as more highly functionalised carbocycles. The latter was achieved by the addition of aldehydes or ketones during the cyclisation step, which allowed the formation of an additional chiral centre in a selective manner [30].

SmI_2 has also been applied to the synthesis of aminocyclopentanes, as reported by Holzapfel and Grové [31]. As can be seen in Scheme 16, SmI_2 was used to reductively cyclise intermediate **54** to yield cyclopentanes **56** and **57** (9:1) in a total yield of 78%. Replacing the benzyl protecting groups with pivaloyl moieties (**55**) was found to have a positive influence on the stereochemical outcome of the reaction, although this reaction was accompanied by pivaloyl group migration. Only one stereoisomer was isolated, namely the *trans*-product **58**.

Sinaÿ and co-workers demonstrated [32] that the use of SmI_2 permitted the ring contraction of cyclic carbohydrate precursors containing an external aldehyde group in "one pot". As shown in Scheme 17, glucoside **59** was oxidised into derivative **60**, which was then subjected to SmI_2 to give product **66** (63% yield

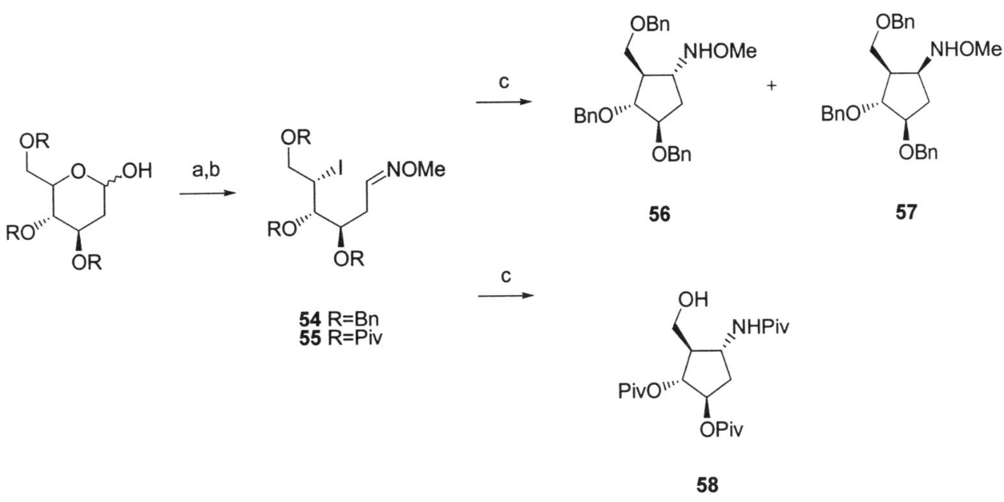

Scheme 16. (a) $H_2NOMe \cdot HCl$, pyridine; (b) PPh_3, imidazole, I_2; (c) SmI_2, THF, HMPA

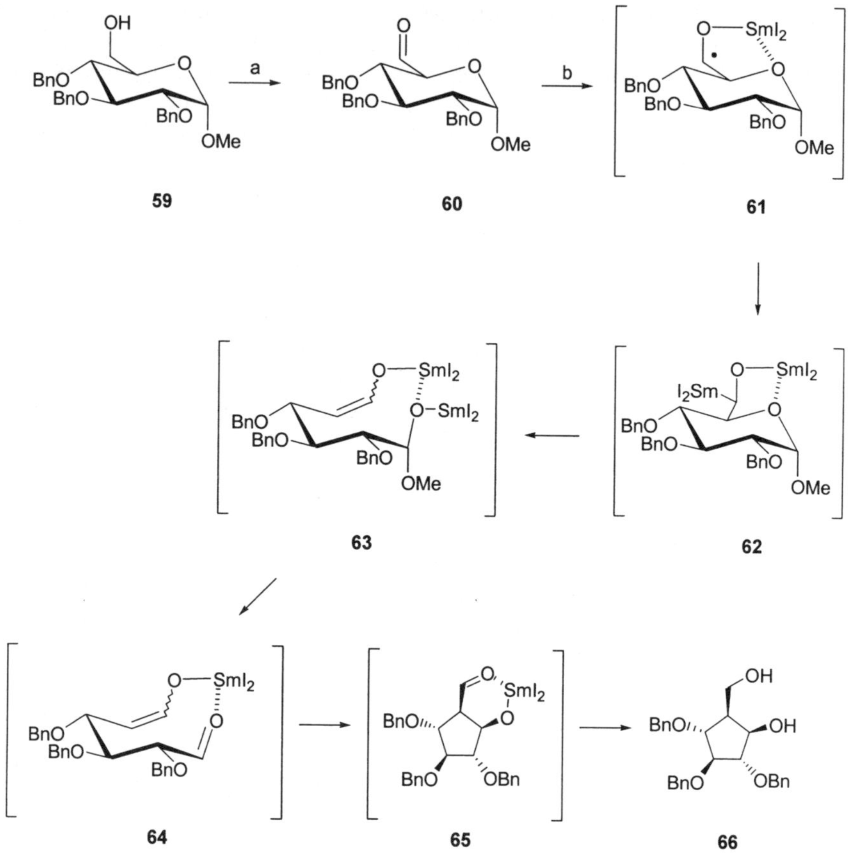

Scheme 17. (a) Swern oxidation; (b) SmI_2, THF, HMPA, *t*-BuOH

calculated from **59**). Interestingly, only the *cis*-product **66** was observed in this reaction. These authors also proposed a possible mechanism. It was envisaged that the first equivalent of SmI_2 reduced aldehyde **60** to samarium ketyl **61**, which was, in turn, reduced by a second equivalent of SmI_2 to species **62**. This disamarium intermediate then ring opened to give **63**, which subsequently eliminated methoxide to give species **64**. This key intermediate cyclised via nucleophilic attack of the samarium enolate onto the aldehyde to furnish **65**. Final reduction of **65** afforded the final product **66**. Employing the *manno*-analog of **60** as starting material in this sequence also gave the *cis*-isomer as the sole product. Other functional groups at C-2 as well as C-3 were also not found to influence the stereochemical outcome of this reaction.

Cyclic substrates containing an external double bond [33] or an α,β-unsaturated ester [34] instead of an aldehyde moiety have also been successfully ring contracted with SmI_2 to give the corresponding cyclopentane derivatives. Recently, Matsuda and co-workers reported the rearrangement of compound **67** into carbocyclic derivative **68** in 91% yield and as the sole product (Scheme 18).

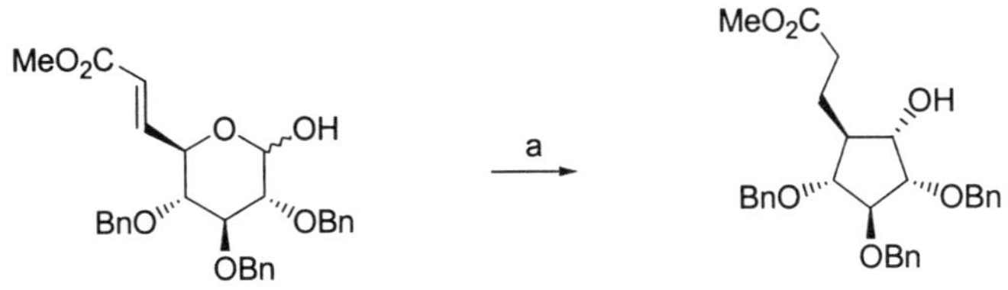

67 68

Scheme 18. (a) SmI_2, THF/MeOH

Here a mixture of SmI_2 (5.0 equiv) and THF/MeOH (15:1) was employed to effect the cyclisation. Interestingly, employing the *galacto*-analog of **67** resulted in a decrease in reaction stereoselectivity.

Zirconium-mediated ring contractions of derivatives possessing an external double bond have also been reported in the literature by Taguchi and co-workers [35, 36]. Zirconacycle "Cp_2Zr" was generated in situ at −78 °C before the vinylic carbohydrate derivative was added followed by $BF_3 \cdot OEt_2$. Compound **69** for example (Scheme 19) was transformed into a single isomer **70** in 65% yield with this reagent. Interestingly, it was found that the stereochemistry of the vinyl group was always *trans* to the adjacent (formerly C-4) benzyloxy or benzyoxymethyl group.

The cyclisation of hex-5-enyl radicals to cyclopentylmethyl radicals and subsequent trapping by various reagents is another synthetically interesting approach to cyclopentane derivatives. In particular, RajanBabu and co-workers have shown [37, 38] that a number of prostanoid intermediates, including the Corey lactone, can be prepared by this approach. Treatment of 3-deoxyglucose derivative **71**, for example, with the Wittig reagent shown in Scheme 20 yielded a mixture of *Z*- and *E*-isomers **72** which were used as a mixture for the next step. Reacting with thiocarbonylbis(imidazole) furnished **73** which, after treatment with tributyltin hydride and AIBN under Barton's deoxygenation conditions, afforded cyclopentane derivative **74** as the sole product. This compound was then converted into the Corey lactone after a number of synthetic steps.

Krohn and Heins [39] reported an unusual ring-contraction reaction following the elimination of water from a *β*-hydroxy-1,3-dithiane intermediate **76**. An attempt to thioacetalize *α*-branched hexopyranoside **75** did not give the expected dithioacetal **76**, but a mixture *C*-glycofuranoside **77** in 80% yield and

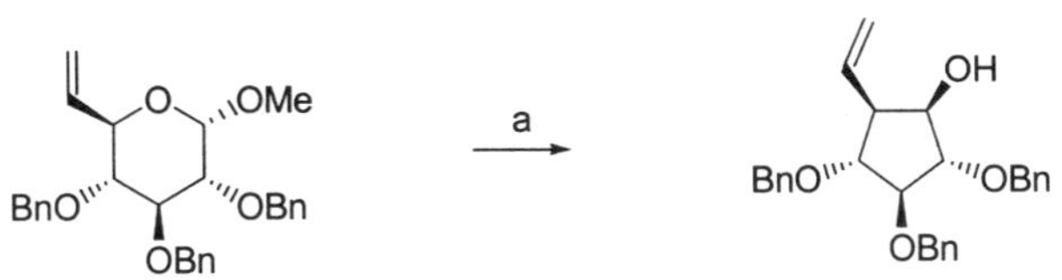

69 70

Scheme 19. (a) "Cp_2Zr", $BF_3 \cdot OEt_2$, toluene

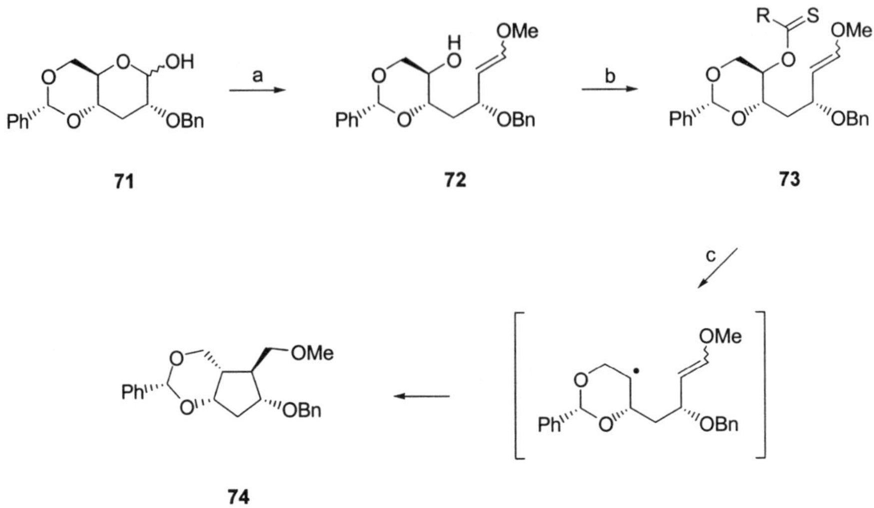

Scheme 20. R = Imidazole (a) $Ph_3P^+CH_2OCH_3Cl^-$, BuLi; (b) thiocarbonylbis(imidazole), $ClCH_2CH_2Cl$, Δ; (c) Bu_3SnH, AIBN, $PhCH_3$, Δ

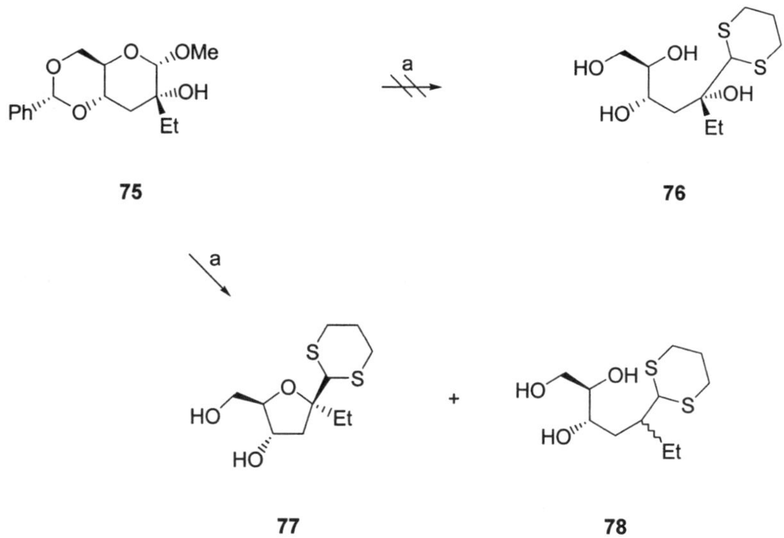

Scheme 21. (a) Propane-1,3-dithiol, $CHCl_3$, HCl

epimeric deoxygenation products **78** (12% yield), as indicated in Scheme 21. Evidently, upon formation of **76**, elimination of the hydroxyl group on C-2 took place and, instead of acid-catalysed addition to the double bond occurring at the dithiane-carbon (C-1), the addition occurred at C-2 to yield *C*-glycoside **77**. This compound clearly lends itself to further transformations on account of the protected aldehyde moiety at C-1 and, in doing so, allows access to other branched deoxy *C*-glycoside derivatives.

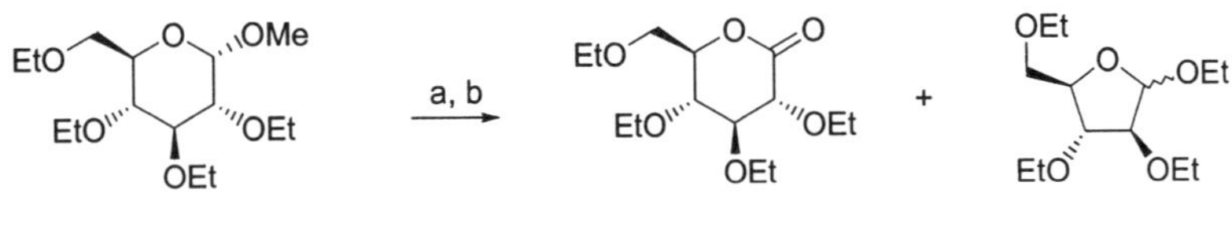

79 80 81

Scheme 22. (a) Me_3SiN_3, $SnCl_4$, CH_2Cl_2; (b) H_2O

Preparing per-*O*-alkylated glycono-δ-lactones from the corresponding glycopyranosides with the aid of trimethylsilyl azide and tin(IV) chloride, Ugi and co-workers [40] also recovered a ring-contracted compound as a minor product. As can be seen from Scheme 22, glycoside **79** afforded two products, the desired lactone **80** as well as the ring-contracted product **81**. An unprecedented sigmatropic rearrangement was thought to account for the formation of **81**.

6 Miscellaneous Rearrangements

Aspinall and Knebl [41] reported an interesting Hofmann rearrangement of amide **82** into carbamate **83** (83% yield, Scheme 23). This reaction was induced with lead tetraacetate in *tert*-butanol and pyridine at reflux temperature. This approach was also applied to disaccharide derivatives with equal success. A number of benzylated glycosides containing an amide substituent were also shown to undergo a Hofmann rearrangement to give the corresponding carbamates [42]. Mixtures of sodium methoxide, bromine and methanol or sodium hydroxide, sodium bromite and methanol were also found to effect this transformation, although in inferior yields (20–30%) as compared with the lead tetraacetate method. In addition, employing a *tert*-butyldimethylsilyl-protected analog permitted the isolation of a bromoamide intermediate.

More recently, Santoyo González and co-workers [43] reported the Hofmann rearrangement of amide **84** into cyclic carbamate **85** (40% yield, Scheme 24). This was achieved with a mixture of mercuric acetate, *N*-bromosuccinimide and *N*,*N*-dimethylformamide.

Ichikawa and co-workers have published a number of papers disclosing an allyl cyanate to isocyanate rearrangement and its application to the synthesis of aminosugars [44, 45]. A nitrogen substituent can be introduced into the pyra-

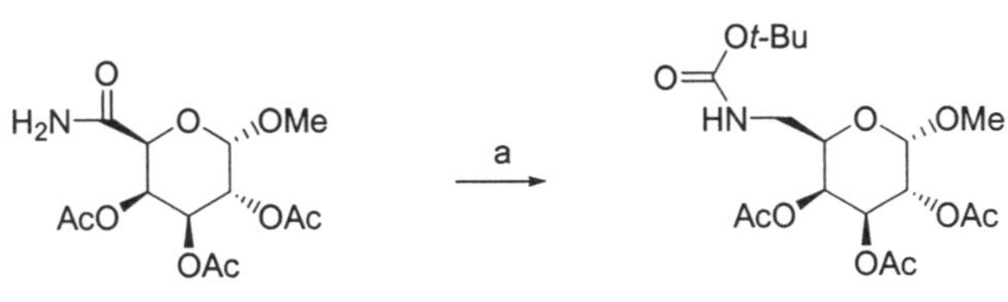

82 83

Scheme 23. (a) $Pb(OAc)_4$, *t*-BuOH, pyridine, reflux

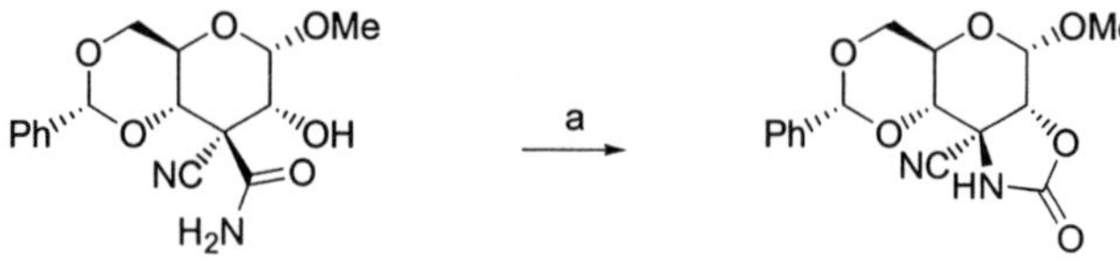

84 85

Scheme 24. (a) $Hg(OAc)_2$, NBS, DMF

nose framework by a [3,3]-sigmatropic rearrangement of an allyl cyanate. For example, allyl cyanate **86** underwent a sigmatropic rearrangement to furnish isocyanate **87**, which was then transformed in situ into acetamide **88** (Scheme 25). Subsequent synthetic steps employing this compound afforded D-perosamine, a substance present in a number of natural products.

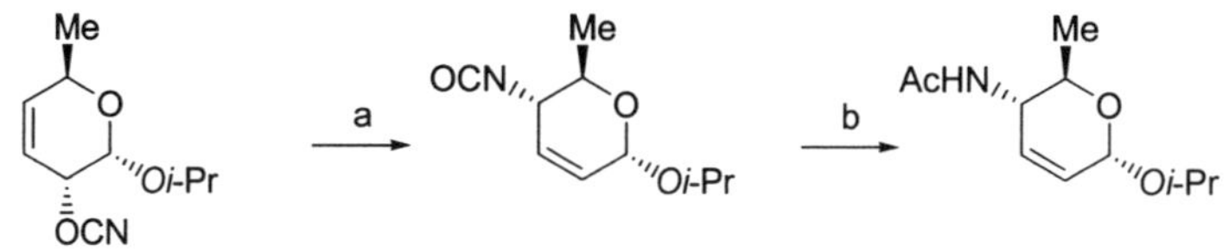

86 87 88

Scheme 25. (a) Bu_3P, CBr_4, Et_3N, CH_2Cl_2; (b) Me_3Al

The rearrangement of carbohydrate thiocarbonates has also been reported by Doane and co-workers [46, 47]. As depicted in Scheme 26, thionocarbonate **89** was treated with potassium iodide in acetonitrile to give monothiolcarbonate **90** in 84% yield. This method was also successfully applied to furanoses bearing other protecting groups as well as similar pyranoside derivatives.

Jenkins and co-workers [48] recently published an interesting report where carbohydrate precursor **91** was converted into enantiomerically pure cyclopentanone aldehydes **95** and **96** in 61 and 14% yield, respectively. This reaction was brought about by the addition of zinc shot in a mixture of isopropyl alcohol and water at reflux temperature, in a similar manner to that reported by Vasella and Bernet [49]. The proposed mechanism of this zinc reduction is given in Scheme 27. Zinc is thought to attack the bromo substituent of **91** leading to ring-cleavage product aldehyde **92**. This compound, which can also be regarded as an enedione, is converted into dianion **93** by the transfer of two electrons from zinc. Intermediate **93** is then protonated to give **94** which can be further protonated from the β or α face to give **95** and **96**, respectively.

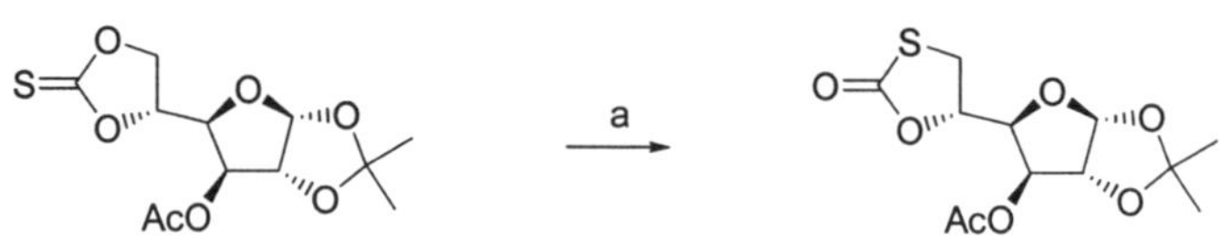

89 90

Scheme 26. (a) KI, CH_3CN, Δ

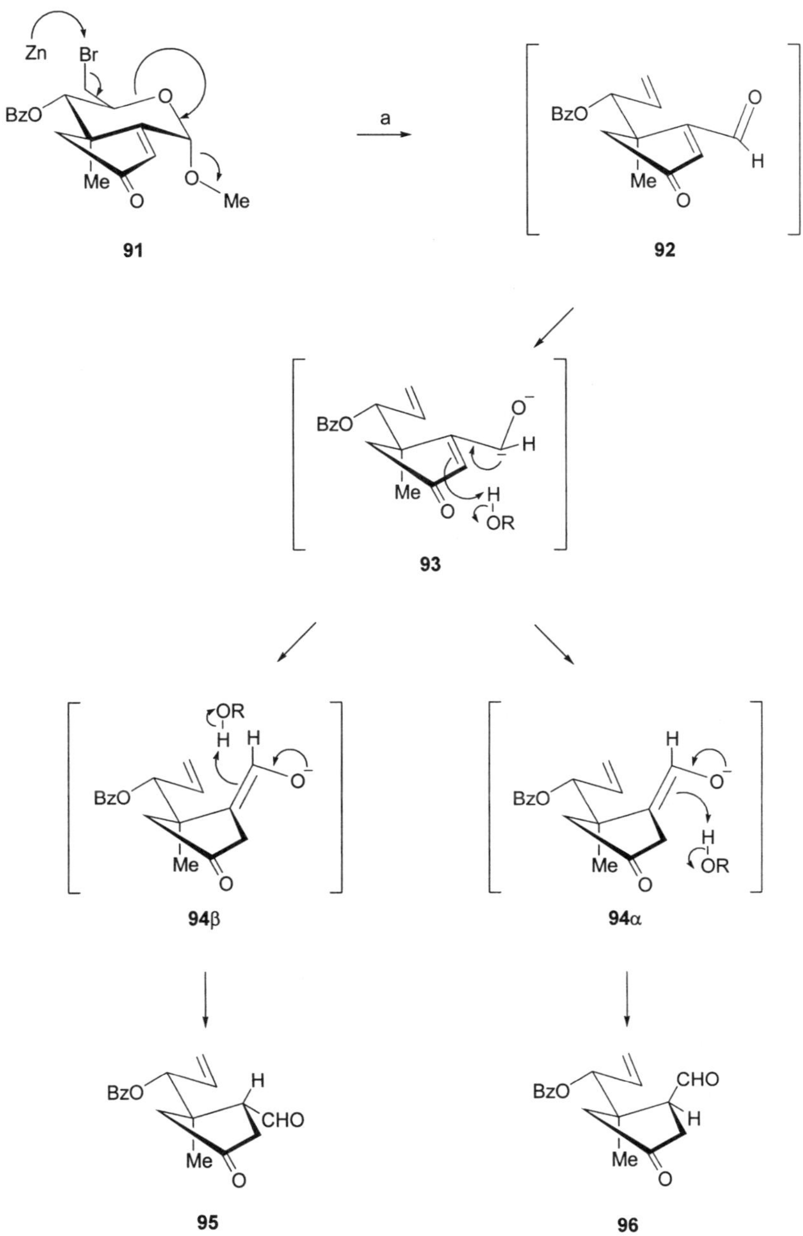

Scheme 27. (a) Zn shot, *i*-PrOH/H_2O (10:1), reflux

7 References

1. de Graauw CF, Peters JA, van Bekkum H, Huskens J (1994) Synthesis 1007
2. Adinolfi M, Iadonisi A, Mangoni L (1996) Tetrahedron Lett 37:5987
3. Adinolfi M, Barone G, Iadonisi A (1998) Tetrahedron Lett 39:7405
4. Kumazawa T, Asahi N, Matsuba S, Sato S, Furuhata K, Onodera J (1998) Carbohydr Res 308:213
5. Balza F, Perlin AS (1982) Carbohydr Res 107:270
6. Frank M, Miethchen R, Degenring D (1999) Carbohydr Res 318:167
7. Miethchen R, Rentsch D, Frank M (1996) J Carbohydr Chem 15:15
8. Miethchen R, Rentsch D, Frank M, Lipták A (1996) Carbohydr Res 281:61
9. Frank M, Miethchen R, Reinke H (1999) Eur J Org Chem 1259
10. Hager C, Miethchen R, Reinke H (2000) Synthesis 226
11. Adinolfi A, Barone G, De Lorenzo F, Iadonisi A (1999) Synlett 336
12. Rabiczko J, Chmielewski M (1999) J Org Chem 64:1347
13. Panfil I, Krajewski J, Gluzinski P, Stefaniak L, Chmielewski M (1994) Tetrahedron 50: 7219
14. Socha D, Rabiczko J, Koneru D, Urbanczyk-Lipkowska Z, Chmielewski M (1997) Carbohydr Lett 2:349
15. Panfil I, Urbanczyk-Lipkowska Z, Chmielewski M (1998) Carbohydr Res 306:505
16. Bols M, Thomsen IB (1998) J Chem Soc Chem Commun 1869
17. Latif F, Malik A, Voelter W (1987) Liebigs Ann Chem 717
18. Ibuka T (1998) Chem Soc Rev 27:145
19. Francisco CG, Martín CG, Suárez E (1998) J Org Chem 63:2099
20. Collins PM, Farnia F, Munasinghe RN, Oparaeche NN, Siavoshy F (1985) J Chem Soc Chem Commun 1038
21. Pedretti V, Mallet J, Sinaÿ P (1993) Carbohydr Res 244:247
22. Tamao K, Ishida N, Tanaka T, Kumada M (1983) Organometallics 2:1694
23. López JC, Gómez AM, Fraser-Reid B (1995) J Org Chem 60:3871
24. Chapleur Y (1998) Carbohydrate mimics: concepts and methods. Wiley-VCH, Weinheim
25. Corey EJ, Albonico SM, Koelliker U, Schaaf TK, Varma RK (1971) J Am Chem Soc 93: 1491
26. Collins PW, Djuric SW (1993) Chem Rev 93:1533
27. (a) Kagan HB, Sasaki M, Collin J (1988) Pure Appl Chem 60:1725; (b) Molander GA, Harris RH (1996) Chem Rev 96:307
28. Enholm EJ, Trivellas A (1989) J Am Chem Soc 111:6463
29. Enholm EJ, Satici H, Trivellas A (1989) J Org Chem 54:5841
30. Enholm EJ, Trivellas A (1994) Tetrahedron Lett 35:1627
31. Grové JJC, Holzapfel CW (1997) Tetrahedron Lett 38:7429
32. Chénedé A, Pothier P, Sollogoub M, Fairbanks AJ, Sinaÿ P (1995) J Chem Soc Chem Commun 1373
33. Aurrecoechea JM, López B (1998) Tetrahedron Lett 39:2857
34. Kann T, Nara S, Ozawa T, Shirahama H, Matsuda F (2000) Angew Chem Intl Ed Engl 39: 355
35. Ito H, Motoki Y, Taguchi T, Hanzawa Y (1993) J Am Chem Soc 115:8835
36. Ito H, Taguchi T, Hanzawa Y (1993) Tetrahedron Lett 34:7639
37. RajanBabu TV (1988) J Org Chem 53:4522
38. RajanBabu TV, Fukunaga T, Reddy GS (1989) J Am Chem Soc 111:1759
39. Krohn K, Heins H (1991) J Carbohydr Chem 10:917
40. Goebel M, Nothofer H, Roß G, Ugi I (1997) Tetrahedron 53:3123
41. Aspinall GO, Knebl MC (1986) Carbohydr Res 157:261
42. Mostowicz D, Belzecki C, Chmielewski M (1991) Synthesis 273

43. Santoyo González F, Berenguel AV, Mateo FH, Mendoza PG, Baer HH (1992) Carbohydr Res 237:145
44. Ichikawa Y, Osada M, Ohtani II, Isobe M (1997) J Chem Soc Perkin Trans 1 1449
45. Ichikawa Y, Kobayashi C, Isobe M (1994) Synlett 919
46. Trimnell D, Doane WM, Russell CR, Rist CE (1971) Carbohydr Res 17:319
47. Shasha BS, Doane WM (1974) Carbohydr Res 34:370
48. Wood AJ, Holt DJ, Dominguez MC, Jenkins PR (1998) J Org Chem 63:8522
49. Vasella A, Bernet B (1979) Helv Chim Acta 62:1990

Author Index Volume 201 – 215

Author Index Vols. 26 – 50 see Vol. 50
Author Index Vols. 51 – 100 see Vol. 100
Author Index Vols. 101 – 150 see Vol. 150
Author Index Vols. 151 – 200 see Vol. 200

The volume numbers are printed in italics

Printing (Computer to Film): Saladruck, Berlin
Binding: Stürtz AG, Würzburg